Climate Change Impact on Groundwater Resources

Balamurugan Panneerselvam
Chaitanya Baliram Pande
Kirubakaran Muniraj
Anand Balasubramanian
Nagavinothini Ravichandran
Editors

Climate Change Impact on Groundwater Resources

Human Health Risk Assessment in Arid and Semi-arid Regions

Springer

Editors
Balamurugan Panneerselvam
Department of Civil, Architectural and Environmental Engineering
University of Naples Federico II
Naples, Nepal

Kirubakaran Muniraj
Department of Civil Engineering
University VOC college of Engineering
Thoothukudi, Tamil Nadu, India

Nagavinothini Ravichandran
Department of Structures for Engineering and Architecture
University of Naples Federico II
Naples, Italy

Chaitanya Baliram Pande
GIS Consultant
Primus Partners Private Limited
Delhi, Delhi, India

Anand Balasubramanian
Department of Agriculture Engineering
KIT-Kalaignarkarunanidhi Institute of Technology
Kannampalayam, Tamil Nadu, India

ISBN 978-3-031-04706-0 ISBN 978-3-031-04707-7 (eBook)
https://doi.org/10.1007/978-3-031-04707-7

This Springer imprint is published by the registered company Springer Nature Switzerland AG
The registered company address is: Gewerbestrasse 11, 6330 Cham, Switzerland

Dedicated to our parents and all our teachers

Preface

Currently, two driving forces, namely climate change and man-made interference, are responsible the water crisis on Earth. Especially in semi-arid regions, hydrological activities are more complex, and they are more sensitive to human activities and climate change. Areas with the most water scarcity are generally defined as arid and semi-arid. Surface and groundwater regimes in these areas are intense and highly diverse, and they face great pressure to provide and manage freshwater resources. In recent years, water circulation has been severely affected by climate change, industrialization, urbanization, and desertification of usable lands (especially in urban areas). The socio-economic development of the population demands proper infrastructure development in and around their habitat resulting in natural water systems being subjected to severe denials. The main purpose of this chapters 1 to 5, is to identify the water source in semi-arid areas and to focus on the response mechanism and geographical modeling of the hydrological activities in those areas. The prime objective of this book is to provide insight into semi-arid regions' climatic and morphometric conditions through rainfall fluctuation and temperature variation studies, land use/landcover change analysis, morphometric analysis, and drought assessment studies and also demonstrate the modern remote sensing and GIS-based quantification and modeling approaches which are used to access the hydrological parameters such as rainfall-runoff, evapotranspiration, groundwater dynamic modeling, water level fluctuation study, water balancing, and aquifer properties.

This book deals with climate change challenges and their impact on groundwater quality, and human health risk assessment due to pollution of surface water and groundwater. To providing a stable outcome presentation by using recent trends in water resource management, pollution mitigation strategies techniques, modeling of groundwater quality and human health risk assessment in arid and semi-arid regions. Population growth has been raising an unparalleled demand for fresh water and also increasing human health risk due to water quality changes in arid and semi-arid regions. Currently, risk to groundwater and human health is very serious in the entire world, where several million people use contaminated surface and groundwater for drinking and irrigation purposes. Nowadays, every sector depends on surface water and groundwater. It is the most significant source of domestic, industrial, and irrigation water and also a limited resource. This book is a collection of chapters providing a multi-disciplinary overview for academics, administrators, scientists, policy makers, social scientists, and professionals involved in the various aspects of climate change impact on groundwater quality, hydrological process, pollution mitigation strategies and techniques, sustainable development, planning, and management. This book combines human health assessment, hydrological process, and groundwater quality monitoring techniques such as remote sensing and GIS techniques, entropy water quality index, weighted arithmetic water quality index, fuzzy logic application for water quality index and irrigation, saline water, and human health risk assessment planning of groundwater resources in arid and semi-arid regions. The main purpose of this book is to develop truthful outlines of major issues under the climate changes impact on groundwater resources and human health risk assessment, mitigation planning and management in arid and semi-arid region. Therefore, this book is organized into four sections.

This book includes research work by professors, planners, scientists, and research scholars from various universities, international organizations, and institutions in India as well as from other countries of world.

Dr. Balamurugan Panneerselvam
Dr. Chaitanya Baliram Pande
Mr. Kirubakaran Muniraj
Dr. Anand Balasubramanian
Dr. Nagavinothni Ravichanran

Contents

About the Editors

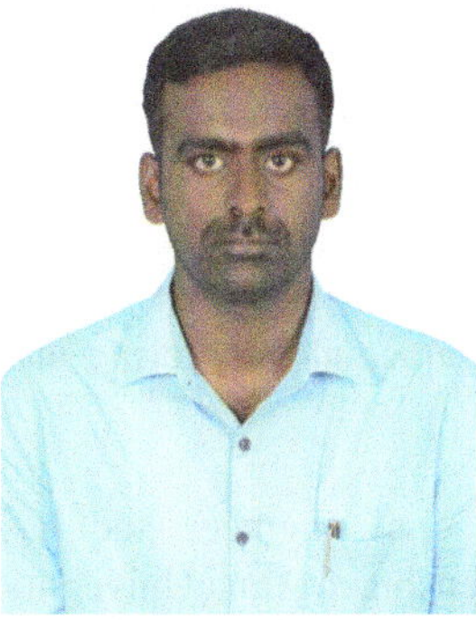

Balamurugan Panneerselvam is a postdoctoral researcher in the Department of Civil, Architectural and Environmental Engineering, University of Naples Federico II, via Claudio 21, 80125, Naples, Italy. His current research interests include groundwater quality assessment for drinking, irrigation, and human health risk associated with major contaminants such as nitrate and fluoride in arid and semi-arid regions, and also groundwater related issues. He extended his research interest in the field of wastewater treatment using moving bed bioreactors and conducting performance evaluation of bioreactor in biogas units. He received his doctoral degree from Anna University, Chennai, in 2019. His thesis was based on the groundwater quality assessment using GIS techniques in Salem district, India. He has published more than 50 papers in international and national journals. He is privileged to act as technical reviewer in journals by Springer, Elsevier, and MDPI. With his immense interest in multidisciplinary research, his recent investigations are focused on parameter estimation using machine learning algorithms, innovative floating structures, and GIS-based potential assessment of photovoltaic systems in India.

Chaitanya B. Pande completed PhD in environment science from Sant Gadge Baba Amravati University, Amravati, and MSc in geoinformatics from Amravati University in 2011. He has more than 10 years of teaching, research, and industrial experience. He is a reviewer for several scientific journals of international repute and is editorial board member of the *American Journal of Agricultural and Biological Sciences*. He has published 57 research papers, 1 textbook, 1 edited book entitled *Groundwater Resources Development and Planning in the Semi-Arid Region* for Springer, 19 conference papers, and 5 book chapters with more than 840 citations. His research interests include remote sensing, GIS, Google Earth Engine, machine learning, watershed management, hydrogeology, hydrological modeling, drought monitoring, land use and land cover analysis, groundwater quality, urban planning, hydrogeochemistry, groundwater modeling, geology, hyperspectral remote sensing, remote sensing and GIS application in natural resources management, watershed management, and environmental monitoring and assessment of subjects.

Kirubakaran Muniraj is a civil engineer and has completed MTech in remote sensing at Anna University Regional Campus, Tirunelveli, TN, India, and is currently pursuing PhD in the Faculty of Civil Engineering, Anna University, Chennai, TN, India, in the area of microwave remote sensing (InSAR). He has published more than 25 research articles in reputed international journals by Springer, Taylor and Francis, and Elsevier, which include *Environmental Research*, *Marine Pollution Bulletin*, *Environmental Geochemistry and Health*, *International Journal of Environmental Analytical Chemistry*, *Environmental Science and Pollution Research*, and *Human and Ecological Risk Assessment*. Kirubakaran's research broadly focuses on remote sensing and GIS applications for environmental issues, water resource management, air pollution, solid waste management, subsidence monitoring, disaster management, climatic change, and urban planning. He is privileged to act as technical reviewer for journals including *Applied Water sciences* (Springer), the *Asia-Pacific Journal of Regional Science*, *Proceedings of National Academy of Sciences Part A Physical Sciences*, and *Marine Georesources and Geotechnology*. He is a life member of the Indian Society of Remote Sensing, member of the International Society for Research and Development (ISRD), associate member of the Institute of Engineers (India), and life member of the Indian National Cartographic Association.

Anand Balasubramanian, MTech, PhD (KIT-Kalaignar Karunanidhi Institute of Technology, Coimbatore), has completed his BTech in geotechnology and geoinformatics in the Centre for Remote Sensing at Bharathidasan University, Tiruchirappalli, and MTech in remote sensing and GIS at the Regional Centre of Anna University, Tirunelveli, and he also completed his PhD in remote sensing and GIS with the specialization in hydro informatics. Anand has worked in the NRDMS-DST Project as a JRF/SRF for 3 years. He has more than 5 years of research and 1 year of teaching experience. He has expertise in the field of remote sensing, GIS, and hydro informatics with respect to water resources management, natural resources mapping, landslide mapping, and agricultural crop management. He has published more 10 research articles in reputed national and international journals. He is privileged to act as a technical reviewer in journals by Springer and Elsevier. Currently, he has been working in the field of agriculture engineering with respect to remote sensing, GIS, and geoinformatics applications.

Nagavinothini Ravichandran is a postdoctoral researcher in the Department of Ocean Engineering at the University of Naples Federico II, Italy. Her current research interests include computer-aided analysis of structures, earthquake engineering and structural control, floating photovoltaic systems, and machine learning applications in engineering problems. She received her PhD in ocean engineering from the Indian Institute of Technology Madras in 2019. Her thesis is based on the dynamic analysis of offshore structures under environmental and accidental loads. She was a university first-rank holder and gold medalist in her undergraduate and postgraduate studies. She has published in approximately 50 international journals and conferences. In addition, she co-authored a textbook titled *Offshore*

Compliant Platforms: Analysis, Design, and Experimental Studies. Nagavinothini is an associate member of the American Society of Civil Engineers and a life member of the Indian Society for Technical Education. With her immense interest in multidisciplinary research, her recent investigations are focused on parameter estimation using machine learning algorithms, innovative floating structures, and GIS-based potential assessment of photovoltaic systems in India.

Chapter 1
Introduction

Chaitanya B. Pande, Balamurugan Panneerselvam, Kirubakran Muniraj, and Nagavinothini Ravichandran

Abstract Climate change and human health risk is a main important for the human and environment, these are factors continue affecting the groundwater resources, water quality and ecosystem in the semi-arid region. This book deals with the climate change impact on groundwater quality and human health risk assessment due to pollution of surface water and groundwater. This book discusses about the climate change factors that affect the quality of surface water and groundwater in the arid and semi-arid regions. Recent trends in management and pollution mitigation strategies, modelling, groundwater quality, and human health risk assessment in arid and semi-arid regions. Population growth has resulted in an unparalleled demand for freshwater and an increase in human health risks due to water quality parameter changes in the arid and semi-arid regions. Groundwater and human health risk condition is a most serious issues in the entire world, where several million people based on contaminated surface and groundwater resources. Nowadays, every sector depends on surface water and groundwater. They are the most significant sources of water for domestic, industrial, and irrigation purposes; however, these resources are limited. This book is a collection of chapters providing a multidisciplinary overview for academics, administrators, scientists, policy-makers, social science researchers, and professionals involved in the various aspects of climate change impact on groundwater quality, hydrological process, pollution mitigation strategies and techniques, sustainable development, planning, and management. This book is a combination of human health assessment, hydrological process, and

C. B. Pande (✉)
GIS Consultant, Primus Partners Private Limited, Delhi, India

B. Panneerselvam
Department of Civil, Architectural and Environmental Engineering, University of Naples Federico II, Naples, Italy

K. Muniraj
Department of Civil Engineering, University VOC college of Engineering, Thoothukudi, Tamil Nadu, India

N. Ravichandran
Department of Structures for Engineering and Architecture, University of Naples Federico II, Naples, Italy

B. Panneerselvam et al. (eds.), *Climate Change Impact on Groundwater Resources*, https://doi.org/10.1007/978-3-031-04707-7_1

groundwater quality monitoring techniques such as remote sensing and geographic information system (GIS) techniques, entropy water quality index, weighted arithmetic water quality index, Fuzzy logic application for water quality index and irrigation purposes, saline water, and human health risk assessment planning of groundwater resources in arid and semi-arid regions. The main purpose of this book is to develop truthful outlines for climate change impact on groundwater resources and for human health risk assessment planning and management on the earth's surface. Therefore, this book will help the readers understand climate change impact on groundwater resources and human health risk assessment in arid and semi-arid regions.

Keywords GIS · Groundwater · Human Health · Hydrological Process · Water Quality Index; Semi-arid

1.1 Introduction

Groundwater is water that exists beneath the earth's surface and makes up 30% of the world's freshwater supply. It is an extremely valuable natural resource that plays a critical role in any country's economy. For a long time, people from all over the world have relied on subsurface water to meet a variety of their needs such as drinking and washing (WWAP 2009; Pande et al. 2021a; Pande et al. 2021b; Rajesh et al. 2021). Due to large-scale human activities, changes in lifestyle, and consumerism, poor drinking water quality has become a severe concern in many countries; thus, groundwater quality evaluation is required, particularly for groundwater sustainable management (Todd and Mays 2005; Pande 2020). The varied properties of water, atmospheric precipitation, inland surface water, and subsurface geochemical activities all contribute to groundwater quality (Khadri and Pande 2016a, b). The majority of research works correlated their observed parameter values with the individual suggested parameter values according to the World Health Organization (WHO) guidelines (Cude 2001). Water quality indices (WQIs) are the most often used instruments for assessing groundwater excellence; they summarize a large number of groundwater quality parameters into a single numerical value (Rawat et al. 2019). The need for freshwater in Tamil Nadu's semi-arid districts has risen in recent years (Pande et al. 2017). Furthermore, due to a decline in groundwater and surface water owing to overextraction, as well as a general lack of public awareness of groundwater quality, systematic management strategies cannot be developed. Groundwater quality is influenced by soil permeability features as well as rock–water interaction, lithological background, and water residence time (Balamurugan et al.2020b; Kirubakaran et al. 2016; Balamurugan et al. 2020c). Increased concentrations of nitrate, fluoride, and arsenic in water are currently responsible for 82.5% of waterborne infections in India (Pande et al. 2019).Therefore, the objectives of this book is measured to variations in the concentration of nitrate under climate change and an evaluated the risk of nitrate contamination on human health, climate change impact on groundwater resources, and human health risk assessment by using GIS, remote sensing, and statistical approaches. The findings will aid in the proper planning and management of groundwater resources as well as in the development of novel techniques to prevent continuous water deterioration in semi-arid areas (Panneerselvam et al. 2021a, 2020b, Shanmugapriya et al. 2021).

1.2 Sections of This Book

This book is divided into four sections: (I) hydrological processes in semi-arid regions; (II) hydrochemical properties of surface water and groundwater; (III) human health risk assessment; and (IV) recent trends in the management and pollution mitigation strategies.

1.2.1 Hydrological Processes in Semi-arid Regions

Currently, the two driving forces, namely, climate change and man-made interference are the main reasons for water crisis on Mother Earth. Especially, hydrological activities are more complex in the semi-arid region as well as more sensitive to the effects of human activities and climate change on it. Areas with the most water scarcity are generally defined as arid and semi-arid. Surface water and groundwater regimes in these areas are intense and highly diverse, and they face great pressure to provide and manage freshwater resources. In recent years, water circulation has been severely affected by climate change, industrialization, urbanization, and desertification of usable lands (especially in urban areas). The socioeconomic development of the population demands proper infrastructure development in and around their habitat, resulting in natural water systems being subjected to severe denials (Batabyal & Chakraborty 2015; Balamurugan et al. 2020d; Kirubakaran et al. 2015; Kumar and Balamurugan 2019, 2018; Panneerselvam et al. 2020a). The main purpose of this chapter is to identify the water source in semi-arid areas and to focus on the response mechanisms and geographical modelling of the hydrological activities in those areas. The prime objective of this section is to provide insights into the semi-arid regions' climatic and morphometric conditions through rainfall fluctuation and temperature variation studies, land use/land cover change analysis, morphometric analysis, drought assessment studies, etc. and also to demonstrate the modern remote sensing and GIS-based quantification and modelling approaches, which are used to access hydrological parameters such as rainfall runoff, evapotranspiration, groundwater dynamic modelling, water-level fluctuation study, water balancing, aquifer properties, etc.

1.2.2 Hydrochemical Properties of Surface Water and Groundwater

In this chapter, the primary mechanisms governing the quality of surface water and groundwater in the arid and semi-arid regions are discussed. Over the past five decades, geochemistry has been significantly contributing to the field of hydrogeology, thus helping in the study of the geological, structural, and behavioural properties of minerals and their lithological effects on water quality. Surface and subsurface sources of water are drastically deteriorated due to rapid increase in population, industrialization, and changes in the methods of the modern agricultural system. The change in climatic conditions, inadequate rainfall, and improper maintenance of water sources are the major reasons for reducing the usage of surface water. Increase in population density and modernization are the primary reasons for excess utilization of groundwater than surface water. The main intention of this study is to evaluate the quality of surface water and groundwater for drinking purposes using various indices, namely, the entropy water quality index, weighted arithmetic water quality index, Fuzzy logic application for the water quality index and for irrigation purposes, sodium adsorption ratio, residual sodium carbonate, percentage sodium, permeability index, magnesium hazard index, residual sodium bicarbonate, potential salinity, synthetic harmful coefficient, and the amount of exchangeable sodium percentage. The result of the study clearly describes the present nature of the groundwater and the sources of contamination in the study area.

1.2.3 Human Health Risk Assessment

The Environmental Protection Agency commonly uses risk assessment to evaluate and characterize the nature and degree of health risks to humans and also ecological receptors such as birds, fish, and wildlife in arid and semi-arid regions. Human health risk exposure the injection of toxic materials such as food and drinking water (Pratap et al. 2021). The main objective of this chapter is to define human exposure to a contaminated environment and its severe effects due to continuous consumption. Nowadays, nitrate and fluoride are the major ions that are easily contaminated by natural and anthropogenic activities and also affect human well-being. Continuous consumption of elevated concentrations of nitrate- and fluoride-containing water causes methemoglobinemia, heart problems, and bone and teeth issues in the human body. It also affects the livestock in the same environment. The health risk evaluation has divulged that more information is needed for policy-makers and government and non-governmental agencies to take remedial measures to avoid contamination and to create awareness on the use of fertilizers and pesticides in agricultural fields.

1.2.4 Recent Trends in Management and Pollution Mitigation Strategies

The water resource-related challenges in the present situation have bestowed a greater responsibility and duty on engineers and scientists dealing with the environment to protect, preserve, and enhance our natural resources in order to deliver a sustainable future. In a market economy, a country's economic growth is forced to move towards industrialization, taking into account the country's gross domestic product (GDP), especially as the resources of developing countries are greatly depleted. Therefore, in recent years, more emphasis has been placed on environmental issues, especially in developed countries (Panneerselvam et al. 2021b, c; Ramalingam et al. 2022). However, due to the economic and technological deficits of many developing countries, the impact of surface/subsurface water pollution and the importance of appropriate control measures are faltering. Therefore, this chapter elaborates on the modern trends of water resources engineering to maintain sustainable water consumption of the available precious resources with more economic concern for a clean environment (Balamurugan et al. 2020a; Colins et al. 2016). Through this chapter, the reader can get a clear explanation of how to deal with the integrated method in the monitoring and mitigation activities of key techniques such as promotion of artificial recharge, identification of urban water body encroachments, deployment of sustainable supply and consumption methods, regulations and legalization activities, industrial pollution source control measures and modern Internet-of-Things (IOT)-based monitoring systems.

This book contains 25 chapters. Chapter 1 discusses about the climate change impact on groundwater resources in semi-arid regions. Chapter 2 discusses the factors that influence the hydrological process, with a climate and land use/land cover perspective in the semi-arid and arid regions. Chapter 3 discusses the climatic changes affecting groundwater quality in the semi-arid region of central India with a special reference to the major cities of Madhya Pradesh.. Climate change views rainfall research as more important for semi-arid regions. Chapter 4 is dedicated to the spatiotemporal analysis of rainfall trends in the semi-arid regions of India over the last 36 years. Chapter 5 discusses the assessment of groundwater quality for drinking and irrigation, citing a case study of the Kattar micro-watershed in Tamil Nadu. Chapters 6 and 7 focus on the hydrogeochemistry and the WQI. Chapter 6 investigates the groundwater hydrochemistry of the shallow aquifers of the Cuddalore coast, South India. Chapter 7 describes Human health risk exposure the injection of toxic materials such as food and drinking water quality index (EWQI) in semi-arid regions in India. Chapter 8 describes a study on the seasonal variation of groundwater quality for irrigation purposes in semi-arid regions in Tamil Nadu, India. The geographic information system is an effectively proved instrument to check the water quality and for WQI research. Chapter 9 discusses the evaluation of groundwater quality and its suitability for drinking and cultivation practices in and around the deltaic regions of South India using the drinking water quality index (DWQI), irrigation water quality index (IWQI), and GIS. Chapter 10 discusses the groundwater quality assessment using GIS-based multi-criteria decision analysis for the Pattukkottai Taluk in Tamil Nadu. Chapter 11 focuses on the study of heavy metal contamination at the Durgapur Barrage site, Damodar River. Chapter 12 explains the heavy metal assessment of groundwater on the periphery of a semi-urban solid waste dumpsite and the mitigation strategy used. This is highly essential for understanding the climate change impacts of nitrate contamination on human health, as discussed in Chap. 13. Chapter 14 describes the catastrophic effects of climate change on child health. Chapter 15 focuses on the mitigation measures of fluoride- and nitrate-contaminated regions. The hydrological system and process are important in semi-arid and arid regions, and, based on this view, Chap. 16 discusses hydrological modelling for ungauged basins, providing an overview of the past, present, and future directions. Chapter 17 focuses on appraising the groundwater potential of the Liddar sub-basin (western Himalayas) using geospatial techniques. Groundwater resources are the key factors in semi-arid and arid regions. In this context, Chap. 18 describes the groundwater potential zone (GPZs) delineation in the Dhamani River basin in Kolhapur district, Maharashtra, India, using remote sensing (RS), geographical information system (GIS), and multi-criteria decision analysis (MCDA) techniques (Chap. 19). Chapter 20 discusses the assessment of groundwater potential zone mapping for development of semi-arid regions through analytical hierarchy process (AHP) and GIS techniques. Chapter 21 describes a study of the assessment of groundwater potential zones in Adigrat town and its surrounding areas using geospatial technology. Rainfall trend analysis is important for arid regions. Chapter 22 describes a study on the computation of rainfall infiltrates into the coastal soil of Andhra Pradesh, India. Chapters 23 and 24 focus on rainwater harvesting activities and rainwater conservation in semi-arid and arid areas. In this view, Chaps. 23 and 24 focus on the feasibility assessment of low-cost filters to be adopted in rooftop rainwater harvesting (RWH) prioritization of sub-watersheds through morphometric analysis of the Amaravathi watershed using geoinformatic techniques. Overall, all these chapters give comprehensive ideas about the importance of climate change impact on groundwater resources for human health risk assessment in arid and semi-arid regions.

References

Arun Pratap Mishra, H. Khali, S. Singh, C.B. Pande, R. Singh, S.K. Chaurasia (2021) An Assessment of In-situ Water Quality Parameters and its variation with Landsat 8 Level 1 Surface Reflectance datasets, International Journal of Environmental Analytical Chemistry, https://doi.org/10.1080/03067319.2021.1954175.

P. Balamurugan, P.S. Kumar, K. Shankar, Dataset on the suitability of groundwater for drinking and irrigation purposes in the Sarabanga River region, Tamil Nadu, India. Data Brief **29**, 105255 (2020a)

P. Balamurugan, P.S. Kumar, K. Shankar, R. Nagavinothini, K. Vijayasurya, Non-carcinogenic risk assessment of groundwater in southern part of Salem District in Tamilnadu, India. J. Chil. Chem. Soc. **65**(1), 4697–4707 (2020b)

P. Balamurugan, K. Shunmugapriya, R. Vanitha, GIS based assessment of ground water for domestic and irrigation purpose in Vazhapadi Taluk, Salem, Tamil Nadu, India. Taiwan Water Conservancy **68**(2), 1–10 (2020c)

P. Balamurugan, P.S. Kumar, K. Shankar, R. Nagavinothini, P.J. Sajil Kumar, Impact of climate and anthropogenic activities on groundwater quality for domestic and irrigation purposes in Attur region, Tamilnadu, India. Desalin. Water Treat. **208**, 172–195 (2020d)

A.K. Batabyal, S. Chakraborty, Hydro-geochemistry and water quality index in the assessment of groundwater quality for drinking uses. Water Environ. Res. **87**(7), 607–617 (2015). https://doi.org/10.2175/106143015X14212658613956

J. Colins, M.C. Sashikkumar, P.A. Anas, M. Kirubakaran, GIS-based assessment of aquifer vulnerability using DRASTIC model: A case study on Kodaganar basin. Earth Sci. Res. J. **20**(1), 1–8 (2016)

C.G. Cude, Oregon water quality index: A tool for evaluating water quality management effectiveness. J. Am. Water Resour. Assoc. **37**(1), 125–137 (2001)

S.F.R. Khadri, C. Pande, GIS-based analysis of groundwater variation in Mahesh River Basin, Akola and Buldhana Districts, Maharashtra, India. IJPRET **4**(9), 127–136 (2016a)

S.F.R. Khadri, C. Pande, Ground water flow modeling for calibrating steady state using MODFLOW software: A case study of Mahesh River basin, India. Model. Earth Syst. Environ **2**, 39 (2016b). https://doi.org/10.1007/s40808-015-0049-7

M. Kirubakaran, C. Ashokraj, J. Colins Johnny, R. Anjali, Groundwater quality analysis using WQI and GIS techniques: A case study of Manavalakurichi in Kanyakumari District, Tamilnadu, India. Int. J. Innov. Sci. Eng. Technol. **2**(11), 341–347 (2015)

M. Kirubakaran, J.C. Johnny, C. Ashokraj, S. Arivazhagan, A geostatistical approach for delineating the potential groundwater recharge zones in the hard rock terrain of Tirunelveli taluk, Tamil Nadu, India. Arab. J. Geosci. **9**(5), 382 (2016)

P.S. Kumar, P. Balamurugan, Evaluation of groundwater quality for irrigation purpose in attur taluk, Salem, Tamilnadu, India. Water Energy Int. **61**(4), 59–64 (2018)

P.S. Kumar, P. Balamurugan, Suitability of ground water for irrigation purpose in. Omalur Taluk, Salem, Tamil Nadu, India. Indian J. Ecol. **46**(1), 1–6 (2019)

C.B. Pande, Introduction, in *Sustainable Watershed Development*, Springer briefs in water, Springer nature (2020)

C.B. Pande, K.N. Moharir, Groundwater Resources Development and Planning in the Semi-Arid Region, Springer, Cham, 1, pp. XIV, 571 (2021a). https://doi.org/10.1007/978-3-030-68124-1

C.B. Pande, K.N. Moharir, B. Panneerselvam, et al. Delineation of groundwater potential zones for sustainable development and planning using analytical hierarchy process (AHP), and MIF techniques. Appl. Water. Sci. 11, 186 (2021b). https://doi.org/10.1007/s13201-021-01522-1.

C.B. Pande, S.F.R. Khadri, N.K. Moharir, R.S. Patode, (2017), Assessment of groundwater potential zonation of Mahesh River basin Akola and Buldhana districts, Maharashtra, India using remote sensing and GIS techniques. Sustain. Water Resour. Manag. 4, 965–979 (2018). https://doi.org/10.1007/s40899-017-0193-5

C.B. Pande, K.N. Moharir, S.K. Singh, A.M. Varade, An integrated approach to delineate the groundwater potential zones in Devdari watershed area of Akola district, Maharashtra, Central India. Environ. Dev. Sustain (2019). https://doi.org/10.1007/s10668-019-00409-1

B. Panneerselvam, S. Karuppannan, K. Muniraj, Evaluation of drinking and irrigation suitability of groundwater with special emphasizing the health risk posed by nitrate contamination using nitrate pollution index (NPI) and human health risk assessment (HHRA). Hum. Ecol. Risk Assess. Int. J., 1–25 (2020a). https://doi.org/10.1080/10807039.2020.1833300

B. Panneerselvam, S.K. Paramasivam, S. Karuppannan, N. Ravichandran, P. Selvaraj, A GIS-based evaluation of hydrochemical characterisation of groundwater in hard rock region, South Tamil Nadu, India. Arab. J. Geosci. **13**(17), 1–22 (2020b). https://doi.org/10.1007/s12517-020-05813-w. a. potential zonation of Mahesh River basin Akola and Buldhana districts, Maharashtra, India

B. Panneerselvam, K. Muniraj, C. Pande, N. Ravichandran, Prediction and evaluation of groundwater characteristics using the radial basic model in semi-arid region, India. Int. J. Environ. Anal. Chem. 1–17 (2021a)

B. Panneerselvam, K. Muniraj, C. Pande, N. Ravichandran, M. Thomas, S. Karuppannan, Geochemical evaluation and human health risk assessment of nitrate-contaminated groundwater in an industrial area of South India. Environ. Sci. Pollut. Res., 1–18 (2021b).

B. Panneerselvam, K. Muniraj, M. Thomas, N. Ravichandran, B. Bidorn, Identifying influencing groundwater parameter on human health associate with irrigation indices using the Automatic Linear Model (ALM) in a semi-arid region in India. Environ. Res. **202**, 111778 (2021c)

J. Rajesh, C.B. Pande, S.A. Kadam, et al. Exploration of groundwater potential zones using analytical hierarchical process (AHP) approach in the Godavari river basin of Maharashtra in India. Appl Water Sci 11, 182 (2021). https://doi.org/10.1007/s13201-021-01518-x

S. Ramalingam, B. Panneerselvam, S.P. Kaliappan, Effect of high nitrate contamination of groundwater on human health and water quality index in semi-arid region, South India. Arab. J. Geosci. **15**(3), 1–14 (2022)

K. Rawat, S. Pradhan, V. Tripathi, L. Jeyakumar, S.K. Singh, Statistical approach to evaluate groundwater contamination for drinking and irrigation suitability. Groundw. Sustain. Dev. (2019). https://doi.org/10.1016/j.gsd.2019.100251. a. science and technology. Cham: Springer. https://doi.org/10.1007/978-3-030-47244-31

K. Shunmugapriya, B. Panneerselvam, K. Muniraj, N. Ravichandran, P. Prasath, M. Thomas, K. Duraisamy, Integration of multi criteria decision analysis and GIS for evaluating the site suitability for aquaculture in southern coastal region, India. Mar. Pollut. Bull. **172**, 112907 (2021)

D.K. Todd, L.W. Mays, *Groundwater Hydrology, Edition*, vol 1625 (Wiley, Hoboken, 2005). a. using remote sensing and GIS techniques. Sustainable Water Resources Management. https://doi.org/10.1007/s40899-017-0193-5

World Water Assessment Programme (United Nations), & UN-Water, *Water in a Changing World*, vol 1 (Earthscan, 2009), UK

Chapter 2
Climate Change Impact on Groundwater Resources in Semi-arid Regions

M. M. Deshmukh, Ahmed Elbeltagi, and Saber Kouadri

Abstract Groundwater is one of the significant freshwater resources and provides essential freshwater supply, chiefly in dry regions where surface water availability is limited. However, groundwater-related problems are among today's most acute and complex technical problems. The climate change outcomes connected to water resources are an increase in temperature, variation in precipitation patterns, snow cover, and a probable increase in the frequency of flooding and droughts. Due to this, climate change intensifies the global hydrological cycle and affects the quality and quantity of groundwater resources. Climate changes also disturb the hydrological cycle, directly increasing the evaporation level of accessible superficial water and vegetation transpiration and affecting the amounts of soil infiltration and deeper percolation. Subsequently, these variations can directly affect evaporative demand over land, rainfall rate, timings, and concentration rates and indirectly affect the flux and storage of water in surface (i.e. lakes, rivers, seas) and subsurface (i.e. groundwater storage (GWS)) reservoirs. In addition, there are other related factors such as groundwater contamination, seawater intrusion, and water scarcity. Climate changes affect surface water resources through climate variants such as air temperature, rainfall, and vegetation transpiration variations, and these variations also affect groundwater storage. Larger rainfall variations directly affect surface water levels and seawater intrusion and result in a drop in water storage. These variations directly affect groundwater storage depending on the percolation rate, volume, and distribution of groundwater recharge. We need to find out the solution on the impact of

M. M. Deshmukh (✉)
Department of Geology, Shri Shivaji Science College, Amravati, India

A. Elbeltagi
Agricultural Engineering Department, Faculty of Agriculture, Mansoura University, Mansoura, Egypt

S. Kouadri
Laboratory of Water and environment Engineering in Sahara Milieu (GEEMS) Department of Civil Engineering and Hydraulics Faculty of Applied Sciences, Kasdi Merbah University, Ouargla, Algeria

B. Panneerselvam et al. (eds.), *Climate Change Impact on Groundwater Resources*, https://doi.org/10.1007/978-3-031-04707-7_2

climate change and groundwater resources and exact estimation of groundwater recharge in the semi-arid region. For understanding climate variations, several global climate models (GCMs) are available. GCMs help detect quantification of groundwater recharge and provide a suitable adaption scheme to protect groundwater storage from climate change. This study shows the impact of climate change on groundwater resources and the methodology used to measure this impact.

Keywords Groundwater · Climate · Groundwater recharge · Hydrological cycle

2.1 Introduction

Water is the most useful and renewable resource for people. Without water, no life can exist, as water is a life-characteristic of air. Two-thirds of the human body are estimated to be made up of water (Priyan 2021). Water is essential not only for the survival of human but also for the survival of animals, plants, and all other ecosystems. Sustainable quality and quantities are threatened by different factors such as agriculture production and polluted water etc. As a result of climate change, there are uncertainties about water supply and management (Lal and Datta 2021; Balamurugan and Balakumaran 2015; Balamurugan and Kumar 2016). Climate change directly affects surface water resources through changes in major long-term climate variables such as precipitation, evapotranspiration, and temperature of the atmosphere. There are more complicated and misunderstood relations between groundwater resources and climate change. Kumar (2016) have said that if greater precipitation change can lead to higher or lower groundwater levels. Furthermore, because it increases sea level and reduces resources on the earth surface. The atmosphere varies by way of soil moisture, percolation, and groundwater systems. Hydrological cycling can influence the amount of soil infiltration, thus resulting in deeper percolation and in the recharge of groundwater (Wu et al. 2020; Kumar and Balamurugan 2018; Panneerselvam et al. 2021b). Effects on precipitation, such as increased precipitation variability and weather, can also lead to prolonged drought and flooding periods that openly affect groundwater availability and dependency (Rane and Jayaraj 2021). According to the International Groundwater Resources Assessment Centre (IGRAC), climate change is becoming a greater concern, and, although this issue is rightly receiving a lot of attention, its impact on groundwater is still underappreciated (Pande et al. 2022). The IGRAC estimates that climate changes also affect groundwater quality, for example, a rise in sea levels may lead to saltwater intrusion into coastal aquifers, affecting groundwater quality (Gaur et al. 2021; Sharief and Zakwan 2021). In order to accurately anticipate future supplies, the groundwater resource management team will increasingly need to factor in the effects of global climate change. Several studies have found that groundwater streamflow is sensitive to climate change in watersheds all around the world (Fathi et al. 2021; Kumar and Balamurugan 2019; Muniraj et al. 2021).

2.2 Background of Research

The findings suggest that a smaller percentage of winter rainfall will fall as snow as the temperature rises. As a result, the spring snowmelt peak is lowered, whereas the winter overflow threat is almost certain to increase. In summer, significant quarterly soils and stream fluxes are reduced by up to 50%, mainly due to water quality, groundwater withdrawals, and hydropower generation problems (Eckhardt and Ulbrich 2003).

An integrated hydrological model was developed in Belgium, aiming to study the influence of climate change on the hydrological cycle of representative basins and to mainly examine groundwater flows modelled using a spatially dispersed system (Bohidar and Ahmad 2021; Balamurugan et al. 2020a; Panneerselvam et al. 2020a). Results on the progress of groundwater and groundwater reserves in relation to climate change were discussed. Using this model, investigate how the levels and resources of groundwater about climatic differences are being reduced. The Geer Basin aquifer has been verified situations and display no effects of seasonal variations on the groundwater levels (Brouyère et al. 2003).

Use for the effect of climate change impact on the streamflow of the upper Mississippi river basin of RCM (regional climate), and Hydrology Model and SWAT (solar and water model). Model could give information about the climate change effects. This survey showed that the repeated fluid variation significantly exceeded the joint modelling system's bias with an 18% bias. Our results also show that the relationship between annual flux and annual precipitation may lead to a change in future climatic conditions, as every per-unit rainfall increase will lead to more significant stream fluctuations (Jha et al. 2004; Balamurugan et al. 2020c; Panneerselvam et al. 2021a). In a case report in East Anglia, Great Britain, researchers explained the direct and indirect impacts of groundwater recharges on climate and socioeconomic change. Many factors will disturb the recovery of groundwater, such as the changing rules of precipitation and temperature, coastal flooding, urbanization and sealing, forest creation, cropping and change in rotation, etc (Shahid et al. 2021). Usually, the direct impacts of climate situations are more important locally than socioeconomic ones. But, despite the several worries have involved in the usage of circumstances, and an exclusively attend to the direct impacts of climate change like those arising from temperature and rainfall variations are negligence the theoretically significant part of social ethics and economic procedures in determining the overhead landscape aquifers (Holman 2006).

The research outcomes specify that the total amount of groundwater recharge is predicted to grow due to climate change. A higher concentration and rainfall rate will also contribute significantly to surface runoff, even though global warming may result in improved evapotranspiration rates. Warmer winter temperatures will decrease the level of ground frost and alter the spring melt from spring towards winter, permitting extra water to penetrate the ground. Although numerous earlier climate change impact investigations have been attentive to the time-based variations in groundwater recharge, outcomes of other studies have recommended that the effects have high spatial changeability (Jyrkama and Sykes 2007; Balamurugan et al. 2020b; Panneerselvam et al. 2020).

Scientists have stated that it is essential to try and identify the problem with persistent research, based on a good set of quality meteorological and hydrological information, which is currently far from satisfactory, to overcome the current and future water and ecological problems. It is crucial that ecosystems such as those that defend recharging areas, wetlands, and mountain forests are secured and reinstated. The gap between the water supply and demand must be reduced, through more energy-efficient irrigation systems, farmers' training, wastewater recycling, public consciousness, and groundwater preservation. More effective irrigation systems, farmers' training, wastewater recycling, water conservation through public awareness, and groundwater regulation are all needed to close the gap between supply and demand for better groundwater management (Dragoni and Sukhija 2008).

The effects of climate change on groundwater recharge as well as baseflow were investigated in the upper catchment area of Ssezibwa in Uganda. For this study, ancient data, which certainly exposed indications of climate change based on fluctuations detected in temperature and discharge rate, were first studied. For this investigation, the Statistical DownScaling Model (SDSM) was used to infer upcoming climate change situations, which were obtained from the UK HadCM3 climate model. The equivalent increase in temperature ranged between 1 and 4 °C, and these temperature changes were expected to strengthen the hydrological cycle. Between the 2020s and 2080s, the existing mean annual daily baseflow of 157 mm/year (69% of discharge) is predicted to increase by 20–80%. From the present 245 mm/year, the comparable increase in recharge ranges from 20% to 100%. The findings lay the groundwork for further research into climate data downscaling and sensitivity analysis for virtual hydrological fluctuations in the catchment area (Nyenje and Batelaan 2009).

An approach for estimating the average changes in groundwater recharge under future climates is described. This approach is used in Australia's Murray–Darling Basin (MDB), which covers 1,060,000 km^2. Climate sequences were created using the outputs of 15 global climate models and 3 scenarios for a 2030 climate compared to a 1990 climate. Owing to the discrepancies between the climate sequences provided by the 15 various global climate models, predicting the direction, much alone the extent, of the shift in recharge for a 2030 climate is problematic (Crosbie et al. 2010; Balamurugan et al. 2020; Panneerselvam et al. 2021).

The results show that recharging is precisely proportional to rainfall in the examined area, even when substantial soil moisture shortages are projected. This means that the water recharging behaviour is controlled by privileged and/or localized flow processes (Abraham and Priyadarshini 2021). In this area underlain by lateritic soils, soil moisture balance and source-responsive models (SMBMs) and unsaturated flow models based on uniform flow guided by the Richards equation are found to be ineffective in estimating recharge. Concerns have been expressed about the use of such models for recharging estimation, and hence for examining future patterns owing to climate or land use change, unless they are backed up by enough hydraulic data to confirm the recharge processes (Cuthbert and Tindimugaya 2010).

Understanding how climate change could influence groundwater systems is an essential component of sound long-term management of our water supplies. On the other hand, calculating how climate change could impact groundwater systems is complicated and forecasting upcoming groundwater situations would still be more complicated because of the multifaceted combinations of processes that have an effect on groundwater recharge, discharge, and quality. Enhanced interpretation, improved understanding of processes, and superior modelling capabilities will be needed to calculate and forecast the upcoming situations of this essential resource in the face of projected climate changes (Earman and Dettinger 2011).

Despite the fact that projections in place and time are burdened by means of doubt, researchers have discovered that the future intensity and frequency of dry periods combined with warming trends must be addressed in the status of groundwater supplies. Finally, the possible effects of groundwater on the global climate system are mostly unknown. All circumstances, changes to global change must include a smart organisation of groundwater as a renewable energy development. Since groundwater storage is already overtapped in many areas, previously untapped subterranean storage could be an option for satisfying the collective demands of agriculture, industry, municipal and domestic water supplies, and ecosystems during times of scarcity (Green et al. 2011; Orimoloye et al. 2022; Pande et al. 2018; Pande et al. 2021; Rajesh et al. 2021).

Frequently, rainfall is downscaled in climate change impact studies; on the other hand, the dependability of the downscaled outcome is frequently poor or variable, as there is a repeatedly small connection among the predictors. A reduced connection has been frequently observed in mesoscale processes occurring at a site scale, which are not represented in provincial models due to their representative spatial and temporal sizes in comparison to large-scale local rainfall. Mesoscale rainfall processes usually occur in the summer period from a variety of convective clouds, which are an outcome of local-scale evapotranspiration from elevated temperatures and solar radiation magnitudes. As an outcome, global-scale models may underestimate the summer precipitation measured at a site (Kumar 2012).

Scientists have stated that the meteoric rise of the nuclear underwater economy demands the development of a groundwater management scheme with convenient supply and demand strategies through brave new ideas and resource allocation policies. In the twenty-first century, the influence of climate change significantly exceeding India's management of hydroclimates is accepted. For instance, the Himalayan snowmelt has heavily loaded the Indo-Gangetic aquifer system. Aquifers react little by little to droughts and climatic variations than surface storage (Shah 2013).

In Eastern Canada, the likely impacts of future climate change on soil water recharge were investigated on a local scale for a watershed of 546 km. Infiltration model recharge was assessed by the Hydrological Depositional Performance Assessment (HELP). The results indicate that higher temperatures do not compensate for the huge increase in winter infiltration, resulting in increased evapotranspiration. Clear changes have been made to the different climate change situations with regard to specific water budget mechanisms (Rivard et al. 2014).

Climate changes have led to disadvantageous phenomena such as high precipitation, which could worsen soil erosion. Therefore, effective and qualitative monitoring of water resources to evaluate groundwater recharges, groundwater flows, and soil erosion is required. The study resulted in a 50% improvement in the usual soil recharge, 42% in baseflow, and a corresponding reduction of 72% in sediments. The recommended procedure thus contributes positively to the hydro-ecosystem and the reduction of muddy water in a steep, sloping water border (Min et al. 2014).

A study evaluated the impact of future climactic variations and sea-level increase on groundwater recharge and groundwater in the low and shallow waters of Hanko on the south coast of Finland and used groundwater flow modelling as an instrument of evaluation. There were strong correlations between groundwater recovery and sea levels with simulated surface leakage. The increase in sea levels will increase groundwater levels in future climates and lead to more surface leakage in low-lying areas, thus increasing the risk of winter and early spring surface floods (Luoma and Okkonen 2014).

The impact of climate change causes doubts about water supply and management. Although the impact of climate on the resources of surface water is directed at the most important long-term climatic variables, such as air temperature, precipitation, and evapotranspiration, it is more difficult and unsuccessful to connect the changing climatic variables to groundwater. Two more frequent periods of high or low groundwater can be caused by higher rainfalls, and saline intrusion into coastal waterfowl can increase and decrease in reserve (Kumar 2016).

For investigating climate change effects on recharges, model studies are helpful. Special effects on recharge for some aquifers have been projected and used for climate change. Predicted temperature and precipitation changes have been reasonably distributed side by side in each charging mechanism, producing approximate, qualitative calculations (not magnitude). Several solution patterns are derived from the research, for example, the standard 10–20% decreases in total refuelling of the southern aquifers are indicated by the available estimated figures but without a vast number of doubts. Second, small alterations are likely to be achieved in the northern aquifers including small increases in total charging. Reasons that contribute to major doubts in the approximation include the following:

1. Limited studies quantitatively coupling climate projections with recharge estimation methods using detailed, process-based numerical models
2. A generally poor understanding of the hydrological flow paths and processes in mountain systems
3. Difficulty in predicting the response of focused recharge to potential changes in the frequency and intensity of extreme precipitation events
4. Unconstrained feedbacks between climate, irrigation practices, and recharge in highly developed aquifer systems (Meixner et al. 2016).

Consequently, it is important to add to this selected standardization plan, where likely, that normal standardization information should include climatic excesses in the directive in order to minimize the standardization model bias. The results strongly recommend that climate forecast bands be fixed with hydrogeological models in order to provide reliable calculations and associated doubts (Moeck et al. 2016).

In the West of Iran, Global Climate Models (GCM) and Hadley Centre is predicting the impact of climate change for soil charging in a dry environment (HADCM3). The study has been shown to assess the impact of climate change in an arid environment on natural recharge. For the aquifer examined, the recharge is reduced by the average of long-term precipitation in comparison with the recharge calculated. Based on the results, a GCM, HADCM3, was carefully selected as the best model to forecast the parameters of the weather conditions in the province of Ilam, west of Iran. The results of HADCM3 have shown that the mean annual precipitation in the study is about 3% lower, whereas the mean annual temperature is around 0.5 °C higher (Ghazavi and Ebrahimi 2018).

The effect of temperature variations on soil storage (GWS) could distress freshwater resources' sustainability. At the time, a fully interconnected climate model was used to study the variation in groundwater storage, and seven dangerous aquifers were found to be suggestively distracted by remote sensing observations. Researchers have assessed the potential impacts on GWS differences in a business-as-usual situation through the twenty-first century (RCP8.5). As a substitute, the trend may result from improved evapotranspiration and decay in snowmelt that together lead to a variety of GWS diagonal fluctuation reactions. The results show that the impacts of GWS on changes due to climate do not necessarily reflect the long-term trend of rains. Finally, the researchers compared the impacts of anthropogenic and climate-driven pumping. The decrease in GWS is mainly due to the mutual impacts of overpumping and climate effects; however, pumping contributions can simply go far beyond natural substitution (Wen-Ying Wu et al. 2020).

Researchers have presented a hypothesis that the quantitative recharging of groundwater and associated water temperature imprinting is primarily determined by infiltrating river-fed aquifers. Investigators showed that regular changes in the methods of groundwater charging could be a significant element of distressing future groundwater temperatures. Moreover, high runoff times are expected to powerfully influence groundwater temperatures through the connections of surface water and the recharging of groundwater. Thus, a shift of precipitation and flood actions from the summer to the winter months could be accompanied by a rise in groundwater recharge during relative winter season for the "business-as-usual" climate change scenario and by a tendency for "cooling down" groundwater resources during the winter season (Epting et al. 2021).

2.3 Impact of Climate Change on Groundwater Resources

Much of the impact of climate change on precipitation and surface water and ultimately on groundwater systems, as the area of groundwater saturation, is mostly charged by precipitation and surface water bodies (lakes, rivers). In addition, the impact on surface water quality and quantity is affected by climate change. Water managers have a major concern about surface water, and management is likely to either reduce or increase the value of groundwater. Groundwater is the main source of drinking water for human consumption and for agriculture. The potential impacts of climate change on groundwater systems are important to consider as part of the hydrological cycle, and groundwater systems are expected to be affected by recharging changes, including changes in precipitation and evapotranspiration and potential environmental changes of connections between groundwater and groundwater systems.

The hydrological cycle represents a continuous and repetitive water process from the earth's surface to the atmosphere and vice versa under the influence of solar heat. The circulation of water from the hydrosphere to the atmosphere and then to the lithosphere is mainly caused by solar heat. So, the impact of climate influence on the hydrological cycle also affects precipitation, soil moisture, evaporation, transpiration, runoff, groundwater recharge, and groundwater movement. Climate change affects evaporation and transpiration; however, warming global temperature increases the rate of evaporation and is causing more precipitation. Higher evaporation and precipitation rates are not uniformly circulated around the world. Some areas may experience heavier-than-normal precipitation, whereas some areas may undergo droughts due to the traditional locations of rain belts and desert shifts in response to a climate change (Fig. 2.1). Soil moisture is stored in the soil and affects precipitation, climatic conditions, solid properties, etc. Soil moisture is good for crop cultivation and farming. The content and output from these models are directly simulated by global climate models to indicate possible changes. However, the impact of climate change on soil moisture will vary not just from climate change to soil capability, soil quality, etc. The soil's water capability will affect possible soil humidity shifts. The lower the capacity, the greater is the climate change and its impact on groundwater resources. Climate change may also affect soil characteristics, perhaps through changes in water logging or cracking, which may affect soil moisture storage properties.

Climate change also affects aquifers, which are usually refilled by efficient rainfall and surface water bodies (Vidya et al. 2021). This water may percolate and reach the aquifer rapidly, through macropores, fissures, cracks, etc. and more gradually by infiltrating through soils and permeable rocks overlying the aquifer (Fig. 2.2). Climate change affects rainfall and will also affect groundwater recharge but so will alteration in the recharge period (Zakwan 2021). Increased rainfall is generally likely to result in increased groundwater recharge. Various types of aquifers include reclaims, such as unrestricted water, which is recharged from lakes, rivers, and plumes over distances ranging from a few kilometres up to thousands of kilometres using local rainfall and surface water bodies.

On the other hand, a confined aquifer features an impermeable overlying bed, and local rainfall has no impact on the aquifer. Charging rate is influenced by the overlying strata's hydrological features. "Rapid" charging can, in principle, occur whenever it rains, so if this process is dominated by charging, then it is more affected by rainfall changes than by soil variability seasonal cycles (Fig. 2.3). Therefore, changes in recharging are determined by changes in local flow times and the

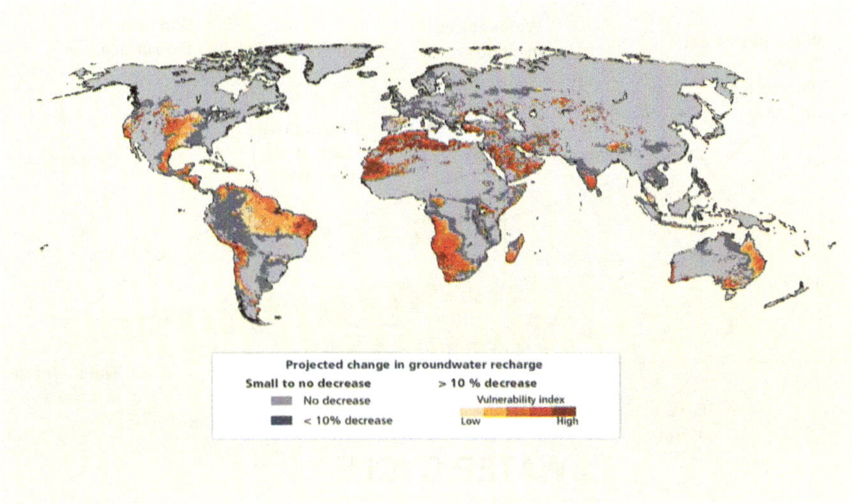

Fig. 2.1 Groundwater recharge. (Source: https://www.metlink.org/resource/impact-of-climate-change-on-water-supplies/)

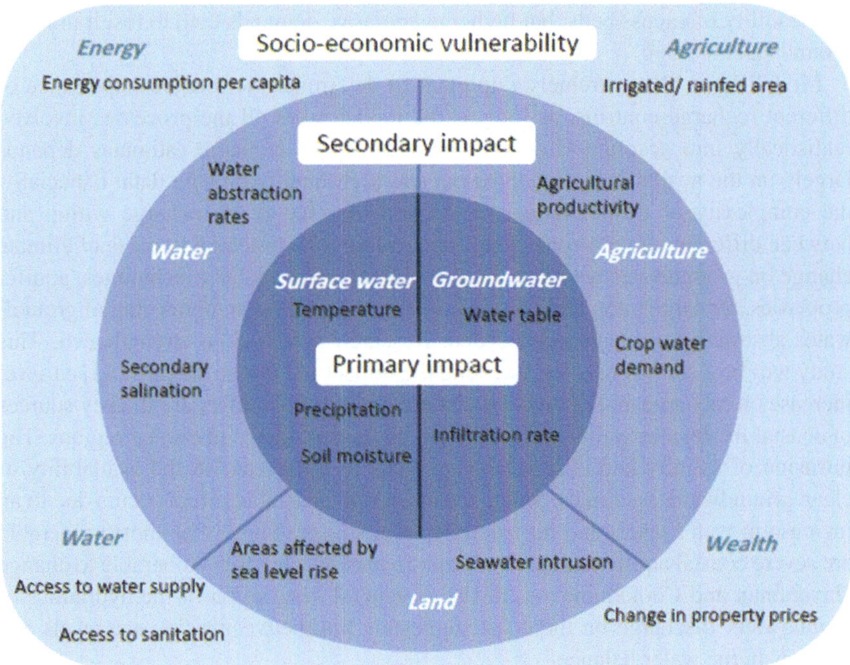

Fig. 2.2 Indicators of climate change impact on water resources in a semi-arid region. (Source: https://water.fanack.com/impact-of-climate-change-on-water-resources/)

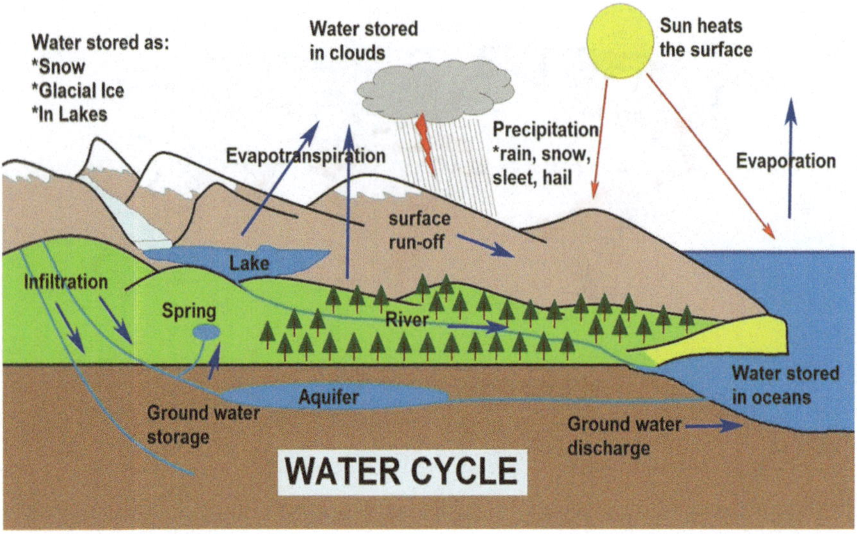

Fig. 2.3 Water cycle

permeability of excess beds, but higher evaporative demands tend to result in lower groundwater storage.

Modelling is also extremely important to determine the relative importance of different recharge controls, as long as the model takes all the processes involved realistically into account. However, the accuracy of recharge estimates depends largely on the available hydrogeological and high-quality climate data. Especially, the complexity of the recharge process and the change in recharge within and between different climate zones make determining the potential impacts of climate change on groundwater resources hard. Research on modelling techniques, aquifer properties, recharge rates, and marine intrusion, as well as on monitoring of groundwater abstraction, is generally needed. Research needs to be strengthened. This study will be a sound basis for assessing the impact of climate change and sea-level increase on recharge and groundwater sources. Coastal aquifers are the key sources of coastal freshwater. Salinity is one of the biggest problems in coastal regions. The intrusion of salinity and climate change leads to a decrease in the availability of clean groundwater resources. Saline intrusion into coastal aquifers occurs due to an increase in water levels and intrusion by local soil gradients. The most vulnerable are severe coastal aquifers, as groundwater in narrow islands is vulnerable to change (Jayaraman and Chokkalingam 2021b). A general description of the hydraulics of groundwater discharge on the coast suggests a link between rising sea levels and changes in the water balance.

2.4 Impact of Climate Change on Groundwater Quality

Impact of Climate is directly impact on the quality of groundwater and improvement of overall climate changes with reduction on annual precipitation (Mishra et al. 2021; Kouadri et al. 2022). The effects may lead to a reduction of the annual recharge. The residence time of water will be higher, degrading the quality of water and increasing the concentration of pollutants (Oliveira). Dragoni and Sukhija stated that climate change is not only expected to affect recharge and discharge but also groundwater quality (Gayathri et al. 2021). Their analysis specifies that water recharged in an arid period has a high concentration of salts and a higher quantity of total dissolved solids. At the same time, as for the duration of a wet period, the converse may happen. The enrichment of seawater pollutes groundwater in coastal aquifers (Mohan and Krishnakumar 2021). Seawater intrusion in coastal aquifers occurs when a permeable formation comes in contact with the ocean. Seawater intrusion is one of the examples of climate influence on the water quality of groundwater resources. Seawater intrusion is due to rising seawater levels due to an increase in sea surface temperatures (Ravi et al. 2020). Seawater intrusion can also be generated by more than just pumping in coastal regions, possibly because of either unfortunate management of water use or increased demands resulting from stress factors related to climate change or non-climatic stressors (Moseki 2017). Seawater intrusion may also be driven by droughts. According to Bates et al. (2008), droughts due to increasing frequency and intensity of deficiencies in catchment areas would lead to more severe and frequent seawater intrusion in estuaries. Therefore, it worsens surface water and groundwater resource quality, decreasing freshwater availability in coastal areas (Jayaraman et al. 2021a, b). The sea level goes up, which, together with climate change, will diminish freshwater supply in many coastal areas. This water is not helpful to humans without filtering and excessive treatment. In addition, any reduction in groundwater recharge will worsen the cause of the sea-level rise. In inland aquifers, reduction in groundwater recharge shows the way to saltwater intrusion from neighbouring saline aquifers (Dyer 2014).

2.5 Concluding Remarks

In conclusion, it is important to conserve natural properties like groundwater, forests, etc., and further research is needed on how to increase and preserve groundwater levels, focusing on factors responsible for climate change, its effect on groundwater quality and quantity on a global scale with importance to groundwater resources and anticipated mitigation strategies. Climate variation is always grated worry in the view ecosystem on the earth. Even though this problem is presently receiving a great deal of attention, its effect on the earth affects water, human life, animal life, etc.

Climate changes are due to deforestation, pollution, etc.; it is important to preserve forests as they reduce gas emissions. If not, the unintentional consequences might make matters worse. The impacts of climate changes on groundwater are as follows:

- Increase in atmospheric water vapour content and changes in rainfall patterns
- Evapotranspiration
- Soil moisture
- Runoff, change in groundwater recharge, ice melting change, snow cover reduction, snowfall, and sea-level increase

Subsequently, these variations can directly impact rainfall amounts, timings, and intensity rate and indirectly impact the flux and storage of water in surface and subsurface reservoirs like lakes, rivers, groundwater, etc. The greater variations in precipitation affect surface water levels, groundwater levels, and seawater intrusion due to sea-level rise and resource reduction. Nevertheless, it is clear that due to climate change, the increasing hazards are real and the results of climate change are numerous and alarming. Geologists must increase their research work and find out a solution to global warming and preserve groundwater resources.

References

M. Abraham, M.K. Priyadarshini, Numerical modeling for groundwater recharge, in *Groundwater Resources Development and Planning in the Semi-Arid Region*, ed. by C. B. Pande, K. N. Moharir, (Springer, Cham, 2021). https://doi.org/10.1007/978-3-030-68124-1_9

P. Balamurugan, S. Balakumaran, Soil quality assessment around magnesite mines and Salem township using GIS techniques. Int. J. Adv. Eng. Technol. **8**(1), 1997 (2015)

P. Balamurugan, P.S. Kumar, Quality of ground water assessment in Salem District using GIS techniques. Adv. Nat. Appl. Sci **10**(3), 22–32 (2016)

P. Balamurugan, P.S. Kumar, K. Shankar, Dataset on the suitability of groundwater for drinking and irrigation purposes in the Sarabanga River region, Tamil Nadu, India. Data Brief **29**, 105255 (2020)

P. Balamurugan, P.S. Kumar, K. Shankar, R. Nagavinothini, K. Vijayasurya, Non-carcinogenic risk assessment of groundwater in southern part of Salem District in Tamilnadu, India. J. Chil. Chem. Soc. **65**(1), 4697–4707 (2020a)

P. Balamurugan, P.S. Kumar, K. Shankar, R. Nagavinothini, P.J. Sajil Kumar, Impact of climate and anthropogenic activities on groundwater quality for domestic and irrigation purposes in Attur region, Tamilnadu, India. Desalin. Water Treat. **208**, 172–195 (2020b)

P. Balamurugan, G. Shunmugapriya, R. Vanitha, GIS based assessment of ground water for domestic and irrigation purpose in Vazhapadi Taluk, Salem, Tamil Nadu, India. Taiwan Water Conservancy **68**(2), 1–10 (2020c)

B. C. Bates, Z. W. Kundzewicz, S. Wu, et al. (eds.), *Climate Change and Water Technical Paper of the Intergovernmental Panel on Climate Change* (IPCC Secretariat, Geneva, 2008) 210 pp

A.K. Bohidar, I. Ahmad, Development of conceptual model and groundwater flow modeling using GMS software: A case study for Dharsiwa Block, Chhattisgarh, India, in *Groundwater Resources Development and Planning in the Semi-Arid Region*, ed. by C. B. Pande, K. N. Moharir, (Springer, Cham, 2021). https://doi.org/10.1007/978-3-030-68124-1_8

S. Brouyère, G. Carabin, A. Dassargues, Climate change impacts on groundwater resources: Modeled deficits in a chalky aquifer, Geer basin, Belgium. Hydrogeol. J. **12**, 123–134 (2003)

R.S. Crosbie, J.L. McCallum, G.R. Walker, H.S. Chiew, Modelling climate-change impacts on groundwater recharge in the Murray-Darling Basin, Australia. Hydrogeol. J. **18**, 1639–1656 (2010)

M.O. Cuthbert, C. Tindimugaya, The importance of preferential flow in controlling groundwater recharge in tropical Africa and implications for modelling the impact of climate change on groundwater resources. J. Water Clim. Change **1**(4), 234–245 (2010)

W. Dragoni, B. S. Sukhija (eds.), *Climate Change and Groundwater: A Short Review*, vol 288 (Geological Society, London, Special Publications., 2008), pp. 1–12

W.L. Dyer, Relationship of climate change to seawater intrusion in coastal aquifers, Unpublished thesis for a Master of Science degree, Graduate College of the Oklahoma State University, USA, 2014

S. Earman, M. Dettinger, Potential impacts of climate change on groundwater resources – A global review. J. Water Clim. Change **2**(4), 213–229 (2011)

K. Eckhardt, U. Ulbrich, Potential impacts of climate change on groundwater recharge and stream flow in a central European low mountain range. J. Hydrol. **284**(1–4), 244–252 (2003)

J. Epting, A. Michel, A. Affolter, P. Huggen berger, Climate change effects on groundwater recharge and temperatures in Swiss alluvial aquifers. J. Hydrol. X **11**, 100071 (2021). Elsevier

S. Fathi, J.S. Hagen, A. Matanó, G.E.H. Nogueira, Review of GIS multi-criteria decision analysis for managed aquifer recharge in semi-arid regions, in *Groundwater Resources Development and Planning in the Semi-Arid Region*, ed. by C. B. Pande, K. N. Moharir, (Springer, Cham, 2021). https://doi.org/10.1007/978-3-030-68124-1_2

S. Gaur, A. Johannet, D. Graillot, P.J. Omar, Modeling of groundwater level using artificial neural network algorithm and WA-SVR model, in *Groundwater Resources Development and Planning in the Semi-Arid Region*, ed. by C. B. Pande, K. N. Moharir, (Springer, Cham, 2021). https://doi.org/10.1007/978-3-030-68124-1_7

S. Gayathri et al., Multivariate statistical tools in assessing the quality of water resources in Netravati River Basin, Karnataka, India, in *Groundwater Resources Development and Planning in the Semi-Arid Region*, ed. by C. B. Pande, K. N. Moharir, (Springer, Cham, 2021). https://doi.org/10.1007/978-3-030-68124-1_16

R. Ghazavi, H. Ebrahimi, Predicting the impacts of climate change on groundwater recharge in an arid environment using modeling approach. Int. J. Clim. Change Strateg. Manag. **11**(1) (2018)., ISSN: 1756-8692

T.R. Green, M. Taniguchi, H. Kooi, J.J. Gurdak, D.M. Allen, K.M. Hiscock, H. Treidel, A. Aureli, Beneath the surface of global change: Impacts of climate change on groundwater. J. Hydrol. **405**(3–4, 5), 532–560 (2011)

I.P. Holman, Climate change impacts on groundwater recharge- uncertainty, shortcomings and the way forward? Hydrogeol. J. **14**(5), 637–647 (2006). https://doi.org/10.1007/s10040-005-0467-0

R. Jayaraman, L. Chokkalingam, Correlation between land surface temperature and vegetation cover of Nagapattinam Coastal Zone, Tamil Nadu, using geospatial techniques, in *Groundwater Resources Development and Planning in the Semi-Arid Region*, ed. by C. B. Pande, K. N. Moharir, (Springer, Cham, 2021a). https://doi.org/10.1007/978-3-030-68124-1_12

R. Jayaraman, L. Chokkalingam, Assessment of aquifer vulnerability for sea-water intrusion in Nagapattinam Coast, Tamil Nadu, using geospatial techniques, in *Groundwater Resources Development and Planning in the Semi-Arid Region*, ed. by C. B. Pande, K. N. Moharir, (Springer, Cham, 2021b). https://doi.org/10.1007/978-3-030-68124-1_10

M. Jha, Z. Pan, E.S. Takle, R. Gu, Impacts of climate change on stream flow in the Upper Mississippi River Basin: A regional climate model perspective. J. Geophys. Res. **109**(D9), 1–12 (2004)

M.I. Jyrkama, J.F. Sykes, The impact of climate change on spatially varying groundwater recharge in the grand river watershed (Ontario). J. Hydrol. **338**(3–4), 237–250 (2007)

S. Kouadri, C.B. Pande, B. Panneerselvam, et al. Prediction of irrigation groundwater quality parameters using ANN, LSTM, and MLR models. Environ Sci Pollut Res **29**, 21067–21091 (2022). https://doi.org/10.1007/s11356-021-17084-3

C.P. Kumar, Chapter 10: Impact of climate change on groundwater resources, in *Handbook of Research on Climate Change Impact on Health and Environmental Sustainability*, (2016), p. 196

P.S. Kumar, P. Balamurugan, Evaluation of groundwater quality for irrigation purpose in attur taluk, Salem, Tamilnadu, India. Water Energy Int. **61**(4), 59–64 (2018)

P.S. Kumar, P. Balamurugan, Suitability of ground water for irrigation purpose in Omalur Taluk, Salem, Tamil Nadu, India. Indian J. Ecol. **46**(1), 1–6 (2019)

C.P. Kumar, Climate change and its impact on groundwater resources. Int. J. Eng. Sci. **1**(5), 43–60 (2012). ISSN: 2278-4721

A. Lal, B. Datta, Robust ensemble modeling paradigm for groundwater salinity predictions in complex aquifer systems, in *Groundwater Resources Development and Planning in the Semi-Arid Region*, ed. by C. B. Pande, K. N. Moharir, (Springer, Cham, 2021). https://doi.org/10.1007/978-3-030-68124-1_3

S. Luoma, J. Okkonen, Impacts of future climate change and Baltic sea level rise on groundwater recharge, groundwater levels, and surface leakage in the Hanko aquifer in southern Finland. Water **6**(12), 3671–3700 (2014)

T. Meixner, A.H. Manning, D.A. Stonestrom, D.M. Allen, H. Ajami, K.W. Blasch, A.E. Brookfield, C.L. Castro, J.F. Clark, D.J. Gochis, A.L. Flint, K.L. Neff, R. Niraula, M. Rodell, B.R. Scanlon, K. Singha, M.A. Walvoord, Implications of projected climate change for groundwater recharge in the western United States. J. Hydrol. **534**, 124–138 (2016)

L.J. Min, J. Younghun, P. Younshik, K. Hyunwoo, L. Kyoung Jae, K. Hungsoo, Assessment of future climate change impact on groundwater recharge, baseflow and sediment in steep sloping watershed. J. Wetlands Res. **16**(2), 173–185 (2014)., 1229-6031(pISSN), 2384-0056(eISSN)

A.P. Mishra, H. Khali, S. Singh, C.B. Pande, R.Singh, S.K. Chaurasia, An Assessment of In-situ Water Quality Parameters and its variation with Landsat 8 Level 1 Surface Reflectance datasets, Int J Environ Anal Chem (2021). https://doi.org/10.1080/03067319.2021.1954175

C. Moeck, P. Brunner, D. Hunkeler, The influence of model structure on groundwater recharge rates in climate-change impact studies Christian. Hydrogeol. J. **24**, 1171–1184 (2016)

U. Mohan, A. Krishnakumar, Seasonal variation of groundwater quality in the Kallada Basin, Southern Western Ghats of India, in *Groundwater Resources Development and Planning in the Semi-Arid Region*, ed. by C. B. Pande, K. N. Moharir, (Springer, Cham, 2021). https://doi.org/10.1007/978-3-030-68124-1_17

M.C. Moseki, Climate change impacts on groundwater: Literature review. Environ. Risk Assess Remediat. **2**(1), 16–20 (2017)

K. Muniraj, B. Panneerselvam, S. Devaraj, C.J. Jesudhas, K. Sudalaimuthu, Evaluating the effectiveness of emissions reduction measures and ambient air quality variability through ground-based and Sentinel-5P observations under the auspices of COVID pandemic lockdown in Tamil Nadu, India. Int. J. Environ. Anal. Chem., 1–12 (2021)

P.M. Nyenje, O. Batelaan, Estimating the effects of climate change on groundwater recharge and baseflow in the upper Ssezibwa catchment, Uganda. Hydrol. Sci. J. **54**(4) (2009)

I.R. Orimoloye, A.O. Olusola, J.A. Belle, et al. Drought disaster monitoring and land use dynamics: identification of drought drivers using regression-based algorithms. Nat Hazards (2022). https://doi.org/10.1007/s11069-022-05219-9

C.B. Pande, K.N. Moharir, S.F.R. Khadri, et al. Study of land use classification in an arid region using multispectral satellite images. Appl Water Sci **8**, 123 (2018). https://doi.org/10.1007/s13201-018-0764-0

C.B. Pande, K.N. Moharir, S.F.R. Khadri, Assessment of land-use and land-cover changes in Pangari watershed area (MS), India, based on the remote sensing and GIS techniques. Appl Water Sci **11**, 96 (2021). https://doi.org/10.1007/s13201-021-01425-1

C.B. Pande, K.N. Moharir, S.K. Singh, et al. Groundwater flow modeling in the basaltic hard rock area of Maharashtra, India. Appl Water Sci **12**, 12 (2022), https://doi.org/10.1007/s13201-021-01525-y

B. Panneerselvam, S. Karuppannan, K. Muniraj, Evaluation of drinking and irrigation suitability of groundwater with special emphasizing the health risk posed by nitrate contamination using nitrate pollution index (NPI) and human health risk assessment (HHRA). Hum. Ecol. Risk Assess. Int. J. **27**(5), 1324–1348 (2020)
B. Panneerselvam, K. Muniraj, C. Pande, N. Ravichandran, Prediction and evaluation of groundwater characteristics using the radial basic model in Semi-arid region, India. Int. J. Environ. Anal. Chem., 1–17 (2021)
B. Panneerselvam, K. Muniraj, M. Thomas, N. Ravichandran, GIS-based legitimatic evaluation of groundwater's health risk and irrigation susceptibility using water quality index, pollution index, and irrigation indexes in semiarid region, in *Groundwater Resources Development and Planning in the Semi-Arid Region*, (Springer, Cham, 2021a), pp. 239–268
B. Panneerselvam, K. Muniraj, M. Thomas, N. Ravichandran, B. Bidorn, Identifying influencing groundwater parameter on human health associate with irrigation indices using the Automatic Linear Model (ALM) in a semi-arid region in India. Environ. Res. **202**, 111778 (2021b)
B. Panneerselvam, S.K. Paramasivam, S. Karuppannan, N. Ravichandran, P. Selvaraj, A GIS-based evaluation of hydrochemical characterisation of groundwater in hard rock region, South Tamil Nadu, India. Arab. J. Geosci. **13**(17), 1–22 (2020a)
K. Priyan, Issues and challenges of groundwater and surface water management in semi-arid regions, in *Groundwater Resources Development and Planning in the Semi-Arid Region*, ed. by C. B. Pande, K. N. Moharir, (Springer, Cham, 2021). https://doi.org/10.1007/978-3-030-68124-1_1
J. Rajesh, C.B. Pande, S.A. Kadam, et al. Exploration of groundwater potential zones using analytical hierarchical process (AHP) approach in the Godavari river basin of Maharashtra in India. Appl Water Sci **11**, 182 (2021). https://doi.org/10.1007/s13201-021-01518-x
N. Rane, G.K. Jayaraj, Evaluation of multiwell pumping aquifer tests in unconfined aquifer system by Neuman (1975) method with numerical modeling, in *Groundwater Resources Development and Planning in the Semi-Arid Region*, ed. by C. B. Pande, K. N. Moharir, (Springer, Cham, 2021). https://doi.org/10.1007/978-3-030-68124-1_5
R. Ravi, S. Aravindan, K. Shankar, P. Balamurugan, Suitability of groundwater quality for irrigation in and around the main Gadilam river basin on the east coast of southern India. Arch. Agric. Environ. Sci. **5**(4), 554–562 (2020)
C. Rivard, C. Paniconi, H. Vigneault, D. Chaumont, A watershed-scale study of climate change impacts on groundwater recharge (Annapolis Valley, Nova Scotia, Canada). Hydrol. Sci. J. **59**(8), 1437–1456 (2014)
T. Shah, Climate change and groundwater, India's opportunities for mitigation and adaptation, in *Water Resources Policies in South Asia*, 1st edn., (2013), p. 31, eBook ISBN9780367818487
M. Shahid, K.U. Rahman, S. Haider, et al. Quantitative assessment of regional land use and climate change impact on runoff across Gilgit watershed. Environ Earth Sci **80**, 743 (2021). https://doi.org/10.1007/s12665-021-10032-x
S.M.V. Sharief, M. Zakwan, Groundwater remediation design strategies using finite element model, in *Groundwater Resources Development and Planning in the Semi-Arid Region*, ed. by C. B. Pande, K. N. Moharir, (Springer, Cham, 2021). https://doi.org/10.1007/978-3-030-68124-1_6
U.K. Vidya, C.B. Pande, J. Rajesh, A. A. Atre, S.D. Gorantiwar, S.A. Kadam Bhau Gavit, Surface water dynamics analysis based on sentinel imagery and Google Earth Engine Platform: a case study of Jayakwadi dam, Sustain. Water Resour. Manag. **7**, 44 (2021). https://doi.org/10.1007/s40899-021-00527-7
W.-Y. Wu, M.-H. Lo, Y. Wada, J.S. Famiglietti, J.T. Reager, P.J.-F. Yeh, A. Ducharne, Z.-L. Yang, Divergent effects of climate change on future groundwater availability in key mid-latitude aquifers. Nat. Commun. **11**(2020), 3710 (2020)
M. Zakwan, Trend analysis of groundwater level using innovative trend analysis, in *Groundwater Resources Development and Planning in the Semi-Arid Region*, ed. by C. B. Pande, K. N. Moharir, (Springer, Cham, 2021). https://doi.org/10.1007/978-3-030-68124-1_20

Chapter 3
Factors That Influence the Hydrological Process: A Climate and Land Use/Land Cover Perspective

Colins Johnny Jesudhas, Kirubakaran Muniraj, and Suresh Devaraj

Abstract The hydrological cycle (HyC) is affected by several factors, but climate and land use/land cover (LU/LC) are the most influential ones. This chapter has tried to show some satellite-based land use/land cover feature extraction methods that are useful for climate studies. Several literature works have claimed that climate is more influential than land use. Land use has an impact on several components of the hydrological cycle. This chapter provides a perspective on climate change, urbanization, land degradation, and other disasters and also on the usage of land use/land cover features in the study of the hydrological cycle. The anomaly in solar radiation due to greenhouse gas (GHG) emissions and its impact on climatic factors and the hydrological cycle with its implication in food production is briefed. Some of the global measurement missions for precipitation and land surface temperature (LST) are also discussed. To investigate the influence of land use/land cover on the hydrological cycle, identification of a particular class or all land use classes of a particular region may be essential. This chapter uses the synoptic view of satellite data and attempts to exercise certain indices to identify certain classes and classification algorithms to classify land use classes. This work has also experimented with certain classification algorithms to delineate some land use/land cover features and has also pointed out some limitations in the application of indices. This chapter discusses the factors that influence the hydrological cycle and highlights the usage of satellite data in regional studies.

Keywords Hydrological cycle · Land use/land cover · Climate · Urbanization and greenhouse gases

C. J. Jesudhas (✉) · K. Muniraj
Department of Civil Engineering, University VOC College of Engineering, Anna University, Thoothukudi Campus, Thoothukudi, India

S. Devaraj
Department of Civil Engineering, Saveetha Engineering College, Chennai, India

B. Panneerselvam et al. (eds.), *Climate Change Impact on Groundwater Resources*, https://doi.org/10.1007/978-3-031-04707-7_3

3.1 Introduction

Many countries in the world give more importance to urban water supply than to rural water supply, and, also, the reliability of surface water (SW) as a source in an urban area is less compared to that in a rural area (JMP 2019). This inconsistency in water supply is due to the improper distribution of water resources both spatially and temporally. The location of countries on Earth plays a major role in their climate. It is the climate that decides the availability of water in a country, so climatic imbalance could disturb the hydrological cycle (HyC). A disturbance in the HyC often affects land surface components such as freshwater and its distribution. The distribution and availability of water in a region depends on the atmosphere, hydrosphere, biosphere, and lithosphere. A change in any of these spheres affects the entire HyC.

In a region, living things are benefitted by the water available to them. Water availability in a watershed (WS) is a factor concerned with the water received, water stored, water consumed, and water lost, all within a time frame. The beneficial quantity of water depends on the storage capacity of a WS. Water stored in a WS can be in the form of SW or groundwater (GW). The HyC is influenced by climatic factors such as solar radiation, Earth's position with respect to the Sun, atmosphere, and climate change as well as WS characteristics such as topography, human intervention, soil and its moisture condition, and land use/land cover (LU/LC). A change in climate is witnessed immediately in the vagaries of precipitation and temperature that affect every component of the HyC and the water stored. Relief affects redistribution of precipitation, orography affects the concentration of runoff, and differences in solar radiation affect evapotranspiration.

Since more components are involved and their interactions are complex, a study of the HyC is always difficult without data. Collection of data becomes imperative to identify the factors that influence the HyC and can be performed only through observations and measurements. A multifaceted view is required to study any cyclic process, and the HyC is not an exemption. All these observations and measurements are carried out through a legacy ground-based system, and, with the advancement of space technology, it is now progressing through satellite-based measurements. The measured components of the HyC are essential to model the atmosphere, land surface, and upper oceans for a better understanding of the changes in the HyC.

3.2 The Hydrological Cycle

Water availability in a region is primarily driven by the components of the HyC, as shown in Fig. 3.1. The hydrological processes that take place inside a WS are precipitation, interception, overland flow, depression storage, surface runoff, infiltration, throughflow, baseflow, evaporation, and evapotranspiration. Any country can

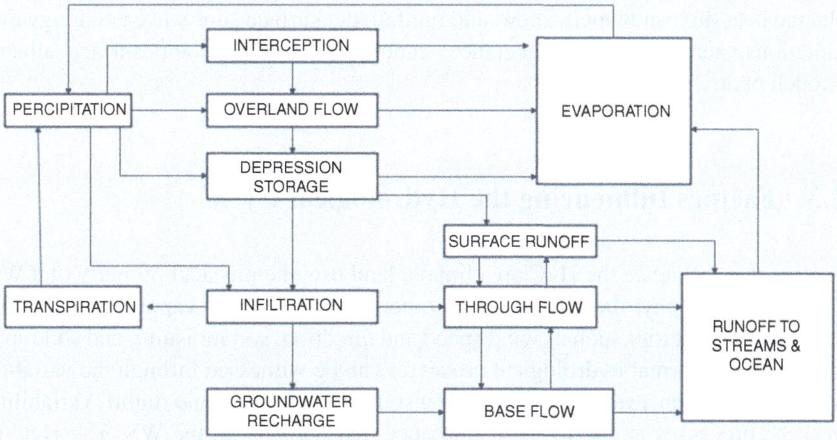

Fig. 3.1 Components of the hydrological cycle

be divided into WSs, and the water stored in them can be effectively managed. A WS is the best hydrological response unit for any hydrological study. It receives water from the atmosphere through precipitation. The water obtained undergoes several transformations and a portion of the water is retained in the basin, whereas the excess water leaves the WS. The stored water supports humans, livestock, and other life forms in the WS. During precipitation, one portion of the water gets evaporated, one portion gets intercepted by the land cover features, and the left out water gets infiltrated into the soil surface or runs down as surface runoff and reaches the water body. The infiltration and the runoff are governed by soil moisture, soil texture, and soil permeability. This infiltrated water is responsible for the saturation of the crop root zone, and, after saturation, water gets percolated through the vadose zone by gravity. This water reaches the water table and gets replenished as GW, and this process is referred to as aquifer recharge.

Although understanding the hydrological components seems simple, quantification of these components is difficult and involves uncertainty because of the poor gauging network. Numerous methods and models have been developed to quantify these components. A data assimilation system is the most popular method to quantify the components from sparse observational data. Global land data assimilation systems use the ground-based monitoring network and the satellite-based network for quantification. These systems receive inputs of forcing fields such as precipitation, downward shortwave radiation, downward longwave radiation, near-surface air temperature, near-surface specific humidity, near-surface wind, and surface pressure to estimate outputs such as soil moisture, snow depth, fractional coverage and water equivalent, plant canopy SW storage, soil temperature, average surface temperature, surface runoff and subsurface runoff, bare soil, snow, canopy SW evaporation, canopy transpiration, latent, sensible, and ground heat flux, snow phase change heat flux, snowmelt, snow, and rainfall, net surface shortwave and longwave radiations, aerodynamic conductance, canopy conductance, and surface albedo (Rodell et al. 2004).

3.3 Factors Influencing the Hydrological Cycle

Factors that influence the HyC are climate, land use, the physical property of a WS such as topography, the atmospheric property such as water vapour content, and circulation properties such as wind speed and direction, soil moisture, and soil type. Variability in normal hydrological processes can be witnessed through the variability in precipitation, evaporation, evapotranspiration, recharge, and runoff. Variability in these processes is due to the disturbance that happens in the WS. The HyC is strongly influenced by the solar influx that disturbs the temperature, temperature, in turn, disturbs the evaporation, evaporation in turn disturbs soil moisture and SW, soil moisture in turn disturbs the runoff and can have a significant impact on the storage of water in a WS and hence affect the local climate. Vegetation on the surface of the earth influences the albedo, surface roughness, storage of energy, and water on the surface of the earth. Although some impacts are slow and have mild influence, some are rapid and harsh, causing disasters that further affect the HyC. Disturbances that cause an alteration in the hydrological function are defined as hydrological disturbances (Mirus et al. 2017). Climate and land use are the two main influencing factors in catchment hydrology (Scilling et al. 2008). Globally, changes in the water vapour content of the atmosphere, clouds, and ice affect the earth's radiation balance and thus play a key role in determining the climate response to accumulative greenhouse gas emissions (Bates et al. 2008). Climate variability influences surface hydrology more than land use. Components that are influenced by climate and land use are shown in Fig. 3.2. Although the literature claims that climatic factors are more influential, land use factors influence most of the components of the HyC. Hence, land use factors have gained importance in HyC studies. Land use change increases evapotranspiration, whereas climate variability decreases evapotranspiration. Land use change and climate variability decrease runoff and soil water content (Li et al. 2009). It can be seen that climate change and LU/LC change have a significant role to play in the HyC. The statistical approach of data sciences is used to analyze the hydroclimatic trends in observatories (Bewket and Sterk 2005); however, the processes of WSs could not captured. With this in mind, the spatially distributed hydrological model is considered to be the most appropriate method for determining the extent of future climate impacts in conjunction with LU/LC changes (Scilling et al. 2008). Physiological hydraulic models have been used to measure the comparative effects of climate change and LU/LC on catchment hydrology (Srivastava et al. 2020).

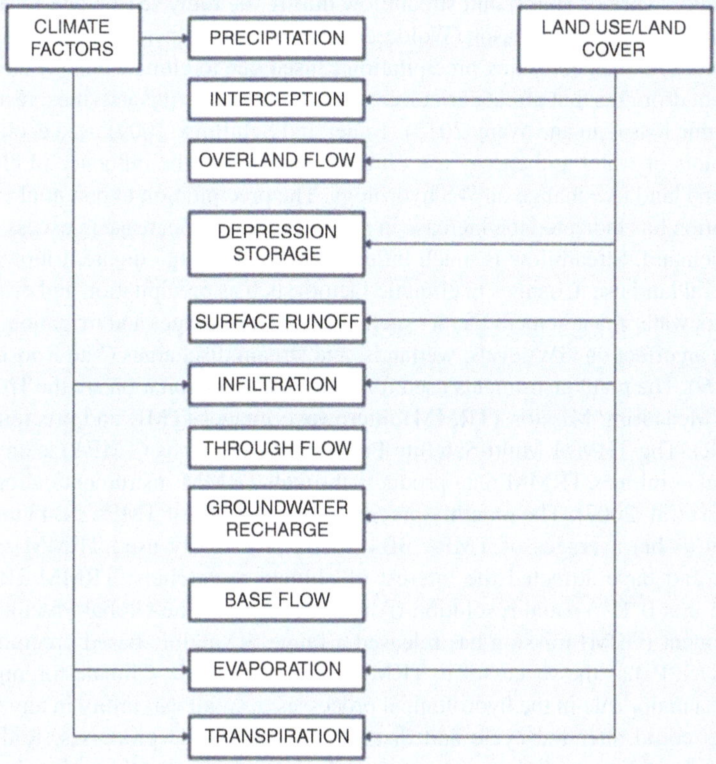

Fig. 3.2 Hydrological components influenced by climate and LU/LC

3.4 Climate

Earth's physical climate system includes the atmosphere, oceans, cryosphere, and land surface, which is complex and changes over time. Climate systems are influenced by the greenhouse effect, global warming, sea-level rise, and radiative forcing of climate due to natural and human activities (NRC 2010). A global rise in greenhouse gas emissions will increase the global temperature and augment evaporation and evapotranspiration, which will increase the water vapour content of the atmosphere and sustain cloud formation, precipitation, discharges, and other components of the HyC. Precipitation and discharges are positively correlated, but temperature and evapotranspiration are negatively correlated with discharges (Ciobotaru 2018). The impact of climatic factors such as solar influx, atmospheric moisture, wind speed, and wind direction affects the HyC and alters the hydrological components. Changes in hydrological factors such as rainfall and actual evapotranspiration and in land use and land cover factors such as forest cover and cultivation on the steep slopes of a basin influence the surface runoff and streamflow in rivers, i.e. forest cover reduces surface runoff and streamflow during the rainy season and raises the water level during the dry season (Woldesenbet et al. 2018). Increasing mean annual air temperature and decreasing precipitation caused due to climate change have led to frequent droughts that affect agriculture and livestock grazing activities, resulting in economic loss (Liu and Wang 2012). Tomer and Schilling (2009) used ecohydrological plots of water and energy use efficiencies to study the influence of climate change and land use change on WS hydrology. The precipitation to potential evapotranspiration has indicated the increase in precipitation and decrease in excess evaporative demand. Streamflow is much influenced by changing climate, followed by agricultural land use. Changes in climatic factors such as precipitation and evaporation and in water management factors such as stream discharges and irrigation water can have an effect on GW levels, wetlands, and stream discharges (Van Roosmalen et al. 2009). The main instruments used for monitoring precipitation are the Tropical Rainfall Measuring Mission (TRMM) microwave imager (TMI) and precipitation radar (PR). The TRMM Multi-Satellite Precipitation Analysis (TMPA) is an algorithm that combines TRMM data products through TRMM instrument calibration (Huffman et al. 2007). The monthly precipitation averages of TMPA 3B43 and the sub-daily (3 hr) averages of TMPA 3B42 daily are widely used TRMM-related products and have attracted the interest of climate researchers. TRMM 3B43 is available in a 0.25° spatial resolution (Almazroui 2011). The Global Precipitation Measurement (GPM) mission has released a range of satellite-based precipitation products (SPPs) as the successor to TRMM (Shi et al. 2020). Climate forcing factors play a major role in the hydrological processes; a small variability in any of the processes could alter the cycle and disturb other hydraulic processes. Radiative forcing is the difference between the rate at which Earth's atmosphere absorbs solar radiation and the rate at which longwave radiation is emitted into space that is influenced by increasing greenhouse gas (GHG) emissions (LMS 2015). Catchment descriptors such as climate, geology/lithology, landscape/morphology, topography, and soil influence hydrology where the climatic factors can be modelled through parameters such as minimum temperature, maximum temperature, precipitation, potential evapotranspiration, aridity, seasonality, and snow elevation (Massmann 2020). A palaeoclimatic study through isotopes suggested that natural and anthropogenic changes in vegetation cover can alter the surface runoff and the GW flow to a lake and control hydrological budgets (Rosenmeier et al. 2002). Hydrological components such as precipitation, surface runoff, GW recharge, evapotranspiration, and streamflow were assessed based on scenarios such as land use change, etc. (Zhang et al. 2016). Anomalies in the climatic factors can affect the HyC, and, hence, monitoring of these factors is essential. Solar radiation that reaches the Earth's surface varies with respect to season, geographic location, time of day, local landscape, and local weather. When this radiation passes through the atmosphere, a portion of it reaches the Earth's surface without disruption; this is called direct solar radiation. Some of it gets altered due to absorption, scattering, and reflection by air molecules, water vapour, clouds, dust, ozone, and other pollutants; this is called diffuse solar radiation. Measurement of this radiation is possible through ground-based instruments and satellite measurements. Irradiance is often measured in terms of

Watts per metre squared (W/m^2). The mean distance from Earth to the Sun is 150 million kilometres, and the distance between them changes due to the elliptical orbit. Due to the variation in the distance between Earth and the Sun and due to the tilt in Earth's axis of rotation by 23.5° relative to the orbital plane, the radiation that reaches the surface of Earth is variable. The top of the atmosphere (TOA) of Earth when closest to the Sun (perihelion) is higher than that when farthest from the Sun (aphelion). The radiation peak that reaches Earth ranges from ultraviolet to near-infrared (0.3–1.1 μm). Remote sensing (RS)-based satellite measurements are used for climate studies. The remote sensing (RS) data extracted from the satellite and aerial data can be used to extract WS properties in the form of a map prepared by geographic information system (GIS) software. Sample satellite images of the South Indian region from the Landsat 8 Operational Land Imager (OLI) and Thermal Infrared Scanner (TIRS) with 11 bands are used for the study. A false colour composite (FCC) prepared from selected bands of the OLI is shown in Fig. 3.3. The FCC was prepared from three bands by assigning the near-infrared band (band 5) to

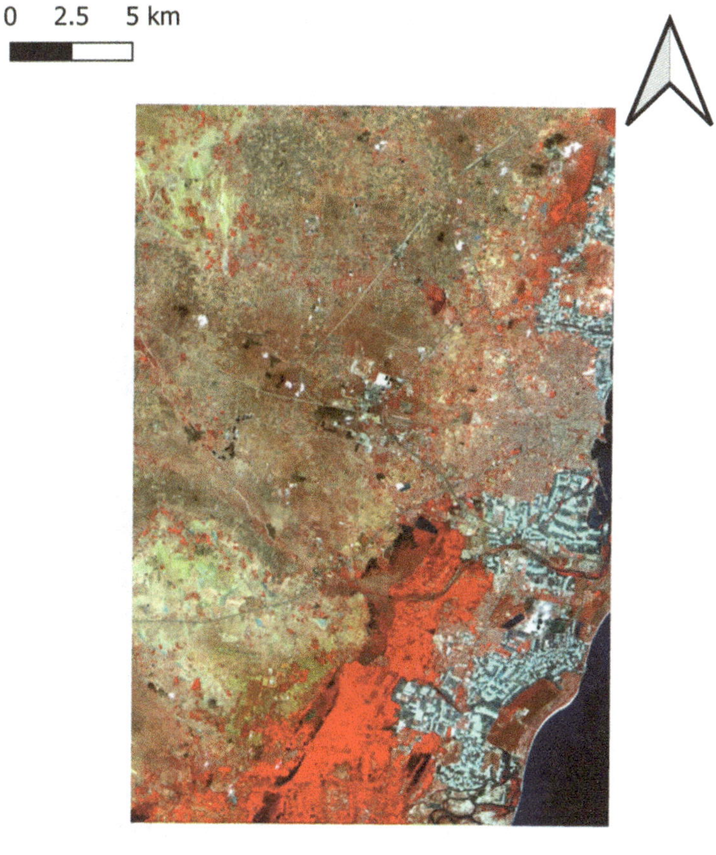

Fig. 3.3 False colour composite prepared from Landsat 8 OLI data

red colour, the red band (band 4) to green colour, and the green band (band 3) to blue colour. For simple understanding, this region will be used in the entire study for experimentation by applying different geospatial techniques. In the FCC image, black represents a water body, red represents vegetation, cyan represents salt pans, grey represents urban settlements, and yellow represents soil. It is obvious that a mixture of LU/LC produces different colours. The study of LU/LC was simplified through RS and GIS technologies.

Climate variables such as land surface temperature can be obtained through the TIRS bands as shown in Fig. 3.4. Landsat 8 band 10 (thermal band) was used to prepare the land surface temperature (LST) map. The influence of temperature on a region can be studied through LST. It is evident from the image that the temperature

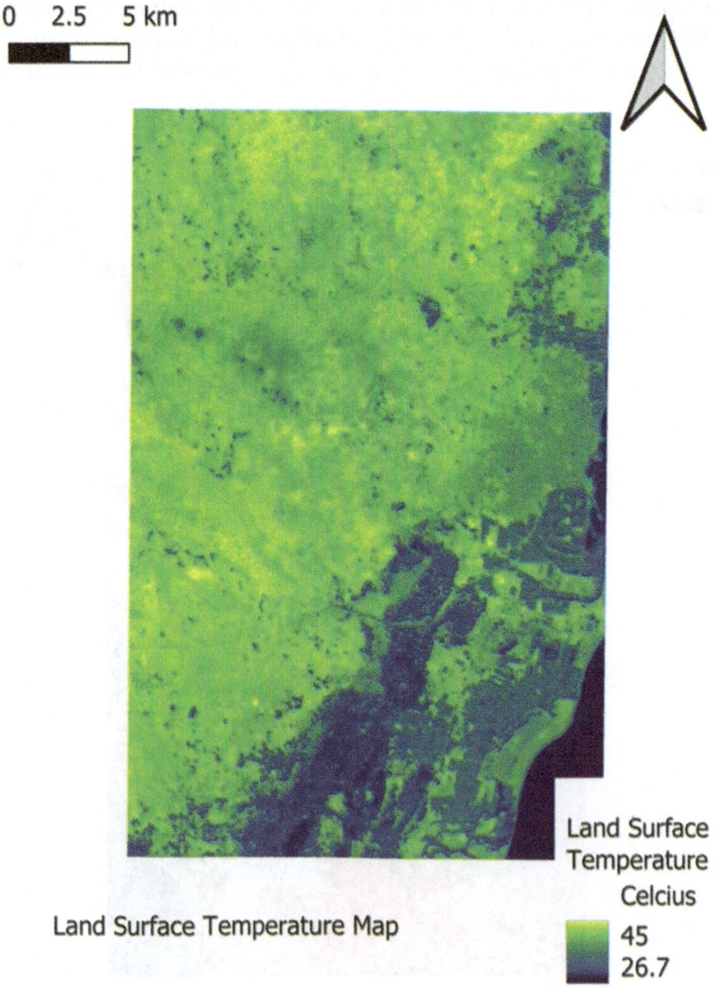

Fig. 3.4 Land surface temperature prepared from Landsat 8 TIRS data

near the vegetated area and near the water body is less compared to those of other areas. LST can also be used to study the urban heat island (UHI). These data can be helpful in studying the influence of temperature on the factors that affect the HyC. Radiation that reaches the Earth's surface is also influenced by the time of the day. When the Sun is directly overhead, i.e. at noon, the atmospheric thickness is minimum, whereas during other times, the thickness will be more. Solar radiation that reaches Earth's surface is partly reflected, with some being absorbed and transmitted. The reflected radiation depends on the albedo of the surface.

The absorbed radiation is radiated back as mid-infrared and far-infrared radiations into the atmosphere where the composition of greenhouse gases such as carbon dioxide, methane, water vapour, etc. absorbs this radiation and heats up the atmosphere. The concentration of CO_2 in the atmosphere increases every year due to the burning of fuel fossils (Anderson et al. 2016). Carbon emission depends on development and production demand, and all countries are responsible for carbon emission on different scales (Abeydeera et al. 2019). The Intergovernmental Panel on Climate Change (IPCC) has used models to project the global surface temperature. The model that was used for evaluation has been improved over the years, and the approach has evolved from atmosphere–ocean-coupled models to climate system models (Randall et al. 2007). Climate warming, by the end of the century, has been projected by an earth system model to be between 1.0 and 3.7 °C by the consideration of future greenhouse gases (IPCC 2019). Models can be developed for different projected scenarios, and their impact on food production can be studied. Such a study shows that climate change has an impact on temperature and affects agriculture growth and productivity (Liu and Wang 2012).

3.5 Land Use/Land Cover

A land use/land cover map helps in understanding the landscape and the current utilization of the land in a region. People living in a WS depend on freshwater sources such as lakes, ponds, rivers, aquifers, and other water bodies for their domestic water needs. This source of freshwater is also used for agricultural applications through irrigation, industrial needs, and other needs. The demand for water supplies and services concurrently amplifies the pressures on water sources. Humans have modified hydrological processes through a variety of behaviours: high water consumption, landscape changes, construction of artificial infrastructures, and introduction of both point and non-point source wastes. As a result, the amount of precipitation fluctuates, runoff varies, and changes occur in flow systems, surface drainage paths, and the composition of water and sediments. All these modifications in a WS alter hydrological functions and, ultimately, are a threat to humans and other ecosystems that depend on it. Therefore, an understanding of the hydrological system is necessary to manage the water resources, and this difficulty shall be overcome through the development of hydrological models. Hydrological models are classified based on functionality and complexity, and, depending on the requirement, an appropriate model can be used. To investigate the influence of climate and LU/LC change on the hydrological response of a WS in different scenarios, integrating different models with the objective to effectively manage the water resources and to forecast future trends can be attempted through the hydrological models (Dwarakish and Ganasri 2015). WS properties that are deciphered through the field-based methods of data collection have been slowly replaced by the synoptic methods of data collection such as satellite and aerial data collection. Satellite remote sensing (SRS) data have been progressively used for automated and semi-automated mapping of various features like vegetation, snow, water, other land covers, etc.

The process of mapping LU/LC features such as vegetation is possible through several methods. Each method has its own advantages and limitations; however, indices dominate other classification methods in terms of time taken for result generation. Various indices have been developed to extract features of interest from SRS imagery. Reduction in SW and GW in a region affects the available water in a WS. SW can be identified through RS data and mapped through GIS. Suppose a researcher is interested in the identification of a water body in a region, a simplified technique called the normalized difference water index (NDWI) can be used as shown in Fig. 3.5. Not all water bodies have higher values of NDWI, and there are other features that have higher NDWI values. Hence, thresholding is the biggest challenge in the identification of a water body's boundary, especially due to mixed pixels and the seasonal variability of the water body. The research of Gupta et al. (2019) shows that water bodies play an important role in the LST dynamics and reduce the temperature of the neighbouring places. Local climate changes caused by urban heat islands (UHIs) differ basically from global climate change (GCC) caused by increases in the Sun's intensity or greenhouse gas concentrations (i.e. in terms of scale and mechanisms, their impacts are often similar) (U.S. Environmental Protection Agency 2014).

Poor availability of water resources in a WS affects the vegetation and its distribution. Vegetation canopies intercept water during rainfall and consume water for their survival. Hence, the characteristics of plant canopies are affected, which is reflected in the leaf area index (LAI) of vegetation. The LAI has a direct impact on interception and consumptive water use. It can be estimated with the help of the normalized difference vegetation index (NDVI) (Tan et al. 2020). The NDVI is the supreme method used for vegetation extraction as shown in Fig. 3.6. Not all vegetation have higher values of NDVI, and there are other features that have higher NDVI values. Hence, thresholding is the biggest challenge in the identification of the boundary of vegetation, especially due to mixed pixels and seasonal variability of the vegetation. There are other indices such as normalized difference snow index (NDSI) for snow and normalized difference built-up index (NDBI) for built-up area extraction (Zha et al. 2003).

Multiple classes of land use/land cover can be studied only through LU/LC maps usually prepared by the classification of satellite data by either supervised or unsupervised methods. A simple pixel-based classified image is shown in Fig. 3.7. Supervised classification is attempted on the satellite imagery using the spectral

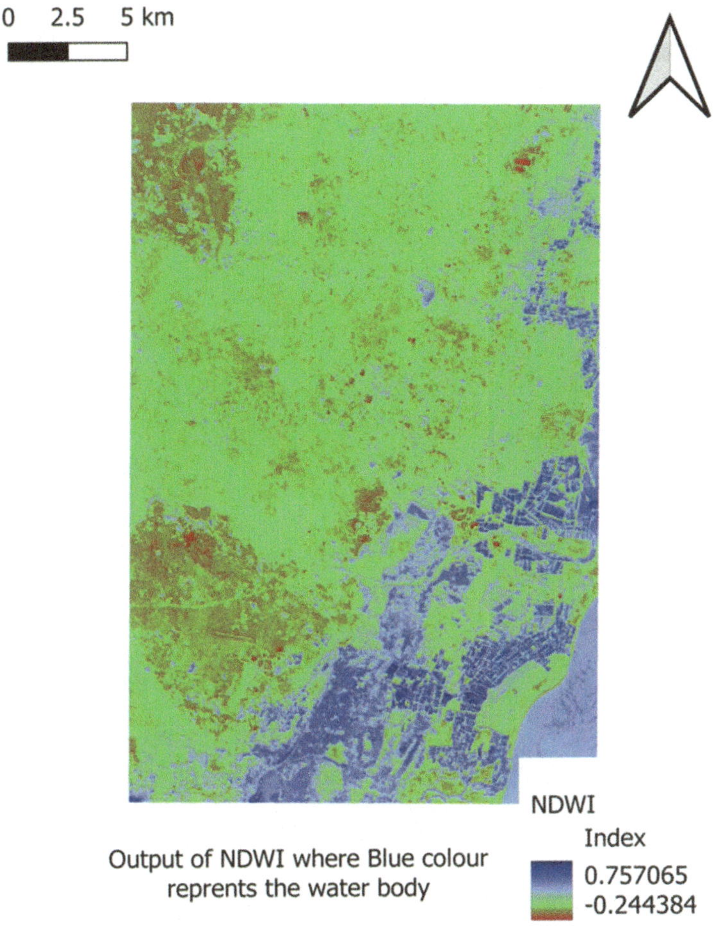

Fig. 3.5 NDWI map prepared from Landsat 8 OLI data

angle mapper algorithm. Six training sites such as soil, vegetation, water body, built-up, salt pan, and soil vegetation mixture were used for the classification. Many techniques have been designed for this purpose, which can be grouped into two common types. The first one is the classification of input imagery, which includes a wide range of pixel- and object-based classifications, whereas the second type involves the direct segmentation of the imagery through indices methods (Kefi et al. 2021). Li et al. (2014) emphasized the need for spatio-contextual analysis and geographic information system in image classification and grouped spatio-contextual analysis techniques into texture extraction, Markov random field modelling, and image segmentation and object-based image analysis. Classification of an impervious area could be challenging and can be attempted by any of these techniques. Object-based image analysis is a popular method, and many researchers claim that this method has achieved maturity, but still more needs to be done such as

Fig. 3.6 NDVI map prepared from Landsat 8 OLI data

development of automation, software, and tools that are user-friendly, integration with machine learning and deep learning, and suitable accuracy assessment (Johnson and Ma 2020).

3.6 Urbanization

Urban areas are only a small part of the total land area of the world; however, compared to the surrounding rural environment, the population density and resource utilization intensity of these areas are extremely high, thus forcing people to adopt

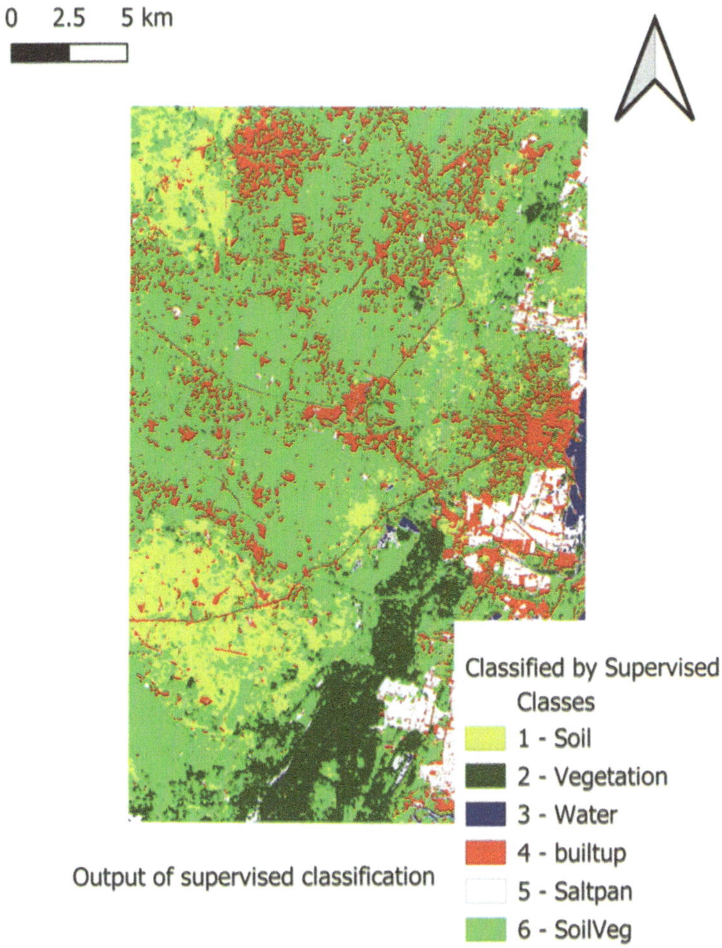

Fig. 3.7 LU/LC map prepared from Landsat 8 OLI data

better resource management practices. Rapid urban sprawl also makes it highly difficult to map the built-up areas of the urban built-up in a timely and accurate manner. Traditional methods such as land surveying and aerial photography, in general, require a lot of resources to be used for survey and mapping operations. RS data, especially from SRS systems, are a vital resource for mapping built-up areas for numerous reasons. They provide a concise and holistic view, making land surveys unrecognizable in large urban areas (Safriel 2007). Another practical benefit of using RS data for urban research is the availability of historical archives, which can help in mapping and understanding urban sprawl over time.

Urbanization and the expansion of impervious areas in a WS affect the surface and subsurface hydrology, stream morphology and ecology, climate, and biogeochemistry (Driscoll et al. 2010). Changes in the HyC caused by urban development have important physical, chemical, and biological impacts on the environment (Edwards 2007). Urban sprawl also significantly alters the amount and delivery of surface runoff (Konrad and Booth 2005). The sources of water in an urban environment are rainwater, grey water, black water, sewer mining, storm water, and treated municipal wastewater (Bichai and Smeets 2015); these waters take part in the urban HyC. Groffman et al. (2003) revealed that the change in urban land use has significantly altered the WS hydrology by lowering the water table that leads to hydrological droughts and affects the soil, vegetation, and microbial process in a riparian zone. Riparian zones are hotspots of interactions between plants, soil, water, microorganisms, and people and may seriously be affected by urbanization. Urban expansion, when not met with insufficient land supply, has led to the conversion of lakes to landfills for construction waste, urban living garbage, farmland waste, as collectors for domestic sewage, industrial and agricultural wastewater, and real estate developments in rapidly growing Wuhan (Wu and Xie 2011). Urbanization influences runoff and decreases the time of concentration. For urban water sustainable management, recognition of water shortages and the limited carrying capacity is essential where water ecological carrying capacity (WECC) could provide a water environment with early warning and control direction (Ding et al. 2015). Studies conducted in rural and urban basins with different population densities and associated streamflow were compared with the climate change scenarios and found to have no difference in the annual streamflow (DeWalle et al. 2000). However, it is believed that there is a differential relation between urban and rural WSs over the runoff. Hence, in these kinds of studies, identification of the built-up area in both an urban and a rural basin is important. The NDBI method of built-up area extraction was used to map built-up areas with good accuracy, and this could serve as an alternative to classification especially for objectively mapping built-up areas (Zha et al. 2003). The NDBI method was adopted to identify the built-up area as shown in Fig. 3.8. However, thresholding is the biggest challenge in the identification of the built-up area's boundary, especially due to mixed pixels. Automated built-up extraction index (ABEI), urban index (UI), and the NDBI were used to extract and test the built-up area, and the result showed better accuracy in the order of ABEI > UI > NDBI (Firozjaei et al. 2019). Bhatti and Tripathi (2014) proposed a new method called the built-up area extraction method (BAEM) that used principal component analysis (PCA) that combined the NDBI, NDVI, and modified normalized difference water index (MNDWI) to extract the built-up area and improved the results of the NDBI. Hydrological changes that result from urban development are increase in the frequency of high flows, redistribution of water to storm flow from baseflow, increase in diurnal variation in streamflow, and reduction in low flow that is potentially significant to stream ecosystems (Konrad and Booth 2005). Increased LST and construction of UHIs can be considered as a negative "by-product" or an impact of urban sprawl. More specifically, UHI formation is affected by changes in the landscape due to urban development in which open lands or impervious surfaces (e.g. buildings, roads, etc.) and by the resulting anthropogenic heat. Therefore, in most cases, terrain elements, impervious surface areas, and green areas determined by SRS data have become an important input for the study of UHIs (Estoque and

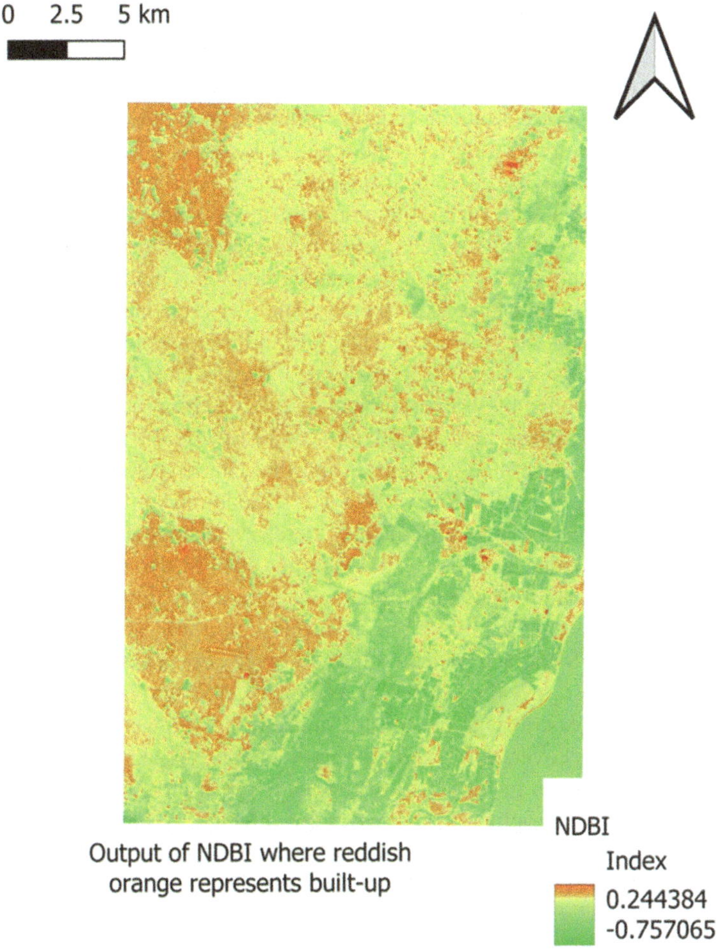

Fig. 3.8 NDBI map prepared from Landsat 8 OLI data

Murayama 2017). Thus, addressing the issues of UHIs will help mitigate the vulnerabilities of GCCs and enhance adaptive capabilities at the local level (U.S. Environmental Protection Agency 2014).

UHIs and global warming due to GCCs can both increase energy consumption and greenhouse gas emission and affect air and water quality as well as humans' health and thermal comfort. Such initiatives particularly contribute to achieving sustainable development goals based on "climate action" and sustainable cities and communities. Expanding population, construction, and urbanization as well as human activities have led to an acceleration in soil erosion, resulting in land degradation (Wu and Xie 2011).

3.7 Land Degradation

Land degradation is defined as the negative trend of the land condition, which is caused by direct or indirect human-induced practices including anthropogenic climate change, manifested as long-term reduction or loss of at least one of the following: biological productivity, ecological integrity, or value to humans. For example, forest degradation is a kind of land degradation that occurs in forest land due to deforestation and other local causes. Deforestation is the transformation of forest land to non-forest land and can result in land degradation (Olsson et al. 2019).

Human activity-related land use changes, such as crop land expansion, have accelerated soil erosion (Borrelli et al. 2017) if land use/land cover changes represent ecological land use conflicts, especially new use deviates from land capability (natural use is determined by WS parameters such as soil properties, slope, drainage density, etc.). In these cases, land degradation can be severe. WS management needs estimation of runoff generated by rainfall and that can be determined by the soil conservation service curve number (SCS-CN) model (Matomela et al. 2019). Change in the urban hydrology subsequently increases the magnitude and frequency of the flooding that enhances soil erosion. Paulinus et al. (2016) used morphometric analysis to show the landform pattern, geological characteristics, soil textural properties, soil erodibility, and the hydrological response of a WS. Morphometric analysis helps evaluate WS characteristics such as drainage network, stream length ratio, bifurcation ratio, etc., revealing different aspects of WSs (Srivastava et al. 2014). Factors that enhance soil degradation are rainfall and rainwater runoff, farming activities, vegetation cover, wind, and topographic slope (Issaka and Ashraf 2017). The affected soil results in low fertility that ends up in reduced biodiversity and human poverty. The agrophysical processes are affected by land degradation, i.e. increases in soil evaporation and runoff and decreases in water holding capacity, infiltration, transpiration, and green water use efficiency (Stroosnijder 2007). RS and GIS techniques are used to perform morphometric analysis and WS prioritization. WS prioritization was used to prioritize and identify a high-priority class exposed to more erosion threat in order to preserve the soil and water in a WS (Farhan et al. 2017). Morphometry can help in flash flood hazard assessment (Youssef et al. 2011) and for the identification of suitable sites for barriers that can conserve excess water and prevent the flood. Morphometrically identified high drainage density in a basin with a highly dissected topography, a low rate of infiltration, and a high surface runoff results in a high flooding potential and is the main reason for flash floods (Farhan et al. 2016). Soil loss in different forms such as interrill, rill, and sheet erosion mainly occurs in farmlands followed by grass/shrub, forest, and built-up areas, where the highest erosion has been observed in farmlands converted from forest, followed by farmlands converted from grass/shrub (Kogo et al. 2020). Morphological parameters can be analyzed to locate artificial recharge structures such as percolating ponds and to check dams that control the recharge in a WS (Sashikkumar et al. 2017). Others have used morphometric analysis to prioritize sub-WSs for erosion control measures. Morphometric analysis can be used to identify areas affected by flash flood hazards (Farhan et al. 2016). Climate change and variability affect land degradation by facilitating soil erosion, especially heavy rainfall accelerates soil erosion in bare vegetation cover. The generation of runoff is an important process in land degradation, resulting in soil erosion and affecting WSs' hydrological regime. One of those regions reflects extensive LU/LC changes in the Upper Blue Nile Basin. The economy of the region depends mainly on crop production, which is mainly dependent on the availability of water resources, and the basin is highly sensitive to rainfall runoff generation. Crop cultivation by rainfall is the major source of food production, but its productivity is declining due to land degradation. The rate of infiltration is governed by the amount of precipitation, topographic slope, and permeability and saturation behaviour of soil media. Generally, the infiltration rate is reduced by site-specific conditions such as steep slopes, tight clay soil, or unfractured bedrock. Recharge is extremely impaired when the vegetation cover is replaced with an impermeable cover. Urbanization significantly affects the ability of the water to circulate through the various components of this cycle and replenish the aquifer. All these accumulated effects on the HyC could cause disasters. Among all these factors that cause soil erosion, poor land management also causes poor soil retention during runoff across the landscape and poor infiltration. Soil and water conservation technologies needed to conserve the soil and water in a WS can be achieved through the traditional land management system.

3.8 Disasters

Disasters are recurrent and happen because of natural and manmade causes. Disasters can happen because of disturbances in the HyC and can thus affect the HyC. The disturbances could be studied for WSs of different scales, through different methods such as field measurements, numerical modelling, remote sensing, and laboratory measurements, and their effects can be examined (Mirus et al. 2017). The disturbances could be insect infestation (Biederman et al. 2015; Penn et al. 2016) and associated drought (Knowles et al. 2017), wildfire (Ebel et al. 2016; Ebel et al. 2016), forest structure change (Jaskierniak et al. 2016), peatland degradation (Menberu et al. 2016), earthquake (Rutter et al. 2016), or landslide (Mirus et al. 2017). As a result of disasters, the HyC gets disturbed and the effects can be witnessed as fluctuations in the climatological factors. For example, the summer monsoon season in South Asia faces fluctuations between heavy rainfall and low rainfall that cause poor regional conditions and adversely affect water resources, agricultural yields, human systems, and infrastructures (Singh et al. 2014). The study conducted by Lasagna et al. (2020) found the importance of the evaluation of climate changes and the impact on GW chemistry and insisted on the need to understand the spatiotemporal trends of GW quality with respect to climate variability. The impact on the quality of GW could adversely affect human health, especially for the rural population where the awareness of the quality of water is low. An analysis of the hydrological components of the pre- and post-scenarios of disasters is extremely essential for understanding the changes in the cycle. A retrospective view of the HyC is needed to investigate the effect of disasters on the HyC.

3.9 Conclusions

This chapter identifies the important factors that influence the HyC such as climate change, land use change, urbanization, land degradation, and other disasters. Many literature works argue that climate and land use have a greater impact on the HyC. The components of the HyC and the role of the influencing factors of climate and land use are discussed. A retrospective view of the HyC is needed to study the impact of every factor on the HyC. LU/LC features can be studied by employing popular indices by knowing their limitations. An experimental study on primary indices has found some complex issues such as judging the appropriate threshold to detect the feature and resolving the issue of mixed pixels. Urbanization and land use change are important concerns in land degradation and the HyC. Land degradation also influences the HyC and affects food productivity. The study, which covers the relationship between geomorphology and riparian vegetation, needs to be explored to help plan WS development and restoration projects. The important factor in a water-stressed urban area of a WS is water security and this should be included in the policy concerns.

References

L.H.U.W. Abeydeera, J.W. Mesthrige, T.I. Samarasinghalage, Global research on carbon emissions: A scientometric review. Sustainability (Switzerland) **11**(14), 1–25 (2019). https://doi.org/10.3390/su11143972

M. Almazroui, Calibration of TRMM rainfall climatology over Saudi Arabia during 1998-2009. Atmos. Res. **99**(3–4), 400–414 (2011). https://doi.org/10.1016/j.atmosres.2010.11.006

T.R. Anderson, E. Hawkins, P.D. Jones, CO_2, the greenhouse effect and global warming: From the pioneering work of Arrhenius and Callendar to today's Earth System Models. Endeavour **40**(3), 178–187 (2016). https://doi.org/10.1016/j.endeavour.2016.07.002

B. Bates, Z. W. Kundzewicz, S. Wu, J. Palutikof, Climate change and water. Technical paper of the intergovernmental panel on climate change IPCC Secretariat, Geneva, Vol. 71 (2008). https://doi.org/10.1029/90EO00112

W. Bewket, G. Sterk, Dynamics in land cover and its effect on stream flow in the Chemoga watershed, Blue Nile basin, Ethiopia. Hydrol. Process. **19**(2), 445–458 (2005). https://doi.org/10.1002/hyp.5542

S.S. Bhatti, N.K. Tripathi, Built-up area extraction using Landsat 8 OLI imagery. GIsci Remote Sens **51**(4), 445–467 (2014). https://doi.org/10.1080/15481603.2014.939539

F. Bichai, P.W.M.H. Smeets, Integrating water quality into urban water management and planning while addressing the challenge of water security. Glob. Issues Water Policy **15**, 135–154 (2015). https://doi.org/10.1007/978-94-017-9801-3_6

J.A. Biederman, A.J. Somor, A.A. Harpold, E.D. Gutmann, D.D. Breshears, P.A. Troch, et al., Recent tree die-off has little effect on streamflow in contrast to expected increases from historical studies. Water Resour. Res. **51**(12), 9775–9789 (2015). https://doi.org/10.1002/2015WR017401

P. Borrelli, D.A. Robinson, L.R. Fleischer, E. Lugato, C. Ballabio, C. Alewell, et al., An assessment of the global impact of 21st century land use change on soil erosion. Nat. Commun. **8**(1) (2017). https://doi.org/10.1038/s41467-017-02142-7

N. Ciobotaru, Factors controlling hydrological processes and characteristics in the Susita Catchment Area - Romania. National Institute for Research and Development, (January 2018)
D.R. DeWalle, B.R. Swistock, T.E. Johnson, K.J. McGuire, DeWalle_et_al-2000-Water_Resources_Research. Water Resour. Res. **36**(9), 2655–2664 (2000)
L. Ding, K.-l. Chen, S.-g. Cheng, X. Wang, Water ecological carrying capacity of urban lakes in the context of rapid urbanization: A case study of East Lake in Wuhan. Phys. Chem. Earth **89–90**, 104–113 (2015). https://doi.org/10.1016/j.pce.2015.08.004
G.S. Dwarakish, B.P. Ganasri, Impact of land use change on hydrological systems: A review of current modeling approaches. Cogent Geosci. **1**(1), 1115691 (2015). https://doi.org/10.1080/23312041.2015.1115691
B.A. Ebel, F.K. Rengers, G.E. Tucker, Observed and simulated hydrologic response for a first-order catchment during extreme rainfall 3 years after wildfire disturbance. Water Resour. Res. **52**(12), 9367–9389 (2016). https://doi.org/10.1002/2016WR019110
S. Edwards (2007). Role of Organic Agriculture in Preventing and Reversing Land Degradation. In: Sivakumar, M.V.K., Ndiang'ui, N. (eds) Climate and Land Degradation. Environmental Science and Engineering. Springer, Berlin, Heidelberg. https://doi.org/10.1007/978-3-540-72438-4_29
R.C. Estoque, Y. Murayama, Monitoring surface urban heat island formation in a tropical mountain city using landsat data (1987 - 2015). ISPRS J. Photogramm. Remote Sens. **133**, 18–29 (2017)
Y. Farhan, O. Anaba, A. Salim, Morphometric analysis and flash floods assessment for drainage basins of the Ras En Naqb Area, South Jordan using GIS. J. Geosci. Environ. Protect. **04**(06), 9–33 (2016). https://doi.org/10.4236/gep.2016.46002
Y. Farhan, A. Anbar, N. Al-Shaikh, R. Mousa, Prioritization of semi-arid agricultural watershed using morphometric and principal component analysis, remote sensing, and GIS techniques, the Zerqa River Watershed, Northern Jordan. Agric. Sci. **08**(01), 113–148 (2017). https://doi.org/10.4236/as.2017.81009
M.K. Firozjaei, A. Sedighi, M. Kiavarz, S. Qureshi, D. Haase, S.K. Alavipanah, Automated built-up extraction index: A new technique for mapping surface built-up areas using LANDSAT 8 OLI imagery. Remote Sens. **11**(1966), 1–20 (2019). https://doi.org/10.3390/rs11171966
P.M. Groffman, D.J. Bain, L.E. Band, K.T. Belt, G.S. Brush, J.M. Grove, et al., Down by the riverside: Urban riparian ecology. Front. Ecol. Environ. **1**(6), 315 (2003). https://doi.org/10.2307/3868092
N. Gupta, A. Mathew, S. Khandelwal, Analysis of cooling effect of water bodies on land surface temperature in nearby region: A case study of Ahmedabad and Chandigarh cities in India. Egypt. J. Remote Sens. Space. Sci. **22**(1), 81–93 (2019). https://doi.org/10.1016/j.ejrs.2018.03.007
G.J. Huffman, R.F. Adler, D.T. Bolvin, G. Gu, E.J. Nelkin, K.P. Bowman, et al., The TRMM Multisatellite Precipitation Analysis (TMPA): Quasi-global, multiyear, combined-sensor precipitation estimates at fine scales. J. Hydrometeorol. **8**(1), 38–55 (2007). https://doi.org/10.1175/JHM560.1
IPCC, Summary for policy makers, in *IPCC Special Report on the Ocean and Cryosphere in a Changing Climate*, ed. by H. O. Portner, D. C. Roberts, V. Masson-Delmotte, P. Zhai, M. Tignor, E. Poloczanska, K. Mintebeck, M. Nicolai, A. Okem, J. Petzold, B. Rama, N. Weyer, (2019) https://report.ipcc.ch/srocc/pdf/SROCC_SPM_Approved.pdf
S. Issaka, M.A. Ashraf, Impact of soil erosion and degradation on water quality: A review. Geol. Ecol. Landsc. **1**(1), 1–11 (2017). https://doi.org/10.1080/24749508.2017.1301053
D. Jaskierniak, G. Kuczera, R. Benyon, Predicting long-term streamflow variability in moist eucalypt forests using forest growth models and a sapwood area index. Water Resour. Res. **52**(4), 3052–3067 (2016). https://doi.org/10.1002/2015WR018029
JMP, *Progress on household drinking water, sanitation and hygiene 2000–2017 Special focus on inequalities New York. WHO/UNICEF Joint Monitoring Programme for Water Supply and Sanitation* (2019)
B.A. Johnson, L. Ma, Image segmentation and object-based image analysis for environmental monitoring: Recent areas of interest, researchers' views on the future priorities. Remote Sens. **12**(1772), 1–9 (2020)

C. Kefi, A. Mabrouk, N. Halouani, H. Ismail, Comparison of pixel-based and object-oriented classification methods for extracting built-up areas in coastal zone. Environ. Sci. Eng. **8**(August), 2151–2155 (2021). https://doi.org/10.1007/978-3-030-51210-1_336

J.F. Knowles, L.R. Lestak, N.P. Molotch, On the use of a snow aridity index to predict remotely sensed forest productivity in the presence of bark beetle disturbance. Water Resour. Res. **53**(6), 4891–4906 (2017). https://doi.org/10.1002/2016WR019887

B.K. Kogo, L. Kumar, R. Koech, Impact of land use/cover changes on soil erosion in western Kenya. Sustainability (Switzerland) **12**(22), 1–17 (2020). https://doi.org/10.3390/su12229740

C.P. Konrad, D.B. Booth, Hydrologic changes in urban streams and their ecological significance. Am. Fish. Soc. Symp. **2005**(47), 157–177 (2005)

M. Lasagna, D. Ducci, M. Sellerino, S. Mancini, D.A. De Luca, Meteorological variability and groundwater quality: Examples in different hydrogeological settings. Water (2020). https://doi.org/10.3390/w12051297

Z. Li, W.z. Liu, X.c. Zhang, F.l. Zheng, Impacts of land use change and climate variability on hydrology in an agricultural catchment on the Loess Plateau of China. J. Hydrol. **377**(1–2), 35–42 (2009). https://doi.org/10.1016/j.jhydrol.2009.08.007

M. Li, S. Zang, B. Zhang, S. Li, C. Wu, A review of remote sensing image classification techniques: The role of Spatio-contextual information. Euro. J. Remote Sens. **47**(1), 389–411 (2014). https://doi.org/10.5721/EuJRS20144723

S. Liu, T. Wang, Climate change and local adaptation strategies in the middle Inner Mongolia, northern China. Environ. Earth Sci. **66**(5), 1449–1458 (2012). https://doi.org/10.1007/s12665-011-1357-5

LMS. (2015). Climate Change Toolkit Additional Notes. Lesotho Meteorological Services Ministry of Energy and Meteorology

C. Massmann, Identification of factors influencing hydrologic model performance using a top-down approach in a large number of U.S. catchments. Hydrol. Process. **34**(1), 4–20 (2020). https://doi.org/10.1002/hyp.13566

N. Matomela, L. Tianxin, L. Morahanye, O.K. Bishoge, H.O. Ikhumhen, Rainfall-runoff estimation of Bojiang lake watershed using SCS-CN model coupled with GIS for watershed management. J. Appl. Adv. Res. (February), 16–24 (2019). https://doi.org/10.21839/jaar.2019.v4i1.263

M.W. Menberu, T. Tahvanainen, H. Marttila, M. Irannezhad, A.-K. Ronkanen, J. Penttinen, B. Kløve, Water-table-dependent hydrological changes following peatland forestry drainage and restoration: Analysis of restoration success. Water Resour. Res. **52**(5), 3742–3760 (2016). https://doi.org/10.1002/2015WR018578

Michael, O'Driscoll Sandra, Clinton Anne, Jefferson Alex, Manda Sara, McMillan (2010) Urbanization Effects on Watershed Hydrology and In-Stream Processes in the Southern United States. Water 2(3) 605-648. https://doi.org/10.3390/w2030605

B.B. Mirus, B.A. Ebel, C.H. Mohr, N. Zegre, Disturbance hydrology: Preparing for an increasingly disturbed future. Water Resour. Res. **53**(12), 10007–10016 (2017). https://doi.org/10.1002/2017WR021084

NRC, (National Research Council), *Advancing the Science of Climate Change* (The National Academies Press, 2010). https://doi.org/10.17226/12782

L. Olsson, H. Barbosa, S. Bhadwal, A. Cowie, K. Delusca, D. Flores-Renteria, et al., Land Degredation, in *Climate Change and Land: An IPCC Special Report on Climate Change, Desertification, Land Degradation, Sustainable Land Management, Food Security, and Greenhouse Gas Fluxes in Terrestrial Ecosystems*, (2019), pp. 345–436

U.U. Paulinus, N.G. Ifedilichukwu, A.C. Ahamefula, Morphometric analysis of sub-watersheds in Oguta and environs, southeastern Nigeria using GIS and remote sensing data. J. Geosci. Geomatics **4**(2), 21–28 (2016). https://doi.org/10.12691/jgg-4-2-1

C.A. Penn, L.A. Bearup, R.M. Maxwell, D.W. Clow, Numerical experiments to explain multiscale hydrological responses to mountain pine beetle tree mortality in a headwater watershed. Water Resour. Res. **52**(4), 3143–3161 (2016). https://doi.org/10.1002/2015WR018300

D.A. Randall, R.A. Wood, S. Bony, R. Colman, T. Fichefet, J. Fyfe, et al., Climate models and their evaluation, in *Climate Change 2007: The Physical Science Basis. Contribution of Working Group I to the Fourth Assessment Report of the Intergovernmental Panel on Climate Change*, ed. by S. Solomon, D. Qin, M. Manning, Z. Chen, M. Marquis, K. B. Averyt, M. Tignor, H. L. Miller, (Cambridge University Press, Cambridge, United Kingdom and New York, 323, 2007), pp. 589–662. https://doi.org/10.1016/j.cub.2007.06.045

M. Rodell, P.R. Houser, U. Jambor, J. Gottschalck, K. Mitchell, C.J. Meng, et al., The global land data assimilation system. Bull. Am. Meteorol. Soc. **85**(3), 381–394 (2004). https://doi.org/10.1175/BAMS-85-3-381

M.F. Rosenmeier, D.A. Hodell, M. Brenner, J.H. Curtis, J.B. Martin, F.S. Anselmetti, et al., Influence of vegetation change on watershed hydrology: Implications for paleoclimatic interpretation of lacustrine δ18 O records. J. Paleolimnol. **27**, 131–141 (2002). https://doi.org/10.1023/A:1013535930777.

H.K. Rutter, S.C. Cox, N.F. Dudley Ward, J.J. Weir, Aquifer permeability change caused by a near-field earthquake, Canterbury, New Zealand. Water Resour. Res. **52**(11), 8861–8878 (2016). https://doi.org/10.1002/2015WR018524

Safriel, U.N. (2007). The Assessment of global trends in land degradation. In: Sivakumar, M.V.K., Ndiang'ui, N. (eds) Climate and land degradation. environmental science and engineering. Springer, Berlin, Heidelberg. https://doi.org/10.1007/978-3-540-72438-4_1

M.C. Sashikkumar, S. Selvam, V.L. Kalyanasundaram, J.C. Johnny, GIS based groundwater modeling study to assess the effect of artificial recharge: A case study from Kodaganar river basin, Dindigul district, Tamil Nadu. J. Geol. Soc. India **89**(1), 57–64 (2017). https://doi.org/10.1007/s12594-017-0558-2

K.E. Scilling, M.K. Jha, Y.-K. Zhang, P.W. Gassman, C.F. Wolter, Impact of landuse and landcover change on water balance of large agricultral watershed: Historical effects and future directions. Water Resour. Res. **44**, 1–12 (2008). https://doi.org/10.1029/2007WR006644

J. Shi, F. Yuan, C. Shi, C. Zhao, L. Zhang, L. Ren, et al., Statistical evaluation of the latest GPM-Era IMERG and GSMaP satellite precipitation products in the Yellow River source region. Water (Switzerland) **12**(4), 1–23 (2020). https://doi.org/10.3390/W12041006

D. Singh, M. Tsiang, B. Rajaratnam, N.S. Diffenbaugh, Observed changes in extreme wet and dry spells during the south Asian summer monsoon season. Nat. Clim. Chang. **4**(6), 456–461 (2014). https://doi.org/10.1038/nclimate2208

O.S. Srivastava, D.M. Denis, S.K. Srivastava, M. Kumar, N. Kumar, Morphometric analysis of a semi urban watershed, trans Yamuna, draining at Allahabad using Cartosat (DEM) data and GIS. Int. J. Eng. Sci. (June), 2319–1813 (2014). www.theijes.com

A. Srivastava, N. Kumari, M. Maza, Hydrological response to agricultural land use heterogeneity using variable infiltration capacity model. Water Resour. Manag. **34**(12), 3779–3794 (2020). https://doi.org/10.1007/s11269-020-02630-4

L. Stroosnijder (2007). Rainfall and land degradation. In: Sivakumar, M.V.K., Ndiang'ui, N. (eds) Climate and land degradation. environmental science and engineering. Springer, Berlin, Heidelberg. https://doi.org/10.1007/978-3-540-72438-4_9

C.W. Tan, P.P. Zhang, X.X. Zhou, Z.X. Wang, Z.Q. Xu, W. Mao, et al., Quantitative monitoring of leaf area index in wheat of different plant types by integrating NDVI and Beer-Lambert law. Sci. Rep. **10**(1), 1–10 (2020). https://doi.org/10.1038/s41598-020-57750-z

M.D. Tomer, K.E. Schilling, A simple approach to distinguish land-use and climate-change effects on watershed hydrology. J. Hydrol. **376**(1–2), 24–33 (2009). https://doi.org/10.1016/j.jhydrol.2009.07.029

U.S. Environmental Protection Agency, *Climate Change Indicators in the United States Metadata*, Third edn. (2014). EPA 430-R-14-004

L. Van Roosmalen, T.O. Sonnenborg, K.H. Jensen, Impact of climate and land use change on the hydrology of a large-scale agricultural catchment. Water Resour. Res. **45**(7), 1–18 (2009). https://doi.org/10.1029/2007WR006760

T.A. Woldesenbet, N.A. Elagib, L. Ribbe, J. Heinrich, Catchment response to climate and land use changes in the Upper Blue Nile sub-basins, Ethiopia. Sci. Total Environ. **644**, 193–206 (2018). https://doi.org/10.1016/j.scitotenv.2018.06.198

J. Wu, H. Xie, Research on characteristics of changes of lakes in Wuhan's main urban area. Procedia Eng. **21**, 395–404 (2011). https://doi.org/10.1016/j.proeng.2011.11.2031

A.M. Youssef, B. Pradhan, A.M. Hassan, Flash flood risk estimation along the St. Katherine road, southern Sinai, Egypt using GIS based morphometry and satellite imagery. Environ. Earth Sci. **62**(3), 611–623 (2011). https://doi.org/10.1007/s12665-010-0551-1

Y. Zha, J. Gao, S. Ni, Use of normalized difference built-up index in automatically mapping urban areas from TM imagery. Int. J. Remote Sens. **24**(3), 583–594 (2003). https://doi.org/10.1080/01431160304987

L. Zhang, Z. Nan, Y. Xu, S. Li, Hydrological impacts of land use change and climate variability in the headwater region of the Heihe River Basin, Northwest China. PLoS One **11**(6), 1–25 (2016). https://doi.org/10.1371/journal.pone.0158394

Chapter 4
Climatic Changes Affecting Groundwater Quality in the Semi-Arid Region of Central India with a Special Reference to the Major Cities of Madhya Pradesh: A Short Review

Dhanya Mary Koshy and Preeti Chincholikar

Abstract Over the past few decades, Madhya Pradesh (M.P.) has moved from being a medium rainfall-receiving state to a semi-arid state due to the fact that the amount of moderate rainfall has considerably fallen. M.P., being a geographically semi-arid land, is crucially troubled by climate changes, viz., variable precipitation trends, droughts, and varying temperatures. From the mid-2000s, the state has been witnessing irregular and erratic monsoon conditions, which have drastically affected the groundwater levels. The major occupation of M.P., i.e. agriculture, has always been rainfed. Since, the state is not receiving the required rainfall for agricultural production, most of the practices are therefore dependent on groundwater, which, in turn, poses a great threat to the quality and level of groundwater. The indiscriminate usage of groundwater has not only led to its overexploitation but also has evidently deteriorated the quality of crop yields and soils. Increased use of groundwater has resulted in unpredictable groundwater-level fluctuations, and recharging it is extremely difficult. Proper management has to be employed to maintain the quality and a satisfactory level of groundwater and to evaluate the rate of withdrawal and find ways to effectively recharge the aquifers. This chapter provides a review of the effect of climatic changes since the past two decades (i.e. from 2000 onwards). The emphasis is on the major cities that are affected by groundwater quality in M.P.

Keywords Climatic change · Groundwater · Major cities of M.P

D. M. Koshy · P. Chincholikar (✉)
Department of Chemistry, IES College of Technology, Bhopal, Madhya Pradesh, India

B. Panneerselvam et al. (eds.), *Climate Change Impact on Groundwater Resources*, https://doi.org/10.1007/978-3-031-04707-7_4

4.1 Introduction

Water resources are imperative for human life as well as for the flora and fauna of any given geographical zone of the world. Surface water, although present in abundance, is not suitable for human consumption (Saxena and Sharma 2017; Saxena et al. 2016; Sharma et al. 2016, 2018; Shivhare et al. 2019; Singh et al. 2017; Singh et al. 2020a, b; Thakur and Bhople 2018; USGS (n.d.-a), USGS (n.d.-b)). Civilization growth and industrial development have led humans to explore water resources near their dwelling place. This has resulted in the exploitation of underground water resources. Based on the rainfall per se, geographical zones have been marked as arid or semi-arid zones (Andrade et al. 2007; IGRAC n.d.; Khan and Khan 2019; Kori et al. 2019, 2020; Montoya 2017; Nigam and Yadav 2018; Parashar and Sharma 2017; Prajapati and Rokde 2016).

In the Indian subcontinent, the semi-arid zone covers an area of around 956,756 km^2, and the states that come under this zone are Punjab, Haryana, Uttar Pradesh, Rajasthan, Gujarat, Madhya Pradesh, Maharashtra, Andhra Pradesh, Karnataka, and Tamil Nadu. The dwelling populations in these states are dependent on the limited natural water resources, viz., lakes, rivers, and streams, whichever is available, but urbanization and industrialization have increased the demand for water supply. The available natural resources of surface water would not suffice to meet the demand, which has led to the dependency on groundwater resources – both for domestic and industrial purposes. The wastes generated through these resources either contaminate or leach into the groundwater table. Agricultural practices like the use of fertilizers, landscape simplification, drainage, and the use of machinery have significantly contributed to the degradation of groundwater quality and have also increased the concentration of pollutants in it (Mishra et al. 2021; Kouadri et al. 2022; Elbeltagi et al. 2022). Madhya Pradesh is no different from the situation and is not untouched by drastic conditions of drought and water scarcity as evidenced by the frequent delays in the commencement of monsoon every year (Bagde 2016). A report from a local newspaper describes the climatic variations on groundwater conditions in M.P. on a gross level:

Hindustan Times, January 14, 2016
The M.P. state coordinator of the Climate Change Knowledge Management Centre on Climate Change said that more than 90% of the water used for drinking and domestic purposes in M.P. was being sourced from groundwater, according to the Action Plan on Climate Change prepared by the state government.

(The Fall in Ground Water Levels in State, Hindustan Times, January 14, 2016)

HT on the issue, Lokendra Thakkar, state coordinator of the Climate Change Knowledge Management Centre on Climate Change, said over 90% of the water used for drinking and domestic purposes in MP was being sourced from groundwater, according to the Action Plan on Climate Change prepared by the state government. M.P.'s climate change report had also expressed concerns over the indiscriminate exploitation of groundwater and its depletion in some areas. In general, the depth of water levels in M.P. ranges from 5 to 20 m below ground level (bgl). In all, 1% of monitoring wells show water levels in the 0–2 m bgl range, and 15% of wells show water levels in the 2–5 m bgl range. Water levels of more than 20 m bgl were observed in 5% of wells in the northernmost parts of the state.

This chapter will shed light on a few districts of M.P. with variations in the groundwater quality based on climatic changes that are reported through various articles of the Ministry of Water Resources, research publications, etc.

It is important to know the climatic conditions prevailing in the major cities of M.P., along with the geographical conditions and soil pattern to some extent. A list of this is briefly described below.

Some of the natural and anthropogenic factors and their effects on groundwater quality are discussed below.

4.2 Natural Factors

4.2.1 Lithology and Soil Characteristics

This refers to the study of the physical characteristics of rocks and their composition. The two main parameters that should be taken into consideration while evaluating groundwater quality are soil porosity and hydraulic conductivity. Soil porosity refers to the open space and pores or gaps between the soil particles for water to pass through, and hydraulic conductivity refers to the ability of the porous soil to transmit water (Chandrasekar et al. 2013). Weathering and other processes are important to increase soil porosity and thus result in the enhancement of hydraulic parameters to transmit and store water. This is significant for the occurrence, extension, and productivity of groundwater.

4.2.2 Climate

Climate change can affect hydrogeological processes and groundwater resources directly and indirectly. Since groundwater is not highly likely to be directly affected by climatic changes, there have been cases where due to changes in the weather conditions, increasing dependency on groundwater has led to its major depletion. Hence, an increase in climatic changes can adversely affect the quality as well as quantity of groundwater. Variability in rainfall and other extreme weather conditions can lead to droughts, floods, or other calamities, which can have a direct impact on the availability and dependency on groundwater. This could in turn result in the depletion of groundwater. In coastal areas, the sea level plays a crucial role in determining the salinity of groundwater, which therefore affects its quality by contaminating the drinking water.

4.2.3 Interaction with Other Water Bodies and Topography

There is a direct interaction between surface water and groundwater, which occurs by water draining from surface water bodies to groundwater or by water escaping from the ground to the surface through pores. The physical features of a particular area, i.e. topography, also play an important role in determining the interaction between two water bodies. Groundwater gets easily recharged in lower slope areas than in higher slope areas, as in the latter case, rainwater can easily runoff and significant infiltration and percolation do not take place.

4.3 Anthropogenic Factors

4.3.1 Industrial Discharge and Disposal of Waste

In the era of industrialization and urbanization, groundwater contamination has become an alarming issue, with various chemical traces found in water bodies. These can not only affect the quality of groundwater but also pose major health hazards. The prevalent health hazards are due to the presence of chemicals such as nitrates, fluorides, arsenic, cadmium, lead, and other toxic substances, which lead to cardiovascular, neurological, and renal problems. Moreover, due to inefficient solid waste management and dumping of municipal solid wastes (MSWs – all the discarded solid wastes from industrial, commercial, municipal, and agricultural activities) into open landfills, both groundwater and surface water are severely affected, leading to contamination of groundwater, soil, and air.

4.3.2 Urban Activities

Urban activities are all those things that human beings do, including what, when, where, and for how long. Development of infrastructure, increase in population, deforestation, and improper sewage system are some of the activities that can have a direct impact on the quality of groundwater. As the population increases, there is an emerging need for new and advanced technologies and more areas for settlement. This affects the rainfall and water drainage patterns adversely, resulting in the degradation of groundwater quality.

4.3.3 Agriculture

The main source of water in arid and semi-arid regions for irrigation is groundwater; due to variability in climatic conditions, the annual rainfall rate in many regions has noticeably declined. Thus, groundwater is being constantly used for irrigation purposes, which affect the quantity of groundwater. Therefore, if excessive pumping of water from aquifers is performed faster than the replenishment or recharging of groundwater, then it can lead to depletion of the groundwater table. It can lead to many problems like drying up of wells, declination of water levels in rivers and lakes, degradation of water quality, increase in the cost of pumping, etc.. Moreover, the use of fertilizers, pesticides, and insecticides can also lead to the deterioration of groundwater quality (Moharir et al. 2019; Mishra et al. 2022). The toxic substances present in these chemicals get infiltrated and seeped through the soil and get dissolved in the water, which makes it harmful for other purposes (Devic et al. 2013).

Some case studies for groundwater quality had been carried out in the major cities of M.P.; the water quality indices (WQIs) based on the analysis conducted by various researchers and government organizations have been compiled to present an overview of the groundwater quality based on the climatic variations occurring in the state of Madhya Pradesh. Here is a glimpse of the major cities.

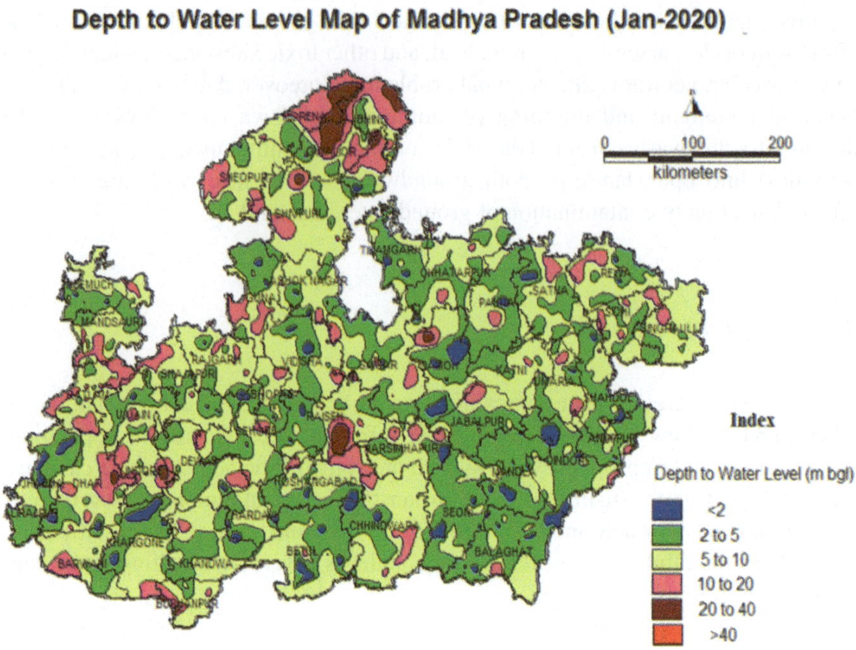

Fig. 1 GW_Year Book_19-20_Madhya Pradesh

4.3.3.1 Bhopal

Bhopal district, with a land area of around 2772 km^2, is located in the central part of Madhya Pradesh, India. Its border is surrounded by the following:

1. On the north – Guna district
2. On the northeast – Vidisha district
3. On the east – Raisen district
4. On the southwest and west – Sehore and Rajgarh districts

Bhopal district lies between N 23°05′ and 23°54′ latitude and E 77°10′ and 77°40′ longitude (SOI, Topography Sheet No. 55 E). Bhopal city is the headquarters of the state. The index map of Bhopal is shown in Fig. 1. As per the 2001 census, the population of Bhopal was about 33,26,228 with the rural population of about 18,43,510 and the urban population of 4,82,718. Thus, the district is largely urban.

The agricultural activity in Bhopal is mainly dependent on the monsoon. At present, the main sources of irrigation in the district are wells and tanks. During the administration of the Nawab of Bhopal state, only two minor irrigation schemes (Chanchal Dam and Ajnal Dam) were in operation with an irrigation potential of 6.59 km^2. After the reorganization of states in 1956, there was a gradual increase in surface water development (CWGE, Bhopal, 2018).

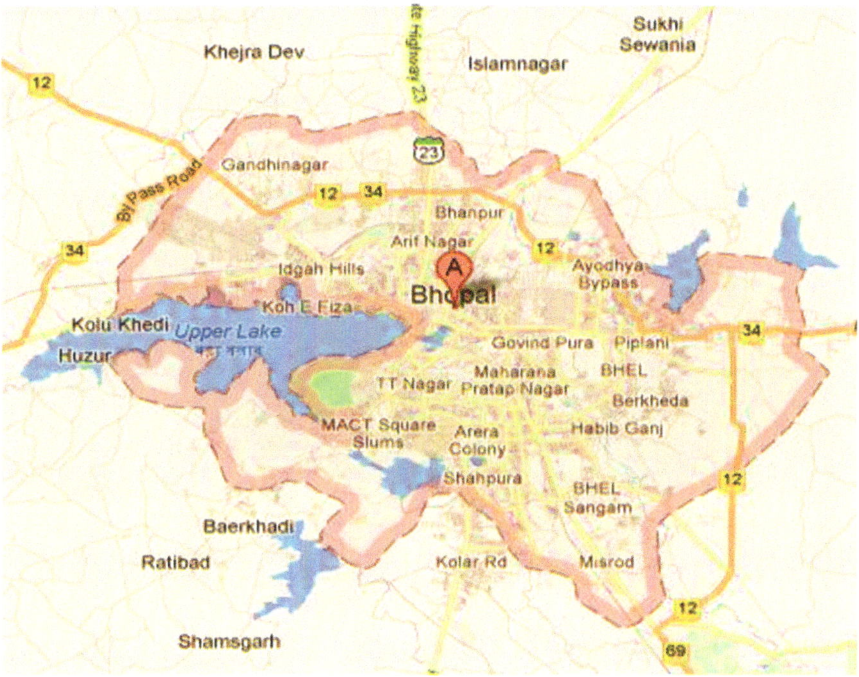

4.3.3.2 Indore

The city of Indore, with an area of 3898 km^2, is located at 22.7196° N and 75.8577° E. Indore lies in the west-central part of M.P. and is the most populated city in M.P. with a population of approximately 31 lakhs. It has a humid tropical climate with an average rainfall of 1033 mm.

4.3.3.3 Jabalpur

The city of Jabalpur lies in the east-central part of M.P. and is located at 23.1815° N and 79.9864° E. It has an area of 5211 km^2 with a population of approximately 15 lakhs. The climate in Jabalpur is mildly warm and temperate. The city lies in the eastern part of M.P. and experiences an average rainfall of about 1279.5 mm (Damle et al. 2020).

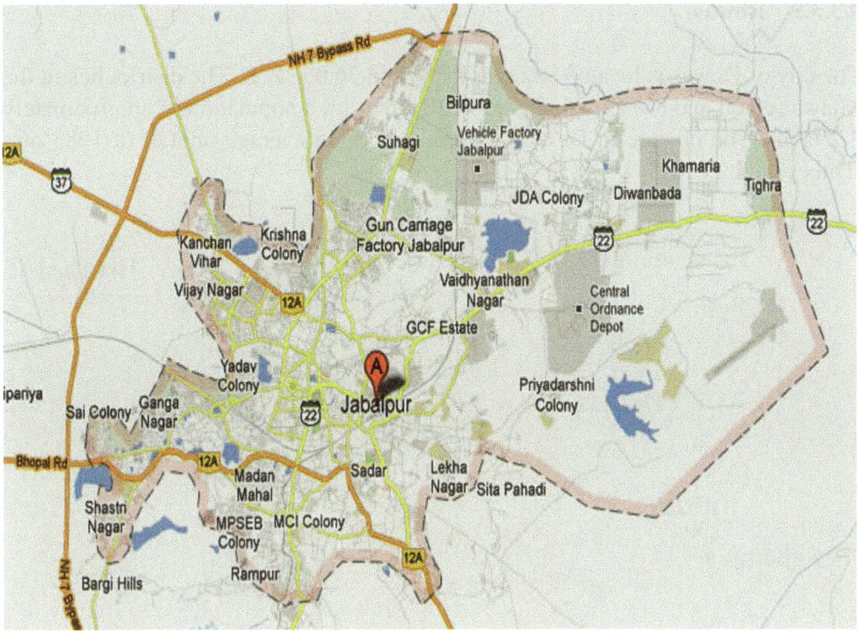

4.3.3.4 Ujjain

Located at 23.1765° N and 75.7885° E, Ujjain has a tropical climate with an average precipitation of 943 mm. The city lies in the western part of M.P. and has an area of 6091 km^2 with a population of approximately 6.2 lakhs.

4.3.3.5 Dewas

The city of Dewas is located at 22.9676° N and 76.0534° E. The district lies in the Malwa region and has a total area of 7020 km² with a population of approximately 3.7 lakhs. The climate in Dewas is tropical with an average rainfall of 1083 mm (Dubey et al. 2015).

4.3.3.6 Gwalior

The city of Gwalior lies in the northern part of M.P. and is located at 26.2183° N and 78.1828° E. It has a subtropical climate and experiences extreme weather conditions. It has an area of 4560 km² with a population of 14.1 lakhs. The average rainfall in Gwalior is 700 mm.

4.3.3.7 Rewa

Located at 24.5362° N and 81.3037° E, Rewa is a north-eastern city in M.P. and has a warm and temperate climate. It covers an area of 6314 km² and has a population of approximately 26.5 lakhs. The average rainfall experienced in Rewa is 890 mm.

4.3.3.8 Chhindwara

The city of Chhindwara is one of the southern districts of M.P. and is located at 22.0574° N and 78.9382° E with an area of 11815 km^2. The climate in Chhindwara is warm and subtropical. It experiences an average rainfall of 1139.3 mm. The population of the city is approximately 23.4 lakhs.

4.3.4 *The Groundwater Variation Report*

The overall climatic effect on the groundwater quality in M.P. depends on the groundwater-level variations occurring throughout the period of study (from 2015 to 2020). A water-level rise of more than 4 m was mainly observed in the Morena, Gwalior, Sheopur, Sagar, Hoshangabad, Umaria, and Balaghat districts and in parts of the Malwa region. In May 2018, it was observed that about 40.93% of the wells showed a rise in groundwater levels and about 59.07% of the wells showed a decline in water levels in the state. The water levels decline in the order to ranges such as 0-2 m (41.61%) 2 and 4 m (11.51%) and more than 4 m (5.94%) of the wells. A decline in water levels of less than 2 m was observed in most of the districts in the north-eastern part of the state compared to the north-western part. A decline in water levels of more than 4 m was mainly observed in the Barwani, Indore, Ratlam, Ujjain, Hoshangabad, Chhindwara, Balaghat, Sagar, Gwalior, Datia, and Sheopur districts of the state. In general, by comparing the mean water level of previous 10 years with that of the current year, it is inferred that there is no significant decline in water levels, but, in general, there is a rise in water levels in the western part of the state (Kathik et al. 2020).

4.3.5 The Monsoon Effect on the Groundwater Levels in M.P.

The groundwater level chart the trend of the pre-season period (May 2015–May 2020) state of Madhya Pradesh, shown a decline of groundwater level (0.10 m/year) in the past ten years, we can say that in all districts of the state. This growth trend of more than 0.20 m/year was observed mainly in the Sheopur, Gwalior, Panna, Revin, Bhopal, Mandsaur, Shajapur, and Ujjain districts, as a small, pocket-sized area. The rising trend is at a higher level (>0.2 m/year) as seen in Sheopur and in the southern part of the district, Ujjain, small spots. A down trend with a speed of up to −0.10 m /year was observed in some areas, especially in the central and southern parts of the state, and a low-speed trend (from −0.10 to −0.20 m/year) was observed mainly in the Damoh, Narsinghpur, Chhindwara, Guna, Shivpuri, Mandsaur, Ratlam, Devas, and Indore districts. However, the trend is more intense (>0.2 m/year), as mentioned above, in the Indore region. The trend of pre-season water level changes, reflecting land management and water management activities in this area. Approximately 58% of well monitoring shows a downward trend in sea levels, and 42% of well monitoring shows a trend in rising groundwater levels. Water table trends in the Post-monsoon period of (November 2015–November 2018) are decreasing water label up to 0.10 m/year over the past ten years in the state. A downward trend in speed (−0.10 to −0.30 m/year) was observed mainly in the Sheopur, Guna, Shivpuri, Vidisha, Indore, Raizen, Damoh, and Betul districts. An increasing trend at a rate of 0.10 m/year was observed in the north, northwest, and east regions of the state. A downward trend in this indicator (from 0.10 to 0.20 m/year) was observed in the spots, Rajgarh, Mandsaur, and Ujjain districts. The upswing of the post-monsoon trend, on the other hand, shows that the additional water stored in the aquifer is due to an increase in rain or leakage of irrigation water used. Approximately 56% of well monitoring shows a downward trend in sea levels, and 44% of well monitoring shows a trend in rising groundwater levels (Damle et al. 2020; Devic et al. 2013; Dubey et al. 2015; Dwivedi and Tripathi 2017; Gedam and Hatiye 2020; Green 2016; Kamboj and Choudhary 2013; Kashyap 2016; Kathik et al. 2020).

4.3.6 Materials and Methods

The groundwater quality of the eight cities, studied over 5 years, i.e. from 2015 to 2020, at different seasonal and climatic variations reported by researchers from meteorological, geochemical, and soil and environmental sciences has been analyzed to interpret the effect of climatic factors in the following cities of Madhya Pradesh. For the year 2015, the methodology used for collecting samples was a standard method according to the American Public Health Association (APHA) Standard Methods for the Examination of Water and Wastewater (1989).

4.3.6.1 Indore

The instruments used for estimating the parameters were a pH meter for calculating the pH, a conductometer for electrical conductivity (EC), and a total dissolved solid (TDS) meter to measure the TDSs. Concentrations of sodium and potassium were measured with the help of a flame photometer, those of sulphate, nitrate, fluoride, iron, and mercury were measured by a spectrophotometer, and those of zinc, lead, copper, and cadmium were estimated by an atomic absorption spectrophotometer.

From 2016 to 2019, the samples were analyzed physically and chemically by the standard methods prescribed in the APHA (2012). The groundwater samples from all 10 cities were collected in pre-washed, rinsed, and labelled polythene/glass containers. Field parameters like temperature, pH, and dissolved oxygen (DO) were examined immediately as these are non-conservative parameters and cannot be preserved.

4.3.6.2 Bhopal

Bhopal has a warm and temperate climate with an average rainfall of 1126.7 mm. The sampling areas were around 10 residential and commercial areas. A total of 20 water samples were collected during the pre- and post-monsoon seasons of 2015. The samples were collected in plastic bottles and analyzed within 24 hours of sample collection. Analysis was conducted based on the standard methods prescribed by the APHA. The instruments/methods used for estimating the parameters were a pH meter for calculating the pH, a conductometer for electrical conductivity, a gravimeter to measure the TDS, and a nephelometer for turbidity analysis. Concentrations of sodium and potassium were measured with the help of a flame photometer, those of sulphate, phosphate, nitrate, fluoride, and iron were measured by an ultraviolet–visible (UV–vis) spectrophotometer, total coliform was calculated by the multiple-tube fermentation technique, and titrimetric methods were employed for the estimation of total alkalinity (TA), total hardness (TH), BOD, COD, DO, chloride, calcium, and magnesium. From 2016 to 2018, a study of all the physicochemical parameters of groundwater was carried out by dividing the city of Bhopal into four zones. In all, 10 locations were selected for groundwater monitoring across the city. Monitoring was conducted as per the standard guidelines of the Central Pollution Control Board. Water samples were collected from bore wells and hand pumps and were analyzed using the standard methods prescribed by the APHA (2017).

4.3.6.3 Jabalpur

Groundwater samples were collected from ten different locations and stored in pre-washed and sterilized 1-litre-capacity polythene bottles during the month of May–June 2015. The collected samples were examined for various physicochemical parameters like pH, temperature, turbidity, electrical conductivity, total alkalinity, total dissolved solids, total hardness, dissolved oxygen, concentrations of calcium and magnesium, chloride, nitrate, fluoride, and phosphate. The climate in Jabalpur is mildly warm and temperate. The city lies in the eastern part of M.P. and experiences an average rainfall of about 1279.5 mm. In 2018, a study was conducted regarding the water quality index of groundwater of nine tribal blocks of M.P., which included Jabalpur. Water samples were collected and stored in non-reactive bottles, which were pre-washed with 2% HNO_3 and then rinsed thrice with distilled water. The samples were collected from 8 a.m. to 11 a.m. The weight of the samples was measured, and then the WQI was calculated by following all the steps and formulae. In 2020, samples were collected from 253 locations and the WQI was calculated for 12 parameters. The methods employed were the standard methods prescribed in the APHA (1994).

4.3.6.4 Ujjain

Located at 23.1765° N and 75.7885° E, Ujjain has a tropical climate with an average precipitation of 943 mm. The city lies in the western part of M.P. and has an area of 6091 km^2 with a population of approximately 6.2 lakhs. From 2015 to 2016, 868 groundwater samples from 124 households were collected and analysis was conducted by standard analytical methods. Collection of samples was conducted during the pre-monsoon (May 2015 to June 2016), monsoon (August 2015 to September 2016), and post-monsoon (January 2015 to January 2016) seasons. The polythene bottles and other glassware were cleaned by soaking in detergent, then soaking in 10% HNO_3 for 48 hours, and rinsing with deionized water. Collected samples were maintained at 4 °C and then brought to the laboratory for physicochemical analysis.

In 2017, groundwater samples were collected as per the standard sample collection techniques given in the APHA (1998), from 10 sampling stations in the Ghatiya block in the Ujjain district. A total of 14 WQI parameters were calculated such as pH, colour, turbidity, conductivity, TDS, chloride, fluoride, TA, TH, calcium, magnesium, nitrate, total coliform, and faecal coliform.

4.3.6.5 Dewas

The climate in Dewas is tropical with an average rainfall of 1083 mm. In 2015, an analysis of groundwater quality was conducted in Dewas to determine whether the groundwater is fit for irrigation. Four samples were collected from five villages with a frequency of one sample a week.

A study on the suitability of groundwater for irrigational purposes was conducted near the Dewas district in 2017. A comparison of the parameters present in the samples during both pre- and post-monsoon seasons is provided. Water samples were collected in pre-washed 1-litre-capacity polythene bottles. Field parameters like colour, temperature, odour, and taste were examined and recorded at the sampling stations. Other parameters were analyzed by standard instruments like a pH meter, conductometer, flame photometer, Hach spectrophotometer, titrations, and UV–vis spectrophotometer.

In 2020, a study on groundwater in various industrial locations in Dewas was conducted. Samples were collected from bore wells and monitored according to the standard guidelines of the Central Pollution Control Board. Analysis was conducted based on the standard methods as per the APHA (2017).

4.3.6.6 Gwalior

The average rainfall in Gwalior is 700 mm. In 2016, an evaluation of WQI parameters was carried out to study the quality of water for drinking purposes in five villages in Gwalior city. The main sources of water are open wells, tubes, or bore wells. Samples were collected from bore wells from March to August in pre-washed polythene bottles. The temperature was measured at the sampling site. Other parameters were examined at the laboratory where the samples were refrigerated at 4°C to minimize the changes in the physicochemical parameters. Various parameters that were analyzed included colour, odour, temperature, pH, electrical conductivity (EC), total dissolved solids (TDSs), total alkalinity (TA), total hardness (TH), calcium, magnesium, chloride, and dissolved oxygen (DO) as per the standard procedures prescribed by the APHA (2003).

A study on the WQI parameters of groundwater was carried out in the city of Gwalior in 2018. During the pre-monsoon season, i.e. from March to May, and the monsoon season, i.e. from June to August, samples from 12 locations were collected in pre-washed plastic bottles. The samples were analyzed within 6–10 hours of collection, and, hence, no preservatives were used. The samples were maintained at a temperature below 4°C.

4.3.6.7 Rewa

The average rainfall experienced in Rewa is 890 mm. Physicochemical analysis of various water samples from various locations of the Rewa district was carried out in the year 2015. The samples were collected from these sites between 9 a.m. and 11 a.m. and stored in polythene bottles.

During the pre- and post-monsoon seasons of 2018, water quality assessment for drinking and irrigation purposes was conducted in the Rewa district. Samples were collected from hand pumps and dug wells and stored in 500-ml-capacity plastic bottles. The bottles were pre-washed with HNO_3 and rinsed with distilled water. In the laboratory, the samples were refrigerated at 4 °C. Standard methods were used for the analysis of physicochemical parameters. Parameters like pH, electrical conductivity, TH, TDS, chloride, calcium, magnesium, sodium, and potassium were determined using their respective instruments.

4.3.6.8 Chhindwara

Chhindwara experiences an average rainfall of 1139.3 mm. The population of the city is approximately 23.4 lakhs. Groundwater quality assessment was conducted in five tehsils/blocks of the Chhindwara district in the year 2015. In all, 10 samples from each block, i.e. a total of 50 samples, were collected from these blocks and chemically examined. Physicochemical parameters like pH, electrical conductivity, salinity, total hardness, fluoride ion, iron, and turbidity were analyzed using standard techniques and were compared with the standard values provided by the World Health Organization (WHO).

In another study in the year 2020, the effect of the concentration of fluoride ions along with the determination of other physicochemical parameters was evaluated. Water samples were collected from 53 bore wells in 5 blocks of the district. The samples were collected in one-litre-capacity plastic bottles during the pre- and post-monsoon seasons. The concentration of ions in the water samples was determined by techniques according to the APHA (1975).

pH estimation of the groundwater was determined by the following protocol. After calibration of the instrument, that is, the electrodes, the cup should be thoroughly rinsed. A coffee cup should be filled with a portion of the sample solution and to obtain a preliminary to the pH value. In general, this will be the value of the drift and should to be treated as a modification. Furthermore, evidence was collected on the additional portions of the same solution that will consistently yield greater values. B For solutions that are well-buffered, 3 portions of it may not be enough to get the pH values of up to ± 0.04 units show errors of less than ± 0.04 units for 1 or 2 minutes. In the case of a very dilute or non-buffered solutions, it can take up to 6 a portion of the test solution, and the pH values in the range ± 0.05 units. As the accuracy, exceeding 0.1 pH units, at a temperature that differs from the standard solutions, and the electrodes can be used for water quality analysis. The water sample should be stored at the measuring temperature for at least 2 hours prior to the measurement to minimize the risk of contamination. The probe should be immersed in the groundwater sample of which the pH is to be measured. The values obtained were noted down for each variant sample of the cities. Calcium, magnesium, bicarbonate, and chloride were determined by complexometric titration methods of hardness (ethylenediaminetetraacetic acid (EDTA) method). This provides a gross estimation of calcium, magnesium, and bicarbonates present in the sample.

4.4 Results and Discussion

Water samples were collected from 1194 Wells during May 2019. A detailed analysis has been carried out on the quality of the measurements, namely, pH, electrical engineering, electrical conductivity (EC), carbonate, bicarbonate, chloride, fluoride, nitrate, sulphate, phosphate, total hardness, calcium, magnesium, sodium, and potassium. The physicochemical parameters of the respective cities are displayed in Tables 1, 2, 3, 4, and 5.

Table 1 Physicochemical parameters in the major areas of Bhopal city

S No.	Location	Physicochemical parameters													
	City: Bhopal	pH	EC	HCO_3	Cl	SO_4	NO_3	F	PO_4	SiO_2	TH	Ca	Mg	Na	K
1	Amlia	7.7	681	239	30	36	58	0.4	0.1	22	275	68	26	21	11
2	Ayodhya Nagar	7.5	1291	514	135	25	3	0.8	0.7	44	410	64	61	99	8.9
3	Bairagarh 1	7.5	1976	453	400	32	1	0.3	0.1	36	800	172	90	78	10
4	Bairagarh 2	7.8	1095	325	192	26	2	0.2	0.1	44	395	62	58	50	25
5	Balampur Ghati	7.2	1492	288	265	54	47	0.1	0.6	49	650	128	80	33	12
6	Barkheda	7.7	360	98	40	35	3	0.3	0.1	12	110	28	10	26	11
7	Berasia	7.4	659	276	37	26	3	0.3	0.1	19	235	66	17	35	9.9
8	Bilkhiriya	7.5	528	214	30	19	15	0.3	0.1	28	160	42	13	39	11
9	Chichli	6.5	322	61	45	33	4	0.1	0	13	95	28	6	21	11
10	Dig Bangla	7.7	1123	490	92	13	1	0.7	0.2	29	275	30	49	120	12
11	E-2 Nursery	7.5	1084	504	61	22	8	0.4	0.3	33	450	102	47	36	4.6
12	Gandhi Nagar	7.4	758	221	106	33	5	0.4	0.1	20	270	70	23	42	8.6
13	Gunga	7.1	869	123	174	48	2	0.3	0.1	31	275	94	10	65	9
14	Lalghati	7.1	643	135	76	61	21	0.2	0.1	25	195	58	12	49	10
15	Nabibagh	7.3	1165	350	172	17	22	0.5	0.3	36	390	80	46	78	12
16	Nagirabad	7.3	1236	178	295	27	13	0.1	0.1	44	375	84	40	101	13
17	Nipaniya Jat	7.8	899	289	129	19	1	0.3	0.1	32	300	66	33	59	11
18	Patel Nagar	7.6	1203	350	177	11	40	0.3	0.2	47	435	114	36	66	12
19	Piplani	7.3	524	191	43	24	3	0.3	0.1	31	185	46	17	28	7.7
20	Ramgarha	7.2	982	381	78	20	40	0.3	0.1	19	330	114	11	67	6.9
21	Sarvar	7.5	646	264	25	15	50	0.4	0.1	27	260	74	18	21	7
22	Shahjahanabad	7.3	966	369	93	21	5	0.3	0.1	40	255	72	18	89	21
23	Shahpura	8.4	674	221	71	22	4	0.2	0.1	31	210	30	33	49	11
24	South T.T. Nagar	7.2	907	357	86	29	2	0.7	0.2	35	260	58	28	79	12
25	Suhaya Kala	7.5	608	271	23	19	26	0.4	0.1	27	215	48	23	33	12
	AVG	**7.4**	**907.6**	**286**	**115**	**27**	**15**	**0.3**	**0.2**	**31**	**312**	**72**	**32**	**55**	**11**

Table 2 Physicochemical parameters in the major areas of Chhindwara city

S No.	Location	Physicochemical parameters													
	City: Chhindwara	pH	EC	HCO_3	Cl	SO_4	NO_3	F	PO_4	SiO_2	TH	Ca	Mg	Na	K
1	Amarwara	7.5	594	178	50	48	11	0.7	0.11	26	141	44	7	65	1.9
2	Bangaon 1	7.4	687	252	55	32	13	1.5	BDL	33	232	63	18	47	1.4
3	Banjari	7.8	410	98	37	42	26	0.9	BDL	22	51	12	5	70	1.6
4	Borgaon 1	7.6	###	406	147	67	189	1.8	0.14	40	551	119	61	112	2.1
5	Chaurai 1	7.5	###	240	145	52	39	1.7	BDL	22	268	65	26	106	2.7
6	Chhindi	8	804	215	105	40	21	1.1	0.15	29	242	59	23	71	1.9
7	Chhindwara	7.4	###	338	132	54	40	0.9	BDL	32	379	93	36	85	2.3
8	Chinchkheda	7.3	743	252	85	20	10	0.5	BDL	21	212	40	27	70	1.9
9	Chincholi Ward	7.4	876	301	70	26	55	0.7	BDL	33	313	77	29	53	2.7
10	Damua	7.7	###	750	222	118	65	1	BDL	46	773	204	64	173	1.4
11	Delakhari	7.1	977	314	107	39	25	0.9	BDL	22	348	85	33	55	11
12	Goni	7.1	901	277	80	41	60	0.7	0.36	33	313	65	37	59	2.9
13	Harrai	7.4	###	314	192	22	49	0.8	0.22	39	359	91	32	115	1.1
14	Jamai	7.1	###	381	60	98	90	0.6	0.11	28	485	127	41	49	3.2
15	Jamunia Ner	7.4	924	289	97	36	41	0.8	0.14	27	313	79	28	66	1.7
16	Kundali	7.1	696	264	45	29	30	0.5	BDL	25	258	75	17	38	2.2
17	Lahgudna	7.3	778	240	90	32	13	1.1	BDL	34	227	65	16	70	1.9
18	Linga Rly. Stn.	7.9	###	338	192	98	112	1		22	510	137	41	119	1.7
19	Mahuljhir	7.9	###	394	132	62	43	1.3		48	369	79	42	118	1.3
20	Markahandi	7.6	794	264	57	48	35	0.4		56	313	73	32	35	0.8
21	Mohi	7.4	590	246	27	22	25	0.8		51	217	63	15	35	1.3
22	Pandhurna	7.9	598	178	57	43	17	1.1		37	182	38	21	51	1.7
23	Piplanarayanwar	7.5	997	350	72	42	49	0.7		20	379	95	34	49	2.8

(continued)

Table 2 (continued)

S No.	Location	Physicochemical parameters													
	City: Chhindwara	pH	EC	HCO_3	Cl	SO_4	NO_3	F	PO_4	SiO_2	TH	Ca	Mg	Na	K
24	Rajna	7.2	###	320	115	58	32	1		29	338	79	34	86	1.4
25	Ramakona New	7.4	###	455	92	34	15	2		24	424	87	50	62	2.8
26	Ramgarh	7.7	###	400	145	83	39	0.7		18	465	152	21	95	1.7
27	Renikhera	7.1	###	240	90	38	152	1.1		22	348	95	27	67	3.1
28	Saori 1	7.3	929	295	80	42	62	0.6		19	348	85	33	50	1.5
29	Sarangbheri	8	669	215	45	31	52	0.9		24	253	75	16	33	1.7
30	Sathiya	7.6	354	117	35	16	8	0.2		33	91	10	16	38	0.9
31	Sausar	7.4	###	461	142	42	24	0.4		44	444	149	17	98	2.2
32	Silwani 1	7.4	758	234	67	27	53	0.7		26	258	77	16	52	1.7
33	Silwanighati	7.3	986	424	52	18	41	0.9		29	359	95	29	59	0.9
34	Singhori	7.3	###	461	172	73	38	0.7		32	601	176	39	69	0.7
35	Sonapipri	7.6	###	320	72	22	125	1.9		26	283	97	10	100	1.4
36	Surla	7.3	997	375	80	36	21	0.3		34	348	77	38	65	1.1
37	Tansara Mal	7.5	596	252	35	14	18	0.4		23	227	71	12	30	1.5
38	Thanvari Kunda	7.6	615	221	47	32	13	0.4		26	197	42	22	48	2.5
	AVG	**7.5**	**976**	**307**	**93**	**44**	**46**	**0.9**	**0.18**	**30**	**327**	**85**	**28**	**70**	**2.1**

Table 3 Physicochemical parameters in the major areas of Dewas city

S No.	Location	Physicochemical parameters													
	city: Dewas	pH	EC	HCO_3	Cl	SO_4	NO_3	F	PO_4	SiO_2	TH	Ca	Mg	Na	K
1	Alari	7.9	845	299	84	22	21	0.5		23	235	42	32	75	13
2	Bagli 1	7.6	712	201	72	65	18	0.1		34	245	46	32	47	4.2
3	Bhamori	7.2	795	92	158	75	5	1.8		36	115	22	15	123	1
4	Bhesuni	7.5	2972	549	686	12	51	0.3		26	755	212	55	333	1.5
5	Bhikupura	7.6	590	171	27	56	42	0.1		42	185	42	19	46	1.9
6	Bhonrasa	7.6	1298	403	126	36	80	0.1		32	355	92	30	127	3.4
7	Bijwad	7.9	712	214	49	22	65	0.6		38	205	52	18	61	1.9
8	Dewas	7.5	1497	256	277	45	53	0.5		22	405	116	28	152	1
9	Dhayli	7.3	1050	397	37	55	73	0.5		35	190	52	15	142	1.5
10	Kannod	7.8	1078	427	67	25	65	0.4		46	195	46	19	150	1.6
11	Kantaphod	7.6	753	311	54	23	6	0.2		33	245	62	22	53	1.7
12	Karnawad	7.6	647	232	49	15	47	0.4		28	235	64	18	35	5.5
13	Kusmaniya	7.6	642	67	139	32	7	1.8		44	150	42	11	72	1.3
14	Matmor New	7.7	777	250	82	18	35	0.8		42	205	44	23	78	1.3
15	Nevri	7.5	875	305	54	15	89	0.6		49	315	78	29	53	1.6
16	Pipilianankar	7.3	1501	451	124	45	145	0.6		36	350	112	17	177	1.7
17	Pipri	7.4	1850	665	235	42	25	0.6		22	420	142	16	200	42
18	Punjapura	7.3	1210	512	42	58	57	0.1		28	260	72	19	128	45
19	Satwas New	7.6	977	244	151	25	42	0.3		27	210	62	13	88	65
20	Udainagar	7.3	1488	488	161	28	97	1		35	405	124	23	148	5.3
	AVG	**7.5**	**1113**	**327**	**134**	**36**	**51**	**0.5**		**34**	**284**	**76**	**23**	**114**	**10**

Table 4 Physicochemical parameters in the major areas of Gwalior city

S No.	Location	Physicochemical parameters													
	City: Gwalior	pH	EC	HCO_3	Cl	SO_4	NO_3	F	PO_4	SiO_2	TH	Ca	Mg	Na	K
1	Aarauli	7.1	740	278	42	30	42	0.5		25	297	89	18	30	1.2
2	Aron	7.3	645	230	60	12	36	0.6		30	257	87	10	18	16
3	Bajna	7.2	868	369	50	23	25	0.6		28	287	93	13	62	1
4	Bajrang Colony	7.6	480	157	57	10	18	0.2		35	183	59	8	24	1.3
5	Behat	7.3	234	91	12	9	15	0.2		22	79	20	7	16	1.2
6	Charai Shyampur	7.3	###	224	212	40	102	0.5		29	421	152	9	85	3.5
7	Dongarpur	7.4	935	411	52	22	40	0.9		24	282	71	25	80	1.6
8	Ghatigaon	7.4	675	303	35	19	12	0.4		33	208	59	14	56	2.3
9	Harsibandh	7.7	278	91	17	21	12	0.3		41	94	30	5	18	1.2
10	Kariyawati	7.6	990	430	65	17	30	1		40	312	53	43	80	1
11	Makoda	7.2	945	321	102	17	49	0.7		29	262	79	16	94	0.9
12	Manpura	7.9	710	327	42	12	8	0.7		25	228	59	19	55	1.1
13	Masudpur New	7.2	###	514	77	45	26	1.4		35	327	71	37	132	0.8
14	Mohanpur	7.3	920	418	47	29	21	1		42	262	53	31	88	2.6
15	Mohana	7.8	###	315	462	90	42	1.1		29	723	216	43	145	3.4
16	Nayagaon 2	7.5	620	303	22	17	10	0.8		35	223	69	12	37	2
17	Panihar	7.6	665	309	32	18	10	0.5		33	208	57	15	52	5
18	Tekanpur	7.8	###	563	72	30	23	1.4		36	213	42	26	185	0.9
	AVG	**7.5**	**864**	**314**	**81**	**26**	**29**	**0.7**		**32**	**270**	**76**	**20**	**70**	**2.6**

4.4.1 Analysis for Salinity

The salt content of the water shall be represented by means of its electrical conductivity. Electrical conductivity is a numerical expression of the ability of water to conduct an electric current. It depends on the total concentration of the ionized substances dissolved in water and the temperature at which the measurement is to be carried out. Electrical conductivity increases by about 2% per degree Celsius increase in temperature. The dissolved ions and their valencies, and the concentration of the influence of the electrical conductivity. If the electrical conductivity is an index of the degree of salinity of the water, its high electric conductivity, and high levels of salinity. The high level of salinity is not the only one it will give a bad taste to the water, but this water is not suitable for irrigation, as well as the results to the deposition of salts in the soil, and its effect on the fertility of the soil, in the absence of adequate soil management practices.

The water, the conductivity ranges in between 1000 μS/cm at 25 °C it can be seen that there is a fresh & fit for the drinking purpose. The quality of the groundwater is a very good, if 58.71% an electrical conductivity of less than 1000 μS/cm at 25 °C and the salinity tolerance of crops to be grown in the region, where the electrical conductivity is more than 2000 μS/cm in 25 °C. The chemical quality of the

Table 5 Physicochemical parameters in the major areas of Indore city

S No.	Location	Physicochemical parameters													
	City: Indore	pH	EC	HCO_3	Cl	SO_4	NO_3	F	PO_4	SiO_2	TH	Ca	Mg	Na	K
1	Bhil Paltan	7.5	1084	303	35	19	12	0.4		33	208	59	14	56	2.3
2	Bijalpur Masjid	7.6	772	91	17	21	12	0.3		41	94	30	5	18	1.2
3	Cable Factory	7.6	2462	430	65	17	30	1		40	312	53	43	80	1
4	Depalpur	7.6	2423	321	102	17	49	0.7		29	262	79	16	94	0.9
5	Dudhia	7.5	1969	327	42	12	8	0.7		25	228	59	19	55	1.1
6	Footi Kothi	7.7	853	514	77	45	26	1.4		35	327	71	37	132	0.8
7	Gandhi Hall	7.7	1232	418	47	29	21	1		42	262	53	31	88	2.6
8	Hatod	7.8	1314	315	462	90	42	1.1		29	723	216	43	145	3.4
9	Mari Mata	7.8	986	303	22	17	10	0.8		35	223	69	12	37	2
10	Mhow	7.4	628	549	686	12	51	0.3		26	755	212	55	333	1.5
11	Mushakhedi	7.5	1145	171	27	56	42	0.1		42	185	42	19	46	1.9
12	Nandpura	7.5	2524	403	126	36	80	0.1		32	355	92	30	127	3.4
13	Polo Ground	7.7	1771	214	49	22	65	0.6		38	205	52	18	61	1.9
14	Prakash Nagar	7.8	544	256	277	45	53	0.5		22	405	116	28	152	1
15	Rangwasa	7.6	1067	397	37	55	73	0.5		35	190	52	15	142	1.5
16	Ranjeet Hanuman Temple	7.5	1047	427	67	25	65	0.4		46	195	46	19	150	1.6
17	Ravindra Nagar	7.6	709	311	54	23	6	0.2		33	245	62	22	53	1.7
18	Sajan Nagar	7.8	1260	232	49	15	47	0.4		28	235	64	18	35	5.5
19	Sanwer	7.5	3880	549	686	12	51	0.3		26	755	212	55	333	1.5
20	Soybean Research Centre	7.6	795	201	72	65	18	0.1		34	245	46	32	47	4.2
21	Telephone Nagar	7.7	1105	92	158	75	5	1.8		36	115	22	15	123	1
22	Ushapura	7.9	3137	549	686	12	51	0.3		26	755	212	55	333	1.5
	AVG	7.6	1486.7	171	27	56	42	0.1		42	185	42	19	46	1.9

groundwater is a very good read the frequency distribution of the electrical conductivity under an unconfined aquifer 86.85%, all this wells are covered range of 0–1500 μS/cm in 25 °C, and it is only 1.01% of the wells, the electrical conductivity is higher than that of 3000 μS/cm in 25 °C.

4.4.2 Fluoride Content

In general, fluoride is found in low concentrations in natural waters. The presence of fluoride in groundwater is extremely important from a health point of view; consumption of a low concentration helps in the prevention of dental caries (up to

1.0 mg/l (ppm). However, high concentrations of fluoride, above the permissible limit of 1.5 mg/l in the drinking water standard (Indian Standard (IS): 10500 drinking water standards) is harmful to humans and causes discoloration of teeth and skeletal fluorosis.

In Madhya Pradesh, the intake of fluoride-rich water in the shallow aquifer occurs in the Alirajpur, Anuppur, Bhind, Chhindwara, Datia, Dewas, Dhar, Dindori, Harda, Jabalpur, Mandla, Neemuch, Raisen, Rajgarh, Ratlam, Singrauli, Tikamgarh, Ujjain, and Umaria districts of the city. The distribution of fluoride in the state is shown. The analysis shows is no particular trend can be observed in the distribution of Fluoride in the State. The concentration of fluoride (4.00 mg/l ppm) of Alirajpur district was observed.

4.4.3 Nitrates

Changes that occur in the quality of land and water as a result of the anthropogenic activities of humans are a matter of great concern. The contribution of the atmospheric and geological sources of nitrates to groundwater is extremely small compared to the sources, in the context of pollution, as a result of the activities of human beings. Sewage and animal waste, which can contain a significant amount of Nitrogen and the emissions of soil, and the subsequent leaching by water. Also in agriculture, such as the excessive use of chemical fertilizers, irrigation with wastewaters, vegetables, and yard waste-the contribution of nitrate to the groundwater.

The concentration of nitrates in excess of 45 mg/l in water is harmful for humans, especially children, since it causes blue baby syndrome. The highest desirable and maximum permissible limit for nitrate is 45 mg/l (IS: 10500 drinking water standards). The locations of the sources of study area which the concentration of the nitrate ion was greater 45 mg/l in the groundwater. In Madhya Pradesh, nitrate concentrations in the shallow aquifer occur in the Agar Malwa, Alirajpur, Anuppur, Ashok Nagar, Balaghat, Barwani, Betul, Bhind, Bhopal, Burhanpur, Chhatarpur, Chhindwara, Damoh, Datia, Dewas, Dhar, Dindori, Gwalior, Harda, Hoshangabad, Indore, Jabalpur, Jhabua, Katni, Khandwa, Khargone, Mandla, Mandsaur, Morena, Narsinghpur, Neemuch, Panna, Raisen, Rajgarh, Ratlam, Rewa, Sagar, Satna, Sehore, Seoni, Shahdol, Shajapur, Sheopur, Shivpuri, Sidhi, Singrauli, Tikamgarh, Ujjain, Umaria, and Vidisha districts. A high concentration of nitrates (458 mg/l) was detected in the groundwater table in Palara in the Damoh region.

Various physicochemical parameters of groundwater quality have been reported by the Central Ground Water Board (CGWB) (NCR), Ministry of Jal Shakti, Department of Water Resources, River Development and Ganga Rejuvenation, India. Some of the cities have been taken into consideration to get a glimpse of the overall underground water quality index in the major cities of M.P.; this covered major areas pertaining to the cities covered under the case study. The physicochemical parameters covered and their analysis for Bhopal, Indore, Dewas, Chhindwara, Ujjain, etc. are shown in a tabulated water quality index (WQI) report.

4.5 Conclusions

Soil and water samples in the study area are slightly alkaline in nature. There are acceptable limits to chloride concentrations and types. We can say that 20% of water samples exceed the usual limits for fluoride, which may be due to phosphorous fertilizers rich in fluoride and mica. In addition, in some places nitrates concentration has permissible limits. The relatively high concentration of nitrates in water may be due to the erratic use of nitrogen fertilizers in agriculture and the collection of fertilizers in the vicinity of water courses. The concentration of Sodium, Potassium, Calcium, and Magnesium standard values is inside. Thus, the overall study shows that, the water in the study area is not suitable for the immediate drinking purposes necessary for the proper operation of medicinal products, in order to minimize risks.

References

E.M.d. Andrade et al., Land use effects in groundwater composition of an alluvial aquifer (Trussu River, Brazil) by multivariate techniques. Environ. Res. **106**(2008), 170–177 (2007)

N. Bagde, Ground water quality assessment and its impact with special reference to Chhindwara District of Madhya Pradesh, India. Int. J. Life Sci. **4**, 1–13 (2016)

T. Chandrasekar et al., Statistical analysis of the hydrogeochemical evolution of groundwater in hard rock coastal aquifers of Thoothukudi district in Tamil Nadu, India. Environ. Develop. Sustain. **71**(1), 1–23 (2013)

N.M. Damle, D. Deolia, M. Jha, S. Tignath, Assessment of water quality index for the groundwater in and around Jabalpur city, Madhya Pradesh, India. Int. J. Innov. Eng. Res. Manag. **7**(3), 1–16 (2020)

G. Devic et al., Natural and anthropogenic factors affecting the groundwater quality in Serbia. Sci. Total Environ. **468–469**(2014), 933–942 (2013)

P. Dubey, D. Killedar, S. Sharma, M.I. Faraz, Ground water quality analysis for irrigation of some villages in Dewas. Int. Res. J. Eng. Technol. **2**(6), 1–10 (2015)

A.P. Dwivedi, I.P. Tripathi, Physico-chemical analysis and mapping of ground water quality in residential area of two different zone of Central India. Int. J. Adv. Res. Chem. Sci. **4**(10), 1–14 (2017)

A. Elbeltagi, C.B. Pande, S. Kouadri, et al. Applications of various data-driven models for the prediction of groundwater quality index in the Akot basin, Maharashtra, India. Environ Sci Pollut Res **29**, 17591–17605 (2022). https://doi.org/10.1007/s11356-021-17064-7

K. Gedam, S.D. Hatiye, Identification of groundwater potential zones using proxy data: Case study of Megech watershed, Ethiopia. J. Hydrol. Reg. Stud. **28**, 100676 (2020)

T.R. Green, Linking climate change and groundwater, in *Integrated Groundwater Management*, (Springer, Cham, 2016), p. 1

IGRAC, Groundwater & climate change (n.d.). [Online] Available at: https://www.un-igrac.org/areas-expertise/groundwater-climate-change

N. Kamboj, M. Choudhary, Impact of solid waste disposal on ground water quality near Gazipur dumping site, Delhi, India. J. Appl. Nat. Sci. **5**(2), 306–312 (2013)

V.R. Kashyap, Physico-chemical analysis of various water samples of Rewa district (M.P.) India. Int. J. Appl. Res. pp. 1–11 (2016)

J. Kathik, P. Kathal, R. Trivedi, Fluoride contamination in ground water in a part of the tribal belt in Chhindwara district, Madhya Pradesh, India. Bhu-Jal Q. J. **27** (2020)

S. Kouadri, C.B. Pande, B. Panneerselvam, et al. Prediction of irrigation groundwater quality parameters using ANN, LSTM, and MLR models. Environ Sci Pollut Res **29**, 21067–21091 (2022). https://doi.org/10.1007/s11356-021-17084-3

H.H. Khan, A. Khan, Groundwater and surface water interaction, in *GIS and Geostatistical Techniques for Groundwater Science*, (Elsevier, 2019), p. 1

R. Kori et al., Study of ground water quality of Bhopal City, Madhya Pradesh, India. Int. J. Curr. Res. **11**(10), 1–12 (2019)

R. Kori, A. Saxena, H. Wankhade, A. Baig, A. Kulshreshtha, S. Mishra, S. Sen, Study on Ground Water of Various Locations in Dewas Industrial Area of Madhya Pradesh India. Int. J. Environ. Agric. Res. **10**(6), 10–18 (2020)

A.P. Mishra, H. Khali, S. Singh, C.B. Pande, R. Singh, S.K. Chaurasia, An Assessment of In-situ Water Quality Parameters and its variation with Landsat 8 Level 1 Surface Reflectance datasets, Int J Environ Anal Chem (2021). https://doi.org/10.1080/03067319.2021.1954175

A.P. Mishra, S. Singh, M. Jani, K.A. Singh, C.B. Pande, A.M. Varade, Assessment of water quality index using Analytic Hierarchy Process (AHP) and GIS: a case study of a struggling Asan RiverInt. J. Environ. Anal. Chem. pp. 1–13 (2022)

K. Moharir, C. Pande, S. Singh, P. Choudhari, R. Kishan, L. Jeyakumar, Spatial interpolation approach-based appraisal of groundwater quality of arid regions J. Water Supply Res. Technol. - Aqua (2019)

S. Montoya, Main factors affecting the groundwater regime (2017). [Online] Available at: https://www.hatarilabs.com/ih-en/main-factors-affecting-groundwater-regime

N. Nigam, S. Yadav, Seasonal Variation of Groundwater Quality in and Around Laharpur Reservoir, Bhopal. In Groundwater, 239–247, Springer, Singapore (2018)

V. Parashar, S.K. Sharma, Suitability of groundwater quality for irrigational use between Silkanth and Kundgaon, north of river Narmada in the upper Alluvial Plains of Narmada Valley, District Sehore and Dewas, M.P., India. Orient. J. Phys. Sci. **2**, 1–15 (2017)

R. Prajapati, R. Rokde, A study of drinking water quality in southern Indore City, M.P. Int. J. Sci. Res. **5**(9), 194–197 (2016)

N. Saxena, A. Sharma, Evaluation of water quality index for drinking purpose in and around Tekanpur area M.P. India. Int. J. Appl. Environ. Sci. **12**, 359–370 (2017)

S. Saxena, N. Singh, A. Bajpai, Physico-chemical assessment of water quality in and around Jabalpur City of Madhya Pradesh (India). IJCSR **6**(3) (2016), 1-13

A. Sharma, P.A. Saxena, D.S. Sharma, Determination of Water Quality Index of groundwater from some locations in Gwalior city. Int. J. Res. Appl. Sci. Eng. Technol. **6**(8), 1–9 (2018)

S. Sharma, G. Sharma, R.K. Yadav, Impact of industrial development on groundwater quality- a case study of impact of effluent from Vishwakarma industrial area, Jaipur on ground water. Bull. Environ. Pharmacol. Life Sci. **5**, 1–14 (2016)

N. Shivhare, A. Gour, A.V. Pandya, R.T. Nair, Water quality index assessment for ground water of Indore City (M.P.), India. Ecol. Environ. Conserv. **5**(2), 1–16 (2019)

S. Singh, R. Chauhan, A. Kumar, V. Vyas, Groundwater quality assessment of Ghattia Block of Ujjain District of Madhya Pradesh, India. Indian J. Appl. Res. **1**(3), 1–16 (2017)

S.K. Singh et al., Water quality assessment for drinking and irrigation purpose of Rewa Block, District-Rewa, Madhya Pradesh, India. Int. J. Curr. Microbiol. App. Sci. **9**, 1–10 (2020a)

S. Singh et al., Physicochemical quality monitoring of groundwater for drinking purposes in rural Ujjain, Central India: Findings of a 2-year longitudinal study. SN Appl. Sci. 1–13 (2020b)

S. Thakur, S. Bhople, Water quality index of groundwater in nine tribal block of Madhya Pradesh. Int. J. Pure Appl. Biosci. 1–16 (2018)

USGS, Groundwater decline and depletion (n.d.-a). [Online] Available at: https://www.usgs.gov/special-topic/water-science-school/science/groundwater-decline-and-depletion?qt-science_center_objects=0#qt-science_center_objects

USGS, Urbanization and water quality (n.d.-b). [Online] Available at: https://www.usgs.gov/special-topic/water-science-school/science/urbanization-and-water-quality?qt-science_center_objects=0#qt-science_center_objects

Chapter 5
Spatiotemporal Analysis of Rainfall Trends in Semi-Arid Regions of India Over the Last 36 Years

Nilufer Sajjad, Arun Pratap Mishra, Chaitanya B. Pande, and Abebe Debele Tolche

Abstract A quantitative and qualitative understanding of the anticipated, climate change-driven, multiscale spatiotemporal shifts in precipitation and attendant river flows is crucial to the development of water resource management approaches, which are capable of sustaining and even improving the ecological and socioeconomic viability of rainfed agricultural regions. A set of homogeneity tests for change point detection, non-parametric trend tests, and Sen's slope estimator were applied to the long-term gridded rainfall records of 26 districts in the state of Maharashtra, India. This study focuses on a key element of climate, i.e. rainfall. Accurately predicting rainfall or precipitation trends is vital for the economic growth and development of a country. The research studied the precipitation variability across the state of Maharashtra using Climate Hazards Group InfraRed Precipitation with Station data (CHIRPS) over a period of 36 years (1982–2018). The data were processed to find seasonal and annual trends using non-parametric analysis. The Mann–Kendall (MK) test was used for the analysis of increasing and decreasing (monotonic) trends together with the non-parametric Sen's slope estimator to estimate the magnitude of trends over time series data. The statistical analysis of the whole reference time series data highlights that (1) the trends look predominantly positive (increasing), both at the annual and seasonal scales, except for the pre-monsoon season (PMS) and (2) the pre-monsoon season (PMS) showed a negative trend for few districts. A total of 26 districts consisting of 130 sub-districts covering 148086.45 km^2 of

N. Sajjad
Department of Natural Resources, TERI School of Advanced Studies, New Delhi, India

A. P. Mishra
Botanical Survey of India, Dehradun, Uttarakhand, India

C. B. Pande (✉)
GIS Consultant, Primus Partners Private Limited, Delhi, India

A. D. Tolche
School of Water Resources and Environmental Engineering, Haramaya Institute of Technology, Haramaya University, Dire Dawa, Ethiopia

B. Panneerselvam et al. (eds.), *Climate Change Impact on Groundwater Resources*, https://doi.org/10.1007/978-3-031-04707-7_5

area or 48.1% of the area of Maharashtra showed a positive trend when the p-value was >0.05. In the pre-monsoon season, 8 sub-districts showed a negative trend, whereas 21 districts showed positive trends.

Keywords Rainfall trend · Mann–Kendall test · Sen's slope estimates · Trend analysis

5.1 Introduction

As a complex and interwoven phenomenon, climate change has pervasive effects on all of Earth's spheres, in particular, the hydrosphere (water resources, glaciers, and oceans), biosphere (the environment, terrestrial ecosystems, and biodiversity), and lithosphere (agriculture, food security) (Orimoloye et al. 2022). Analysis and detection of historical changes in climatic parameters such as rainfall, temperature, etc. are the first and most important steps towards a systematic study of climate change (Fankhauser and Tol 1997). The characterization and quantification of climate change and its effects on rainfall are important in helping the sustainable development of India's largely monsoon-dependent agriculture and management of its associated water resources (Gadgil 1986; Mondal et al. 2014; Balamurugan et al. 2020a, b; Kumar and Balamurugan 2018, 2019). Such studies are important, given that India's agricultural sector generates about 60% of the jobs and 20% of the national gross domestic product (GDP). Moreover, beyond affecting rainfall patterns or agriculture directly, the rapid retreat of glaciers triggered by climate change-driven local warming can lead to significant soil erosion and flooding. Climatic change impact on the world climate due to increasing greenhouse gas (GHG) emissionsgreenhouse gas (GHG) emissions, this probelm is a global issue that has implications in the regional climate with temperature and precipitation important elements of climate. Global climate change has an impact on the rainfall patterns amongst many other phenomena, and, since rainfall and its variability is the driving factor behind the economies and livelihood of a country, it is important to understand the changes at a fine resolution level. Precipitation throughout the season is important as the total amount of monthly or annual precipitation changes when understanding its effects on ecology, hydrology, water use, and agriculture (Kumar et al. 2022; Pande et al. 2021a, b; Panneerselvam et al. 2020a, b, 2021a, b, c; Kouadri et al. 2021; Shunmugapriya et al. 2021). This distribution of seasonal precipitation has ultimately resulted in the unequal heating of Earth's surface over the years, and, as a final step of the process, this has led to what we know as atmospheric general circulation on Earth.

Rainfall is essential to all life forms on Earth, but climate change is posing a threat to this entire system of self-sustenance. Since 1991, the percentage of rainfall in extremely heavy precipitation instances has been largely above average. Heavy rainfall, which was a one-in-a-20-year event, has now been predicted to occur about every 4–10 years by the end of this century (Elbeltagi et al. 2022a, b). In 2016, the amount of annual rainfall drastically decreased due to the intense La Niña conditions induced by climate change, causing Asian nations, including Thailand, Myanmar, India, and other highly populated regions, to face water insecurity.

India has a huge climatic and cultural variation across the country. The primary economic activity is still dominated by agriculture, which is vulnerable to the slightest change in rainfall trends. The Indian climate is characterized by monsoons, which is a major source of water supply (75% of the annual rainfall supply) for the country as well as its agriculture. From 67 floods over a period of 10 years (1996–2005), India has now faced more than 90 floods (2006–2015), according to the UN Office for Disaster Risk Reduction. As per the data obtained from the European Centre for Medium-Range Weather Forecast of 2018, "the humidity in the mid-troposphere increased by about 2% over the sub-continent over the last 39 years, which resulted in a rise in extreme rainfall events over central India". A study showed a massive rise in precipitation events across central India from 1950 to 2015. The nature of the monsoons causes variations in annual rainfall. Certain regions receiving low rainfall such as parts of Gujarat, Rajasthan, and the leeward side of the Western Ghats are facing high variability. On the other hand, areas experiencing high rainfall are susceptible to floods, whereas areas receiving less rainfall are drought-prone. Maharashtra is one such state that is highly affected due to its geography and population.

Having analyzed the daily precipitation records for India from 1951 to 2000, Goswami et al. (2006) found a rising occurrence and magnitude of extreme rainfall events but a decline in the occurrence of rainfall events over central India during monsoon seasons. After an extensive analysis of the Ganges, Brahmaputra, and Meghna River basins, Mirza et al. (1998) reported precipitation in the Ganges River basin to be stable. Singh et al. (2008) found that one of the Brahmaputra River's three sub-watersheds showed a rising rainfall pattern, whereas another showed a diminishing pattern. Although India-wide precipitation and surface runoff showed no significant patterns, an increasing trend in precipitation, in the order of 2–19% of the 100-year annual mean, was noted for nine basins in north-western and central India (Sinha Ray and De 2003). Sinha Ray and Srivastava (1999) noted a rising recurrence of extreme precipitation events in the southwest (SW), with a downward trend in winter and pre- and post-monsoon rainfall.

Geographically, Maharashtra is widely spread that gives it different types of climatic features. Due to the variation in climate and topography, the state is divided into four meteorological subdivisions. The meteorological subdivision of Konkan and Goa is the extreme western part spreading from north to south along the western coast of India. As a result of these topographical features, the region experiences extremely high rainfall during the monsoon season. "Climate change is acting as a spark for changes in rainfall pattern and temperature which can significantly impact the hydrological cycle and crop calendar of the area" (Abrol et al. 1996). Monocropping is generally observed, with rice production during the Kharif season while horticultural crops like cashew, mango, cotton, and coconut are the major crops of the region. Over the last few years, "the production & quality of mangoes has been affected in a negative way due to the change in meteorological parameters of the region" (Munj et al. 2017).

The main objective of this study is to comprehend the temporal variability in rainfall in the state of Maharashtra. Thus, studying these changes, patterns, and intensity is important to assess the net change in climate over the years to equip us to make informed decisions and prepare for future events. This chapter further looks into the background of the issue using various literature reviews of published papers, journals, and reports followed by the methodology that explains the methods followed to obtain the results of the analysis. The results of study area are disscused and presented with the help of various maps and graphs for understading of the rainfall pattern entire in the Maharashtra.

5.2 Background of This Study

Over the last five decades, studies are being conducted on rainfall variation using various methods. There are various research works on trend analysis of rainfall in India. Rainfall is a vital part of the Indian economy as, at present, 70% of the agriculture produced is dependent on rainfall. So, a change in rainfall patterns affects the economy of the country. Therefore, it is necessary to study the trend and predict the pattern for future convenience. So, in this context, a brief literature review of research works has been mentioned to understand the rainfall trend in Maharashtra.

Patil et al. (2013) have presented a study on rainfall trend analysis for the district of Pune in Maharashtra. According to their paper, the study was to assess the trend of rainfall in Pune using two methods, namely, the non-parametric Mann–Kendall (MK) test and the modified Mann–Kendall (MMK) trend test along with Sen's slope estimator. Data for 47 years from 1958 to 2004 were acquired. The data were in the form of annual, seasonal, and monthly rainfall estimates for the Kharif season. The trend analysis for the study area showed changes in spatial and temporal dimensions. During Kharif, the season rainfall showed a decreasing trend, whereas in June, it showed an increasing trend and July's rainfall showed a decreasing trend at a significant level. Sen's slope estimates also pointed at the decreasing and increasing trends of the slope concerning other MK parameters.

Mondal et al. (2012) reviewed the rainfall trend for the north-eastern part of Cuttack, Odisha, using the MK test. They used the data for 40 years from 1971 to 2010. In addition, the MMK test was used together with Sen's slope estimator for determining the trend and slope magnitude of the study region. The research concluded that some months have an increasing trend and some have a decreasing rainfall trend and that, spatially, there are no significant changes. Sridhar and Raviraj (2017), in their study, aimed to assess the trends in annual and seasonal rainfall and magnitude in the Amaravati basin of Tamil Nadu. In this study, the data for the mean seasonal and annual rainfall for the Amaravati basin were collected from 10-gauge stations, which were analyzed to determine the rate of change and its magnitude for

32 years (1982–2014). The Mann–Kendall test and Sen's slope estimator were used for this analysis. The results showed variations in all the seasons. This trend has been found to be aggressively rising during the northeast monsoon season in comparison to other seasons.

Basha et al. (2017) analyzed an area located in the Western Ghats of Maharashtra. The yearly rainfall conditions differ from region to region. In spite of the developments in enabling improved crop varieties and better management tactics in the district of Satara, agriculture is still a gamble as production depends on the monsoon. The average annual rainfall is 1277 mm. The analysis of the annual rainfall of 11 tehsils involved the use of the mean, standard deviation (SD), and coefficient of rainfall variability. This attempt was made for periods 1998–2016 (annual rainfall) in the Satara district. The research aimed to study the related trends in rainfall in the eastern part of the Satara district in Maharashtra. There were variations in the results of the taluka-wise studies, and a clear and rational picture of the rainfall trend showed variability in drought-prone regions. In a study on taluka-wise trend analysis, seven tehsils showed a decreasing trend in annual rainfall. Among two tehsils that showed an increasing trend, the Mahabaleshwar tehsil showed the highest rainfall trend. The remaining two tehsils had the same direction as the trend in annual rainfall and seasonal scale.

Barakade and Sule (2011) in their *Rainfall Variability in Solapur District of Maharashtra: A Geographical Study* worked on rainfall variability for district of Solapur for the period 2000–2001. In the study, they found that the area monitored received 561.47 mm annual average rainfall. The rainfall variation in the tehsil was recorded at 107 mm, and the Mangalwedha tehsil recorded the highest rainfall during the year 2007, which showed that the annual rainfall and its variation are non-uniform throughout the Solapur district. The Barshi tehsil received the highest mean annual rainfall of 642.65 mm, whereas the Malshiras tehsil received 455.64 mm mean annual rainfall. This variance in rainfall had a massive impact on the agricultural practices in the Solapur district. This is why underirrigated area crops and Bhima River basin perennial crops are cultivated in the central part of the district. Low rainfall crops cultivated are bajra, horticulture, jawar, and pulses in the Sangola, Malshiras, and Madha tehsils.

India Meteorological Department have observed the rainfall variability and changes in the Maharashtra State (Malik et al., 2019). The day-to-day rainfall statistics from 1989 to 2018 were considered for analysis of trend variability in this study. The study analyzed the "rainfall pattern and its variation and also the changes based on the recent 30 years of data". The analysis was carried out at monthly, seasonal, and annual scales. As per the results obtained from the study, the state gets its 33% of SW monsoon rainfall (maximum rainfall) in July, which is followed by 28% of SW monsoon rainfall in August. The maximum percentage of annual rainfall occurs during the southwest monsoon. The study also mentioned that from June to September, there was a significant increase in heavy rainfall days in the Nandurbar, Kolhapur, Jalgaon, Raigarh, and Bhandara districts. However, districts like Kolhapur, Pune, Solapur, Aurangabad, Ahmednagar, Jalna, Beed, Yavatmal, Hingoli,

Wardha, and Nanded saw a significant decrease in heavy rain days, whereas the other remaining districts had no significant changes.

5.3 Objectives of This Study

On the basis of its background, this study aims to investigate the temporal rainfall pattern in Maharashtra using gridded data of the last 36 years (1982–2018). The objectives are to investigate the following:

1. The trends in annual rainfall and its spatial variability across the districts
2. The trends in seasonal rainfall and its spatial variability across the districts

5.4 Materials and Methods

5.4.1 Study Area

The area selected for the study (Fig. 5.1) was the state of Maharashtra (India), as 61% of this state falls under the semi-arid category that demonstrates large variation in rainfall, thus affecting food production (Barron et al. 2003; Wani et al. 2009). Maharashtra is also an agriculturally and socioeconomically vulnerable state of India as almost every year farmers commit suicide in large numbers due to climate-related stressors, namely, monsoon variability, droughts, hail storms, etc. (Basu and Bazaz 2016).

Maharashtra's expansion (both longitudinal and latitudinal) is between 72°E to 83°E and 15°N to 23°N, respectively. Moreover, Maharashtra has been categorized into five different regions: Konkan in the west coastal area, between the Western Ghats and the Arabian Sea; Marathwada (Aurangabad region) in the southeast; Vidarbha (Amravati and Nagpur division) in the eastern side; Nashik in the north-west; and Pune toward the west-central area. The span of monsoons in Maharashtra is around June to September for approximately 122 days. The rainfall distribution pattern is diverse across the state. For instance, while the coastal Maharashtra, which is Konkan, receives 6000 mm rainfall, the plain areas receive 2500 mm and the central areas receive just 500 mm of rainfall, all of which depict the variation in the rainfall that the state receives. If we talk about the number of rainy days in general, then Konkan gets 75–85 days, Vidarbha gets 40–50 days, and the central areas get 30–40 days. The state undergoes great variation in annual rainfall, for instance, Konkan gets 23%, Vidarbha gets 26%, Marathwada gets 31%, and the central areas get 30% of the annual rains as per the National Institute of Disaster Management (NIDM), 2013.

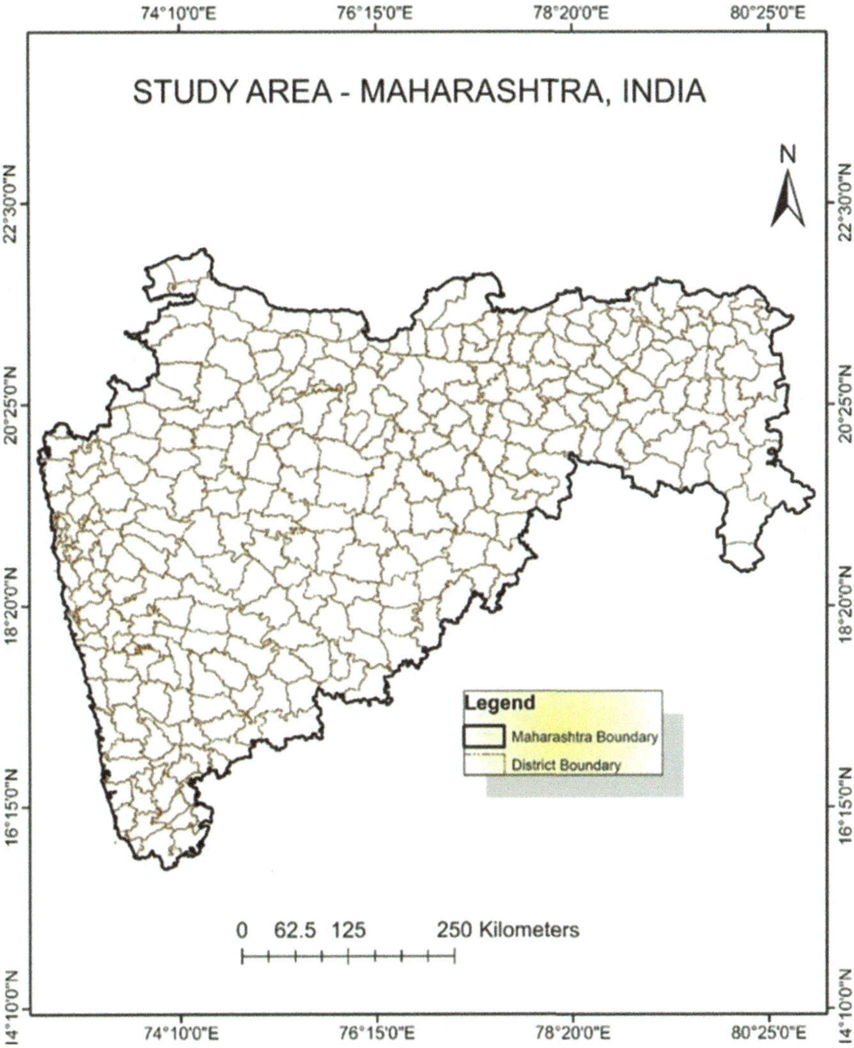

Fig. 5.1 Extent of the study area: Maharashtra

5.4.2 Data and Methodology

CHIRPS, which means Climate Hazards Group InfraRed Precipitation with Station data, is a worldwide rainfall dataset spanning over 35 years. Starting from the year 1981 to the current scenario and ranging from 50°S to 50°N, CHIRPS involves our in-house climatology, CHPclim, 0.05° (5.5 km × 5.5 km) resolution satellite image, and station data to produce gridded rainfall statistics for trend analysis and seasonal drought monitoring. CHIRPS has recently generated satellite-based rainfall images with high spatial and temporal resolutions. The distinctive feature of this dataset is its various temporal scales (daily, monthly, decadal, annual) that make it highly comprehensive, and, therefore, the spatial resolution of 5 km (approximately) makes it suitable for this analysis. The annual and seasonal data of 36 years of rainfall for Maharashtra were used in the study obtained from CHIRPS. The trend was assessed for annual and seasonal rainfall, i.e. winter, pre-monsoon, monsoon, and post-monsoon.

5.4.3 The Mann–Kendall Test

The Mann–Kendall test is a non-parametric test used to find trends in non-parametric data. It is a robust measure to analyze long-term or short-term trends over an area. One of the significances of this particular test is that it accommodates missing values and the data do not need to follow any specified statistical distribution. This test is particularly useful for a monotonous trend. The trend we derive at the end does not have any effect on the outliers, which arise due to extreme rainfall events. This test involves assumption of the zero hypothesis as there is zero trend in the dataset and no existence of an alternate hypothesis as a trend (Fig. 5.2).

The Mann–Kendall test (S) can be derived as:

$$S = \sum_{\substack{1 \le i \le n-1 \\ i+1 < j < n}} sign\left(X_j - X_i\right)$$

where n = sample size and

X_j-X_i = data for ith and jth terms, respectively, and sign (X_j-X_i) can be calculated as:

$$sign\left(X_j - X_i\right) = \begin{cases} +1; if\ X_j - X_i > 1 \\ 0; if\ X_j - X_i = 0 \\ -1; if\ X_j - X_i < 1 \end{cases}$$

The statistic S is assumed to be Gaussian when $n = 18$ or $n > 10$. In general, Var(S) with $E(S) = 0$ can be calculated as:

$$E(S) = 0 \text{and}\, Var(S) = \frac{n(n-1)(2n+5)}{18}$$

$$Var(S) = \frac{n(n-1)(2n+5) - \sum_{p=1}^{q}(t_p - 1)(2t_p + 5)}{18}$$

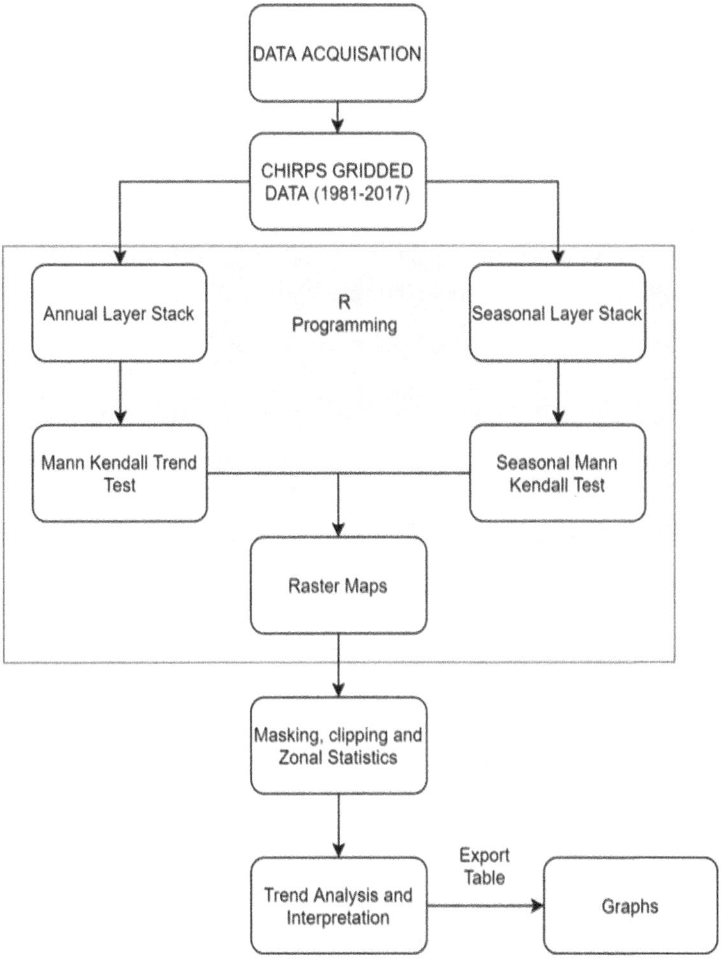

Fig. 5.2 Flowchart of methodology

Within the above-mentioned formula, if ties exist, then Var(*S*) can be calculated as:

where *q* and *tp* are the tied groups within the above-mentioned formula.

The one-tailed *Z* test, which helps define the trend (positive or negative) and also which helps in rejecting or accepting the null hypothesis, can be derived as:

$$Z = \begin{cases} \frac{S-1}{\sqrt{Var(S)}}; if\ S > 0 \\ 0; \quad if\ S = 0 \\ \frac{S+1}{\sqrt{Var(S)}}; if\ S < 0 \end{cases}$$

With the value of *Z* being positive, the trend is said to be increasing and vice versa.

5.4.4 *Sen's Slope*

Sen's slope estimator analyzes the magnitude of the trend. For the duo of *i* and *x*(*i*) being a time series, Sen's slope is defined as:

$$d(k) = (x(j) - x(i)) / (j - i)$$

for (1 < = *i* < *j* < = *n*), where *d* stands for slope, *x* stands for the variable, *n* is the number of data, and *i* and *j* are both indices.

Sen's slope is then calculated as the median of all slopes:

$$b = \text{median}(d(k))$$

Here, *j* and *i* are time points (*j* > *i*), and *x*(*j*) and *x*(*i*) are rainfall values. The median of these values Ti A positive value of β depicts that the trend is an increasing trend and a negative value of β portrays a decreasing trend.

The tau and slope for the annual and seasonal rainfall data of Maharashtra was calculated using an R script on RStudio. The main packages used were Raster and spatialEco. Furthermore, ArcGIS was used to process the raster and calculate significant slope values for different seasons and years.

5.5 Results and Discussion

The analysis of the annual and seasonal trends of rainfall over Maharashtra for 36 years (1982–2018) was conducted using the Mann–Kendall test and the Theil–Sen slope method. The Mann–Kendall test was used to obtain significance in the trend results, and, further, Sen's slope was used to get the magnitude of the trend. The significant changes are shown as maps wherein the increasing and decreasing trends for seasonal data are shown (Fig. 5.3).

The Mann–Kendall test was used to test the long-term rainfall as it is considered a better option than the other parametric tests for a monotonic trend and handles outliers effectively. These methods are widely used and referred by the hydrological and meteorological trend detection.

A total of 26 districts showed a positive trend, out of which Sindhudurg, Ratnagiri, Thane, Raigarh, and Greater Bombay showed the highest trends (Fig. 5.4). A total of 26 districts consisting of 130 sub-districts, which cover 148086.45 km^2 of area or 48.1% of the area of Maharashtra, showed a positive trend when the *p*-value was >0.05 for the period 1982–2018. Sangrampur, Sillod, Sindkhed Raja, and Kavathe Mahankal showed a low positive trend (Fig. 5.5).

During winter and post-monsoon seasons only, a few districts showed a positive trend. Figures 5.6 and 5.7 show seasonal significant slopes.

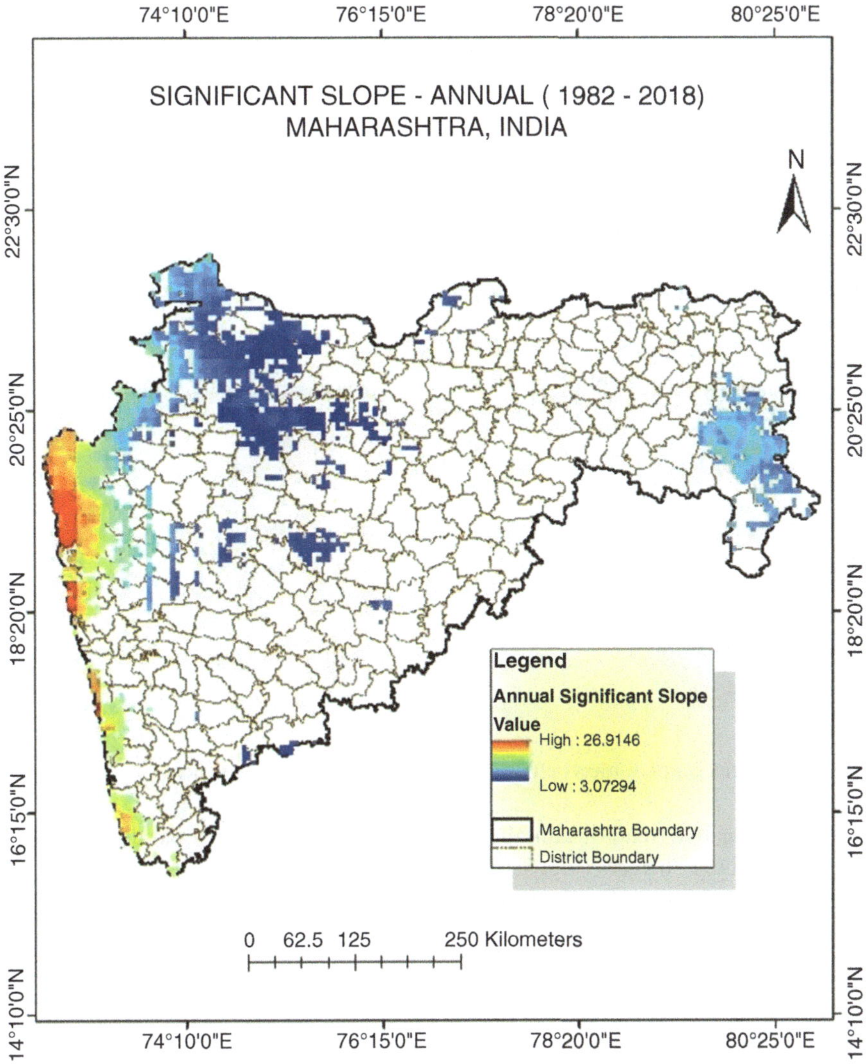

Fig. 5.3 Significant slope in the annual rainfall trend

In the pre-monsoon season, 8 sub-districts showed a negative trend, whereas 21 districts showed positive trends, which can be observed in Figs. 5.8 and 5.9. In the monsoon season, a majority of the areas of Maharashtra showed a positive trend, and a total of 147 sub-districts showed a positive trend (Figs. 5.10 and 5.11).

In the monsoon season, most of the districts showed an increasing rate of change. In all, 27 districts in which 147 sub-districts are accounted for showed a positive trend in an area of about 160415 km^2. In the pre-monsoon season, five districts

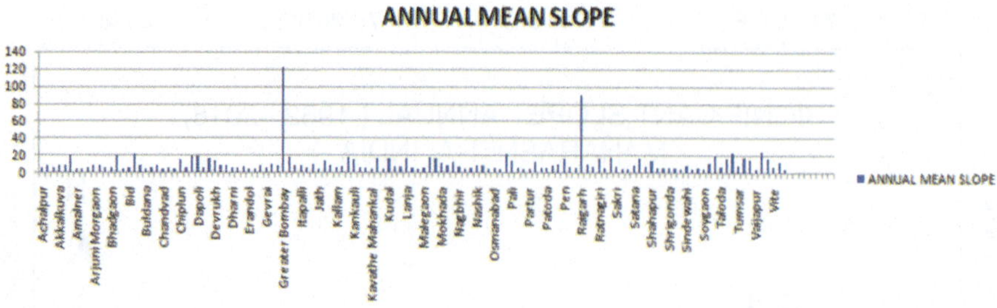

Fig. 5.4 District-wise slope of the annual rainfall trends

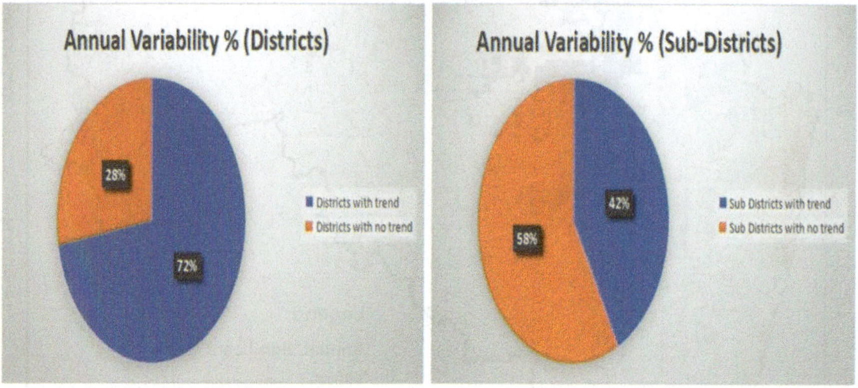

Fig. 5.5 Percentages of districts (left) and sub-districts (right) showing trends

showed a negative trend of rainfall, which is the only negative trend that the study found over these 36 years (Fig. 5.12).

In the post-monsoon season, only three districts showed a significant rate of change and the trends are increasing.

5.6 Conclusions and Recommendations

This study was carried out to investigate the spatial variability in the long-term pattern of rainfall in Maharashtra. The results of the annual and seasonal rainfall trend analysis using the Mann–Kendall test at the district and sub-district levels for a period of 36 years from 1982 to 2018 indicate that about 76% of the districts (26 districts) in Maharashtra consisting of 130 sub-districts and covering 48.1% of the area of the state have significant increases in rainfall over the years. Among all these districts, Sindhudurg, Ratnagiri, Thane, Raigarh, and Greater Bombay showed the maximum increase in rainfall with a higher magnitude of the rate of change. Furthermore, trends of rainfall for the winter season were only significant in four districts of the state and only in three districts during the post-monsoon season. These findings in the changes

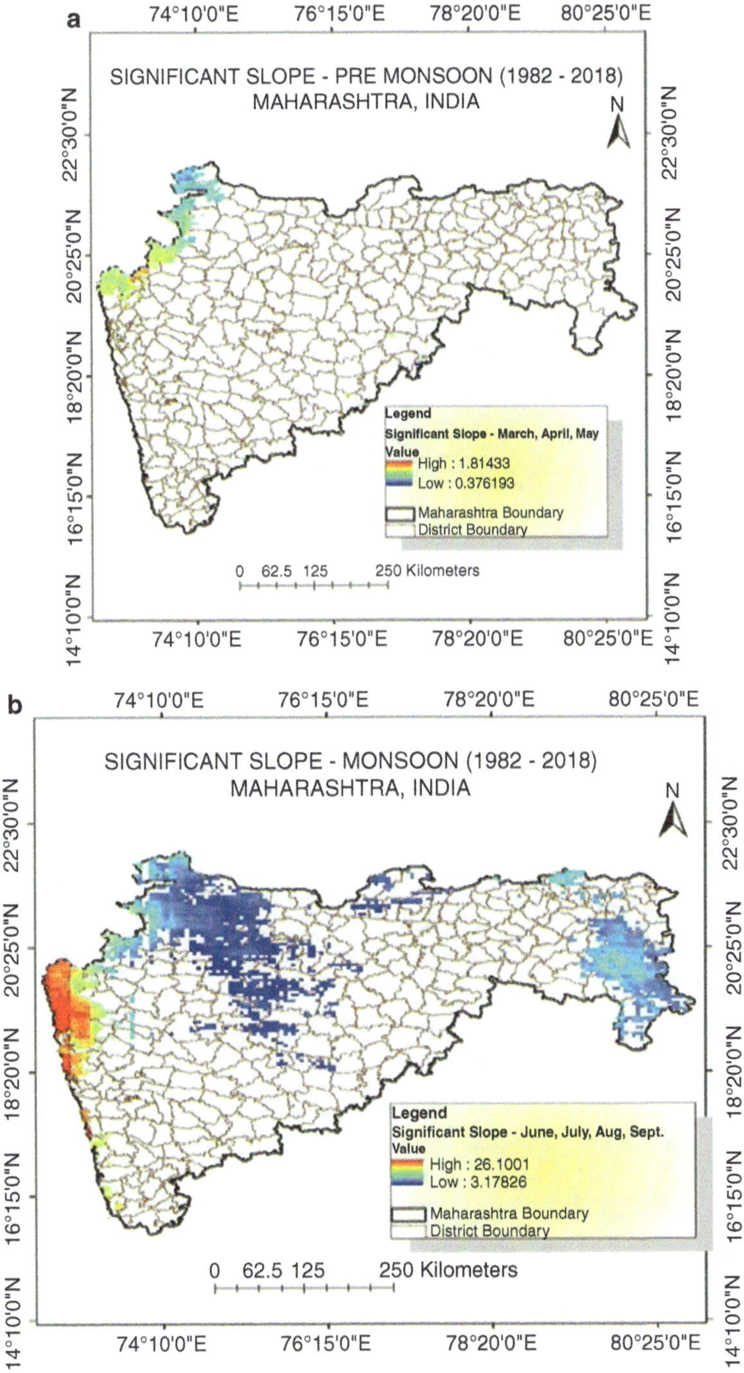

Fig. 5.6 Regions with a significant slope in the pre-monsoon (**a**) and monsoon (**b**) seasons

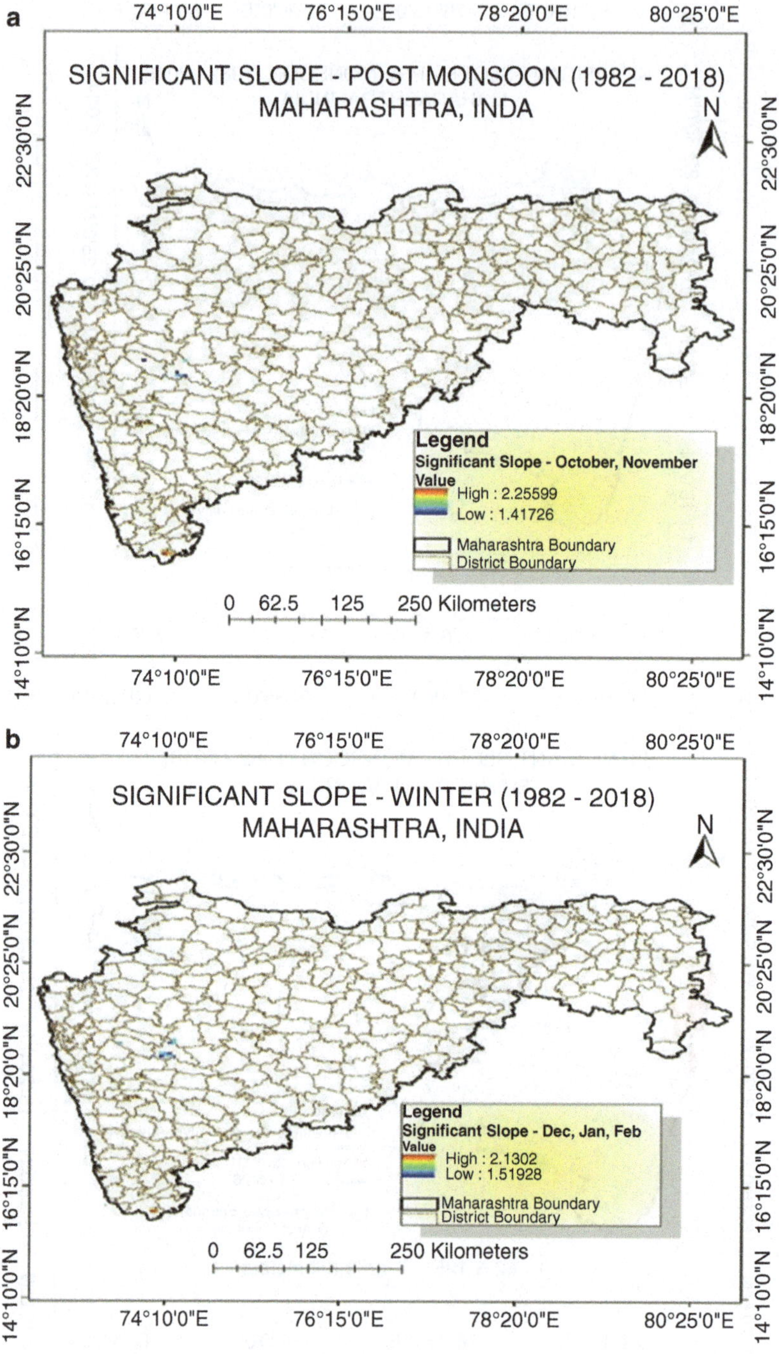

Fig. 5.7 Regions with a significant slope in the (**a**) post-monsoon (left) and (**b**) winter (right) seasons

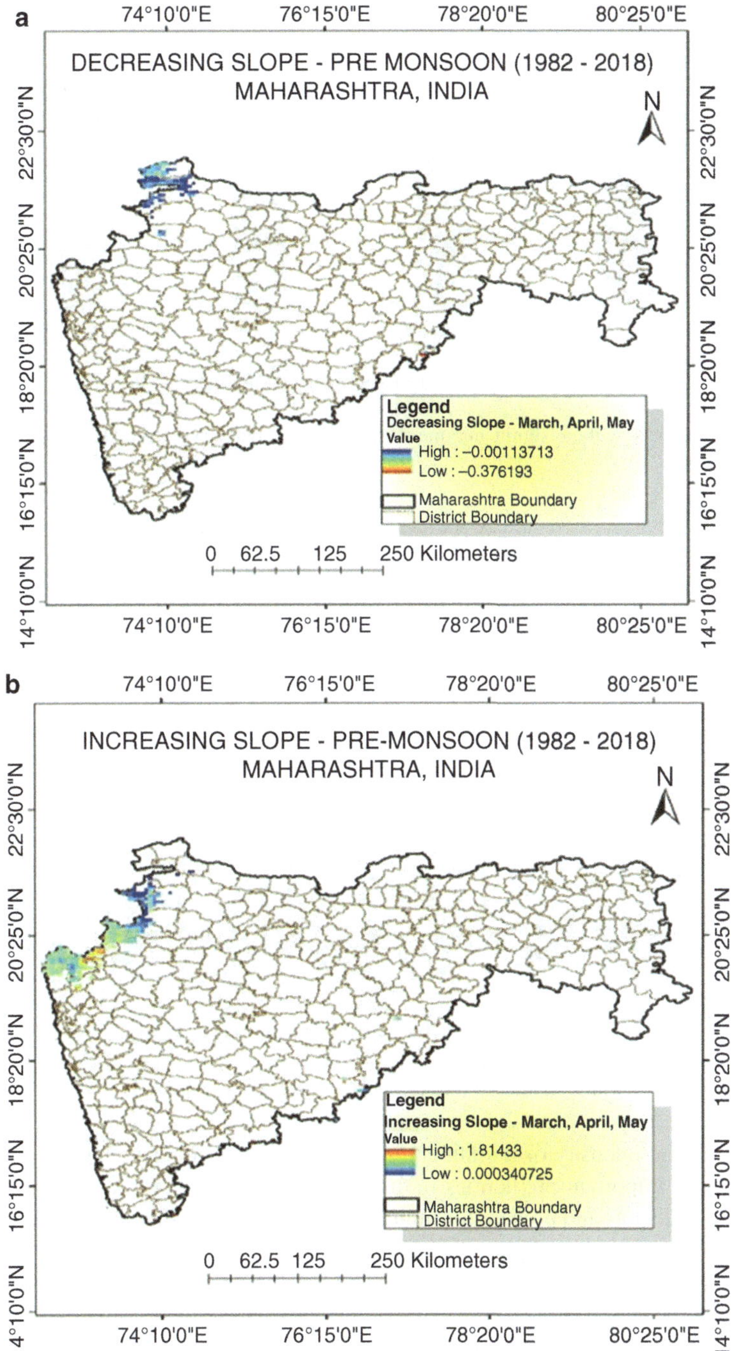

Fig. 5.8 Regions with a decreasing slope (**a**) and an increasing slope (**b**) in the pre-monsoon season

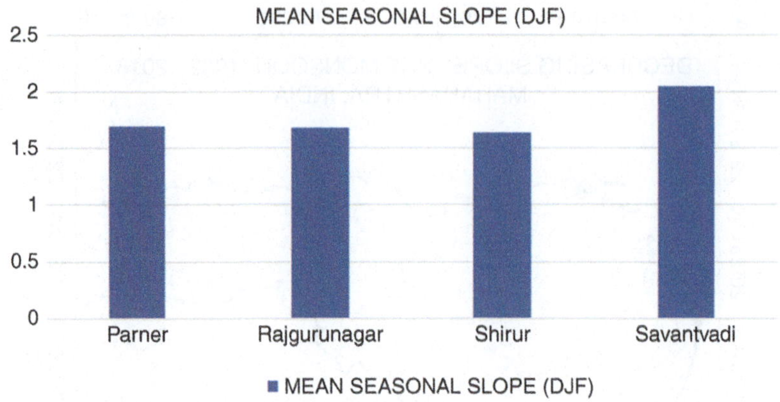

Fig. 5.9 Districts with a significant mean slope in the winter season December, January, February (DJF)

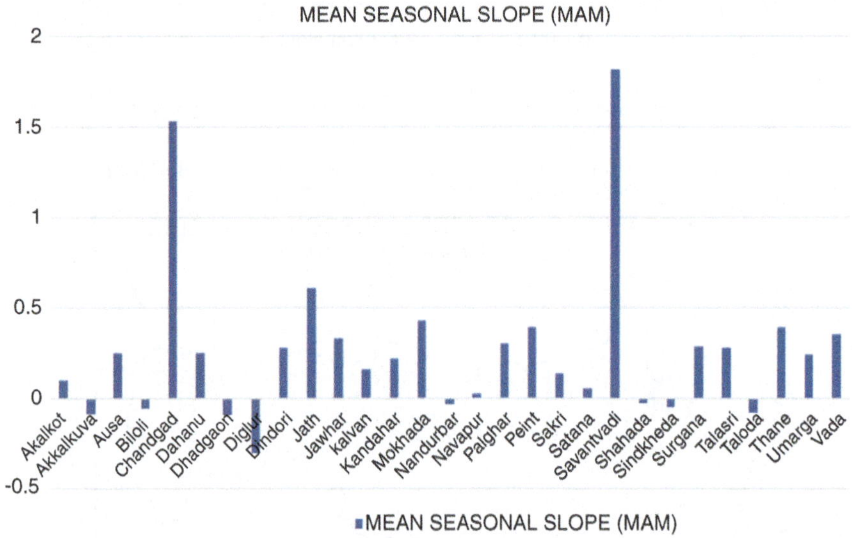

Fig. 5.10 Districts with a significant mean slope in the pre-monsoon season

in patterns and intensity of rainfall are important to assess the net change in climate and to make important predictions in the region over the years to equip us to make informed decisions and prepare for future events. The results showing climatic variations indicate the vulnerability of multiple socioeconomic factors, majorly affecting the economic practices of the region. Therefore, the study leaves ample scope for examining the various underlying factors of the observed climate variations and their impacts on the social and economic practices of the region. The rainfall trend analysis is most important for groundwater and surface water conservation in aquifer zones. The results of this study can be helpful in the development of agricultural practices and in the understanding of metrological drought conditions.

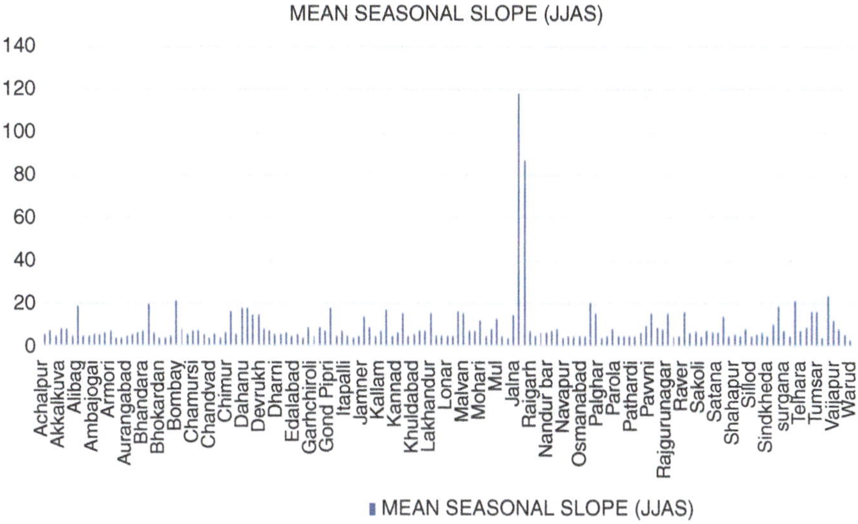

Fig. 5.11 Districts with a significant mean slope in the monsoon season

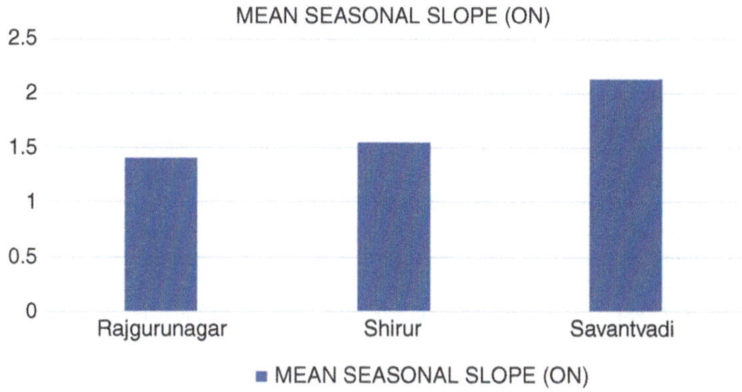

Fig. 5.12 Districts with a significant mean slope in the post-monsoon season

References

Y.P. Abrol, S. Gadgil, G.B. Pant, *Climate Variability and Agriculture* (Narosha Publishing House, New Delhi, 1996)

P. Balamurugan, P.S. Kumar, K. Shankar, R. Nagavinothini, P.J. Sajil Kumar, Impact of climate and anthropogenic activities on groundwater quality for domestic and irrigation purposes in Attur region, Tamilnadu, India. Desalin. Water Treat. **208**, 172–195 (2020a)

P. Balamurugan, P.S. Kumar, K. Shankar, R. Nagavinothini, K. Vijayasurya, Non-carcinogenic risk assessment of groundwater in southern part of Salem District in Tamilnadu, India. J. Chil. Chem. Soc. **65**(1), 4697–4707 (2020b)

A.J. Barakade, B.M. Sule, Rainfall variability in Solapur District of Maharashtra: A geographical study. Rev. Res. **1**(2), 1–3 (2011)

J. Barron, J. Rockström, F. Gichuki, N. Hatibu, Dry spell analysis and maize yields for two semi-arid locations in East Africa. Agric. For. Meteorol. **117**(1–2), 23–37 (2003)

S.J. Basha, A.S.R. Sarma, P. Munirathnam, Probability analysis of weekly rainfall for crop planning at Nandyal of Andhra Pradesh. Trends Biosci. **10**(2), 956–958 (2017)

R. Basu, A. Bazaz, Assessing Climate Change Risks and Contextual Vulnerability in Urban Areas of Semi-Arid India the Case of Bangalore, (2016)

A. Elbeltagi, A. Nagy, S. Mohammed, B. Chaitanya Pande, Manish Kumar, Shakeel A. Bhat, József Zsembeli, László Huzsvai, János Tamás, Elza Kovács, Endre Harsányi, and Csaba Juhász. (2022a) "Combination of Limited Meteorological Data for Predicting Reference Crop Evapotranspiration Using Artificial Neural Network Method" Agronomy 12, no. 2: 516. https://doi.org/10.3390/agronomy12020516.

A. Elbeltagi, N. Kumar, A. Chandel, et al. (2022b) Modelling the reference crop evapotranspiration in the Beas-Sutlej basin (India): an artificial neural network approach based on different combinations of meteorological data. Environ Monit Assess 194, 141. https://doi.org/10.1007/s10661-022-09812-0.

S. Fankhauser, R.S. Tol, The social costs of climate change: The IPCC second assessment report and beyond. Mitig. Adapt. Strat. Global Change **1**(4), 385–403 (1997)

A. Gadgil, Annual and weekly analysis of rainfall and temperature for Pune: A multiple time series approach. Inst. Indian Geograph. **8**(1p), 4–20 (1986)

B.N. Goswami, V. Venugopal, D. Sengupta, M.S. Madhusoodanam, P.K. Xavier, Increasing trends of extreme rain events over India in a warming environment. Science **314**(5804), 1442–1445 (2006)

S. Kouadri, C.B. Pande, B. Panneerselvam, K.N. Moharir, A. Elbeltagi, Prediction of irrigation groundwater quality parameters using ANN, LSTM, and MLR models. Environ. Sci. Pollut. Res., 1–25 (2021)

P.S. Kumar, P. Balamurugan, Evaluation of groundwater quality for irrigation purpose in attur taluk, Salem, Tamilnadu, India. Water Energy Int. **61**(4), 59–64 (2018)

P.S. Kumar, P. Balamurugan, Suitability of ground water for irrigation purpose in Omalur Taluk, Salem, Tamil Nadu, India. Indian J. Ecol. **46**(1), 1–6 (2019)

Kumar, M., Elbeltagi, A., Pande, C.B. et al. (2022) Applications of Data-driven Models for Daily Discharge Estimation Based on Different Input Combinations. Water Resour Manage 36, 2201–2221, https://doi.org/10.1007/s11269-022-03136-x

A. Malik, A. Kumar, P. Guhathakurta, O. Kisi, Spatial-temporal trend analysis of seasonal and annual rainfall (1966–2015) using innovative trend analysis method with significance test. Arab. J. Geosci. **12**(10), 1–23 (2019)

M.Q. Mirza, R.A. Warrick, N.J. Ericksen, G.J. Kenny, Trends and persistence in precipitation in the Ganges, Brahmaputra and Meghna River basins. Hydrol. Sci. J. **43**(6), 845–858 (1998)

A. Mondal, S. Kundu, A. Mukhopadhyay, Rainfall trend analysis by Mann-Kendall test: A case study of north-eastern part of Cuttack district, Orissa. Int. J. Geol. Earth Environ. Sci. **2**(1), 70–78 (2012)

A. Mondal, D. Khare, S. Kundu, Spatial and temporal analysis of rainfall and temperature trend of India. Theor. Appl. Climatol. **122**(1–2), 143–158 (2014)

A.Y. Munj, B.S. Rana, S.K. Godase, Population dynamics of mango hopper, Idioscopus niveosparsus Leth and its natural enemies under coastal Konkan conditions of Maharashtra. J. Agrometeorol. **19**(4), 372–374 (2017)

I.R. Orimoloye, A.O. Olusola, J.A. Belle, et al. Drought disaster monitoring and land use dynamics: identification of drought drivers using regression-based algorithms. Nat Hazards. (2022) https://doi.org/10.1007/s11069-022-05219-9

C.B. Pande, N. Kanak Moharir, Sudhir Kumar Singh, Abhay M.Varade ,Ahmed Elbeltagie, S.F.R. Khadri, Pandurang Choudhari (2021a), Estimation of crop and forest biomass resources in a semi-arid region using satellite data and GIS, Journal of the Saudi Society of Agricultural Sciences, **20**(5), pp. 302–311.

C.B. Pande, K.N. Moharir,B. Panneerselvam et al. Delineation of groundwater potential zones for sustainable development and planning using analytical hierarchy process (AHP), and MIF techniques. Appl. Water. Sci. **11**, 186. (2021b) https://doi.org/10.1007/s13201-021-01522-1

B. Panneerselvam, S. Karuppannan, K. Muniraj, Evaluation of drinking and irrigation suitability of groundwater with special emphasizing the health risk posed by nitrate contamination using nitrate pollution index (NPI) and human health risk assessment (HHRA). Hum. Ecol. Risk Assess. Int. J. **27**(5), 1324–1348 (2020a)

B. Panneerselvam, S.K. Paramasivam, S. Karuppannan, N. Ravichandran, P. Selvaraj, A GIS-based evaluation of hydrochemical characterisation of groundwater in hard rock region, South Tamil Nadu, India. Arab. J. Geosci. **13**(17), 1–22 (2020b)

B. Panneerselvam, K. Muniraj, C. Pande, N. Ravichandran, Prediction and evaluation of groundwater characteristics using the radial basic model in semi-arid region, India. Int. J. Environ. Anal. Chem., 1–17 (2021a)

B. Panneerselvam, K. Muniraj, M. Thomas, N. Ravichandran, B. Bidorn, Identifying influencing groundwater parameter on human health associate with irrigation indices using the Automatic Linear Model (ALM) in a semi-arid region in India. Environ. Res. **202**, 111778 (2021b)

B. Panneerselvam, K. Muniraj, C. Pande, N. Ravichandran, M. Thomas, S. Karuppannan, Geochemical evaluation and human health risk assessment of nitrate-contaminated groundwater in an industrial area of South India. Environ. Sci. Pollut. Res., 1–18 (2021c)

J.P. Patil, A. Sarangi, D.K. Singh, D. Chakraborty, M.S. Rao, S. Dahiya, Rainfall trend analysis: A case study of Pune district in Western Maharashtra region. J. Soil Water Conserv. **12**(1), 35–43 (2013)

K. Shunmugapriya, B. Panneerselvam, K. Muniraj, N. Ravichandran, P. Prasath, M. Thomas, K. Duraisamy, Integration of multi criteria decision analysis and GIS for evaluating the site suitability for aquaculture in southern coastal region, India. Marine Pollut. Bull. **172**, 112907 (2021)

P. Singh, V. Kumar, T. Thomas, M. Arora, Changes in rainfall and relative humidity in river basins in northwest and central India. Hydrol. Proc. **22**(16), 2982–2992 (2008)

K.C. Sinha Ray, U.S. De, Climate change in India as evidenced from instrumental records. Bull. World Meteorol. Org. **52**(1), 53–58 (2003)

K.C. Sinha Ray, A.K. Srivastava, *Is There Any Change in Extreme Events Like Droughts and Heavy Rainfall? Intropmet-97* (IIT, New Delhi, 1999), pp. 2–5

S.I. Sridhar, A. Raviraj, Statistical trend analysis of rainfall in Amaravathi river basin using Mann-Kendall test. Curr. World Environ. **12**(1), 89 (2017)

S.P. Wani, T.K. Sreedevi, J. Rockström, Y.S. Ramakrishna, Rainfed agriculture–past trends and future prospects. Rainfed Agr.: Unlock. Potent. **7**, 326 (2009)

Chapter 6
Assessment of Groundwater Quality for Drinking and Irrigation: A Case Study of the Kattar Micro-Watershed, Tamil Nadu

Jothiramalingam Killivalavan, Ravichandran Venkatesh, Periasamy Thilagaraj, Jothimani Lokeshwari, Jegankumar Rajagopal, and Masilamani Palanisamy

Abstract This study attempted to evaluate the groundwater quality of the Kattar micro-watershed (~255 km^2), located in the Tiruchirappalli and Pudukottai districts of Tamil Nadu by analyzing 38 samples collected in and around the study area in January 2019. The hydrochemical parameters (total dissolved solids (TDSs), electrical conductivity (EC), percentage of hydrogen (pH), calcium, magnesium, potassium, sodium, fluoride, nitrate, sulphate, chloride, bicarbonate) were analyzed, and their limits were categorized based on World Health Organization standards to calculate the water quality index. Parameters like magnesium ratio, sodium adsorption ratio, residual sodium carbonate, total hardness, soluble sodium percentage, permeability index, TDSs, EC, and the United States Salinity Laboratory (USSL) classifications were considered for irrigation suitability analysis. The hydrochemical analysis inferred that the predominant water type present in the study area was of mixed Ca–Mg–Cl and Ca–Cl types and calcium and bicarbonate were the dominant anion and cation, respectively. The result of drinking suitability illustrated that the study area mostly fell under the poor to very poor category. On the other hand, irrigation suitability showed that most of the stations fell under the moderately suitable to marginally suitable category. Thus, the methods used in this study were undoubtedly significant for identifying the areas with poor groundwater quality that needed immediate attention of proper water management strategies.

J. Killivalavan (✉) · P. Thilagaraj · J. Lokeshwari · J. Rajagopal · M. Palanisamy
Department of Geography, Bharathidasan University, Tiruchirappalli, Tamil Nadu, India
e-mail: jegankumar@bdu.ac.in; masilamani@bdu.ac.in

R. Venkatesh
Department of Civil Engineering, Indian Institute of Technology, Guwahati, Assam, India
e-mail: r.v@rnd.iitg.ac.in

B. Panneerselvam et al. (eds.), *Climate Change Impact on Groundwater Resources*, https://doi.org/10.1007/978-3-031-04707-7_6

Keywords Groundwater quality · Kattar micro-watershed · Water quality index · Irrigation suitability

6.1 Introduction

Next to surface water, groundwater, an essential natural resource in present-day circumstances, is utilized for drinking, domestic, and industrial purposes (Shunmugapriya et al. 2021; Jesudhas et al. 2017, 2021; Saleem 2007). It is considered as a major part of water resource in semi-arid regions, where, in recent years, both its quality and quantity have been directing many environmental problems and threatening planners and policy-makers managing the resources (Bain et al. 2014). In the global scenario, nearly 1.5 billion people depend on groundwater for drinking purposes (Shen et al. 2008). It is estimated that India is consuming about ~ 230 km^3 of groundwater every year and thus is facing a groundwater crisis due to the overexploitation of the resources; 85% of the rural population in India depends on groundwater for both irrigation and drinking purposes (Adimalla and Qian 2019; World Bank 2010; CGWB 2017). In the past decade, the groundwater resources have been overexploited due to increasing demand, with 65% utilized for drinking, 20% for irrigation, and 15% for industries (Aravinthasamy et al. 2020; Karunanidhi et al. 2021; Salehi et al. 2018). Groundwater quality is influenced by various point sources (i.e. industrial, domestic, agricultural activities, etc.; Johnny et al. 2018; Panneerselvam et al. 2020, 2021a) and non-point sources (i.e. pesticides, fertilizers, etc.) (Nag and Lahiri 2012). Both natural and anthropogenic activities determine the quality of water, whereas its characteristics are influenced by physical, chemical, and biological factors (WHO 2011; BIS 2012; Akter et al. 2016). In recent times, rapid urbanization, development of industries, and increased use of fertilizers for agricultural activities have rendered changes in the quality of surface water as well as groundwater (Jha et al. 2020; Vasanthakumari Sivasankara Pillai et al. 2020). Moreover, changing rainfall patterns and current land use changes due to increased cultural or anthropogenic activities will also have an adverse effect on both the quality and quantity of groundwater (Zaki et al. 2019).

6.2 Background of This Study

Globally, groundwater contamination has become the most threatening issue as the quality of water has become poor and the quantity has also reduced, leading to water crises (Khan and Jhariya 2017; Ameen 2019). In developing countries, it has been estimated that ~250 million people are affected due to consumption of highly polluted water and about 10–20 million have lost their lives (Dzwairo et al. 2006). The quality of groundwater has become poor in India, which is impacting the soil quality (Adomako et al. 2011). This contamination will reduce agricultural production due to increased salinity and will lead to many health issues when consumed (Mohamed et al. 2018; Lanjwani et al. 2020; Arslan 2012; Kamaraj et al. 2021). Unlike surface water, groundwater pollution is hard to assess as it is beneath the surface and its contamination is evaluated based on the hydrochemical properties of cations and anions (Vadiati et al. 2016; Jha et al. 2020). This will trigger planning towards optimum utilization of the resources and development of conservation practices. Accordingly, the analysis of groundwater quality has become a primary concern and plays a major role in determining its suitability for drinking and agricultural and industrial processes (Selvam 2012; Singaraja et al. 2012). However, it is important to assess groundwater quality in river basins or watersheds as they are more prone to the influence of domestic, agricultural, and industrial wastes (Simeonova et al. 2003). The region of the Kattar micro-watershed positioned in the hard rock formation of the Tiruchirappalli district, South India, was selected as the study region. Due to lack of a surface water system in the study area, the dependency on groundwater has increased, leading to serious issues like groundwater contamination and declined water table levels. Thus, the need for groundwater quality assessment in the Kattar micro-watershed is essential. Eventually, several research studies have been conducted to assess groundwater quality for drinking suitability and irrigation suitability in different terrain and climatic conditions (Samantray et al. 2009; Abdul et al. 2010; Vasanthavigar et al. 2010; Srinivasamoorthy et al. 2011; Krishna kumar et al. 2015; Akter et al. 2016; Khan and Jhariya 2017; Adimalla et al. 2018; Abbasnia et al. 2018; Singh et al. 2020; Kamaraj et al. 2021; Karunanidhi et al. 2021). Initially, Horton (1965) formulated the water quality indices (WQIs) to evaluate water quality, and several researchers (Bhat and Pandit 2014; Lobato et al. 2015; Kangabam et al. 2017; Adimalla and Qian 2019; Ameen 2019; Jha et al. 2020) used the modified and updated WQI to identify the improved results. From research studies (Srinivasamoorthy et al. 2011; Kangabam et al. 2017; Selvakumar et al. 2017; Adimalla et al. 2018; Karunanidhi et al. 2021; Panneerselvam et al. 2021b, c), the essential parameters used to evaluate the groundwater quality for drinking suitability are percentage of hydrogen (pH), electrical conductivity (EC), total dissolved solids (TDSs), magnesium (Mg), potassium (K), calcium (Ca), sodium (Na), chloride (Cl), fluoride (F), sulphate (SO_4), nitrate (NO_3), and bicarbonate (HCO_3). Then, in studies related to irrigation suitability (Srinivasamoorthy et al. 2011; Thapa et al. 2017; Singh et al. 2018; Kamaraj et al. 2021) states, parameters like magnesium ratio (MR), sodium adsorption ratio (SAR), residual sodium carbonate (RSC), total hardness (TH), soluble sodium percentage (SSP), permeability index (PI), and USSL are the best indicators of irrigation suitability as individual or combined index. Hence, the objective of this study is to evaluate the groundwater quality parameters (i.e. the major cations and anions) based on World Health Organization's (WHO 2011) guidelines and standards and to determine drinking suitability through the WQI and index-based irrigation suitability assessment.

6.3 Study Area

The Kattar micro-watershed, a part of the Cauvery basin distributed over the Tiruchirappalli and Pudukottai districts of Tamil Nadu, South India, is located between the north latitudes 10°50′06″ and 10°35′23″ and the east longitudes 78°49′36″ and 78°40′56″. The total geographical extent of the study area is 255 km^2. The location of the study area is shown in Fig. 6.1. The topographical variation of the study area ranges at an altitude between 60 m and 129 m above the sea level. The Cauvery and Kollidam Rivers are the only sources for a surface water system. The northern portion of the study area is distributed over the core city region of the Tiruchirappalli Corporation, whereas the southern part is distributed with the presence of industries and agricultural lands. The study region is a hard rock formation with an unconfined aquifer system and hornblende biotite gneiss and granite as the dominant geological features covering 70% and 12%, respectively. With regard to the soil texture, gravelly loam soil occupies most parts of the study area and next to that loamy soil and calcareous cracking clay soil cover the remaining part. Based on the grid sampling method, groundwater samples were collected from 38 locations in equal intervals. Those water samples were used to evaluate the groundwater suitability for drinking and irrigation purposes.

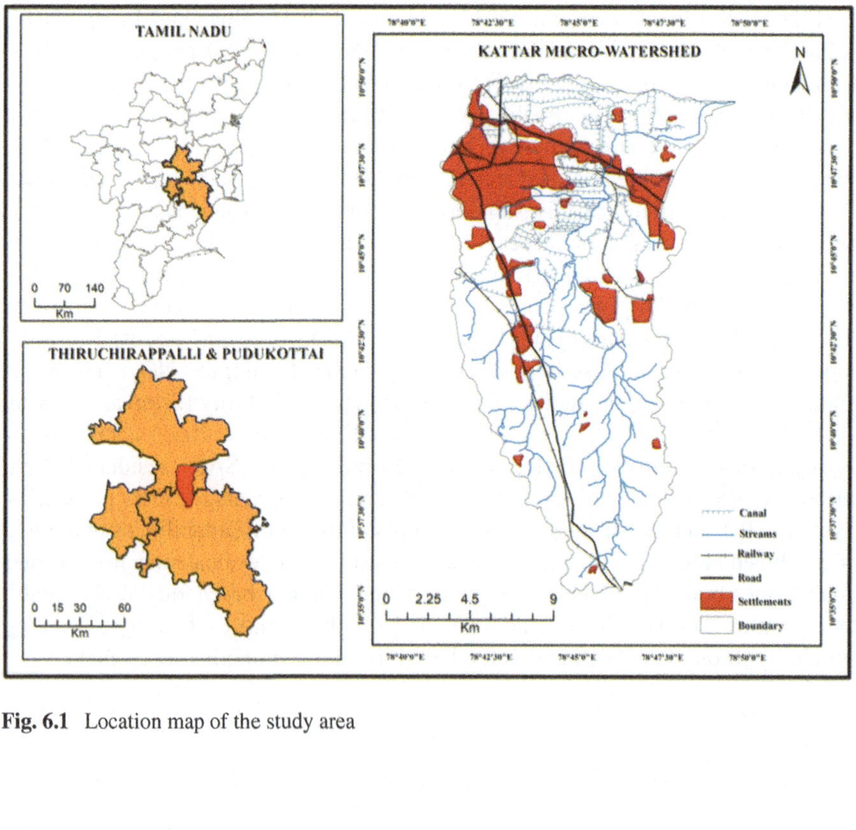

Fig. 6.1 Location map of the study area

6.4 Materials and Methods

6.4.1 Data Acquisition

Around 38 groundwater sampling locations were selected through the grid method within the study region during the post-monsoon season (January 2019) (Fig. 6.2). Groundwater samples were collected in 500-ml high-density polyethylene (HDPE) bottles from open wells and bore well locations at a depth level of 20–120 m. Then, the water samples were stored in a dry-cleaned bottle without any contamination factors. The collected water samples were further subjected to laboratory analysis to

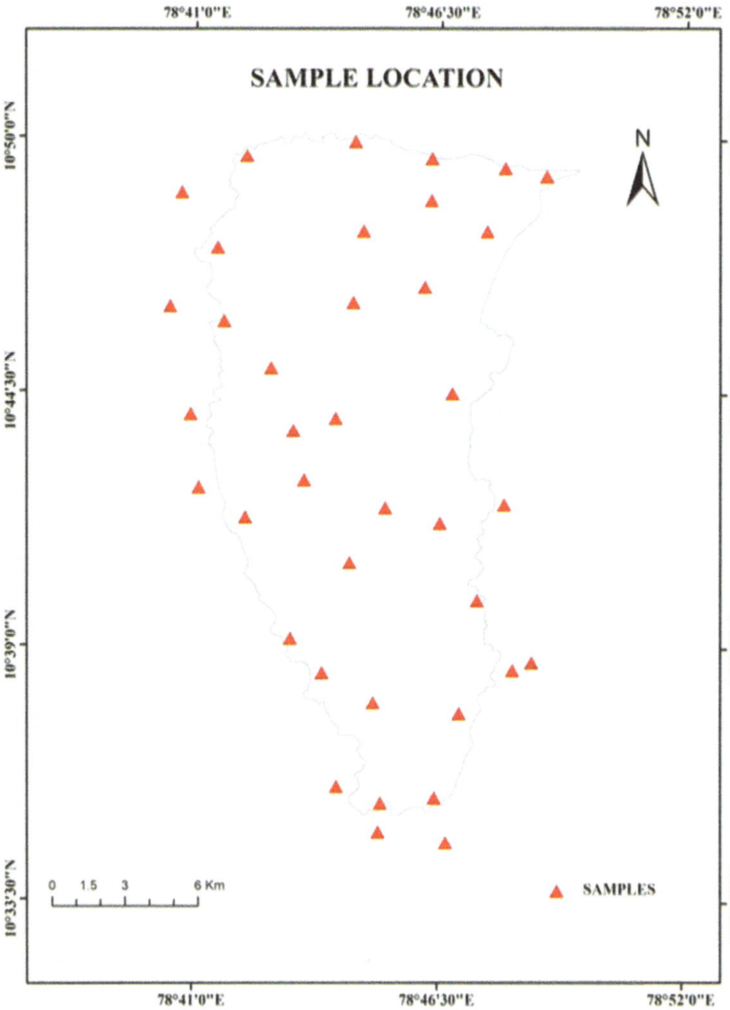

Fig. 6.2 Sample location of the study area

identify the major anions (i.e. sulphate (SO_2^{4-}), nitrate (NO_3^-), bicarbonate (HCO_3^-), chloride (Cl^-), and fluoride (F)), cations (i.e. sodium (Na^+), magnesium (Mg_2^+), potassium (K), and calcium (Ca)), and physical parameters (i.e. percentage of hydrogen (pH), electrical conductivity (EC), total dissolved solids (TDSs)). Each hydrochemical parameter was analyzed through distinct methods and instruments (Table 6.1) based on standard procedures (APHA 1995). Ion balancing was performed for the major anions and cations, and the estimated percentage of error was less than 10%. Later, the parameters were used as inputs for groundwater suitability analysis for drinking and irrigation purposes.

6.4.2 *Water Quality Index (WQI)*

In geochemistry, the water quality index (WQI) is an effective tool for quantifying groundwater quality (Khan and Jhariya 2017; Adimalla and Qian 2019), which is utilized in numerous research studies by researchers and scientists worldwide (Vasanthavigar et al. 2010; Wu et al. 2015; Akter et al. 2016; Kangabam et al. 2017; Karunanidhi et al. 2021). The WQI measures the level of suitable quality for human consumption, by considering various hydrochemical parameters. In this study, the standards and guidelines provided by the World Health Organization (WHO 2011) were followed to compute the WQI for drinking purposes. The major influencing parameters like pH, TDSs, EC, F, NO_3, SO_4, Na, K, Ca, Mg, HCO_3, and Cl were selected for WQI calculation. A weightage (w_i) (i.e. ranging from 1 to 5) was assigned to each parameter, where a high weightage of "5" is assigned to TDS and a low weightage to NO_3 (Table 6.2). To compute the WQI, the following steps were considered: relative weightage (W_i) and quality rating scale (q_i) were calculated for

Table 6.1 Hydrochemical parameters and their analytical methods/instruments

S No.	Parameter	Units	Methods	Instruments
1	pH	pH unit	pH meter	Systronics pH system System 361
2	EC	µs/cm	Electrometric	Systronics conductivity TDS meter 308
3	TDS	mg/l	Electrometric	
4	Ca	mg/l	Titrimetric	Borosilicate burette 100 ml
5	Cl	mg/l	Titrimetric	
6	Mg	mg/l	Titrimetric	
7	HCO_3	mg/l	Titrimetric	
8	K	mg/l	Flame photometric	Elico flame photometer CL 378
9	Na	mg/l	Flame photometric	
10	F	mg/l	Spectrophotometric	Systronics PC-based double-beam spectrometer 2202
11	SO_4	mg/l	Spectrophotometric	
12	NO_3	mg/l	Spectrophotometric	

Table 6.2 Relative weightage (W_i), weight (w_i), and standard for drinking water suitability (S_i) (WHO 2011)

Parameters	WHO standards	Weight (w_i)	Relative weight (W_i)
pH	6.5–8.5	3	0.081081
EC	780	4	0.108108
TDS	500	5	0.135135
HCO_3	300	4	0.108108
Ca	75	4	0.108108
Mg	30	2	0.054054
Cl	200	3	0.081081
SO_4	200	3	0.081081
Na	200	4	0.108108
K	10	2	0.054054
NO_3	45	1	0.027027
F	1.5	2	0.054054
		$\sum w_i = 37$	$\sum W_i = 1$

12 parameters through Eqs. 6.1 and 6.2; sub-index values (SI_i) were calculated by multiplying (W_i) and (q_i) (Eq. 6.3); then, the WQI was computed by adding the sub-index value (SI_i) of the selected parameters (Eq. 6.4) (Ramakrishnaiah et al. 2009; Khan and Jhariya 2017; Kangabam et al. 2017; Adimalla et al. 2018).

$$W_i = \frac{w_i}{\sum_{i=1}^{n} w_i} \tag{6.1}$$

$$q_i = \frac{C_i}{S_i} \times 100 \tag{6.2}$$

$$SI_i = W_i \times q_i \tag{6.3}$$

$$WQI = \sum_{i=1}^{n} SI_i \tag{6.4}$$

where w_i is the assigned weightage and n is the number of selected hydrochemical parameters. C_i is the concentration of the ith parameter, and S_i is the standard of drinking (WHO 2011). SI_i is the sub-index value, and W_i and q_i are the relative weightage and quality rating scale for the ith parameter, respectively.

The WQI value is computed through Eq. 6.4, and those values were classified based on proper standard values as mentioned in Table 6.3 (Ramakrishnaiah et al. 2009; Vasanthavigar et al. 2010; Kangabam et al. 2017).

Table 6.3 WQI value range and types of drinking water suitability

Range	Type of water	Number of stations	Percentage
<50	Excellent	–	–
50–100	Good	8	21
100–200	Poor	25	66
200–300	Very poor	4	10
>300	Unsuitable for drinking	1	3

Table 6.4 Irrigation suitability class

Suitability class	Category	Number of stations	Percentage
Highly suitable	<60	7	18.42
Moderately suitable	60–75	26	68.42
Marginally suitable	>75	5	13.15

6.4.3 *Irrigation Water Quality (IWQ)*

A measure of water quality suitable for irrigation purposes is referred to as irrigation water quality (Table 6.4). Meanwhile, in this study, the irrigation water quality (IWQ) was evaluated by considering various parameters like total hardness (TH), sodium adsorption ratio (SAR), residual sodium carbonate (RSC), magnesium ratio (MR), permeability index (PI), soluble sodium percentage (SSP), USSL salinity, electrical conductivity (EC), and total dissolved solids (TDSs) (Vasanthavigar et al. 2010; Srinivasamoorthy et al. 2011; Singh et al. 2020; Kamaraj et al. 2021). The water quality fluctuated relatively to the concentration of each parameter. To calculate the IWQ, the selected nine parameters were assigned to the individual class weightage (R_i) (Table 6.5) based on the concentration level. Then, the summation of all parameter weightage represents the IWQ value $\left(\sum_{i=1}^{n} R_i\right)$, and the values are categorized into three classes as mentioned in Table 6.4.

Total hardness (TH) is a measure of the concentration of multivalent cations (i.e. calcium and magnesium) in a given water sample (Raghunath 1987a; Kamaraj et al. 2021). This is calculated through the following equation (Eq. 6.5):

$$TH = \left(Ca + \mathrm{Mg}\right) \times 50 \tag{6.5}$$

The sodium adsorption ratio (SAR) is an excellent indicator of alkali hazards to measure irrigation water suitability. The SAR, calculated using Eq. 6.6, is defined as the amount of Na concentration over Ca and Mg (Richards 1954; Wilcox 1955; Vasanthavigar et al. 2010).

$$SAR = \frac{Na^{+}}{\frac{\sqrt{Ca^{2+} + \mathrm{Mg}^{2+}}}{2}} \tag{6.6}$$

Table 6.5 Irrigation suitability indices and weightage (R_i)

Parameters	Unit	Category	Class	Weightage (R_i)
USSL class	–	C1-S1	Very good	10
		C2-S1/C1-S2	Good	8
		C3-S1/ C4-SI/C2-S2/ CI-S3	Moderate	6
		C3-S2/C2-S3/C4-S2	Poor	4
		C4-S3/C3-S3	Very poor	2
Permeability index	epm	>75	Suitable	10
		25–75	Moderately suitable	8
Magnesium ratio	epm	0–25	Suitable	10
		25–50	Permissible	6
		50–75	Doubtful	4
		>75	Unsuitable	2
Sodium adsorption ratio	epm	0–5	Excellent	10
		10 May	Very good	8
Residual sodium carbonate	epm	<1.25)	Good	10
		1.25–2.50	Harmful	4
		>2.50	Serious	2
Soluble sodium percentage	%	<40	Good	10
		40–60	Permissible	8
Conductivity	µS/cm	<750	Desirable	10
		750–1500	Permissible	8
		1500–2250	Doubtful	6
		2250–4000	Not permissible	4
Total dissolved solids	mg/l	<500	Desirable	10
		500–2000	Permissible	8
		2000–3000	Not permissible	6
Total hardness	mg/l	<300	Soft	10
		300–600	Moderate soft	8
		600–900	Moderate hard	6
		900–1200	Hard	4
		>1200	Very hard	2

Residual sodium carbonate (RSC) is a widely used index to estimate the sodium permeability hazard (Richards 1954; Siddiqui et al. 2005). The following equation (Eq. 6.7) is used for RSC calculation:

$$RSC = (CO_3 + HCO_3) - (Ca + \mathrm{Mg}) \tag{6.7}$$

The magnesium ratio (MR) is a quantitative estimation of magnesium hazard by calculating the excess concentration of magnesium over Ca and Mg (Szabolcs and Darab 1964; Paliwal 1972; Singh et al. 2020). Eq. 6.8 is used to compute the MR.

$$MR = \frac{Mg^{2+}}{Ca^{2+} + Mg^{2+}} \times 100 \quad (6.8)$$

The permeability index (PI), calculated through the formula mentioned below (Eq. 6.9), is used to measure the permeability (i.e. water flux efficiency) in a soil system (Doneen 1964; Kamaraj et al. 2021).

$$PI = \frac{\left(Na + \sqrt{HCO_3}\right)}{\left(Ca + Mg + Na\right)} \times 100 \quad (6.9)$$

Soluble sodium percentage (SSP) is essential for evaluating the percentages of all dissolved cations (i.e. Ca, Mg, Na, K) (Doneen 1962; Raghunath 1987b; Kamaraj et al. 2021). The formula mentioned in eq. 6.10 is used to calculate SSP.

$$SSP = \frac{Na^{+} + K^{+}}{Na^{+} + Ca^{2+} + Mg^{2+} + K^{+}} \times 100 \quad (6.10)$$

6.5 Results and Discussion

6.5.1 Hydrochemical Characteristics

A basic statistical analysis (i.e. maximum, minimum, average, median, mode, standard deviation) was attempted for selected hydrochemical parameters (pH, EC, TDSs, Na, K, Ca, Mg, Cl, SO_4, NO_3, F, CO_3, HCO_3, TH) to understand the statistical characteristics. Table 6.6 summarizes the statistical parameters, and Fig. 6.3 shows

Table 6.6 Statistical parameters of the samples

Parameters	Maximum	Minimum	Average	Median	Mode	Standard deviation
pH	8.91	7.01	7.79	7.735	7.56	0.41
EC	4283	270	1412.47	1187.09	1208	947.33
TDS	2870	181	946.35	795.35	810	634.71
Na	426.9	11.4	140.97	98.4	–	106.26
K	61.1	8.3	11.33	9.1	9.7	8.76
Ca	708	66	245.68	222	278	145.39
Mg	207.6	8.4	47.05	42.6	52.8	30.31
Cl	1113.13	42.54	260.65	209.15	219.79	224.58
SO_4	552	93	239.36	208	–	119.78
NO_3	64.42	0.85	19.52	18.08	1	14.63
F	1.97	0.15	0.57	0.47	0.47	0.34
HCO_3	884.5	18.3	422.18	448.35	451.4	198.02

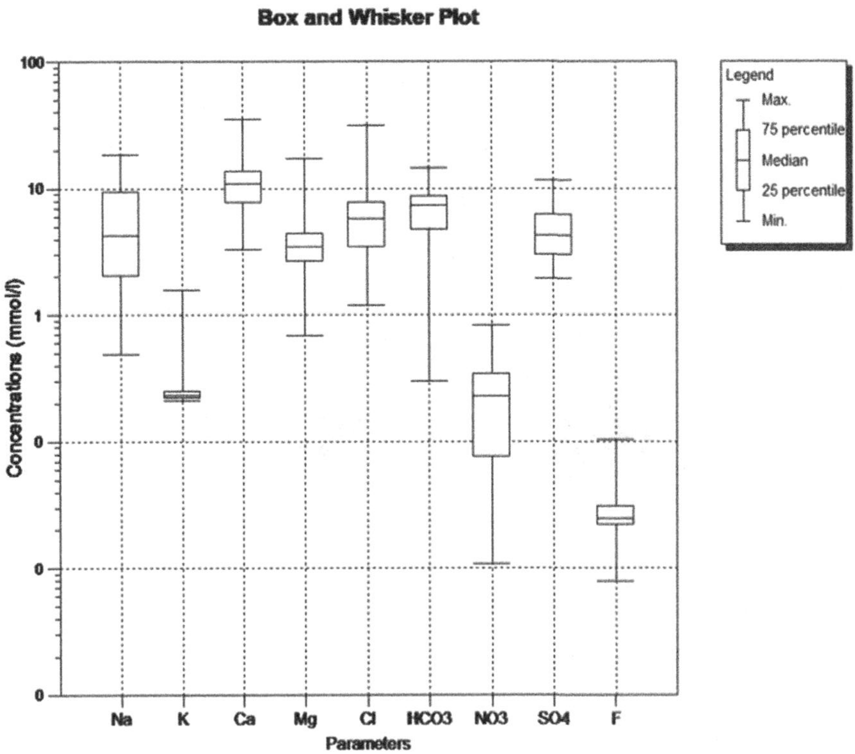

Fig. 6.3 Box and Whisker plot of the major cations and anions

the graphical representation of the major cations and anions, and it also represents the maximum and minimum values and the median, 75th percentile, and 25th percentile.

6.5.2 Piper Plot

A Piper plot is shown in Fig. 6.4. In this diagram, the left triangle represents the cation concentrations and the right triangle represents the anion concentrations (Piper 1994). A majority of the samples fell under the mixed Ca–Mg–Cl and Ca–Cl types, and few samples represented the Ca–HCO_3 and Na–Cl– SO_4 types. On the basis of cations, there were about 25 samples under the calcium type, 3 samples under the sodium and potassium types, and a further 10 samples under the no dominant type. In anions, about eight samples were under the bicarbonate type and five samples under the chloride type. Furthermore, 25 samples were under the no dominant type.

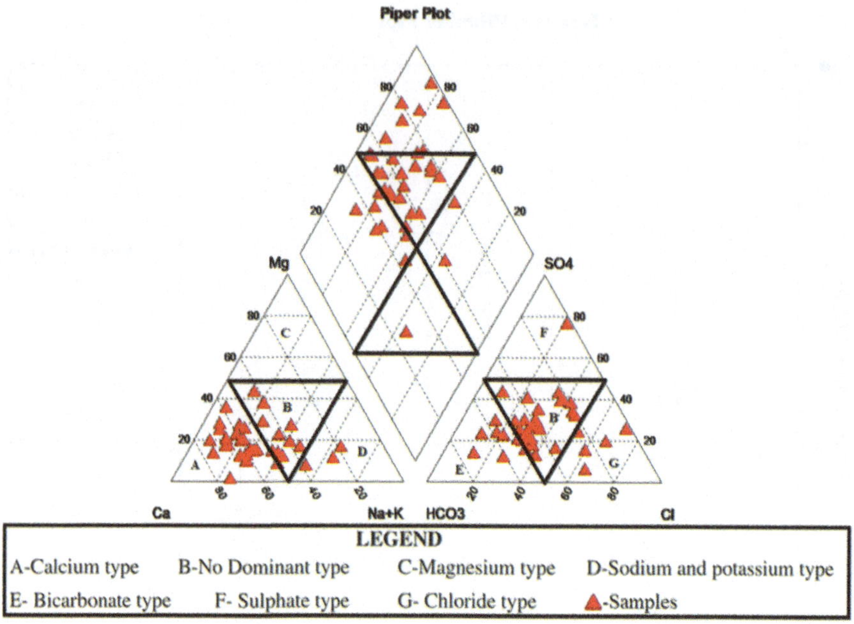

Fig. 6.4 Piper plot diagram of the hydrochemical facies of samples

6.6 Groundwater Quality Parameters for Drinking Purposes

6.6.1 Physical Parameters

Physical parameters were calculated according to the WHO standards (WHO 2011). pH signifies the nature of water (i.e. acidic to alkaline). In this study, the pH values varied between 7.01 and 8.91. About 95% of the samples were under the permissible limit, and the nature of water in the study area was neutral to alkaline. Total dissolved solids (TDSs) represent the total amount of organic and inorganic matters dissolved in water. When groundwater moves in a direction for a long period, the TDS level will increase accordingly (Freeze and Cherry 1979). The TDSs ranged between 181 and 2870 mg/l. About 80% of the study area was under the permissible limit and 15% was under the desirable limit in the south-eastern part. Notably, the samples near the tannery industry reached the not-permissible TDS limit. Electrical conductivity (EC) ranged from 270 to 4283 μS/cm. Although electrical conductivity is directly proportional to total dissolved solids, the spatial attributes portray similar patterns. The spatial distribution of all the physical parameters is shown in Fig. 6.5.

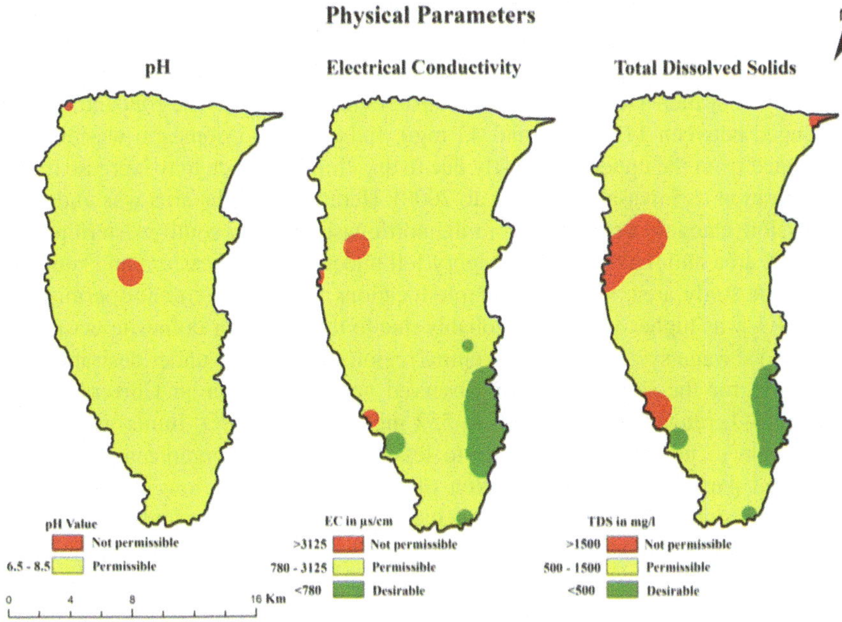

Fig. 6.5 Spatial distributions of the physical parameters in the study area

6.6.2 Major Cations and Anions

Na ranged between 11 mg/l and 426 mg/l. Weathering and overexploitation of plagioclase-bearing rocks may cause higher concentrations of sodium (Hem 1985). Nearly three-fourth of the study area was under the permissible category, and some minor portions in the northern part were under the not-permissible limit. The concentration of Ca was higher during Post-monsoon (POM) (708 mg/l), and it ranged from 66 mg/l to 708 mg/l. $CaCO_3$ and $CaMgCO_3$ dissolution during groundwater recharge increases the concentration of Ca (Datta and Tyagi 1996; Lakshmanan et al. 2003). About 65% of the study area was under the not-permissible limit, and only one sample location was under the desirable limit, which was in the Melaputhuvayal village. K was higher (61 mg/l), especially near agricultural lands due to usage of more chemical fertilizers and soil leaching. The northern, north-eastern, and central parts were under the not-permissible category, whereas the rest of the area was under the permissible limit. The concentration of Mg was about desirable to permissible limit all over the study area, but one sample location (i.e. Vadugappatti village) was marked as not permissible (207 mg/l) due to the extensive use of chemical fertilizers like $MgSO_4$. The highest concentration of HCO_3 during POM was 884 mg/l and the lowest was 18 mg/l. In 32 sample locations, HCO_3 fell under the not-permissible limit. Only six sample locations were under the permissible limit. HCO_3 is higher due to precipitation of CO_3, where it exceeds the WHO limit for drinking; further, it can affect pipelines and pumps and thus can affect the economy of farmers (Rengarajan and Balasubramanian 1990). The concentration of Cl ranged between 1113 mg/l and 42 mg/l. Industrial and domestic wastes were percolated from the upper soil mostly due to dry climate, which may increase the Cl concentration (Srinivasamoorthy et al. 2008). Here, most of the area was under the permissible category extending from the north-eastern to the south-western part of the study area and the desirable category fell under the south-eastern and southern part of the study area, and three sample locations were under the not-permissible limit. NO_3 was higher (64 mg/l) probably due to the improper domestic waste disposal and sewage system. Here, the entire region was covered under desirable limits, except for the sample location observed near Bharathidasan University. The value of SO_4 ranged from 93 mg/l to 552 mg/l, and these SO_4 limits were under permissible to not-permissible. Due to less breakdown of organic matters from weathered particles, the concentration of sulphate was also low (Miller 1979). Fluoride (F) concentration in 36 sample locations was noted to be under the permissible category, and the sample locations of Kizhakalkandarkottai and Arasangudi were under the not-permissible limit due to dumping of industrial wastes in the nearby river and usage of more fluoride-containing pesticides in agricultural fields. Fig. 6.6 shows the spatial distribution of all cations and anions.

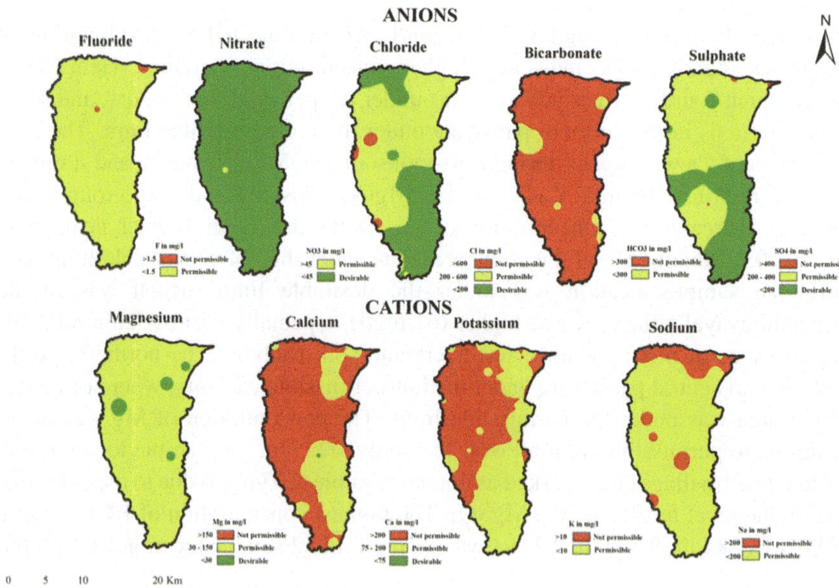

Fig. 6.6 Spatial distribution of the major cations and anions

6.7 The Water Quality Index (WQI)

The water quality index (WQI) integrated the value of the 11 hydrochemical parameter into a single index value to define the overall groundwater quality. The groundwater quality categorized based on the WQI value represents the chemical suitability of groundwater for drinking purposes. Its values ranged from 43 to 323 and were classified into five classes: excellent (<50), good (50–100), poor (100–200), very poor (200–300), and unsuitable (>300) (Ramakrishnaiah et al. 2009; Vasanthavigar et al. 2010; Kangabam et al. 2017). In the study area, none of the sample locations was noted in the excellent category. Around 21% (8 sample locations) of the study area fell under the good suitability class. The poor and very poor suitable classes covered almost 75% (29 sample locations) of the study area, which is harmful for drinking. The unsuitable category was marked only in one sample location (i.e. Vadugappatti village) (Fig. 6.7). The WQI showed that the water quality in the study area was not suitable for drinking and consumption for a more extended period, resulting in severe health issues. Some parts of the northeast and mid-central part of the study area were marked as unsuitable and very poor categories, and smaller areas in the southern and southeast parts were marked as good category. The rest of the three-fourth regions were marked as poor categories.

6.8 Irrigation Water Quality (IWQ)

6.8.1 Physical Parameters

6.8.1.1 Electrical Conductivity and Total Dissolved Solids

EC and TDSs are important factors for assessing the irrigation suitability of groundwater and range from high to poorly suitable. Since EC and TDSs are directly proportional to each other, the southern part was under highly suitable, most of the central and northern parts were under good category, and the industrial region was under poor category for irrigation.

6.8.1.2 Magnesium Ratio (MR)

Increase in the magnesium ratio can severely affect the soil and reduce agricultural yield (Paliwal 1972). The magnesium ratio was divided into four classes: suitable (<25 epm), permissible (25–50 epm), doubtful (50–75 epm), and unsuitable (>75 epm). Around 35% of the samples were under suitable, 47% permissible, 12% doubtful, and 6% unsuitable categories. Especially, the southern and northernmost parts of the study area were under permissible to unsuitable limits of the magnesium ratio.

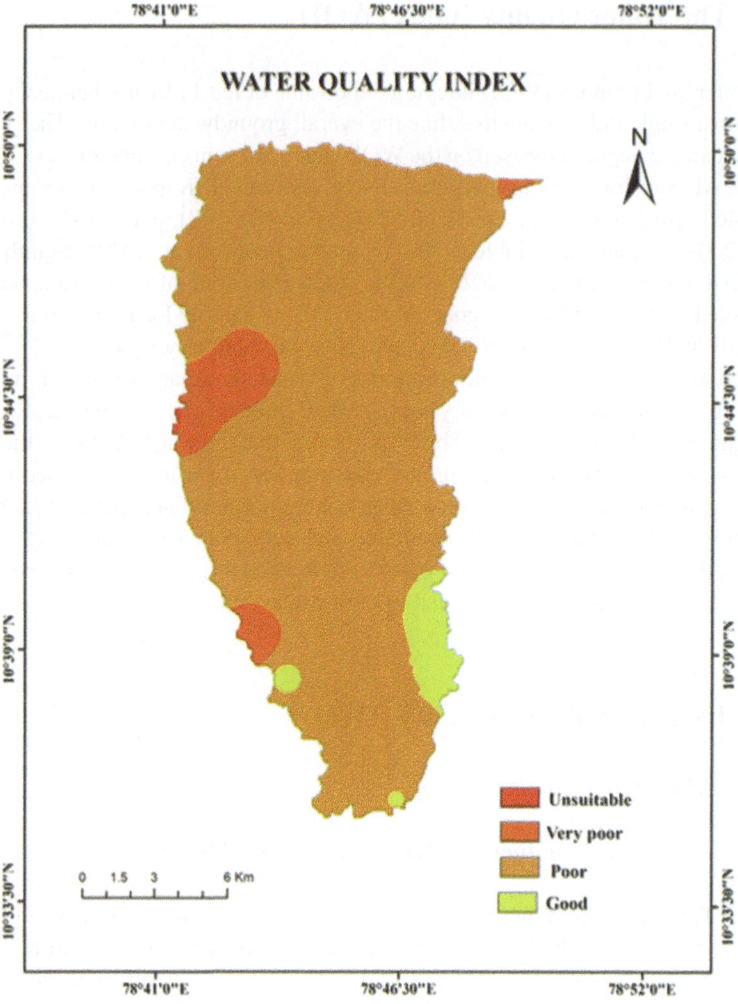

Fig. 6.7 Spatial distribution of the WQI

6.8.1.3 Sodium Adsorption Ratio (SAR)

If the combined concentration of calcium and magnesium is higher than the concentration of sodium, then sodium pollution should be low (Balasubramani 2015). Based on the USSL limits (USSL 1954), the SAR level was divided into two categories such as excellent (<5 epm) and very good (5–10 epm). In the study area, the SAR level was either very good or excellent. The calculated SAR values ranged between 0.16 epm and 5.8 epm.

6.8.1.4 Total Hardness

Total hardness is the quantity of the mineral content in water that is irreversible by boiling. Hence, total calcium and magnesium hardness is equivalent to the total hardness of water. Total hardness ranged between 298 mg/l and 2622 mg/l, which was classified as highly suitable (<300 mg/l) to unsuitable (>1200 mg/l). A major part of the study area fell under the good to moderate category. However, poor and unsuitable categories covered only the minor portions.

6.8.1.5 Residual Sodium Carbonate (RSC)

RSC values were classified into three classes such as good (<1.25 epm), harmful (1.25– 2.5 epm), and serious (>2.5 epm) as recommended by the USSL (1954). In the study area 82% of the samples were suitable for irrigation and rest of the samples were harmful to serious state.

6.8.1.6 Soluble Sodium Percentage (SSP)

The calculated SSP values of groundwater samples were divided into good (<40%) and permissible (40–60%) as suggested by Wilcox (1955). The SSP values indicated that the groundwater samples in the study area fell under the permissible to good category.

6.8.1.7 Permeability Index (PI)

The values of the permeability index present in the study area were categorized into highly suitable (>75%) and moderately suitable (25–75%). The PI values indicated that the groundwater for irrigation suitability ranged from moderate to highly suitable. The northern, middle, and south-eastern parts fell under the highly suitable category and all other areas under the moderately suitable class.

6.8.1.8 USSL Classification

The USSL classification is considered as an important factor in evaluating irrigation suitability. The USSL diagram will plot the results of SAR and EC, in which salinity hazard is categorized from C1 (low saline) to C4 (high saline) and sodium hazard is categorized from S1 (low sodium) to S4 (high sodium). Then, the major saline–alkaline hazard classes are represented as good (C2-S1/C1-S2), moderate (C3-S1/C4-SI/C2-S2/CI-S3), and poor (C3-S2/C2-S3/C4-S2). The results were interpreted,

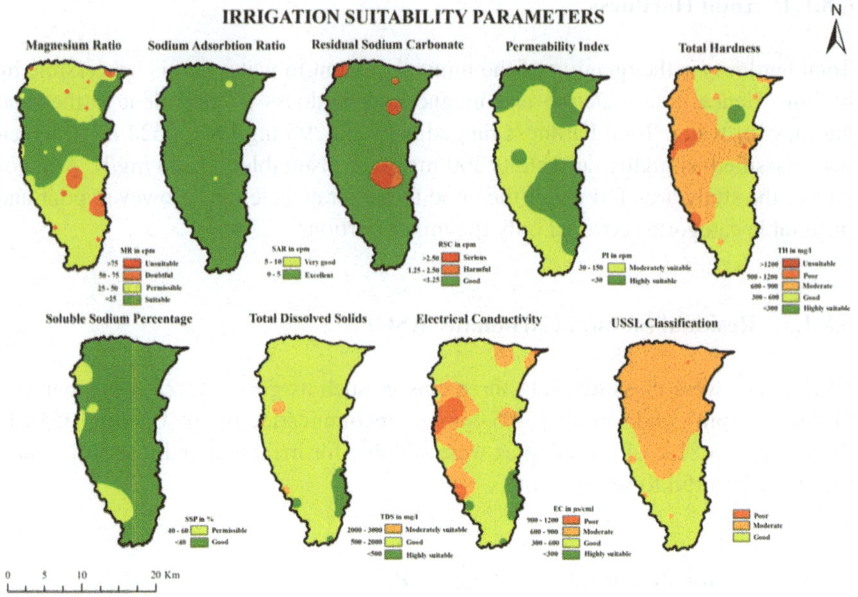

Fig. 6.8 Spatial distribution of the major parameters used to calculate irrigation suitability

and the C2-S1 of the good class was found only in eight sample locations covering 21% of the study area. Then, the C3-S1 of the moderate class covered almost 57% of the study area with 22 sample locations. The poor class of C4-S2 and C3-S2 was marked only in three sample locations covering more than 7.6% of the area. In this study area, the salinity hazard was noted as vulnerable, which fell under the medium to very high category. Figure 6.8 clearly shows the spatial distribution of all the parameters used to calculate irrigation suitability.

6.9 Irrigation Suitability

The irrigation suitability of groundwater was obtained through the weighted overlay analysis and was helpful in understanding the comprehensive risk associated with groundwater. The study area was categorized as marginally suitable (55–65), moderately suitable (65–75), and highly suitable (>75). The major portion (63%) of the watershed fell under the moderately suitable class covering 26 sample locations. The highly suitable class covered seven sample locations, which covered almost 18% of the study area, and the marginally suitable class covered 13% of the study area (Fig. 6.9). The results clearly revealed that nearly one-fourth of the study area had poor quality of water for irrigation purposes, and so the Kattar micro-watershed needs proper management of groundwater quality.

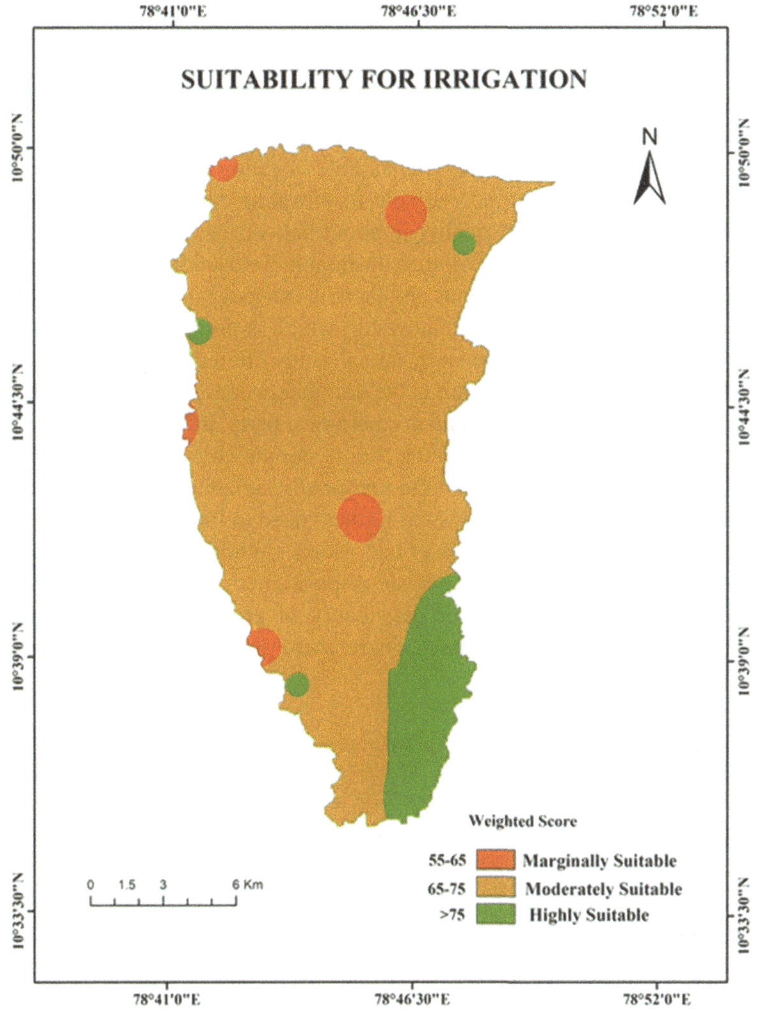

Fig. 6.9 Spatial distribution of irrigation suitability

6.10 Conclusions

This study mainly used field-collected primary data to examine the groundwater quality for drinking and irrigation purposes. Initially, the major cations and anions were analyzed, and further the WQI and IWQ were used to demarcate the suitability area for drinking and irrigation purposes. These methods were proved to be useful to this study region because the results exactly reflected the observations made from field surveys. The major types of water present in the study area were the mixed Ca–Mg–Cl and Ca–Cl types, and Ca–HCO_3 and Na–Cl–SO_4 types were present in minor quantities. The major cation and anion present in the study area were calcium and bicarbonate, respectively. The WQI clearly indicated that the major parts of the study area have poor to very poor quality of groundwater due to the higher contribution of calcium, bicarbonate, and potassium. The northern part, although located in a riverine environment, has poor water quality due to the dumping of industrial and domestic wastes into the river. The sample locations like Vadugappatti and Kilaputhutheru fell under the not-suitable-for-drinking category. If this trend continues for a decade, the water quality in about half of the study area will become extremely poor and a few areas will turn unsuitable for drinking. The study area has a vast extension of agricultural lands and there is extensive usage of chemical fertilizers that affects the quality of groundwater, which is noted in the southern parts. Improper sewage disposal in the study area directly altered the chemical composition of the groundwater. Agriculture is the dominant economic activity in the study region, especially in the southern and northernmost parts. In those areas, most of the samples were under moderately suitable, some samples were under good category, and 13% of the samples were under the marginally suitable for irrigation category. To maintain the quality of groundwater, farmers need to limit their usage of chemical fertilizers and pesticides, rainwater harvesting method should be implemented, and industries must practice proper waste disposal methods. Policy-makers should implement proper plans to recover the quality of groundwater or find alternate sources of drinking water in areas where there is extremely poor quality of groundwater.

Acknowledgements All authors thank the Department of Geography, Bharathidasan University, for providing essential support and UGC SAP DRS-II lab facilities to carry out the entire work. In addition, JL thank the Tamil Nadu State Council for Science and Technology (TNSCST) for their indispensable support.

References

A. Abbasnia, N. Yousefi, A.H. Mahvi, R. Nabizadeh, M. Radfard, M. Yousefi, M. Alimohammadi, Evaluation of groundwater quality using water quality index and its suitability for assessing water for drinking and irrigation purposes : Case study of Sistan and Baluchistan province (Iran). Hum. Ecol. Risk. Assess. **25**(4), 1–18 (2018). https://doi.org/10.1080/10807039.2018.1458596

H.M. Abdul, H.S. Abid, B.K. Maulood, Application of water quality index for assessment of Dokan Lake Ecosystem, Kurdistan Region, Iraq. J. Water Resour. Protect **2010**(September), 792–798 (2010). https://doi.org/10.4236/jwarp.2010.29093

N. Adimalla, H. Qian, Groundwater quality evaluation using water quality index (WQI) for drinking purposes and human health risk (HHR) assessment in an agricultural region of Nanganur South India. Ecotoxicol. Environ. Saf. **176**(126), 153–161 (2019). https://doi.org/10.1016/j.ecoenv.2019.03.066

N. Adimalla, P. Li, S. Venkatayogi, Hydrogeochemical evaluation of groundwater quality for drinking and irrigation purposes and integrated interpretation with water quality index studies. Environ. Proc. **5**, 363–383 (2018). https://doi.org/10.1007/s40710-018-0297-4

D. Adomako, A. Gibrilla, T.T. Akiti, R. Fianko, P. Maloszewski, Hydrogeochemical evolution and groundwater flow in the Densu River Basin, Ghana. J Water Resour Protect **03**(07), 548–561 (2011). https://doi.org/10.4236/jwarp.2011.37065

T. Akter, F.T. Jhohura, F. Akter, T.R. Chowdhury, S.K. Mistry, D. Dey, M.K. Barua, M.A. Islam, M. Rahman, Water quality index for measuring drinking water quality in rural Bangladesh: A cross-sectional study. J. Health Popul. Nutr. **35**(4) (2016). https://doi.org/10.1186/s41043-016-0041-5

H.A. Ameen, Spring water quality assessment using water quality index in villages of Barwari Bala Duhok Kurdistan Region Iraq. Appl Water Sci **9**(8), 1–12 (2019). https://doi.org/10.1007/s13201-019-1080-z

APHA, *Standard Methods for the Examination of Water and Wastewater* (American Public Health Association, Washington, DC, 1995)

P. Aravinthasamy, D. Karunanidhi, T. Subramani, K. Srinivasamoorthy, B. Anand, Geochemical evaluation of fluoride contamination in groundwater from Shanmuganadhi River basin, South India : implication on human health. Environ. Geochem. Health **42**(7), 1937–1963 (2020). https://doi.org/10.1007/s10653-019-00452-x

H. Arslan, Spatial and temporal mapping of groundwater salinity using ordinary kriging and indicator kriging: The case of Bafra Plain, Turkey. Agr. Water Manag. **113**, 57–63 (2012). https://doi.org/10.1016/j.agwat.2012.06.015

R.E.S. Bain, J.A. Wright, E. Christenson, J.K. Bartram, Rural: Urban inequalities in post 2015 targets and indicators for drinking-water. Sci. Total Environ. **490**, 509–513 (2014). https://doi.org/10.1016/j.scitotenv.2014.05.007

K. Balasubramani, Application of Geospatial Technologies in Land and Water Resources Evaluation for Sustainable Development of Andipatti Watershed, Tamil Nadu, India. Ph.D Thesis, Madurai Kamaraj University (2015)

S.A. Bhat, A.K. Pandit, Surface water quality assessment of Wular Lake, A Ramsar Site in Kashmir Himalaya. Using Discrim. Anal. WQI, 1–18 (2014). https://doi.org/10.1155/2014/724728

Bureau of Indian Standards (BIS), *Specification for Drinking Water* (Food and Agricultural Division Council, New Delhi, 2012)

CGWB, *Dynamic Groundwater Resources of India (As on March 2013). Central Ground Water Board (CGWB), Ministry of Water Resources* (Government of India, New Delhi, 2017)

P.S. Datta, S.K. Tyagi, Major ion chemistry of groundwater in Delhi area: Chemical weathering processes and groundwater flow regime. J. Geol Soc. India **47**, 179–188 (1996)

L.D. Doneen, The influence of crop and soil on percolating water, in *Proceedings of the Biennial Conference on Ground water Recharge*, (1962), pp. 156–163

L.D. Doneen, *Notes on Water Quality in Agriculture* (Published as a Water Science and Engineering, Paper 4001, Department of Water Sciences and Engineering, University of California, Davis, 1964)

B. Dzwairo, Z. Hoko, D. Love, E. Guzha, Assessment of the impacts of pit latrines on groundwater quality in rural areas : A case study from Marondera district, Zimbabwe. Phys. Chem. Earth, Parts A/B/C **31**, 779–788 (2006). https://doi.org/10.1016/j.pce.2006.08.031

R.A. Freeze, J.A. Cherry, *Groundwater*, vol 7632 (Prentice-Hall Inc., Englewood Cliffs, 1979), p. 604

J.D. Hem, *Study and Interpretation of the Chemical Characteristics of Natural Water. 3rd Edition, US Geological Survey Water-Supply Paper 2254* (University of Virginia, Charlottesville, 1985)

R.K. Horton, An index-number system for rating water quality. J. Water Pollut. Cont. Federat. **37**(3), 300–306 (1965)

C.J. Jesudhas, S.M. Chidambaram, K. Muniraj, Isolation of wells contaminated by tannery industries using principal component analysis. Arab. J. Geosci. **10**(14), 1–8 (2017)

C.J. Jesudhas, A. Chinnasamy, K. Muniraj, A. Sundaram, Assessment of vulnerability in the aquifers of rapidly growing sub-urban: A case study with special reference to land use. Arab. J. Geosci. **14**(1), 1–20 (2021)

M.K. Jha, A. Shekhar, M.A. Jenifer, Assessing groundwater quality for drinking water supply using hybrid fuzzy-GIS-based water quality index. Water Res. **179**, 115867 (2020). https://doi.org/10.1016/j.watres.2020.115867

J.C. Johnny, M.C. Sashikkumar, M. Kirubakaran, L.M. Mathi, GIS-based assessment of groundwater quality and its suitability for drinking and irrigation purpose in a hard rock terrain: A case study in the upper Kodaganar basin, Dindigul district, Tamil Nadu, India. Desalin. Water Treat. **102**, 49–60 (2018)

J. Kamaraj, S. Sekar, P.D. Roy, V. Senapathi, S.Y. Chung, M. Perumal, A.V. Nath, Groundwater pollution index (GPI) and GIS-based appraisal of groundwater quality for drinking and irrigation in coastal aquifers of Tiruchendur, South India. Environ. Sci. Pollut. Res., 18212232061030 (2021). https://doi.org/10.1007/s11356-021-12702-6

R.D. Kangabam, S.D. Bhoominathan, S. Kanagaraj, M. Govindaraju, Development of a water quality index (WQI) for the Loktak Lake in India. Appl Water Sci **7**(6), 2907–2918 (2017). https://doi.org/10.1007/s13201-017-0579-4

D. Karunanidhi, P. Aravinthasamy, T. Subramani, G. Muthusankar, Revealing drinking water quality issues and possible health risks based on water quality index (WQI) method in the Shanmuganadhi River basin of South India. Environ. Geochem. Health **43**(2), 931–948 (2021). https://doi.org/10.1007/s10653-020-00613-3

R. Khan, D.C. Jhariya, Groundwater quality assessment for drinking purpose in Raipur city, Chhattisgarh using water quality index and geographic information system. J. Geol. Soc. India **90**(1), 69–76 (2017). https://doi.org/10.1007/s12594-017-0665-0

S. Krishna Kumar, A. Logeshkumaran, N.S. Magesh, P.S. Godson, N. Chandrasekar, Hydrogeochemistry and application of water quality index (WQI) for groundwater quality assessment, Anna Nagar, part of Chennai City, Tamil Nadu, India. Appl Water Sci **5**(4), 335–343 (2015). https://doi.org/10.1007/s13201-014-0196-4

E. Lakshmanan, K. Kannan, M. Senthil Kumar, Major ion chemistry and identification of hydrogeochemical process of groundwater in part of Kancheepuram district, Tamilnadu, India. J. Environ. Geosci. **10**(4), 157–166 (2003)

M.F. Lanjwani, M.Y. Khuhawar, T.M.J. Khuhawar, Groundwater quality assessment of Shahdadkot, Qubo Saeed Khan and Sijawal Junejo Talukas of District Qambar Shahdadkot, Sindh. Appl. Water Sci. **10**(1), 1–18 (2020). https://doi.org/10.1007/s13201-019-1098-2

T.C. Lobato, R.A. Hauser-davis, T.F. Oliveira, A.M. Silveira, H.A.N. Silva, M.R.M. Tavares, A.C.F. Saraiva, Construction of a novel water quality index and quality indicator for reservoir water quality evaluation: A case study in the Amazon region. J. Hydrol. **522**, 674–683 (2015). https://doi.org/10.1016/j.jhydrol.2015.01.021

G.T. Miller, *Living in the Environment*, vol 470 (Wadsworth, Belmond, 1979)

A.K. Mohamed, D. Liu, M.A.A. Mohamed, K. Song, Groundwater quality assessment of the quaternary unconsolidated sedimentary basin near the Pi river using fuzzy evaluation technique. Appl Water Sci **8**(2), 1–12 (2018). https://doi.org/10.1007/s13201-018-0711-0

S.K. Nag, A. Lahiri, Hydrochemical characteristics of groundwater for domestic and irrigation purposes in Dwarakeswar Watershed Area, India. Am. J. Clim. Chang. **1**, 217–230 (2012). https://doi.org/10.4236/ajcc.2012.14019

K.V. Paliwal, *Irrigation with Saline Water. Monogram No. 2 (New Series)* (IARI, New Delhi, 1972), p. 198

B. Panneerselvam, S. Karuppannan, K. Muniraj, Evaluation of drinking and irrigation suitability of groundwater with special emphasizing the health risk posed by nitrate contamination using nitrate pollution index (NPI) and human health risk assessment (HHRA). Hum. Ecol. Risk Assess. Int. J. **27**(5), 1324–1348 (2020)

B. Panneerselvam, K. Muniraj, M. Thomas, N. Ravichandran, GIS-based legitimatic evaluation of groundwater's health risk and irrigation susceptibility using water quality index, pollution index, and irrigation indexes in semiarid region, in *Groundwater Resources Development and Planning in the Semi-Arid Region*, (Springer, Cham, 2021a), pp. 239–268

B. Panneerselvam, K. Muniraj, M. Thomas, N. Ravichandran, B. Bidorn, Identifying influencing groundwater parameter on human health associate with irrigation indices using the Automatic Linear Model (ALM) in a semi-arid region in India. Environ. Res. **202**, 111778 (2021b)

B. Panneerselvam, K. Muniraj, C. Pande, N. Ravichandran, Prediction and evaluation of groundwater characteristics using the radial basic model in semi-arid region, India. Int J Environ. Anal. Chem., 1–17 (2021c)

A.M. Piper, A graphic procedure in the geochemical interpretation of water analysis. Am. Geophys. Union Trans. **25**, 914–923 (1994)

H.M. Raghunath, *Groundwater* (Wiley Eastern, New Delhi, 1987a), p. 563

H.M. Raghunath, *Geochemical Survey and Water Quality, Groundwater* (Wiley Eastern Limited, New Delhi, 1987b), pp. 343–347

C.R. Ramakrishnaiah, C. Sadashivaiah, G. Ranganna, Assessment of water quality index for the groundwater in Tumkur Taluk, Karnataka State, India. J. Chem. **6**(2), 523–530 (2009). https://doi.org/10.1155/2009/757424

R. Rengarajan, A. Balasubramanian, Corrosion and scale formation characteristics of groundwater in and around Nangavalli, Salem District, Tamil Nadu. J. Appl. Hydrol. **3**, 15–22 (1990)

L.A. Richards, Diagnosis and Improvement of Saline and Alkaline Soils. U.S. Department of Agriculture Hand Book (1954)

R. Saleem, Groundwater management—emerging challenges. Water Digest., 12–28 (2007)

S. Salehi, M. Chizari, H. Sadighi, M. Bijani, Assessment of agricultural groundwater users in Iran : A cultural environmental bias. Hydrogeol. J. **26**, 285–295 (2018). https://doi.org/10.1007/s10040-017-1634-9

P. Samantray, B.K. Mishra, C.R. Panda, S.P. Rout, Assessment of water quality index in Mahanadi and Atharabanki Rivers and Taldanda Canal in Paradip Area, India. J. Hum. Ecol. **26**(3), 153–161 (2009). https://doi.org/10.1080/09709274.2009.11906177

S. Selvakumar, K. Ramkumar, N. Chandrasekar, N.S. Magesh, S. Kaliraj, Groundwater quality and its suitability for drinking and irrigational use in the Southern Tiruchirappalli district, Tamil Nadu, India. Appl. Water Sci. **7**, 411–420 (2017). https://doi.org/10.1007/s13201-014-0256-9

S. Selvam, Use of remote sensing and GIS techniques for land use and land cover mapping of Tuticorin Coast, Tamilnadu. Univers. J. Environ. Res. Technol. **2**(4), 233–241 (2012)

Y. Shen, T. Oki, N. Utsumi, S. Kanae, N. Hanasaki, Projection of future world water resources under SRES scenarios: Water withdrawal. Hydrol. Sci. J. **53**(1), 11–33 (2008). https://doi.org/10.1623/hysj.53.1.11

K. Shunmugapriya, B. Panneerselvam, K. Muniraj, N. Ravichandran, P. Prasath, M. Thomas, K. Duraisamy, Integration of multi criteria decision analysis and GIS for evaluating the site suitability for aquaculture in southern coastal region, India. Mar. Pollut. Bull. **172**, 112907 (2021)

A. Siddiqui, S. Naseem, T. Jalil, Groundwater quality assessment in and around Kalu Khuhar, super highway, Sindh, Pakistan. J. Appl. Sci. **5**(7), 1260–1265 (2005). https://doi.org/10.3923/jas.2005.1260.1265

P. Simeonova, V. Simeonov, G. Andreev, Water quality study of the Struma River Basin, Bulgaria (1989–1998). Central Eur. Chem. **1**, 136–212 (2003)

C. Singaraja, S. Chidambaram, P. Anandhan, M.V. Prasanna, C. Thivya, R. Thilagavathi, A study on the status of fluoride ion in groundwater of coastal hard rock aquifers of South India. Arab. J. Geosci. (2012). https://doi.org/10.1007/s12517-012-0675-6

S. Singh, N.C. Ghosh, S. Gurjar, G. Krishan, S. Kumar, P. Berwal, Index-based assessment of suitability of water quality for irrigation purpose under Indian conditions. Environ. Monitor. Assess. **190**(29), 1–14 (2018). https://doi.org/10.1007/s10661-017-6407-3

K.K. Singh, G. Tewari, S. Kumar, Evaluation of groundwater quality for suitability of irrigation purposes: A case study in the Udham Singh Nagar, Uttarakhand. J. Chem., 1–15 (2020). https://doi.org/10.1155/2020/6924026

K. Srinivasamoorthy, M. Chidambaram, M.V. Prasanna, M. Vasanthavigar, A. John Peter, P. Anandhan, Identification of major sources controlling groundwater chemistry from a hard rock terrain— A case study from Mettur taluk, Salem district, Tamilnadu, India. J. Earth Syst. Sci. **117**(1), 49–58 (2008)

K. Srinivasamoorthy, C. Nanthakumar, M. Vasanthavigar, K. Vijayaragavan, R. Rajivgandhi, S. Chidambaram, P. Anandhan, R. Manivannan, S. Vasudevan, Groundwater quality assessment from a hard rock terrain, Salem district of Tamilnadu, India. Arab. J. Geosci. **4**, 91–102 (2011). https://doi.org/10.1007/s12517-009-0076-7

I. Szabolcs, C. Darab, The Influence of Irrigation Water of High Sodium Carbonate Content of Soils. Proceedings of 8th International Congress of ISSS **2**, 803–812 (1964)

R. Thapa, S. Gupta, D.V. Reddy, H. Kaur, An evaluation of irrigation water suitability in the Dwarka river basin through the use of GIS-based modelling. Environ. Earth Sci. **76**(14), 1–12 (2017). https://doi.org/10.1007/s12665-017-6804-5

USSL, *Diagnosis and Improvement of Saline and Alkali Soils. United States Development Agency Handbook 60* (Government Printing Office, Washington, DC, 1954), p. 147

M. Vadiati, A. Asghari-Moghaddam, M. Nakhaei, J. Adamowski, A.H. Akbarzadeh, A fuzzy-logic based decision-making approach for identification of groundwater quality based on groundwater quality indices. J. Environ. Manag. **184**, 255–270 (2016). https://doi.org/10.1016/j.jenvman.2016.09.082

A. Vasanthakumari Sivasankara Pillai, C. Sabarathinam, T. Keesari, T. Chandrasekar, T. Rajendiran, V. Senapathi, P.M. Viswanathan, D.R. Samayamanthu, Seasonal changes in groundwater quality deterioration and chemometric analysis of pollution source identification in South India. Environ. Sci. Pollut. Res. **27**, 20037–20054 (2020). https://doi.org/10.1007/s11356-020-08258-6

M. Vasanthavigar, K. Srinivasamoorthy, K. Vijayaragavan, R. Rajiv Ganthi, S. Chidambaram, P. Anandhan, R. Manivannan, S. Vasudevan, Application of water quality index for groundwater quality assessment: Thirumanimuttar sub-basin, Tamilnadu, India. Environ. Monit. Assess. **171**, 595–609 (2010). https://doi.org/10.1007/s10661-009-1302-1

WHO, *Guidelines for Drinking-Water Quality*, 4th edn. (World Health Organization, Geneva, 2011)

L.V. Wilcox, Classification and use of irrigation water. U.S. Geological Department Agri. Arc. **969**, 19 (1955)

World Bank, Deep Wells and Prudence: Towards Pragmatic Action for Addressing Groundwater Overexploitation in India. World Bank Report No. 51676, Washington, DC (2010)

J. Wu, P. Li, H. Qian, Hydrochemical characterization of drinking groundwater with special reference to fluoride in an arid area of China and the control of aquifer leakage on its concentrations. Environ. Earth Sci. **73**(12), 8575–8588 (2015). https://doi.org/10.1007/s12665-015-4018-2

S.R. Zaki, M. Redwan, A.M. Masoud, A. Ahmed, A. Moneim, Chemical characteristics and assessment of groundwater quality in Halayieb area, southeastern part of the Eastern Desert, Egypt. Geosci. J. **23**, 149–164 (2019). https://doi.org/10.1007/s12303-018-0020-5

Chapter 7
An Investigation of Groundwater Hydrochemistry in the Shallow Aquifers of the Cuddalore Coast, South India

R. Ayyandurai and S. Venkateswaran

Abstract Coastal aquifers are affected by different sources from both coast and land, which alter the quality of groundwater in nature. Hence, this study was conducted to investigate the groundwater hydrochemistry in the shallow aquifers of the Cuddalore coast, South India. Groundwater samples were collected from 66 bore wells (out of 132 specimens) with respect to 2 seasons (pre- and post-monsoon). The samples were analyzed by standard procedures, and the data were presented with ±5% error. Depending on the chemical concentration, the ions were arranged based on the box plot in the following order: $Na^+ > Ca^{2+} > Mg^{2+} > K^+$ and $Cl^- > HCO_3^- > SO_4^{2-} > NO^{3-} > PO_4^{3-} > Br^- > F^-$ for the pre-monsoon and $Na^+ > Ca^{2+} > K^+ > Mg^{2+}$ and $Cl^- > HCO_3^- > SO_4^{2-} > NO^{3-} > PO_4^{3-} > Br^- > F^-$ for the post-monsoon seasons. The Hill–Piper plot demonstrated that most of the groundwater samples fell under the seawater intrusion (Na–Cl) category during the pre-monsoon rather than the post-monsoon season due to rainfall recharge. Multivariate statistical results indicated the different hydrogeochemical sources in the study area, which are mainly seawater intrusion, rock–water interaction, and anthropogenic sources. Ionic ratio plots were used to discover the hydrogeochemical evolution and ion exchange processes in nature. The hydrochemistry of the groundwater samples varied with respect to the monsoon due to its diversity in processes and activities.

Keywords Groundwater · hydrogeochemistry · Hill–Piper · statistics and ionic ratio plots

---

R. Ayyandurai (✉) · S. Venkateswaran
Department of Geology, Periyar University, Salem, Tamil Nadu, India

B. Panneerselvam et al. (eds.), *Climate Change Impact on Groundwater Resources*, https://doi.org/10.1007/978-3-031-04707-7_7

7.1 Introduction

Groundwater is the world's most important source of water for domestic, industrial, and agricultural purposes (Jesudhas et al. 2017, 2021; Shunmugapriya et al. 2021; Panneerselvam et al. 2020, 2021). It is estimated that one-third of the world's population uses groundwater for drinking purposes to meet their drinking needs (Nickson et al. 2005). It was found in the capacity of those reservoirs to store and circulate significant amounts of water for drinking and irrigation purpose (Todd 1980). In general, the quality of groundwater depends on the precipitation of recharged water, inland surface water, and groundwater chemistry processes (Twarakavi and Kaluarachchi 2006; Kumar et al. 2008). Groundwater quality models in coastal areas are complex due to inputs from various water sources, including rainfall, seawater, deep groundwater, and anthropogenic resources (Steinich et al. 1998). Problems in coastal areas generally contaminate freshwater resources, including salt water and well areas. Assessing groundwater quality provides scientifically rational judgment and evidence to play a key role in environmental management, monitoring and appropriate use and protection in product production (Karro et al. 2004). Such as natural processes, human activities, urbanization, and agricultural activities are the main causes of seawater intrusion and water turbulence, declining quality in coastal aquifers (Mondal et al. 2011a, Selvam et al. 2013). Complete dependence on groundwater for drinking, agricultural and industrial sector processes, and the diversity of climatic and natural phenomena all pollute coastal groundwater (Srinivas et al. 2013).

7.2 Background of This Study

In addition, the chemistry of groundwater and the various geochemical processes that regulate the properties of groundwater have been well documented by many studies worldwide. Moreover, hydrochemistry maps are potentially useful in depicting water resources and their history (Elango et al. 2003). It is therefore important to understand their hydrochemistry and pollution effects to assess groundwater. Many researchers have tried to identify the process of salinization of coastal groundwater. Mandal et al. (2011 b) identified a saltwater mixture using ionic ratio plots in a coastal reservoir in Tuticorin in southern India. Ghiglieri et al. (2010) identified salinization sources in coastal carbonate aquifers in Italy and groundwater chemistry regulated by seawater mixing and cation conversion processes. Separate saline infiltration and water processes in the coastal waters of Greece using geochemical modelling indicate water quality variation in flow paths due to their chromatographic origin. Wang and Jiao (2012) found salinity and subsequent freshness in limited quaternary aquifers in the Pearl River Delta through China due to the combination of seawater and aquaculture recharge rainwater. Using geophysical and hydrochemical approaches, Praveena et al. (2010) identified groundwater salinity sources in Malaysian coastal reservoirs. Other sources of groundwater salinization have been attempted by Srinivasamoorthy et al. (2011) Coastal Tamil Nadu, the Cuddalore region of India and revealed that the major factors responsible for salinization of the source of groundwater are sewage dispsoal, agricultural activities and groundwater overextraction. Giménez-Forcada (2010) attempted to identify the seawater interface using a hydrochemical facies evolution diagram and then the freshwater affected by sedimentation. The region is highly dependent on groundwater for drinking and agricultural and industrial purposes, and various climatic and natural phenomena cause coastal groundwater pollution (Srinivas 2015). Cuddalore is a major industrial city of Tamil Nadu. The Cuddalore district, covered with alluvial sediments of river basins, has unorganized structures in its coastal and deltaic areas. Deeply exploited groundwater has disturbed the natural balance between groundwater and salt water. This can lead to drainage of saltwater resources and progressive pollution of groundwater. Understanding the evolution of groundwater salinity can help improve hydrochemical processes that promote the development and management of sustainable water resources.

7.3 Objectives of This Study

The aim of this study is to test groundwater and identify its purpose and major geological features. It also aims to explain the factors that affect the process. For this purpose, we conducted and integrated a regional hydrochemical survey. The applied Hill–Piper plot diagrams, statistics, and description ratios on the hydrochemical datasets obtained from the ionic order illustrate the effect of water pollution on the plot to assess its suitability in terms of water quality as well as for a better understanding and future management perspectives.

7.4 Materials and Methods

7.4.1 Study Area

Cuddalore is a large industrial city in the southern Indian state of Tamil Nadu, which has experienced rapid coastal growth. The coast of Cuddalore stretches from the Gadilam River mouth in the north to the Pichavaram mangrove forests in the south, and the study area consists of three major rivers, the Ponnaiyar, the Vellar, and the Coleroon (Fig. 7.1). The Cuddalore district is covered with alluvial sediments of river valleys and coastal and deltaic areas and has unorganized structures. The study area ranges from 117°12′20.67″ to 11°52′48.79″ N latitude and 79°29′25.06″ to 79°48′41.70″ E longitude with an area of 1772.4 km^2. The study area, with an average elevation of 1 m (3 feet) above sea level, has fewer relief features. The annual

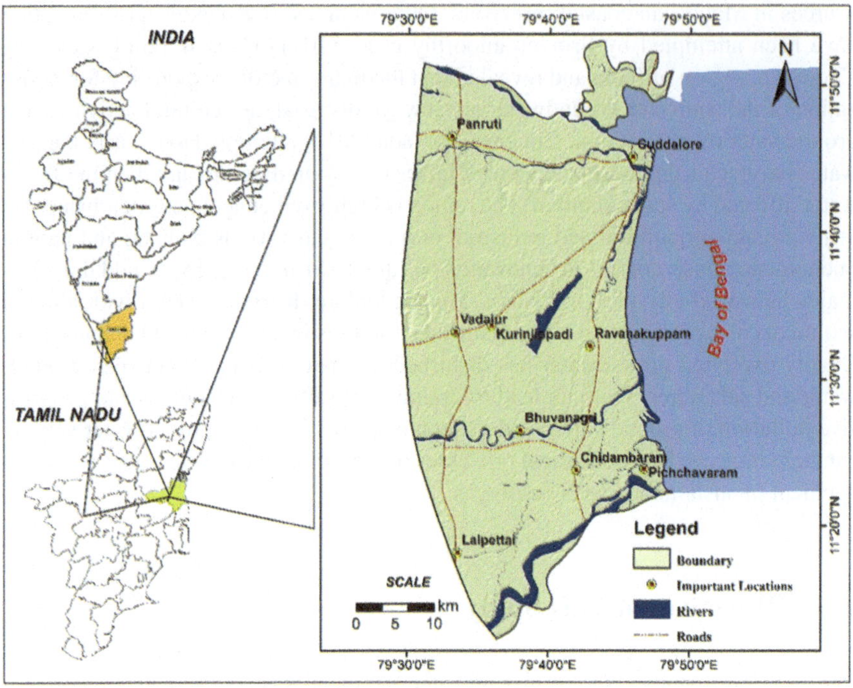

Fig. 7.1 Study area map

average rainfall of the study area is 1446.3 mm/year, and the highest rainfall (80% of the total rainfall) occurs during the northwest monsoon as it is located along the Bay of Bengal.

7.4.2 *Sample Collection and Field Site*

Groundwater samples were collected over two different seasons to understand the hydrochemical properties and evolution. This pattern occurred in the year 2017, which represents the pre- and post-monsoon seasons of 2018 (Fig. 7.2). Groundwater samples were collected from 66 bore wells (out of a total of 132 specimens) for 2 seasons (before and after the monsoon). Groundwater samples were collected in low-density polythene bottles, followed by air ventilation of the wells for 10–15 min to remove accumulated water. Prior to sampling, the bottles were thoroughly cleaned two to three times with groundwater samples. Electrical conductivity (EC) and pH were measured in the field using a portable conductivity and pH meter. The major ions were analyzed according to the American Public Health Association's (APHA 2012) standard procedures. Calcium (Ca^{2+}) and magnesium (Mg^{2+}) were analyzed by titration with ethylenediaminetetraacetic acid (EDTA) as the standard and

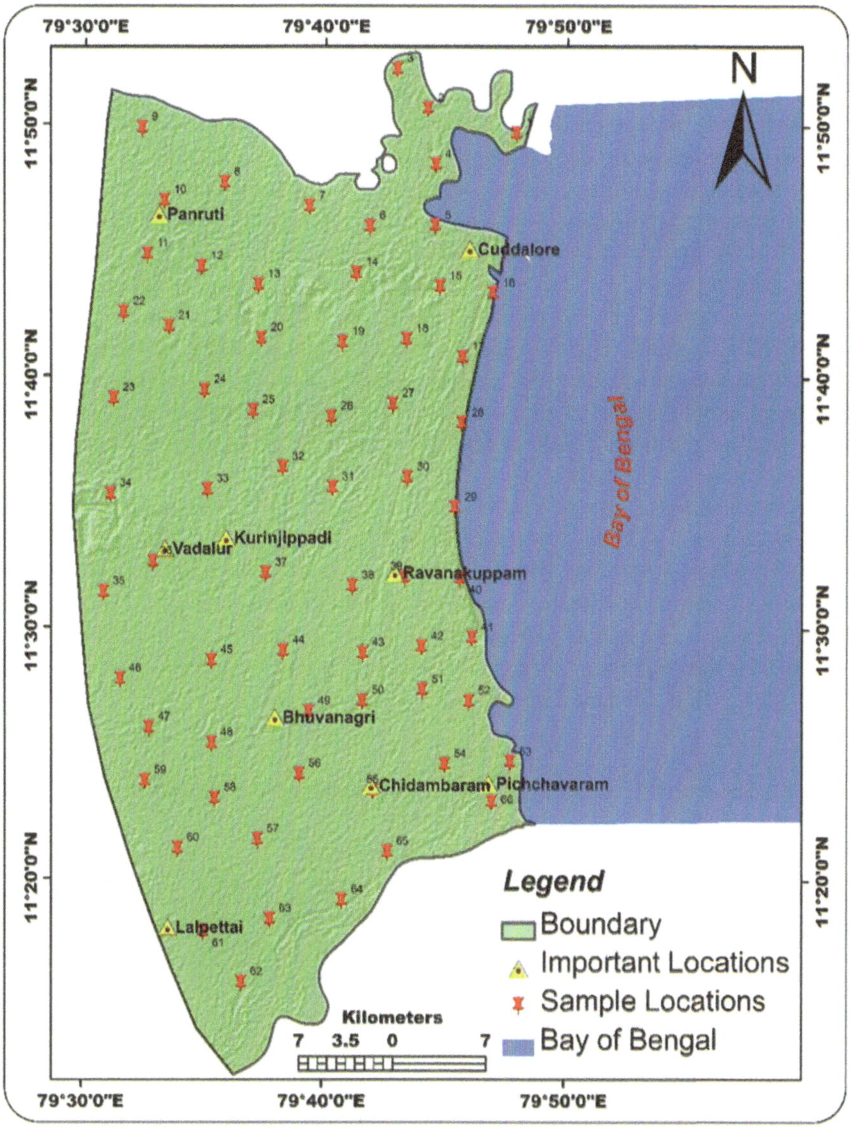

Fig. 7.2 Groundwater location map

chloride (Cl^-) and bicarbonate with silver nitrate ($AgNO_3$) and hydrochloric acid (HCl) were used as the standard for HCO_3^-. Sulphur (SO_4^{2-}), nitrate (NO_3^-), phosphate (PO_4^{3-}), fluoride (F^-), bromide (Br^-), and silica (SiO_2) were analyzed by a spectrophotometer and sodium (Na^+) and potassium (K^+) by flame photometric and colorimetric methods. Data were cross-checked by charge error balance, with an error of within 5%.

7.5 Results and Discussion

Statistical analysis of the hydrochemical composition of the groundwater samples was performed, and the results are shown in Table 7.1. Concentrations of important chemical parameters are depicted in the box and whisker plots (Figs. 7.3 and 7.4). Large standard deviations occur for a wide range of parameters before the monsoon. The chemical compositions of ground water is representative of valuable process that affect the quality of groundwater in the specific region. The pre-monsoon EC values ranged from 1224.22 to 8300.93 μS/cm, with an average of 2919.12 μS/cm, and the monsoon EC values ranged from 1172.36 to 6983.533 μS/cm, with an average of 2356.32 μS/cm. High EC levels are mainly due to specific processes such as the flow of pollutants from saline sources, mineral dissolution, and human activity. The study found that the pre-monsoon levels of total dissolved solids (TDSs) in the study area ranged from 722.29 to 5063.57 mg/l, with an average of 1770.50 mg/l over the course of the study.

TDSs (total dissolved solids) after monsoon in the study area ranged from 692.42 to 4614.28 mg/l, with an average of 1664.65 mg/l. A general increase in TDSs (>1000 mg/l) was observed mainly along the coasts of the study area due to the activities of semi-arid climatic conditions, a deep groundwater table, and natural/ human activities (Richter and Kreitler 1993). The dominance of cations of the study area is essential: $Na^+ > Ca^{2+} > Mg^{2+} > K^+$ and $Cl^- > HCO_3^- > SO_4^{2-} > NO_3^- > PO_4^{3-} > Br^- > F^-$ (Figs. 7.5 and 7.6).

Table 7.1 Statistics of the chemical parameters of groundwater samples during Premonsoon(PRM) and Post monsoon (POM)

	PRM				POM			
Parameter	MIN	MAX	AVG	SD	MIN	MAX	AVG	SD
pH	6.12	7.77	6.95	0.43	6.53	8.96	7.68	0.51
EC	1224.22	8300.93	2919.12	1538.42	1172.36	6983.53	2356.32	1246.78
TDS	722.29	5063.57	1770.50	946.40	692.42	4614.28	1664.65	1106.16
Salinity	816.19	5823.11	2013.42	1084.83	754.74	5121.85	1853.83	1231.97
Ca^{2+}	71.00	350.00	178.93	64.27	75.00	392.00	156.55	62.52
Mg^{2+}	11.36	120.00	43.51	23.33	11.00	101.00	32.00	17.62
Na^+	50.00	1266.00	315.06	291.67	30.00	1033.00	235.71	226.53
K^+	11.00	100.00	34.70	19.14	7.00	221.00	35.34	33.27
Cl^-	112.00	2307.21	624.20	538.99	82.00	2178.00	466.28	469.24
HCO_3^-	127.00	651.00	293.46	103.41	124.00	585.00	322.52	91.76
SO_4^{2-}	10.50	280.90	79.58	58.77	7.00	148.00	53.75	31.24
NO_3^-	22.00	140.00	64.13	27.43	10.00	107.00	48.27	21.35
PO_4^{3-}	9.60	151.00	60.52	27.72	5.76	96.00	36.00	24.79
F^-	0.22	1.40	0.72	0.32	0.04	1.20	0.48	0.29
Br^-	0.14	8.33	1.88	1.78	0.10	8.23	1.63	1.70

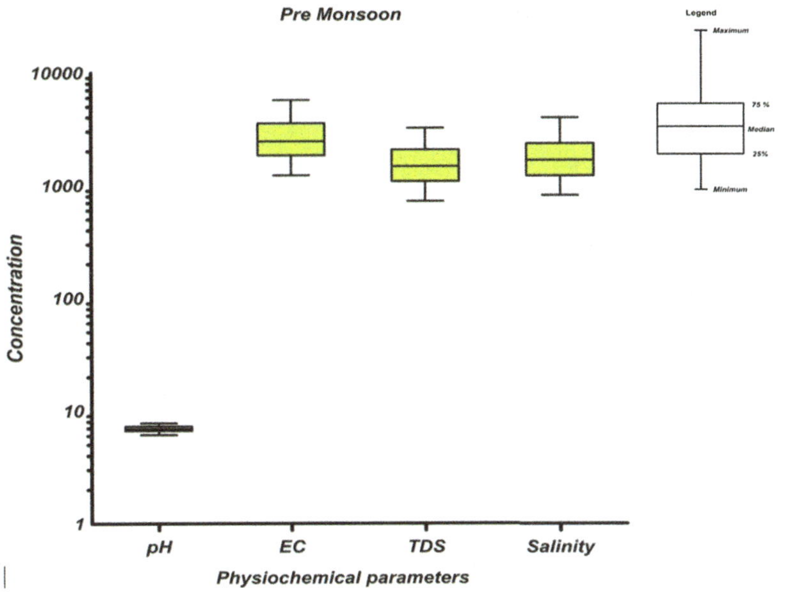

Fig. 7.3 Box and whisker plot for physical parameter during pre-monsoon

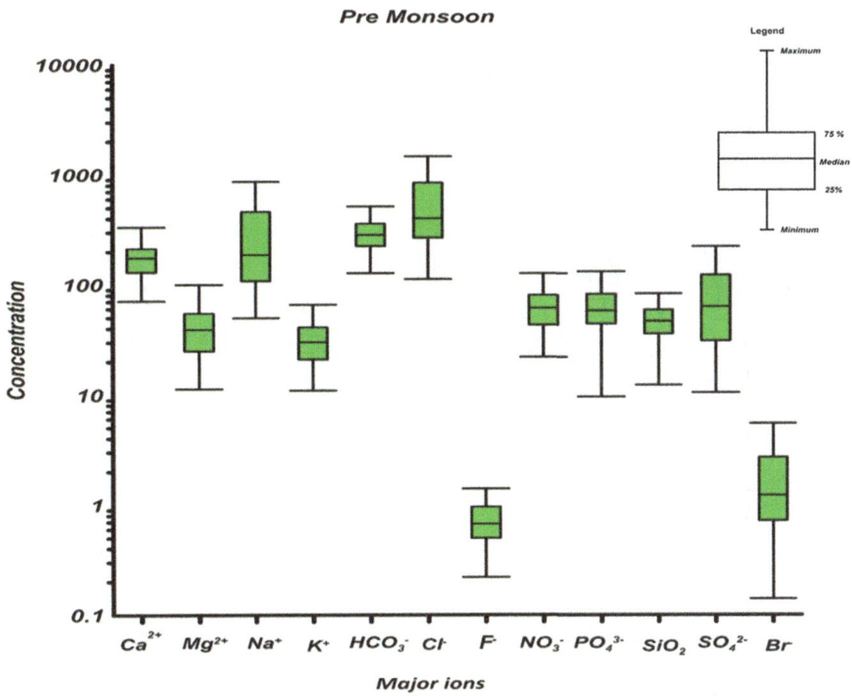

Fig. 7.4 Box and whisker plot for major ions during pre-monsoon

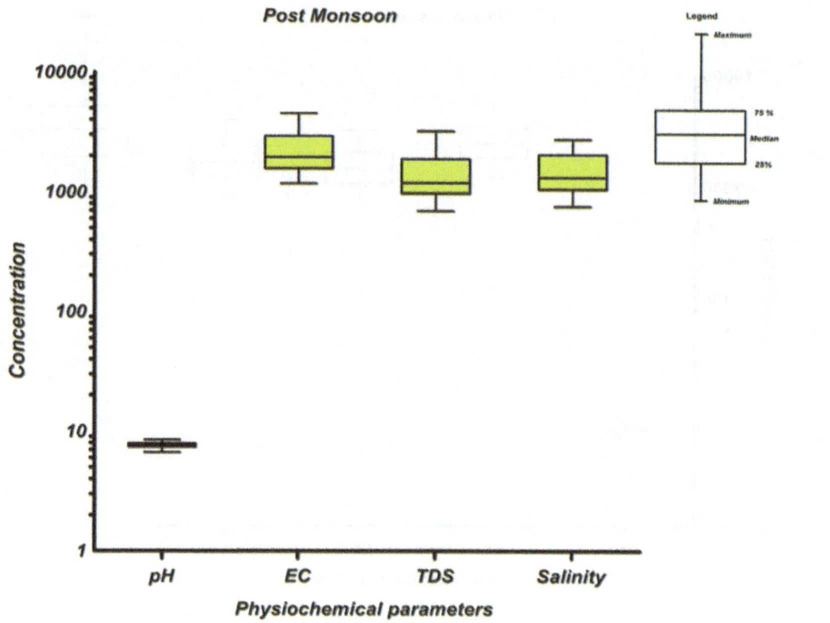

Fig. 7.5 Box and whisker plot for physical parameter during post monsoon

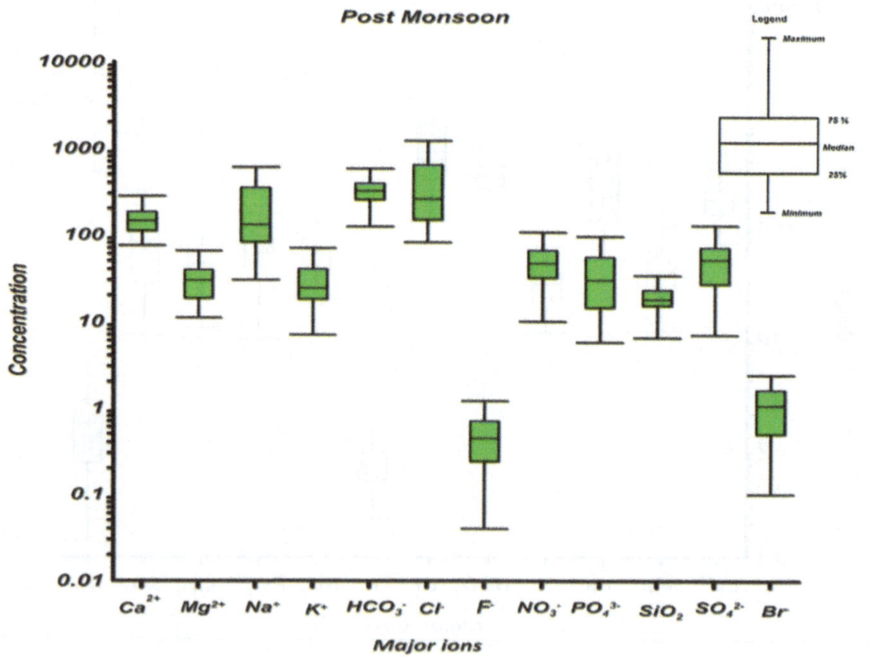

Fig. 7.6 Box and whisker plot for major ions during post monsoon

7.5.1 Hydrogeochemical Facies

The Piper diagram was introduced to investigate the groundwater chemistry and hydrogeochemical species in the study area. Hyperchemical features as well as the chemical evolution of groundwater can be understood with a help of the Piper trilinear diagram (Piper 1944). The diagram shows the chemical relationship between groundwater samples. It represents the cations and ions in various ternary plots, and the ternary plots are extrapolated to a diamond field that detects different types of water (Panneerselvam et al. 2021). Hydrochemical phase analysis is a valuable tool for establishing the flow pattern and chemical evolution of groundwater mass (Srinivasamoorthy et al. 2013). Groundwater samples were spread within 3% of the $CaHCO_3$-type sample during the pre-monsoon period, 19.7% of the $CaHCO_3$-type sample after monsoon, and 37.9% of the CaCl-type water sample during pre-monsoon. About 51.5% of water CaCl type, 48.5% is Ca-Mg-Cl type during the pre-monsoon season and it indicates that groundwater recharge system in few locations increases the temporary hardness of water. Samples in the Na–Cl region represent salt water, whereas those in regions 5 and 6 represent a 12.12% and 7.57% post-monsoon mixture in the study area, respectively (Fig. 7.7).

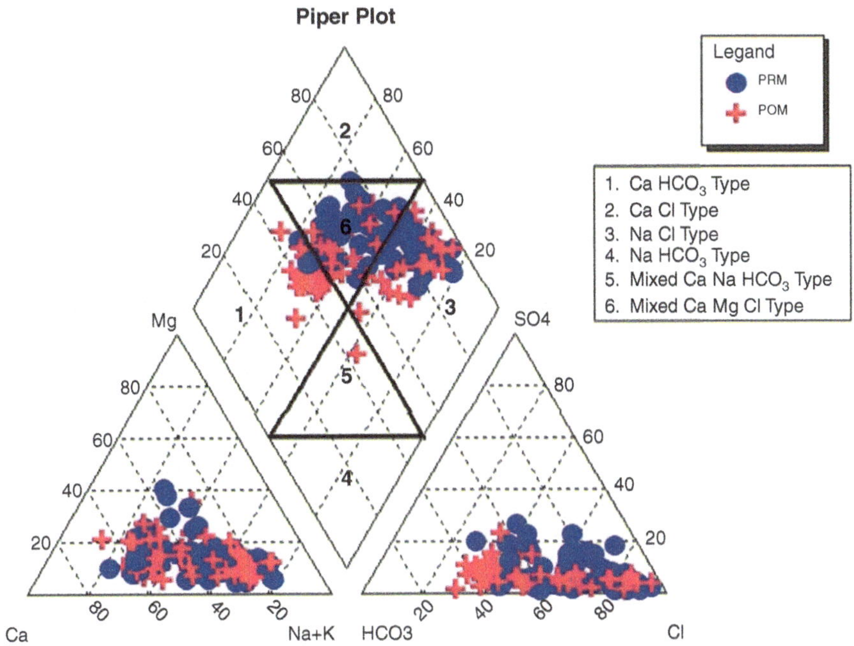

Fig. 7.7 Piper diagram

7.5.2 Gibbs Diagram

Gibbs diagram (Gibbs 1970) is a commonly used technique to establish the relationship between groundwater composition and aquifer lithological features. The boomerang-shaped Gibbs diagram shows three different regions of evaporation – crystallization, atmospheric/rock–water interactions, and precipitation dominance. The distribution of sample points indicates that the major ion chemistry of groundwater is regulated by the chemical environment and human activity that accelerates the rate of rock mineral formation. The Gibbs diagram is drawn separately for cations and anions by plotting the TDS on the logarithmic axis as opposed to the Gibbs ratio on the linear axis. The Gibbs ratio is obtained using [(Na + K)/(Na + K + Ca)] and ion [Cl/(Cl + HCO_3)] against TDS. All samples in the Gibbs plot study area are from rock domination. In alluvial plains, the rock–water interface is the main method of regulating groundwater chemistry.

Hydrogeochemical data from water samples before the monsoon are displayed in the Gibbs diagram (Fig. 7.8). Most specimens show rock dominance, whereas small representations of the groundwater specimens fall under the dominance of evaporation crystallization. The distribution of sample points indicates that the major ion chemistry of groundwater is regulated by the chemical environment and human activity that accelerates the rate of rock mineral formation. Groundwater is hydrogeochemical before and after the rainy season. However, some points fall into the evaporation zone, which indicates a small effect of evaporation in regulating the groundwater chemistry. It can be speculated that the shift of the sample points to the

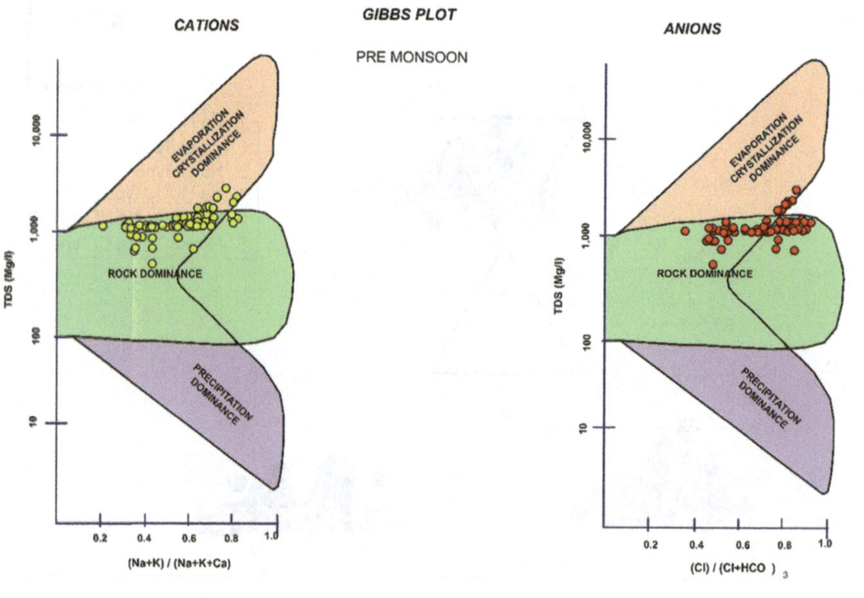

Fig. 7.8 Gibbs diagram for pre-monsoon season

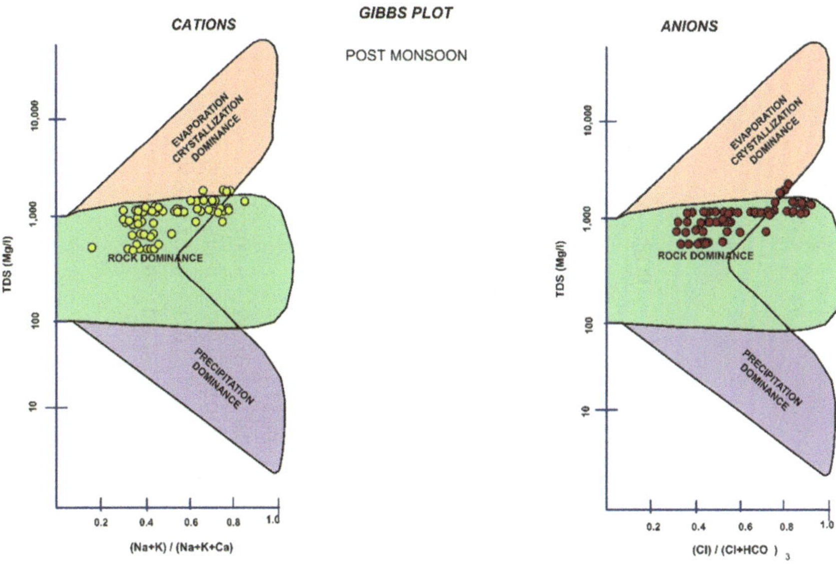

Fig. 7.9 Gibbs diagram for post monsoon season

dominance of evaporation may be due to the contamination of the salinity sources that release more Na^+ and Cl^- into the groundwater (Fig. 7.9).

7.6 Statistics Analysis

7.6.1 Correlation Analysis

Statistical analysis for the hydrochemical description of groundwater is generally applied in some manner. Important procedures are followed for groundwater quality assessment – one is cluster analysis and the other is factor analysis. Hierarchical cluster analysis is an excellent tool for determining groundwater samples. Hydrochemical data collection is subjected to multivariate statistical analysis such as Pearson's correlation and core component analysis for better understanding. Pearson's correlation matrix of different parameters of groundwater samples over two periods is calculated and displayed in Tables 7.2 and 7.3. Post-monsoon models show positive correlations of TDSs, salinity, Ca^{2+}, Mg^{2+}, Na^+, K^+, Cl^-, and Br^- with the effects of multiple processes such as ion exchange and ionic ratios using different combinations of chemical components.

The most positive correlation between EC and TDSs before and after the monsoon is that groundwater contains only charge-soluble components. A high correlation between the various components indicate the mixing of salts with the other water quality parameters along with the coast minerals present in the region. Factor

Table 7.2 Statistics of correlation analysis during PRM

Correlation analysis: pre-monsoon																
	pH	EC	TDS	Salinity	Ca^{2+}	Mg^{2+}	Na^+	K^+	HCO_3^-	Cl^-	F^-	NO_3^-	PO_4^{2-}	SiO_2	SO_4^{2-}	Br^-
pH	1.00															
EC	0.102	1.00														
TDS	0.099	**0.999**	1.00													
Salinity	0.107	**0.998**	**0.999**	1.00												
Ca^{2+}	0.086	**0.697**	**0.703**	**0.701**	1.00											
Mg^{2+}	0.202	**0.656**	**0.657**	**0.660**	0.326	1.00										
Na^+	0.083	**0.974**	**0.974**	**0.974**	**0.563**	**0.591**	1.00									
K^+	0.009	**0.596**	**0.601**	**0.603**	0.420	**0.513**	**0.532**	1.00								
HCO_3^-	−0.077	0.332	0.325	0.323	**0.543**	0.291	0.165	0.274	1.00							
Cl^-	0.099	**0.974**	**0.977**	**0.976**	**0.621**	**0.642**	**0.985**	**0.558**	0.162	1.00						
F^-	0.166	−0.084	−0.075	−0.067	0.014	0.050	−0.081	−0.003	−0.051	−0.035	1.00					
NO_3^-	−0.010	0.044	0.028	0.028	0.133	−0.047	−0.029	0.248	0.177	−0.067	−0.153	1.00				
PO_4^{2-}	0.200	0.300	0.305	0.310	0.379	0.304	0.220	0.228	0.198	0.204	0.021	0.289	1.00			
SiO_2	−0.094	−0.117	−0.119	−0.119	−0.001	−0.041	−0.194	−0.087	0.149	−0.205	−0.302	0.092	0.145	1.00		
SO_4^{2-}	0.186	0.395	0.392	0.396	0.262	0.200	0.368	0.253	0.115	0.288	−0.202	0.043	0.303	0.063	1.00	
Br^-	0.220	**0.861**	**0.871**	**0.869**	**0.612**	**0.564**	**0.860**	0.419	0.205	**0.865**	−0.020	−0.111	0.365	−0.125	0.346	1.00

Table 7.3 Statistics of correlation analysis during POM

Correlation analysis: post-monsoon																
	pH	EC	TDS	Salinity	Ca^{2+}	Mg^{2+}	Na^+	K^+	HCO_3^-	Cl^-	F^-	NO_3^-	PO_4^{2-}	SiO_2	SO_4^{2-}	Br^-
pH	1.00															
EC	0.130	1.00														
TDS	0.151	**0.999**	1.00													
Salinity	0.150	**0.994**	**0.999**	1.00												
Ca^{2+}	0.083	**0.659**	0.480	0.474	1.00											
Mg^{2+}	0.114	**0.661**	**0.515**	**0.515**	0.243	1.00										
Na^+	0.121	**0.973**	**0.801**	**0.798**	**0.517**	**0.606**	1.00									
K^+	−0.014	**0.765**	**0.574**	**0.566**	0.351	**0.600**	**0.721**	1.00								
HCO_3^-	0.032	−0.223	−0.220	−0.225	0.135	−0.213	−0.328	−0.273	1.00							
Cl^-	0.115	**0.985**	**0.798**	**0.794**	**0.624**	**0.652**	**0.977**	**0.748**	−0.356	1.00						
F^-	0.181	0.158	0.172	0.174	−0.129	0.213	0.194	0.219	−0.070	0.163	1.00					
NO_3^-	0.235	0.302	0.295	0.288	0.106	0.278	0.276	0.404	−0.073	0.261	0.223	1.00				
PO_4^{2-}	−0.211	0.044	0.008	0.002	0.177	−0.055	0.011	0.006	0.105	−0.021	−0.168	0.068	1.00			
SiO_2	0.052	−0.145	−0.118	−0.115	−0.210	−0.130	−0.112	−0.175	0.095	−0.155	0.027	−0.212	−0.018	1.00		
SO_4^{2-}	0.124	0.442	0.294	0.296	0.293	0.388	0.407	0.378	−0.223	0.417	−0.030	−0.002	−0.131	−0.106	1.00	
Br^-	0.101	**0.886**	**0.723**	**0.723**	**0.629**	**0.540**	**0.877**	**0.640**	−0.309	**0.903**	0.192	0.271	−0.032	−0.190	0.354	1.00

Table 7.4 Statistics of the principal component analysis during PRM

Principal component analysis: pre-monsoon			
	Factors		
Parameters	I	II	III
pH	0.058	−0.116	−0.200
EC	**0.981**	0.125	0.050
TDS	**0.985**	0.119	0.043
Salinity	**0.984**	0.119	0.038
Ca^{2+}	**0.651**	0.482	−0.026
Mg^{2+}	**0.668**	0.139	−0.111
Na^{+}	**0.975**	−0.059	0.038
K^{+}	**0.598**	0.380	−0.067
HCO_3^-	0.265	**0.713**	0.020
Cl^-	**0.986**	−0.047	−0.037
F^-	−0.069	0.039	−0.845
NO_3^-	−0.093	**0.674**	0.136
PO_4^{2-}	0.207	0.487	0.103
SiO_2	−0.198	0.266	**0.643**
SO_4^{2-}	0.358	−0.010	0.450
Br^-	**0.880**	−0.025	0.006
Total variance	**7.12**	**1.73**	**1.43**
% of variance	**44.50**	**10.83**	**8.91**
Cumulative %	**44.50**	**55.33**	**64.24**

analysis results revealed that shallow reservoirs are highly affected due to presence of minerals in the aquifers and dissolution of ion into water.

7.6.2 *Factor Analysis*

Factor analysis is more conducive to finding the factors that affect groundwater quality. A varimax rotation process helps in easier understanding of the factors by increasing the difference between the variables. The first factor provides as much variation as possible in the dataset. The second factor provides as few residual variations as possible. Analysis is the main procedure to collect four factors from a rotating dataset according to varimax with Kaiser normalization. The results of the principal component analysis (PCA) are summarized in Table 7.4.

Before the monsoon, factor I shows a 44.50% difference with high loads of EC, TDSs, salinity, Ca^{2+}, Mg^{2+}, Na^+, K^+, Cl^-, and Br^-. A high positive loading also indicates that there is an absolutely long linear relationship between the parameters, demonstrating the impact of anthropogenic pollution by fertilizers. Potassium and silica are geographically formed by the chemical environment of silicate-bearing rocks, whereas some components of phosphate and sulphate can be formed by

Table 7.5 Statistics of the principal component analysis during PRM

Principal component analysis: post-monsoon			
	Factors		
Parameters	I	II	III
pH	0.112	0.303	**0.652**
EC	**0.978**	0.009	0.030
TDS	**0.880**	0.138	0.007
Salinity	**0.877**	0.136	0.012
Ca^{2+}	**0.662**	−0.343	−0.082
Mg^{2+}	**0.647**	0.164	0.160
Na^{+}	**0.951**	0.058	0.040
K^{+}	**0.737**	0.186	−0.030
HCO_3^{-}	−0.247	−0.039	−0.111
Cl^{-}	**0.970**	0.002	0.060
F^{-}	0.151	**0.723**	0.183
NO_3^{-}	0.264	**0.643**	−0.054
PO_4^{2-}	0.068	−0.050	−0.780
SiO_2	−0.078	0.055	0.016
SO_4^{2-}	0.436	−0.395	0.463
Br^{-}	**0.895**	0.015	0.049
Total variance	**6.91**	**1.41**	**1.39**
% of variance	**43.20**	**8.82**	**8.37**
Cumulative %	**43.20**	**52.02**	**60.39**

fertilizers. Factor II shows a 10.83% difference with a positive loading of HCO_3^- and NO_3^-, indicating mineral leaching and ion exchange. Factor III exhibits a difference of 8.91% with the highest weight of SiO_2^-, taken from the chemical environment of silicate-bearing rocks as part of the area. Synthetic fertilizers used in agriculture contain phosphorus (Srinivasamoorthy et al. 2014).

The groundwater parameters after monsoon (Table 7.5) simplify various issues without dropping data. Positive values or values close to zero indicate areas affected by the factor and negative values indicate areas not affected by the factor analysis process. The difference after monsoon is 60.39%. For factor I, EC shows a 44.50% difference in TDSs, salinity, Ca^{2+}, Mg^{2+}, Na^+, K^+, Cl^-, and Br^- . Moreover, a high positive loading means that the absolute long linear relationship between the parameters represents the anthropogenic effects, ion exchange and mineral dissolution, and representations of saltwater infiltration along the coast. Factor II shows a difference of 8.82% with a positive weighting of NO_3^- and PO_4^{2-}, which can be obtained from agricultural activities of the area. Synthetic fertilizers may be used for phosphate and sulphate cultivation. Factor III contributes to 8.37% variability, with the highest weight of pH, indicating the process of natural accumulation in the study area.

7.6.3 Ionic Ratio Plots

A specific ionic ratio was used to determine the potential provided by seawater intrusion and salt sources (Vengosh and Rosenthal 1994). This study calculates the sum of ionic ratios of Ca^{2+} and Mg^{2+} vs Ca^{2+}, Mg^{2+}, Na^{+}, K^{+}, and Na^{+}/Cl^{-} vs EC, Na^{+} vs Cl^{-} and Ca^{2+}, and Mg^{2+} vs $HCO_3^{-} + SO_4^{2-}$ ions (Fig. 7.10) of the major ions. Groundwater quality data were analyzed to classify seawater intrusion and calculate different ionic ratios to isolate the source of salinity in Cuddalore's coastal reservoirs. Ca^{2+} and Mg^{2+} were plotted against Ca^{2+}, Mg^{2+}, Na^{+}, and K^{+} to calculate the ionic ratios after monsoons; Ca^{2+}, Mg^{2+}, Na^{+}, and K^{+} showed natural processes in the main study area (Fig. 7.11). Ca^{2+} and Mg^{2+} refer to the conversion of fresh groundwater into salt water in coastal reservoirs. Na + K + The major ions in seawater are Na + and K + – ions and Na ions (Appelo and Postma 2005). Pre-monsoon calculated and planned Na^{+}/Cl^{-} vs ECs are good proxies for (Cl^{-}) chloride (Fig. 7.12), which is used in seawater intrusion (Batayneh et al. 2013; Nwankwoala and Udom 2011; Lee and Song 2006). The ratio of Cl to seawater was greater than 0.0032 in almost all samples analyzed. However, this proportion of specimens is gradually increasing, with the prevalence of pre-monsoon prevalence in the region, the proportion of high-level seawater (Fig. 7.13). This may be due to the fresh recharge of groundwater in coastal reservoirs due to recharging of groundwater after the monsoon season. The ratio of Na^{+} to Cl^{-} (Fig. 7.14) shows samples with high Na, indicating the ion exchange process, and samples with high

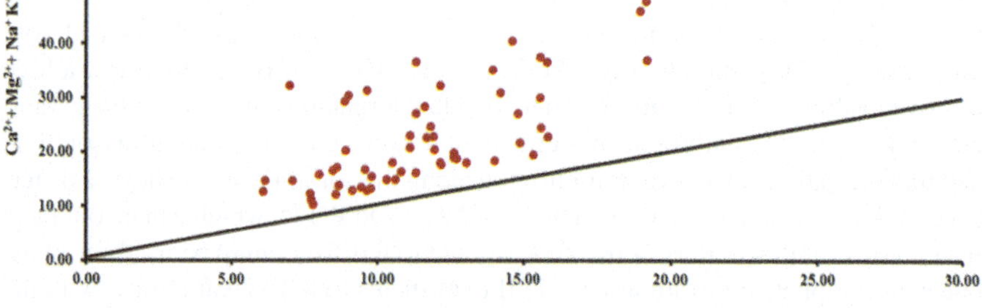

Fig. 7.10 Ionic ratio plots of $Ca^{2+} + Mg^{2+}$ meq/l vs $Ca^{2+} + Mg^{2+} + Na^{+}$ K^{+} meq/l PRM

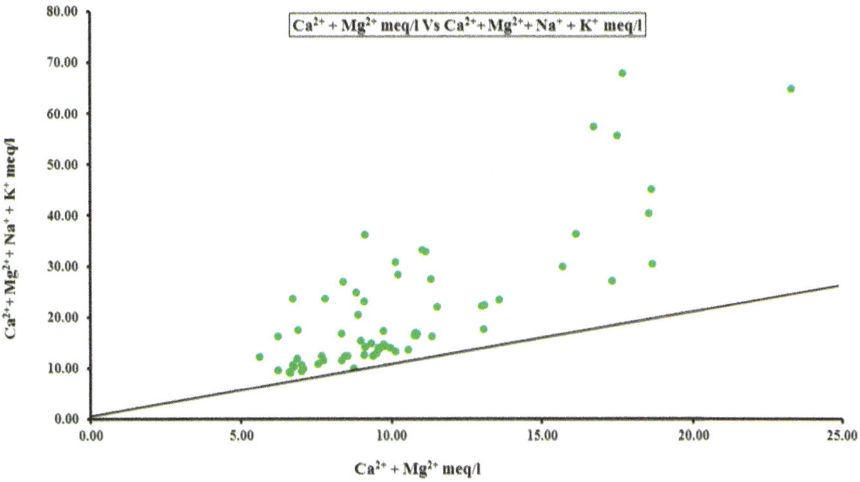

Fig. 7.11 Ionic ratio plots of $Ca^{2+} + Mg^{2+}$ meq/l vs $Ca^{2+} + Mg^{2+} + Na^{+} + K^{+}$ meq/l POM

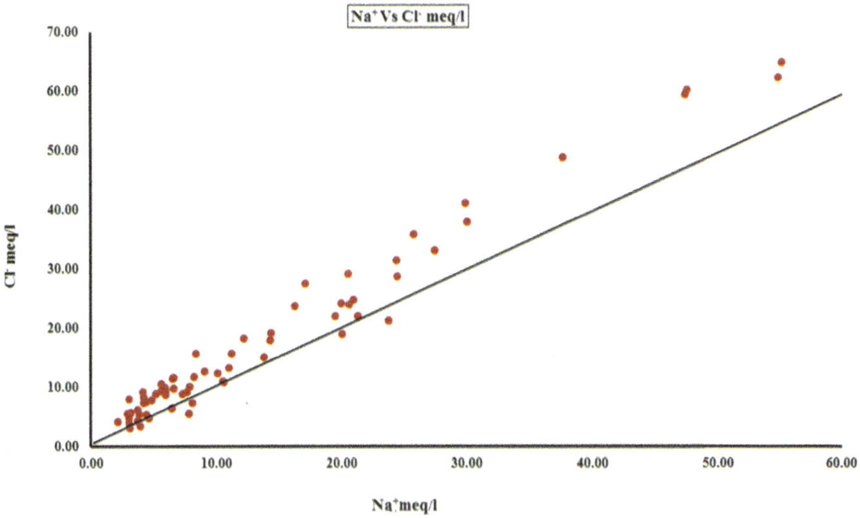

Fig. 7.12 Ionic ratio plots of Na^{+}/Cl^{-} (meq/l) vs EC (micro S/cm) during PRM

chloride ions, indicating the dominance of each saltwater infiltration during the rainy season.

The chemical composition of calcium–magnesium minerals from the geological components of the study area shows a pattern indicating a specific water infiltration, and the samples below the conventional mixing line indicate the atmosphere of the calcium–magnesium minerals.

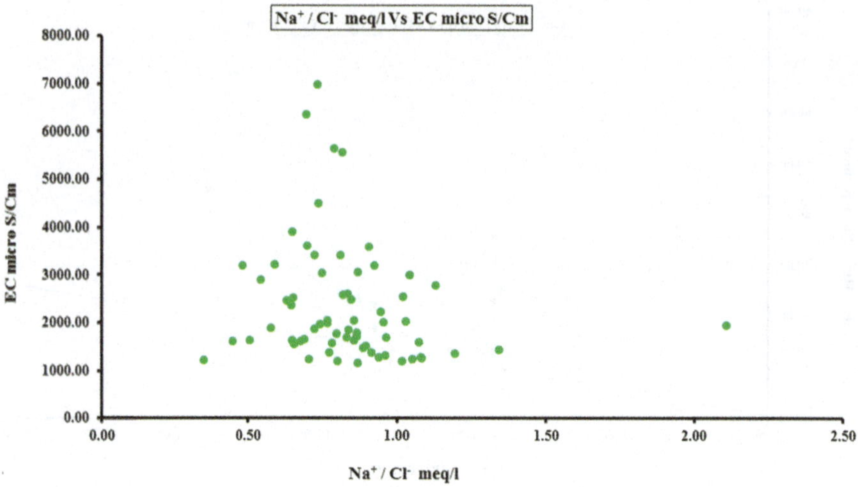

Fig. 7.13 Ionic ratio plots of Na^+/Cl^- (meq/l) vs EC (micro S/cm) during POM

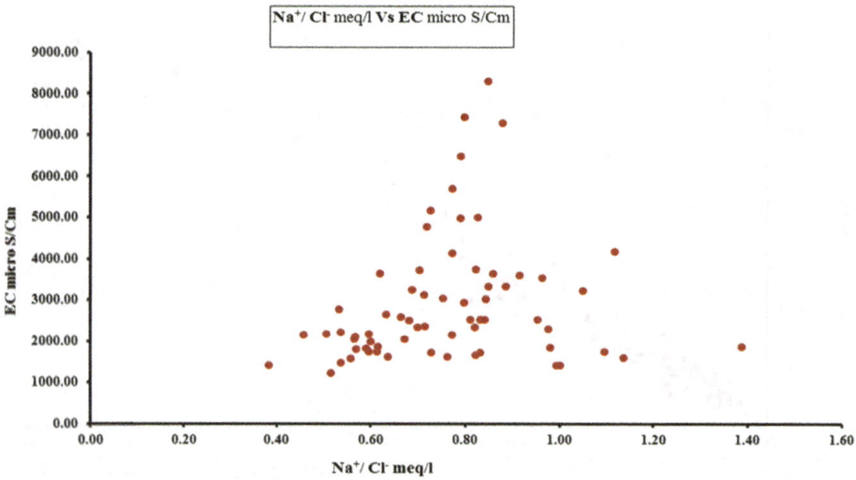

Fig. 7.14 Ionic ratio plots of Na^+ vs Cl^- meq/l PRM

The chemical composition of groundwater, including other groundwater samples, is characterized by the major inorganic compounds Ca^{2+} + Mg^{2+} vs HCO_3^- + SO_4^{2-}, the result of the natural chemical atmosphere, seawater infiltration, and atmospheric and anthropogenic inputs (Fig. 7.15). Concentration is divided into three partial contributions for each compound, one related to the atmospheric reactions to seawater intrusion. This ratio of patterns gradually approaches the

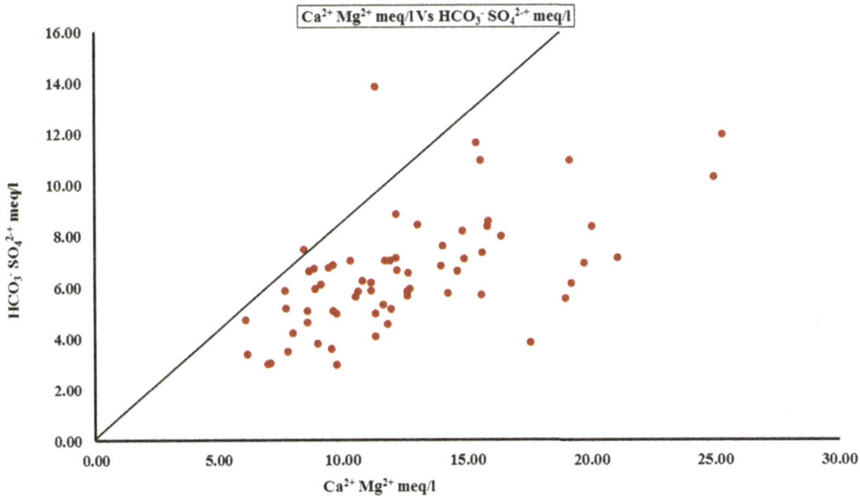

Fig. 7.15 Ionic ratio plots of Ca^{2+} + Mg^{2+} (meq/l) vs HCO_3^- + SO_4^{2-} (meq/l) during PRM

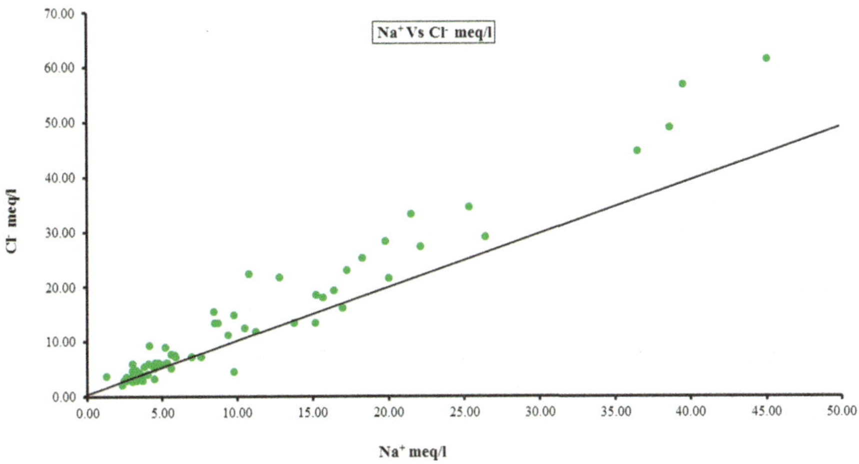

Fig. 7.16 Ionic ratio plots of Na^+ vs Cl^- meq/l POM

proportion of seawater with increasing levels (Fig. 7.16), indicating seawater intrusion before the rainy season. Ca^{2+} + Mg^{2+} vs HCO_3^- + SO_4^{2-} refers to post-groundwater recharge. The mixed nature of $HCO_3^-$and SO_4^{2-}groundwater is influenced by human activities and/or the cation exchange process (Fig. 7.17). Groundwater is confined to nearby coastal areas.

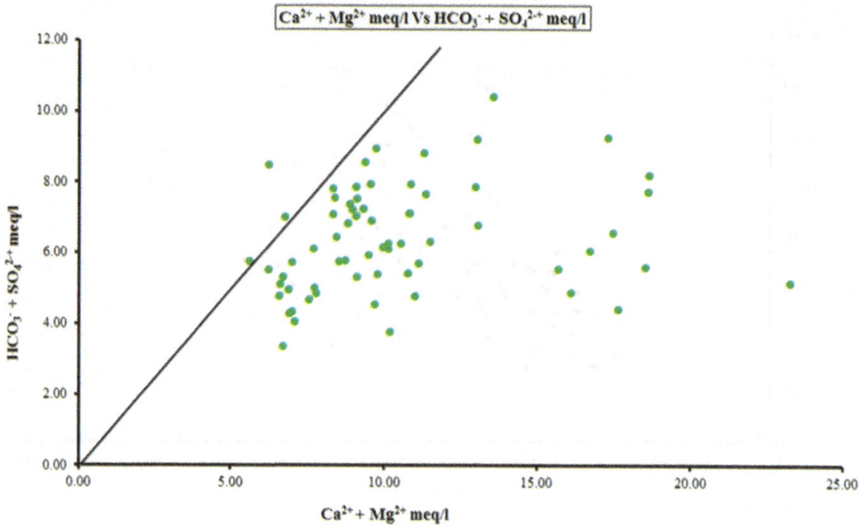

Fig. 7.17 Ionic ratio plots of Ca^{2+} + Mg^{2+} (meq/l) vs HCO_3^- + SO_4^{2-} (meq/l) during POM

7.7 Conclusions

The aim of this study is to combine hydrochemical and statistical methods to understand saltwater intrusion in Cuddalore's coastal aquifers. The subsequent are the findings of the study area. The groundwater in the study area shows an acidic to alkaline nature with a maximum pH of 8.96 after the rainy season. The seawater mixing index is more consistent with the groundwater and seawater mix. The analysis, 48.5% of the samples during pre monsoon and 52.8% of the sample during post monsoon are type of mixed Ca, Mg, Cl, groundwater in the study area. A correlation analysis establishes the positive correlation of EC with Cl^- (pre-monsoon = 0.974; post-monsoon = 0.985), Na (pre-monsoon = 0.974; post-monsoon = 0.973), Mg (pre-monsoon = 0.656; post-monsoon = 0.661), and Ca (pre-monsoon = 0.697; post-monsoon = 0.559), indicating the role of these parameters in regulating groundwater chemistry. The analysis collected three items with a difference of 44.50% before monsoon and 43.20% after monsoon. The first factor explains the effect of natural (silicate rock climate) and human resources (fertilizers) on the overall difference. The second item explains the effect of mineral leaching and ion exchange process of 10.83% and 8.82% on % difference before and after monsoons, respectively. The third factor identifies the effects of evaporation and high salinity in the study area by pre-monsoon, with a variability of 8.91%. Finally, the fourth factor reflects the effects of fertilizers on human wastes as well as the difference of 8.37% after monsoons. Ionic ratio plots reveal a water pattern affected by a combination of factors. The area studied by the ratio plot-based calculations is dominated by agricultural, household, and sewage wastes, which are influenced by saltwater infiltration, rock–water interaction, and ion exchange in flat areas.

Acknowledgements The authors express their gratitude to the University Grants Commission (UGC) and DST-FIST for the necessary financial support to carry out this study.

References

APHA, *Standard Methods for the Examination of Water and Wastewater. American Public Health Association*, 23rd edn. (Washington, DC, 2012)

C.A.J. Appelo, D. Postma, *Geochemistry, Groundwater and Pollution*, 2nd edn. (AA Balkema, Rotterdam, 2005), 649 pp

A. Batayneh, H. Zaman, T. Zumlot, H. Ghrefat, S. Mogren, Y. Nazzal, E. Elawadi, S. Qaisy, I. Bahkaly, A. Al-Taani, Hydrochemical facies and ionic ratios of the coastal groundwater aquifer of Saudi Gulf of Aqaba: Implication for seawater intrusion. J. Coast. Res. **30**(1), 75 (2013). https://doi.org/10.2112/jcoastres-d-13-00021.1

L. Elango, R. Kannan, M. Senthil Kumar, Major ion chemistry and identification of hydrogeochemical processes of groundwater in a part of Kancheepuram district, Tamil Nadu. Environ. Geosci. **4**, 157–166 (2003)

G. Ghiglieri, R. Balia, G. Oggiano, D. Pittalis, Prospecting for safe (low fluoride) groundwater in the Eastern African Rift: The Arumeru District (Northern Tanzania). Hydrol. Earth Syst. Sci. **14**, 1081–1091 (2010). https://doi.org/10.5194/hess-14-1081-2010

R.J. Gibbs, Mechanisms controlling world water chemistry. Science **170**(3962), 1088–1090 (1970)

E. Giménez-Forcada, Dynamic of sea water interface using hydrochemical facies evolution diagram. Ground Water **48**(2), 212–216 (2010)

C.J. Jesudhas, S.M. Chidambaram, K. Muniraj, Isolation of wells contaminated by tannery industries using principal component analysis. Arab. J. Geosci. **10**(14), 1–8 (2017). https://doi.org/10.1126/science.170.3962.1088

C.J. Jesudhas, A. Chinnasamy, K. Muniraj, A. Sundaram, Assessment of vulnerability in the aquifers of rapidly growing sub-urban: A case study with special reference to land use. Arab. J. Geosci. **14**(1), 1–20 (2021)

E. Karro, A. Marandi, R. Vaikma¨e, The origin of increased salinity in the Cambrian-Vendian aquifer system on the Kopli Peninsula, northern Estonia. Hydrogeol. J. **12**, 424–435 (2004)

M. Kumar, A.L. Ramanathan, A.K. Keshari, Understanding the extent of interactions between groundwater and surface water through major ion chemistry and multivariate statistical techniques. Hydrol. Process. **23**, 297–310 (2008)

J.-Y. Lee, S.-H. Song, Evaluation of groundwater quality in coastal areas: Implications for sustainable agriculture. Environ. Geol. **52**(7), 1231–1242 (2006). https://doi.org/10.1007/s00254-006-0560-2

N.C. Mondal, V.P. Singh, S. Singh, V.S. Singh, Hydrochemical characteristic of coastal aquifer from Tuticorin, Tamil Nadu, India. Environ. Monit. Assess. **175**(1–4), 531–550 (2011a)

B.C. Mondal, T. Mukherjee, L. Mandal, C.J. Evans, S.A. Sinenko, J.A. Martinez-Agosto, U. Banerjee, Interaction between differentiating cell- and niche-derived signals in hematopoietic progenitor maintenance. Cell **147**(7), 1589–1600 (2011b)

D. Nickson, C. Warhurst, E. Dutton, The importance of attitude and appearance in the service encounter in retail and hospitality. Manag. Serv. Qual. **15**(2), 195–208 (2005)

H.O. Nwankwoala, G.J. Udom, Investigation of hydrogeochemical characteristics of groundwater in Port Harcourt City, Nigeria: Implications for use and vulnerability. J. Appl. Sci. Environ. Manag. **15**(3), 479–488 (2011)

B. Panneerselvam, S. Karuppannan, K. Muniraj, Evaluation of drinking and irrigation suitability of groundwater with special emphasizing the health risk posed by nitrate contamination using nitrate pollution index (NPI) and human health risk assessment (HHRA). Hum. Ecol. Risk Assess. Int. J. **27**(5), 1324–1348 (2020)

B. Panneerselvam, K. Muniraj, M. Thomas, N. Ravichandran, GIS-based legitimatic evaluation of groundwater's health risk and irrigation susceptibility using water quality index, pollution index, and irrigation indexes in semiarid region, in *Groundwater Resources Development and Planning in the Semi-Arid Region*, (Springer, Cham, 2021), pp. 239–268

B. Panneerselvam, K. Muniraj, M. Thomas, N. Ravichandran, GIS-Based Legitimatic Evaluation of Groundwater's Health Risk and Irrigation Susceptibility Using Water Quality Index, Pollution Index, and Irrigation Indexes in Semiarid Region. In: Pande, C.B., Moharir, K.N. (eds) Groundwater Resources Development and Planning in the Semi-Arid Region. Springer, Cham. 239-268 (2021)

A.M. Piper, A graphic procedure in the geochemical interpretation of water analyses. Trans. Am. Geophys. Union **25**(6), 914 (1944). https://doi.org/10.1029/tr025i006p00914

S.M. Praveena, A.Z. Aris, M. Radojevic, Heavy metals dynamics and source in intertidal mangrove sediment of Sabah, Borneo Island. Environ. Asia **3**, 72–81 (2010)

B.C. Richter, W.C. Kreitler, *Geochemical Techniques for Identifying Sources of Groundwater Salinization* (CRC Press, New York, 1993) ISBN 1-56670-000-0

S. Selvam, I. Jeba, R. Dhana Mala, V. Muthukakshmi, A hydrochemical analysis and evaluation of groundwater quality index in Thoothukudi district, Tamilnadu, South India. Int. J. Adv. Eng. Appl. **2**(3), 25–37 (2013)

K. Shunmugapriya, B. Panneerselvam, K. Muniraj, N. Ravichandran, P. Prasath, M. Thomas, K. Duraisamy, Integration of multi criteria decision analysis and GIS for evaluating the site suitability for aquaculture in southern coastal region, India. Mar. Pollut. Bull. **172**, 112907 (2021)

N.R. Srinivas, Biochanin A: Understanding the complexities in the paradoxical drug–drug interaction potential. Eur. J. Drug Metab. Pharmacokinet. **40**, 119–125 (2015)

Y. Srinivas, D. Hudson Oliver, A. Stanley Raj, N. Chandrasekar, Evaluation of groundwater quality in and around Nagercoil town, Tamil Nadu, India: An integrated geochemical and GIS approach. Appl. Water Sci. **3**, 631–651 (2013)

K. Srinivasamoorthy, C. Nanthakumar, M. Vasanthavigar, et al., Groundwater quality assessment from a hard rock terrain, Salem district of Tamilnadu, India. Arab. J. Geosci. **4**(1), 91–102 (2011)

K. Srinivasamoorthy, M. Vasanthavigar, K. Vijayaraghavan, R. Sarathidasan, S. Gopinath, Hydrochemistry of groundwater in a coastal region of Cuddalore district, Tamilnadu, India: Implication for quality assessment. Arab. J. Geosci. **6**(2), 441–454 (2013)

K. Srinivasamoorthy, M. Gopinath, S. Chidambaram, M. Vasanthavigar, V.S. Sarma, Hydrochemical characterization and quality appraisal of groundwater from Pungar sub-basin, Tamilnadu, India. J. King Saud Univ. – Sci. **26**(1), 37–52 (2014). https://doi.org/10.1016/j.jksus.2013.08.001

B. Steinich, O. Escolero, L. Marin, Salt-water intrusion and nitrate contamination in the Valley of Hermosillo and El Sahuaral coastal aquifers, Sonora, Mexico. Hydrogeol. J. **6**, 518–526 (1998)

D.K. Todd, *Groundwater Hydrology*, 2nd edn. (Wiley, New York, 1980)

N.K.C. Twarakavi, J.J. Kaluarachchi, Arsenic in ground waters of conterminous United States: Assessment, health risk, and cost. J. Am. Water Resour. Assoc. **42**(2), 1–24 (2006)

A. Vengosh, E. Rosenthal, Saline groundwater in Israel: Its bearing on the water crisis in the country. J. Hydrol. **156**, 389–430 (1994)

Y. Wang, J.J. Jiao, Origin of groundwater salinity and hydrogeochemical processes in the confined quaternary aquifer of the Pearl River Delta, China. J. Hydrol. **438–39**, 112–124 (2012)

B. Panneerselvam, K. Muniraj, C. Pande, N. Ravichandran, Prediction and evaluation of groundwater characteristics using the radial basic model in semi-arid region, India. Int. J. Environ. Anal. Chem., 1–17 (2021)

Chapter 8
Evaluation of Groundwater Quality for Drinking Purposes Using the WQI and EWQI in Semi-Arid Regions in India

Shunmuga Priya Kaliyappan, Balamurugan Panneerselvam, Kirubakaran Muniraj, Gopinath Rajendran, Butsawan Bidorn, and Maciej Thomas

Abstract Groundwater is the major source of water for drinking, irrigation, and industrial purposes in the arid and semi-arid regions of South India. This study aims to evaluate the quality of groundwater for drinking purposes using water quality index (WQI) and entropy water quality index (EWQI) values. A total of 53 samples were collected from densely populated areas and industrial zones in the semi-arid region of Tamil Nadu, India. The collected samples were analyzed for basic groundwater parameters by following the standards recommended by the American Public Health Association (APHA, 2005). The results revealed that calcium, magnesium, chloride, and nitrate are the major ions that highly dominate the quality of groundwater in the study region. The WQI results indicated that around 10% of the sample locations were highly contaminated in both the pre- and post-monsoon seasons. The

S. P. Kaliyappan (✉)
Department of Civil Engineering, Nehru Institute of Technology, Coimbatore, India

B. Panneerselvam
Department of Construction Architecture and Environmental Engineering, University of Naples Federico II, Naples, Italy

K. Muniraj
Department of Civil Engineering, University VOC College of Engineering, Thoothukudi, India

G. Rajendran
Department of Civil Engineering, Hindustan Institute of Technology & Science, Chennai, India

B. Bidorn
Department of Water Resources Engineering, Chulalongkorn University, Bangkok, Thailand

M. Thomas
Cracow University of Technology, Faculty of Environmental Engineering and Energy, Cracow, Poland
e-mail: maciej.thomas@pk.edu.pl

B. Panneerselvam et al. (eds.), *Climate Change Impact on Groundwater Resources*, https://doi.org/10.1007/978-3-031-04707-7_8

EWQI values revealed that about 15% of the sample locations were polluted in both the pre- and post-monsoon seasons due to excessive concentrations of chloride and nitrate. This study identified the primary sources of contamination such as rock–water interaction, the nature of an aquifer, and other anthropogenic activities. The results help decision-makers enhance the quality of groundwater and reduce contamination in a specific region in order to maintain a sustainable practice in the study region.

Keywords Groundwater · Water quality index · entropy water quality index · anthropogenic activities

8.1 Introduction

Groundwater is the most valuable natural resource and is essential for all living organisms on the earth's surface. It is a more essential source of water for drinking and agricultural and industrial purposes (Abbasnia et al. 2019; Adimalla et al. 2019; Li et al. 2016). Rapid increases in industrialization, urbanization, and population density are the major threats that affect the quality of groundwater used for drinking and irrigation purposes. Assessment of groundwater quality for sutability of drinking and agriculture purpose is important research to understand the geochemical characteristics and chemical composition. The major factors that affect the quality of groundwater such as temperature, climatic condition, mineral dissolution, rock–water interaction, and anthropogenic activities (Panneerselvam et al. 2020a; Adimalla et al. 2019; Adimalla et al. 2019). Sources of groundwater have been identified almost everywhere on the earth's surface and not just in a single specified aquifer. The characteristics of groundwater mainly depend on the nature of soil layers, minerals present in the aquifer, and time for transfer of the earth layers. The basic information such as occurrence, recharge, and flow of groundwater has received more attention in arid and semi-arid regions due to inadequate rainfall, climatic conditions, high temperatures during the summer reason, insufficient surface water, and excess utilization of groundwater resources. The quality of groundwater is based on the recharge well, quality of rainfall, inland surface water, and the geochemical process of subsurface water. The seasonal changes of groundwater in arid and semi-arid regions highly depend on climatic conditions and rainfall (Aravinthasamy et al. 2019b; Shukla et al. 2020; Wagh et al. 2020).

However, the quality of groundwater for drinking and irrigation purposes is based on the locally available groundwater recharge system. Arid and semi-arid regions are more densely populated, and synthetic fertilizers are the major threat to groundwater quality (Adimalla et al. 2018a; Singh et al. 2020; Singh et al. 2020; Solangi et al. 2019; Verma et al. 2020a). A recent research related to the geochemical studies of groundwater has revealed a better understanding of the sources of groundwater contamination and the possible chemical reaction that affects the quality of groundwater and also the developmental system of groundwater quality.

Panneerselvam et al. (2020b) conducted a geographic information system (GIS)-based investigation of the hydrochemical characteristics of groundwater in a hard rock region in southern Tamil Nadu, India, and found that physical processes such as weathering, rock–water interaction, and human activities are the major sources of groundwater contamination in a semi-arid region. The study also suggested that primary treatments, namely, filtration and coagulation, are needed to treat the contaminated groundwater. Balamurugan et al. (2020a) investigated the impact of climatic conditions and anthropogenic activities on the quality of groundwater for drinking and irrigation purposes in the Attur region in Tamil Nadu, India, and stated that excess utilization of synthetic fertilizers and disposal of human wastes in open areas and dumping yards are the major reasons for groundwater contamination in the study region. The concentration of total hardness (TH) of the samples collected from the study region exceeded by 90.7%. Rock–water interaction and weathering are the major reasons for elevated hardness in groundwater. The study also recommended managing the quality of groundwater and creating awareness regarding the use of natural fertilizers for agricultural purposes. Panneerselvam et al. 2021a identified the influencing factors to affects the groundwater for human health using automatic linear model in a semi-arid region in India and revealed that 76% of the sample locations are exceeds the magnesium and salinity concentration in the study region due to usage of synthetic fertilizers for high crop yield in the agriculture filed. Major chemical parameters such as sodium, magnesium, and bicarbonate highly influence the nature and quality of groundwater (Zhang et al. 2020b; Zhou et al. 2020; Zhu et al. 2020; Adimalla et al. 2018b; Panaskar et al. 2016). Panneerselvam et al. (2021a, b) predicted and assessed the groundwater characteristics using a machine learning platform known as a radial basin model in a semi-arid region in India and determined the correlation coefficient to be $R^2 = 0.95$ and 0.901; the results show that leachates from dumping yards and agricultural fields are the primary sources of groundwater contamination and also the reason for the elevated concentration of anions in the groundwater geochemistry. Balamurugan et al. (2020b) carried out a non-carcinogenic risk assessment of groundwater in a semi-arid region of the southern part of the Salem district in Tamil Nadu, India, and showed that 37.31% of the sample locations were of the Ca–Mg–SO_4 type and 37.34% of the sample locations were of the Na–Cl–SO_4 type. In all, 72% of the sample locations were affected by rock–water interaction and 28% by evaporation caused by anthropogenic activities. The study also identified that climatic conditions, anthropogenic activities, and industrial disposal are the primary sources of groundwater contamination (Verma et al. 2020b; Kumar et al. 2019; Panneerselvam et al. 2021a; Panneerselvam et al. 2021b).

Dindigul is one of the fastest developing districts in Tamil Nadu, India, and a high percentage of the population in the district depends on groundwater for drinking and agricultural purposes. This study aims to assess the groundwater quality for drinking purposes using various indices such as the water quality index and the entropy water quality index. This study also aims to compare the results of the WQI and EWQI using GIS techniques. The results of this research will help suggest remedial measures to reduce groundwater contamination. The results will help improve sustainable practices for future groundwater management and improve the water supply system of wholesome water in the study region.

8.2 Materials and Methods

8.2.1 Study Area

Vedasandur is one of the densely populated taluks in the Dindigul district. According to the population estimates of 2011, the total population was around 3.5 lakhs in a total of 63 villages. The weather condition in the study region is the driest during the month of January, and the coldest months are October and November. The average temperature recorded in the study region is 30.4 °C and the lowest temperature is 23.8 °C. The recorded normal rainfall is 1480 mm throughout the year. The relative humidity of the study region varies between 40% and 85% from morning to afternoon. The drainage pattern in the taluk is radial and parallel. A major part of the study region is covered by red and black cotton soil. The red sandy soil was highly observed and identified in the sample collection stations. The source of groundwater in the study region was groundwater and open wells for irrigation and drinking purposes. The average yield of open well was recorded as 95 to 125 lpm for draw down ranging of 2.5–4 m with a pumping of three to four hours in a day. Based on the observation and field survey, groundwater is the major source of freshwater for drinking and domestic purposes for people in the study region.

8.2.2 Geology

The geological formation of the district is shown in Fig. 8.1. The major minerals in the district are granite, quartz vein, charnockite, granite gneiss, dark-grey biotite gneiss, etc. In the present study region, the major part of the taluk was covered by qartz vein and it was noted that chettinad cement copportation was actively processing for manufacturing of cement. A higher percentage of the study region is covered by an Archaean crystalline metamorphic group of rocks, and the southern part of the study region is covered by dark-grey biotite gneiss. During sample collection, shallow buried pediments were identified in the study region. The spatial representation of the study region is shown in Fig. 8.1., which shows that 868.41 km^2 of the area is covered by quartz vein, 96.38 km^2 by dark-grey biotite gneiss, 43.81 km^2 by granite, 14.04 km^2 by anorthosite, 13.87 km^2 by charnockite, 9.85 km^2 by fissile hornblende biotite gneiss, 9.26 km^2 by granitoid gneiss, and the remaining part of the region was covered by champion gneiss, garnet biotite gneiss, purple sandstone, and shale.

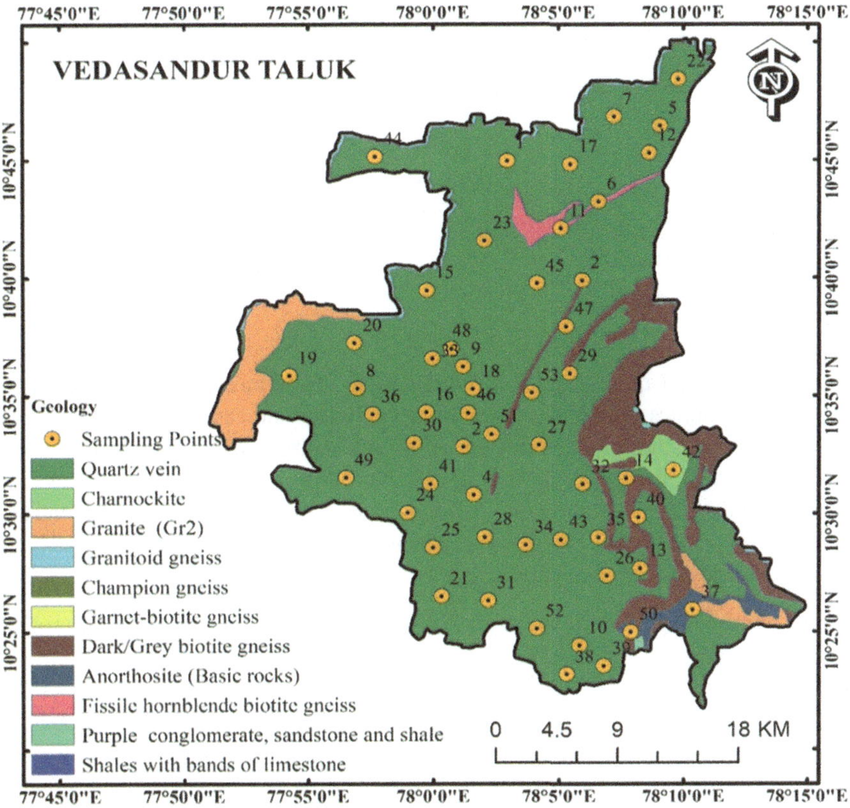

Fig. 8.1 Sampling location and geological classification of the study region

8.2.3 The WQI

The water quality index (WQI) is one of the effective methods to identify the quality of water and also to evaluate the suitability of groundwater for drinking and other domestic purposes. The WQI classification of groundwater is as follows: 0–25 is excellent, 25–50 is good, 50–75 is moderate, 75–100 is bad, and greater than 100 is unsuitable for drinking purposes Adimalla and Qian (2019). The WQI can be calculated using three major parameters: (i) relative weightage, (ii) quality rating, and (iii) sub-index value. In this study, the weighted arithmetic WQI method was used to assess the water quality and demonstrate the major ion influencing the quality of groundwater in a specific region. The following formulas are used to calculate the WQI values of groundwater:

$$Wi = \frac{w_i}{\sum_{i=1}^{n} w_i} \tag{8.1}$$

$$Q_i = \frac{C_i}{S_i} \times 100 \tag{8.2}$$

$$SI_i = W_i \times Q_i \tag{8.3}$$

$$WQI = \Sigma SI_i \tag{8.4}$$

8.2.4 The EWQI

The entropy water quality index (EWQI) is one of the most effective methods to identify the sources of contamination in groundwater and can used to integrate the chemical composition of groundwater in a study region (APHA, 2005). For calculating the EWQI, the following five steps were followed: (1) estimating the evaluation matrix (X), (2) calculating the standard grade matrix (y), (3) calculating entropy information (e_j), (4) calculating weighted entropy (w_j), and (5) calculating the quality rating scale (q_i) for each parameter.

Step:1 Estimating the evaluation matrix (X)

$$X = \begin{bmatrix} x_{11} & x_{12} & \cdots x_{1n} \\ x_{21} & x_{22} & \cdots x_{2n} \\ \vdots & \vdots & \vdots \\ x_{m1} & x_{m2} & \cdots x_{mn} \end{bmatrix}$$

where m is the number of samples (i = 1,2, 3, ..., m) and n is the number of water quality parameters (j = 1,2,3, .., n).

Step:2 Calculating the standard grade matrix (y)

$$y_{ij} = \frac{x_{ij} - \left(x_{ij}\right)_{\min}}{\left(x_{ij}\right)_{\max} - \left(x_{ij}\right)_{\min}} \tag{8.5}$$

Step:3 Calculating entropy information (e_j)

$$e_j = -\frac{1}{\ln m} \sum_{i=1}^{m} P_{ij} \ln P_{ij} \tag{8.6}$$

$$P_{ij} = \frac{(1+y_{ij})}{\sum_{i=1}^{m}(1+y_{ij})} \tag{8.7}$$

where, P_{ij} is the ratio of the index value of the j index in the i sample.

Step:4 Calculating weighted entropy (w_j)

$$w_j = \frac{(1-e_j)}{\sum_{i=1}^{n}(1-e_j)} \tag{8.8}$$

Step:5 Calculating the quality rating scale (q_i) for each parameter

$$q_i = \frac{C_j}{S_j} \times 100 \tag{8.9}$$

$$\text{EWQI} = \sum_{j=1}^{n} w_i q_i \tag{8.10}$$

8.2.5 *Spatial Analysis*

Spatial analysis is one of the most significant ways to represent the level of contamination in groundwater and helps identify the polluted zones in the study region (Balamurugan et al. 2020a, b; Panneerselvam et al. 2021a, b; Kirubakaran et al. 2021). Spatial analysis was conducted using QGIS (version 3.18). The inverse distance weighting (IDW) method was used to integrate the groundwater geochemistry in the study region in order to interpolate findings such as the concentration of chemical parameters and the WQI and EWQI values of groundwater (Marghade et al. 2020; Moharir et al. 2019). The results of the spatial analysis were extremely helpful to decision-makers and policy-makers working on groundwater resources.

8.3 Result and Discussion

8.3.1 *pH*

pH is the primary factor that influences the quality of groundwater and also decides the suitability of groundwater for drinking and irrigation uses (Yadav et al. 2019; Yang et al. 2020; Younas et al. 2019; Badeenezhad et al. 2020). The World Health

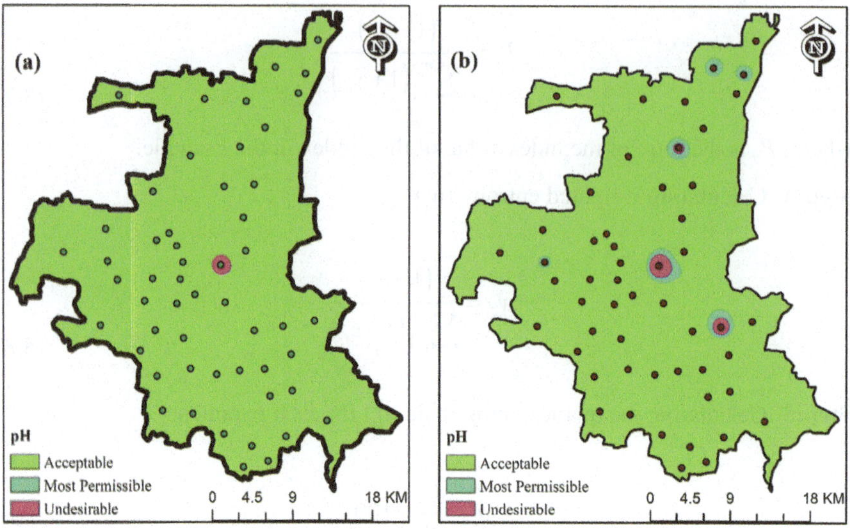

Fig. 8.2 Spatial distribution of pH concentration in the study region

Organization (WHO) and the IS 10500:2016 recommend that the value of pH ranging from 6.5 to 8.5 is the desirable limit for drinking purposes. The spatial distribution of pH in the study region revealed that 1054.60 km^2 and 1029.21 km^2 are the suitable limits and 0.952 km^2 and 20.251 km^2 are the unsuitable limits for drinking and irrigation purposes, respectively, in the study region (Fig. 8.2). The results indicate that the concentration of pH increased during the post-monsoon period, influenced by rainfall infiltration and rock–water interaction, which are the primary factors that governed the quality of groundwater in the study region.

8.3.2 Electrical Conductivity (EC)

Electrical conductivity is an important groundwater parameter to evaluate the quality of groundwater for drinking purposes. EC divulged that the presence of salt concentration and capable of electricity transformation through the water in the sampling locations. It also highly dominates the chemical composition of groundwater for drinking purposes. The WHO 2004 recommended that the permissible value of EC in groundwater is 1500 µmhos/cm. The spatial distribution of EC in the study region revealed that 194.32 km^2 is considered as excellent, 858.67 km^2 as good, and 6.22 km^2 as doubtful during the pre-monsoon period. About 74.06 km^2 is considered as excellent, 314.207 km^2 as good, and 4.94 km^2 as doubtful for drinking during the post-monsoon period (Fig. 8.3). The results indicate that the salt concentration during the pre-monsoon season increased due to the increase in temperature and evaporation of groundwater during the summer season.

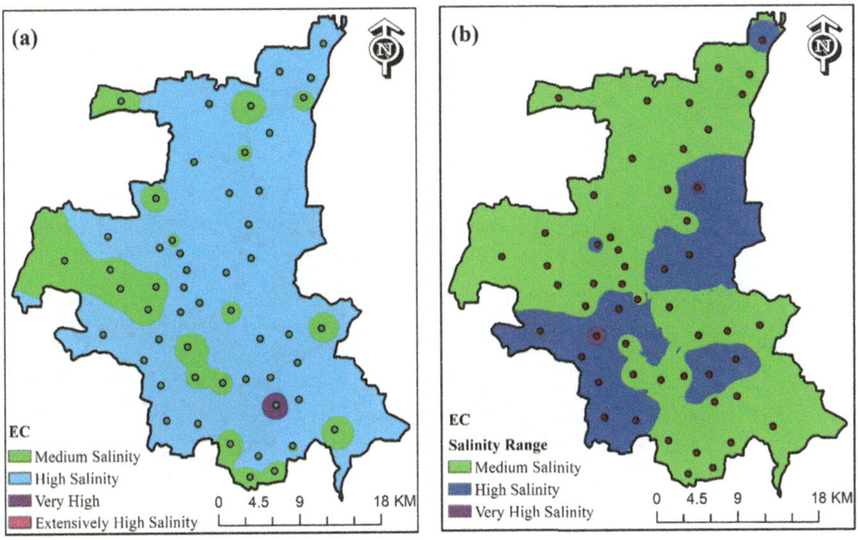

Fig. 8.3 Spatial distribution of EC concentration in the study region

8.3.3 *Total Dissolved Solids (TDSs)*

Total dissolved solids and EC are interrelated in groundwater chemistry. The concentration of TDSs highly influences the quality of groundwater for drinking purposes. The WHO 2004 and IS 10500:2012 B.I.S. (2012) recommended that a TDS value of 500 mg/L is desirable and 1500 mg/L is the maximum permissible limit for drinking purposes. The spatial distribution of TDSs in the study region revealed that 917.57 km^2 is desirable and 141.64 km^2 is the maximum permissible limit during the pre-monsoon period. About 650.24 km^2 is desirable and 403.71 km^2 is the maximum permissible limit for drinking during the post-monsoon period (Fig. 8.4). Anthropogenic activities such as usage of detergents, synthetic fertilizers, and road salt, rainfall runoff, and animal waste are the major sources of excess concentration of TDSs in groundwater.

8.3.4 *TH*

Total hardness is generally measured by calculating the sum of the concentrations of calcium and magnesium in groundwater. The WHO 2011 and IS 10500:2012 recommended that the permissible value of total hardness is 200 mg/L and 600 mg/L is the maximum permissible limit for drinking purposes. The spatial analysis of the total hardness in the study region revealed that 904.50 km^2 is desirable and 154.26 km^2 is the maximum permissible limit for drinking purposes during the

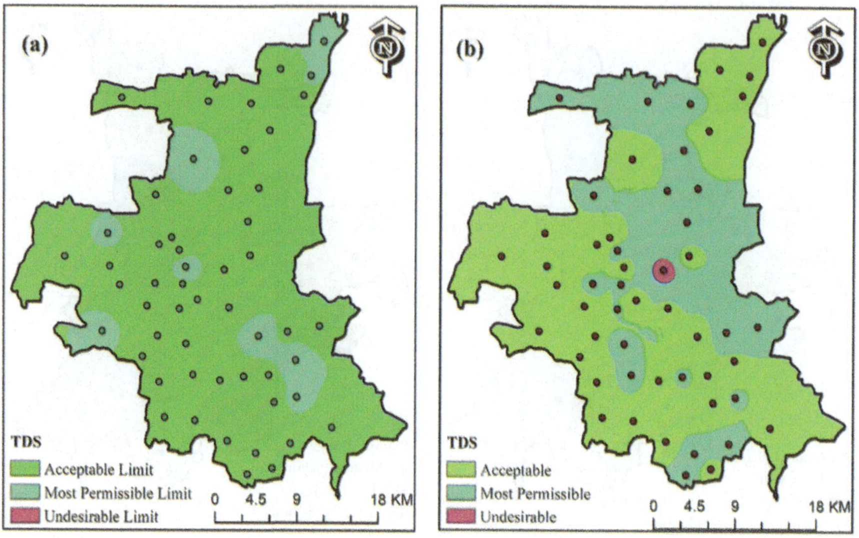

Fig. 8.4 Spatial distribution of TDS concentration in the study region

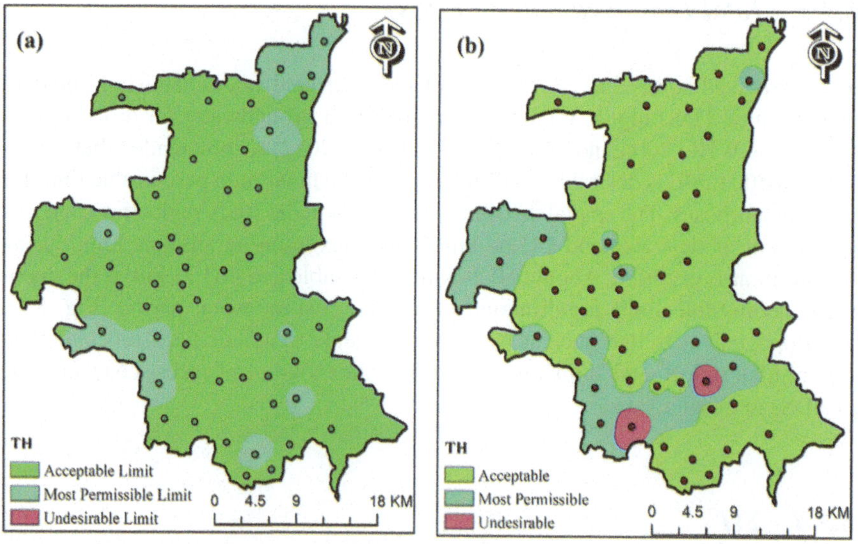

Fig. 8.5 Spatial distribution of TH concentration in the study region

pre-monsoon period. About 806.39 km^2 is desirable and 203.96 km^2 is the maximum permissible limit for drinking purposes during the post-monsoon period (Fig. 8.5). Primary factors such as rock–water interaction, water stagnated in aquifers, and the presence of minerals in aquifers highly influenced the quality of groundwater in the study region.

8.3.5 Calcium

The presence of calcium largely influences the total hardness in groundwater and is one of the basic groundwater parameters. In order to evaluate the quality of groundwater for both domestic and industrial purposes, calcium plays a vital role in the examination of the suitability of water (Panneerselvam et al. 2020a, b; Ravi et al. 2020). The WHO 2011 and IS 10500:2012 recommended that the maximum permissible limit for calcium is 200 mg/L and 75 mg/L, respectively. In this study, the spatial analysis of the calcium concentration showed that 1024.53 km^2 is suitable and 34.68 km^2 is unsuitable during the pre-monsoon period. About 980.47 km^2 is suitable and 67.88 km^2 is unsuitable during the post-monsoon period for drinking purposes (Fig. 8.6). The leaching effects on rocks and other earth crust materials are the major reasons identified in the study region for a high concentration of calcium.

8.3.6 Magnesium

Magnesium plays a vital role in groundwater for drinking, industrial, and irrigation purposes. It also decides the concentration of the total hardness in water. Magnesium, naturally present in water, depends on the rocks, minerals, and other sources in the study region. The WHO 2011 and IS 10500:2012 recommended that the maximum permissible limit for magnesium is 150 mg/L and 30 mg/L, respectively. In this study, the spatial analysis of the magnesium concentration divulged that 179.59 km^2 is suitable and 874.44 km^2 is unsuitable during the pre-monsoon period. About

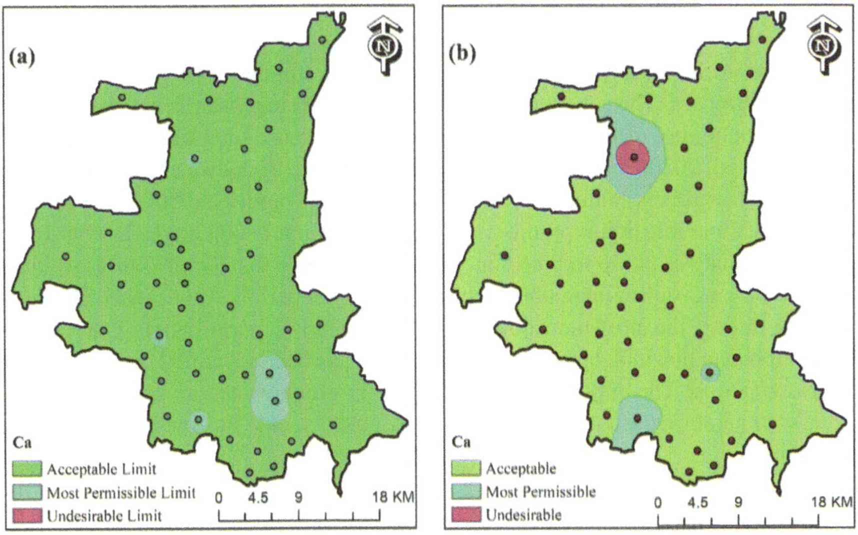

Fig. 8.6 Spatial distribution of calcium concentration in the study region

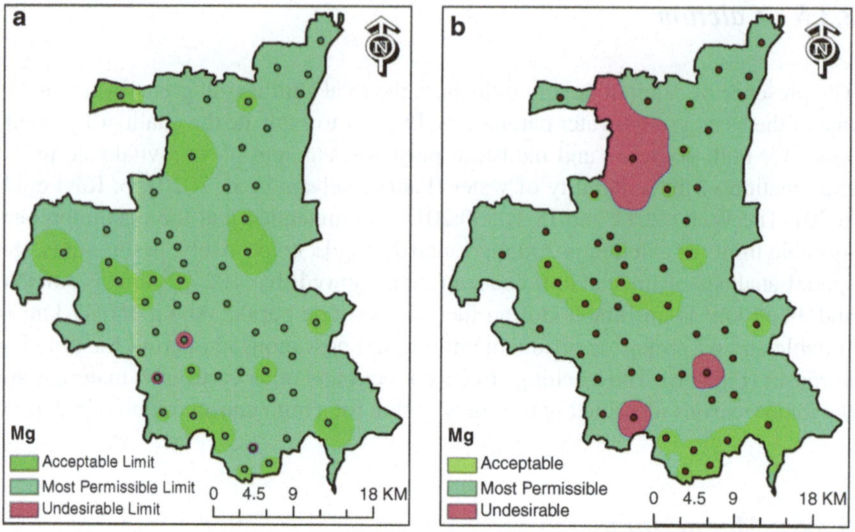

Fig. 8.7 Spatial distribution of magnesium concentration in the study region

118.51 km^2 is suitable and 842.11 km^2 is unsuitable during the post-monsoon period for drinking purposes (Fig. 8.7). The leaching effects on rocks and other earth crust materials are the major reasons identified in the study region for a high concentration of magnesium.

8.3.7 Sodium

Sodium is an important parameter to evaluate the suitability of groundwater for drinking purposes. The concentration of calcium in groundwater is essential for the normal functioning of human health and also for aquatic life (Kumar et al. 2018; Adimalla et al. 2020, 2020). The presence of sodium in groundwater is another form of intake that affects human health. The WHO 2011 and IS 10500:2012 recommended that the maximum permissible limit for sodium is 200 mg/L. In this study, the spatial analysis of the sodium concentration divulged that 793.59 km^2 is suitable and 265.62 km^2 is unsuitable during the pre-monsoon period. About 1059.21 km^2 is suitable during the post-monsoon period for drinking purposes (Fig. 8.8). The results indicate that a high sodium concentration has been identified during the pre-monsoon period. The major sources of sodium in groundwater are saline intrusion, excess mineral deposits, sewage effluents, and cracks in sewage-transferring pipelines.

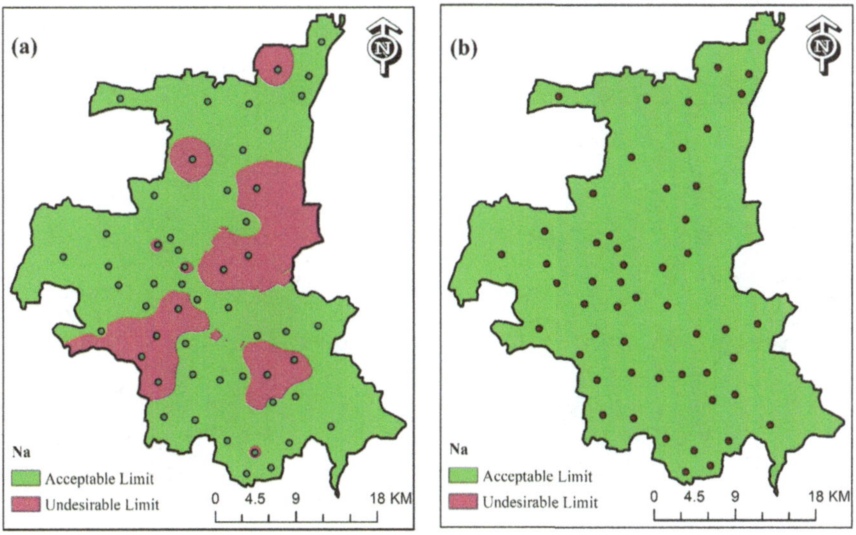

Fig. 8.8 Spatial distribution of sodium concentration in the study region

8.3.8 Potassium

Another important groundwater parameter for drinking and irrigation purposes is potassium. The presence of potassium in groundwater helps identify the number of dissolved salts and also the reactive ions in water (Zango et al. 2019; Zhang et al. 2020a). The WHO 2011 recommended that the maximum permissible limit for potassium is 200 mg/L. In this study, the spatial analysis of the potassium concentration showed that 101.17 km^2 is suitable and 49.04 km^2 is unsuitable during the pre-monsoon period. About 943.04 km^2 is suitable and 116.21 km^2 is unsuitable during the post-monsoon period for drinking purposes (Fig. 8.9). The results indicate that a high potassium content was found in the post-monsoon period and it affects human health. An excess concentration of potassium causes heart diseases, hypertension, and other health-related issues.

8.3.9 Bicarbonate

The primary source of bicarbonate in groundwater is carbon dioxide present in the soil layers and in aquifers. Bicarbonate ions are released through the dissolution of carbonate minerals (Balamurugan et al. 2020a, b). The WHO 2011 recommended that the maximum permissible limit for bicarbonate is 300 mg/L. In this study, the spatial analysis of the bicarbonate concentration divulged that 732.93 km^2 is

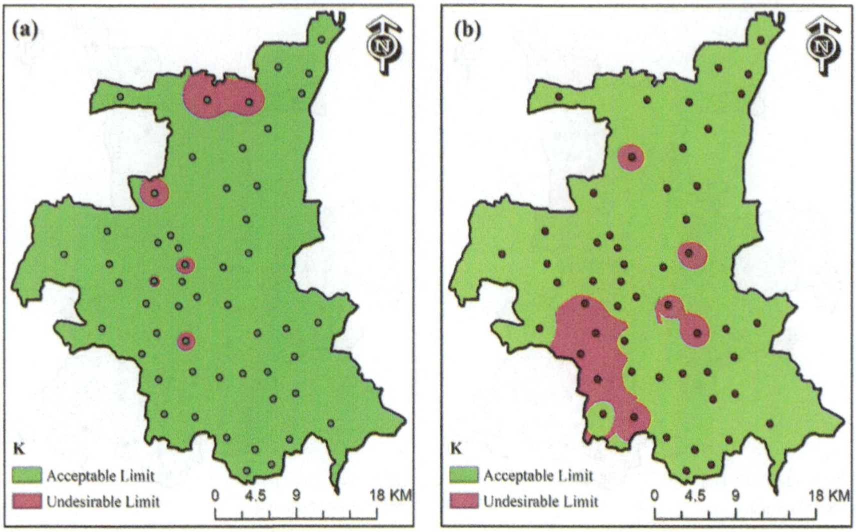

Fig. 8.9 Spatial distribution of potassium concentration in the study region

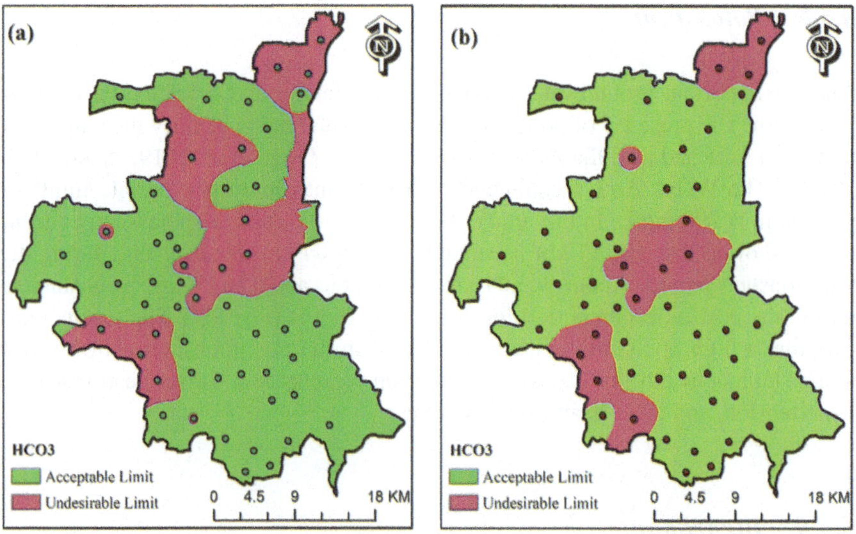

Fig. 8.10 Spatial distribution of bicarbonate concentration in the study region

suitable and 326.28 km^2 is unsuitable during the pre-monsoon period. About 853.46 km^2 is suitable and 205.75 km^2 is unsuitable during the post-monsoon period for drinking purposes (Fig. 8.10). Primary factors such as rock–water interaction, presence of carbon dioxide in aquifers, and chemical reactions under the subsurface highly influenced the excess concentration of bicarbonate in groundwater.

8.3.10 Chloride

Chloride is a major water quality parameter that decides the suitability of water for drinking purposes. An excessive concentration of chloride increases the electrical conductivity of water and results in corrosivity of metal pipelines (Alghamdi et al. 2020; Aravinthasamy et al. 2019a). The WHO 2011 and IS 10500:2012 recommended that the maximum permissible limit for chloride is 200 mg/L and 250 mg/L, respectively. In this study, the spatial analysis of the chloride concentration showed that 379.24 km^2 is suitable and 679.97 km^2 is the maximum permissible level during the pre-monsoon period. About 232.36 km^2 is suitable and 826.85 km^2 is the maximum permissible level in the post-monsoon period for drinking purposes (Fig. 8.11). Anthropogenic activities such as disposal of human waste, usage of synthetic fertilizers and road salt, rainfall runoff, and animal waste are the major sources of excess concentration of chloride in groundwater.

8.3.11 Sulphate

The concentration of sulphate is the primary parameter to evaluate the suitability of groundwater for drinking purposes. It occurs naturally in various chemical compositions in the groundwater chemistry (Ahada et al. 2018; Alam et al. 2020; Asgari, et al. 2019; Xu et al. 2019). The WHO 2011 and IS 10500:2012 recommended that

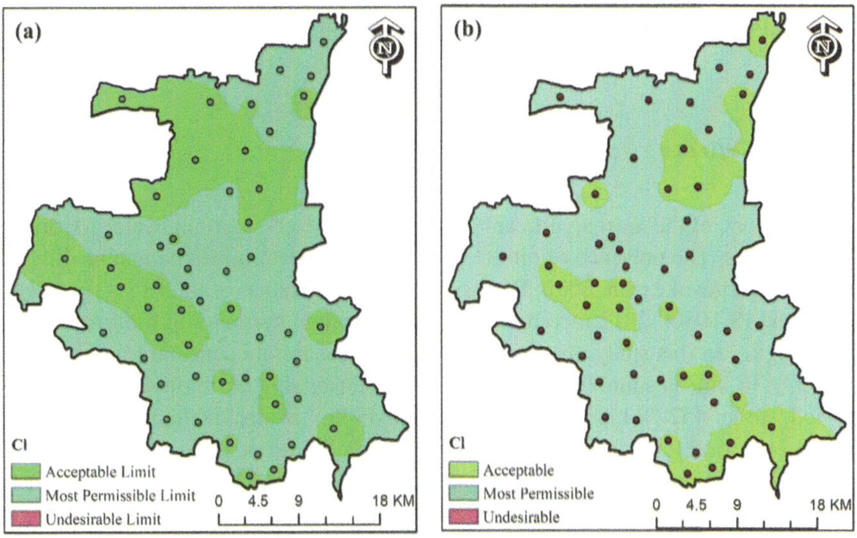

Fig. 8.11 Spatial distribution of chloride concentration in the study region

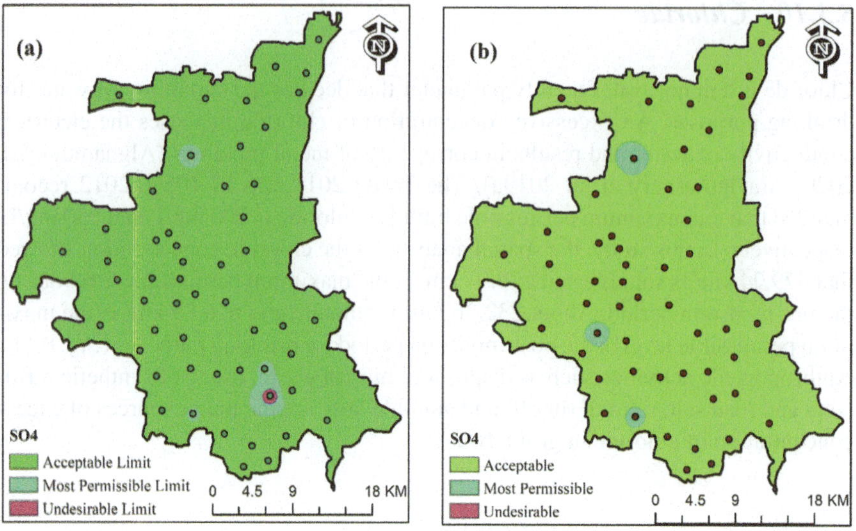

Fig. 8.12 Spatial distribution of sulphate concentration in the study region

the maximum permissible limit for sulphate is 200 mg/L. In this study, the spatial analysis of the sulphate concentration divulged that 1035.36 km^2 is suitable, 21.76 km^2 is the maximum permissible level, and 2.12 km^2 is unsuitable during the pre-monsoon period. About 1039.35 km^2 is suitable and 19.86 km^2 is unsuitable during the post-monsoon period for drinking purposes (Fig. 8.12). The identified man-made sources of sulphate contamination in groundwater in the study region are sewage effluents and discharge of industrial effluents.

8.3.12 Nitrate

Nitrate is one of the most significant chemical parameters in groundwater for drinking purposes. The presence of nitrate in groundwater determines human health and livelihood (Bawoke et al. 2020; Bouksila et al. 2013; Kumar et al. 2017). The WHO 2011 and IS 10500:2012 recommended the maximum permissible limit for nitrate is 45 mg/L . In this study, the spatial analysis of the nitrate concentration divulged that 539.36 km^2 is suitable and 519.85 km^2 is unsuitable during the pre-monsoon period. About 452.78 km^2 is suitable and 606.43 km^2 is unsuitable during the post-monsoon period for drinking purposes (Fig. 8.13). Anthropogenic activities such as disposal of human waste, usage of synthetic fertilizers, cracks in sewage pipelines, and animal waste are the major sources of excess concentration of nitrate in groundwater.

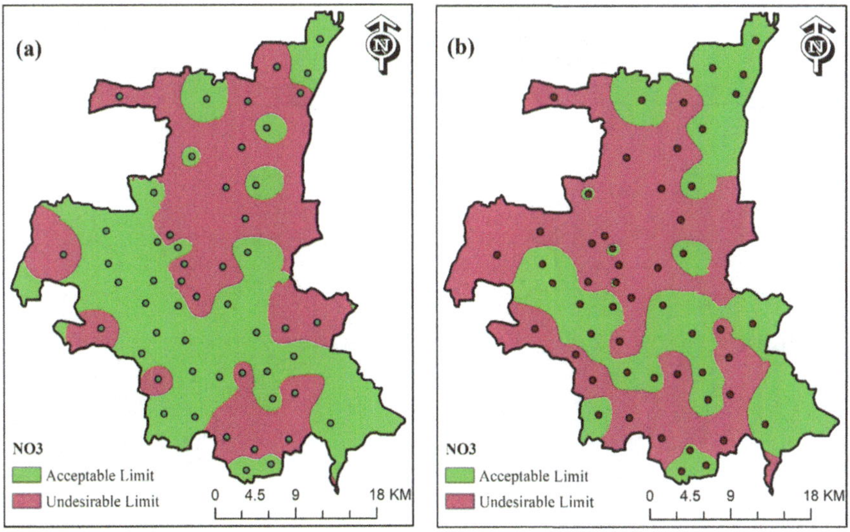

Fig. 8.13 Spatial distribution of nitrate concentration in the study region

8.3.13 Fluoride

A major concentration of fluoride has been identified in groundwater than in surface water. Leachates from dumping yards and other anthropogenic activities are the primary sources of high fluoride contamination in groundwater (Bahir et al. 2020). The WHO 2011 and IS 10500:2012 recommended that the maximum permissible limit for calcium is 1.5 mg/L and 1.2 mg/L, respectively. In this study, the spatial analysis of the fluoride concentration showed that 974.74 km^2 is suitable, 64.66 km^2 is the maximum permissible level, and 19.81 km^2 is unsuitable during the pre-monsoon period. About 943.04 km^2 is suitable, 89.50 km^2 is the maximum permissible level, and 26.68 km^2 is unsuitable during the post-monsoon period for drinking purposes (Fig. 8.14). A long-term excessive ingestion of fluoride causes serious dental and bone problems in humans. In the study region of this research, anthropogenic activities such as disposal of human waste and industrial effluents are the major sources of excessive fluoride in groundwater.

8.3.14 The WQI

In this study, the WQI value was calculated for selected parameters that majorly influenced the quality of groundwater in the study region. The calculated value of the WQI revealed that 32.08% and 26.42% of the samples were good, 49.06% and 49.06% of the samples were moderate, 13.21% and 15.09% of the samples were doubtful, and 5.66% and 9.43% of the samples were unsuitable for drinking and

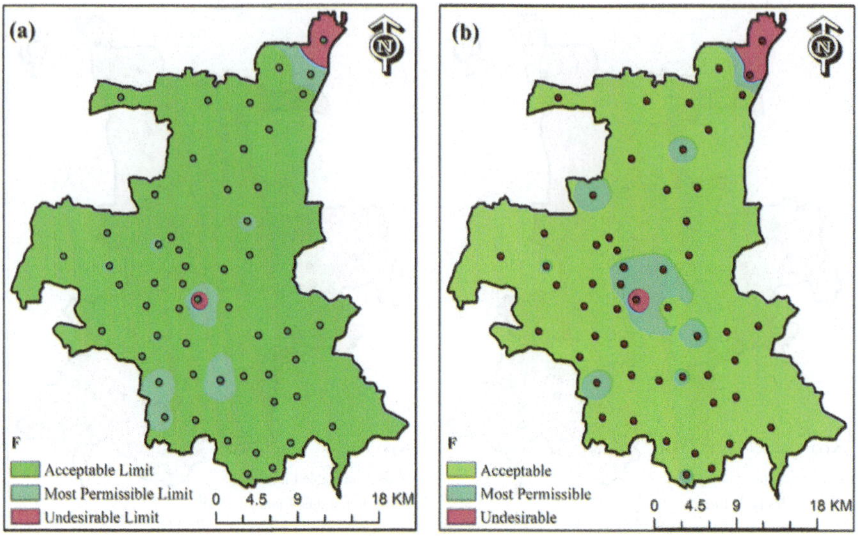

Fig. 8.14 Spatial distribution of fluoride concentration in the study region

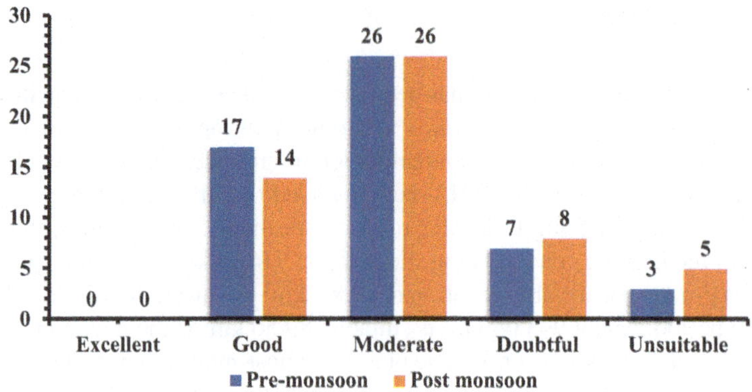

Fig. 8.15 WQI classification of groundwater in the study region

irrigation purposes, respectively, during the pre- and post-monsoon periods. The results indicate that around 10% of the samples were contaminated during the post-monsoon compared to the pre-monsoon season (Fig. 8.15). It was identified that the excessive concentrations of calcium, magnesium, and nitrate are the major chemical parameters that highly influenced the quality of groundwater in the study region (Fig. 8.16). Natural sources of pollution such as rock–water interaction, water retention in aquifers, evaporation, and increase in temperature are the identified sources and anthropogenic activities such as sewage flow, municipal waste disposal, and animal waste are the primary sources of groundwater contamination in the study region (Table 8.1).

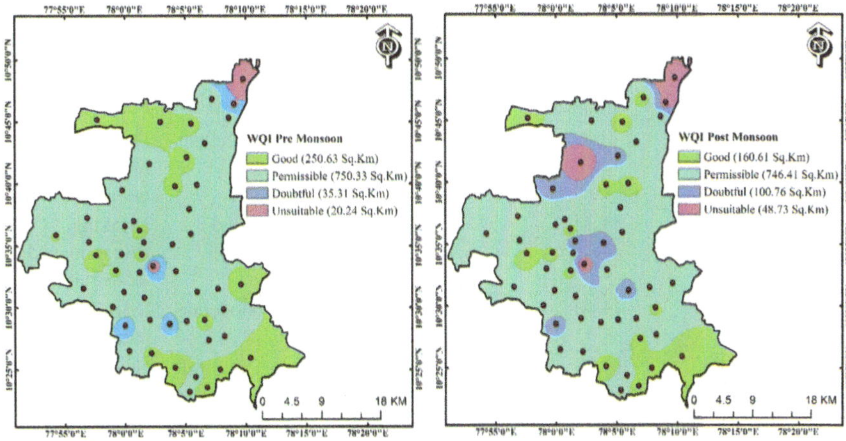

Fig. 8.16 Spatial analysis of the WQI value of groundwater in the study region

Table 8.1 WQI value-based classification of groundwater in the study region

		Pre-monsoon		Post-monsoon			
S No.	WQI value	No. of samples	Percentage of samples	No. of samples	Percentage of samples	Class	Purpose of use
1	0–25	0	0.00	0	0.00	Excellent	Drinking, irrigation, and industrial
2	25–50	17	32.08	14	26.42	Good	Drinking, irrigation, and industrial
3	50–75	26	49.06	26	49.06	Moderate	Irrigation and industrial
4	75–100	7	13.21	8	15.09	Doubtful	Need treatment before use
5	>100	3	5.66	5	9.43	Unsuitable	Need primary and secondary treatments

8.3.15 The EWQI

The EWQI values of the groundwater samples were calculated and are presented in Table 8.2. The calculated values of the EWQI of groundwater show that 15 and 7 samples are good, 23 and 26 samples are moderate, 9 and 12 samples are doubtful, and 6 and 8 samples are unsuitable for drinking purposes in both pre- and post-monsoon periods, respectively. The results indicate that about 20% of the samples exceed the permissible limit for drinking during the post-monsoon compared to the pre-monsoon season (Fig. 8.17). Based on the results, nitrate and magnesium are the major chemical parameters that highly influenced the quality of groundwater in the study region. The primary sources of nitrate contamination in the study region

Table 8.2 Classification of groundwater based on the EWQI in the study region

S No.	WQI value	Pre-monsoon No. of samples	Pre-monsoon Percentage of samples	Post-monsoon No. of samples	Post-monsoon Percentage of samples	Class	Purpose of use
1	0–25	0	0.00	0	0.00	Excellent	Drinking, irrigation, and industrial
2	25–50	15	28.30	7	26.42	Good	Drinking, irrigation, and industrial
3	50–75	23	43.40	26	49.06	Moderate	Irrigation and industrial
4	75–100	9	16.98	12	15.09	Doubtful	Need treatment before use
5	>100	6	11.32	8	9.43	Unsuitable	Need primary and secondary treatments

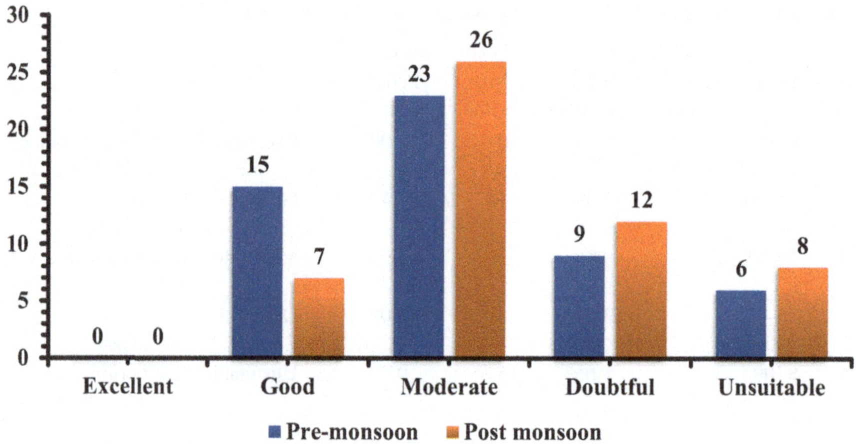

Fig. 8.17 EWQI classification of groundwater in the study region

are residents' waste disposal, municipal waste dumping yards, industrial effluents in some specific locations, and sewage leakage through pipelines.

8.4 Conclusions

This research work was carried out to investigate the quality of groundwater in detail for drinking purposes by comparing the results of the WQI and EWQI using GIS techniques. This study concluded that the primary chemical parameters such as calcium, magnesium, chloride, and nitrate are the major ions that influenced the quality of groundwater for drinking purposes in the study region. The WQI results revealed that about five sample locations were contaminated due to excessive concentrations of nitrate and magnesium in the groundwater composition. The EWQI results indicated that nine sample locations were commonly affected in both pre- and post-monsoon periods due to anthropogenic activities such as sewage effluents, man-made waste disposal, animal waste, and synthetic fertilizers. The graphical representation of each chemical parameter and the WQI and EWQI values will lead to an enhanced understanding of the level of groundwater contamination in the study region. It can also help find the area (km^2) of the total amount of contamination in the specific region. The study results help policy-makers and management people working in groundwater management in the study region.

References

A. Abbasnia, N. Yousefi, A.H. Mahvi, R. Nabizadeh, M. Radfard, M. Yousefi, M. Alimohammadi, Evaluation of groundwater quality using water quality index and its suitability for assessing water for drinking and irrigation purposes: Case study of Sistan and Baluchistan province (Iran). Hum. Ecol. Risk Assess. Int. J. **25**(4), 988–1005 (2019)

N. Adimalla, Groundwater quality for drinking and irrigation purposes and potential health risks assessment: A case study from semi-arid region of South India. Exposure Health **11**(2), 109–123 (2019)

N. Adimalla, Spatial distribution, exposure, and potential health risk assessment from nitrate in drinking water from semi-arid region of South India. Hum. Ecol. Risk Assess. Int. J. **26**(2), 310–334 (2020)

N. Adimalla, H. Qian, Hydro geochemistry and fluoride contamination in the hard rock terrain of central Telangana, India: Analyses of its spatial distribution and health risk. SN Appl. Sci. **1**(3), 202 (2019). https://doi.org/10.1007/s42452-019-0219-8

N. Adimalla, P. Li, S. Venkatayogi, Hydrogeochemical evaluation of groundwater quality for drinking and irrigation purposes and integrated interpretation with water quality index studies. Environ. Process. **5**(2), 363–383 (2018a). https://doi.org/10.1007/s40710-018-0297-4

N. Adimalla, P. Li, H. Qian, Evaluation of groundwater contamination for fluoride and nitrate in semi-arid region of Nirmal Province, South India: A special emphasis on human health risk assessment (HHRA). Hum. Ecol. Risk Assess. Int. J. (2018b). https://doi.org/10.1080/10807039.2018.1460579

N. Adimalla, H. Qian, P. Li, Entropy water quality index and probabilistic health risk assessment from geochemistry of groundwaters in hard rock terrain of Nanganur County, South India. Geochemistry, 125544 (2019)

N. Adimalla, R. Dhakate, A. Kasarla, A.K. Taloor, Appraisal of groundwater quality for drinking and irrigation purposes in Central Telangana, India. Groundwater Sustain Develop **10**, 100334 (2020)

C.P.S. Ahada, S. Suthar, Assessing groundwater hydrochemistry of Malwa Punjab, India. Arabian J. Geosci. **11**(2), 17 (2018). https://doi.org/10.1007/s12517-017-3355-8

W. Alam, K.S. Singh, Y. Gyanendra, R.J. Laishram, N. Nesa, Hydrogeochemical assessment of groundwater quality for few habitations of Chandel District, Manipur (India). Appl Water Sci **10**(5), 1–13 (2020)

A.G. Alghamdi, A.A. Aly, S.A. Aldhumri, F.N. Al-Barakaha, Hydrochemical and quality assessment of groundwater resources in Al-Madinah City, Western Saudi Arabia. Sustainability **12**(8), 3106 (2020)

American Public Health. APHA, *Standard Methods for the Examination of Water and Wastewater*, 21st edn. (American Public Health Association, Washington, 2005)

P. Aravinthasamy, D. Karunanidhi, T. Subramani, B. Anand, P.D. Roy, K. Srinivasamoorthy, Fluoride contamination in groundwater of the Shanmuganadhi River basin (South India) and its association with other chemical constituents using geographical information system and multivariate statistics. Geochemistry **125555** (2019a)

P. Aravinthasamy, D. Karunanidhi, T. Subramani, K. Srinivasamoorthy, B. Anand, Geochemical evaluation of fluoride contamination in groundwater from Shanmuganadhi River basin, South India: Implication on human health. Environ. Geochem. Health, 1–27 (2019b)

G. Asgari, S. Ghalehaskar, V.K. Moghaddam, M. Radfard, Z. Heidarinejad, A.A. Mohammadi, H. Faraji, Monitoring and health risk assessment of fluoride in drinking water in Babol, Mazandaran Province, Iran. Desalination Water Treat. **165**, 141–147 (2019)

B.I.S, *Drinking Water-Specification* (Bureau of Indian Standards, New Delhi IS, 2012), p. 10500

A. Badeenezhad, H. Tabatabaee, H.A. Nikbakht, M. Radfard, A. Abbasnia, M.A. Baghapour, M. Alhamd, Estimation of the groundwater quality index and investigation of the affecting factors their changes in Shiraz drinking groundwater, Iran. Groundwater Sustain. Develop., 100435 (2020)

M. Bahir, S. Ouhamdouch, D. Ouazar, A. Chehbouni, Assessment of groundwater quality from semi-arid area for drinking purpose using statistical, water quality index (WQI) and GIS technique. Carbonates Evaporites **35**(1), 1–24 (2020)

P. Balamurugan, P.S. Kumar, K. Shankar, R. Nagavinothini, P.J. Sajil Kumar, Impact of climate and anthropogenic activities on groundwater quality for domestic and irrigation purposes in Attur region, Tamilnadu, India. Desalin. Water Treat. **208**, 172–195 (2020a)

P. Balamurugan, P.S. Kumar, K. Shankar, R. Nagavinothini, K. Vijayasurya, Non-carcinogenic risk assessment of groundwater in southern part of Salem District in Tamilnadu, India. J. Chil. Chem. Soc. **65**(1), 4697–4707 (2020b)

G.T. Bawoke, Z.L. Anteneh, Spatial assessment and appraisal of groundwater suitability for drinking consumption in Andasa watershed using water quality index (WQI) and GIS techniques: Blue Nile Basin, Northwestern Ethiopia. Cogent Eng. **7**(1), 1748950 (2020)

F. Bouksila, A. Bahri, R. Berndtsson, M. Persson, J. Rozema, S.E. Van der Zee, Assessment of soil salinization risks under irrigation with brackish water in semiarid Tunisia. Environ. Exp. Bot. **92**, 176–185 (2013). https://doi.org/10.1016/j.chemer.2019.125544

M. Kirubakaran, P. Balamurugan, D. Suresh, C. J. Jesudhas, S. Karuppasamy, Evaluating the effectiveness of emissions reduction measures and ambient air quality variability through ground-based and Sentinel-5P observations under the auspices of COVID pandemic lockdown in Tamil Nadu India, Int. J. Environ. Anal. Chem. 1–12 (2021)

P.S. Kumar, P. Balamurugan, Evaluation of groundwater quality for irrigation purpose in attur taluk, Salem, Tamilnadu, India. Water Energy Int. **61**(4), 59–64 (2018)

P.S. Kumar, P. Balamurugan, Suitability of ground water for irrigation purpose in Omalur Taluk, Salem, Tamil Nadu, India. Indian J. Ecol. **46**(1), 1–6 (2019)

M.S. Kumar, R. Dhakate, G. Yadagiri, K.S. Reddy, Principal component and multivariate statistical approach for evaluation of hydrochemical characterization of fluoride-rich groundwater of Shaslar Vagu watershed, Nalgonda District, India. Arabian J. Geosci. **10**(4), 83 (2017). https://doi.org/10.1007/s12517-017-2863-x

P. Li, J. Wu, H. Qian, et al., Hydrogeochemical characterization of groundwater in and around awastewater irrigated forest in the southeastern edge of the Tengger Desert. Northwest China. Expo. Health **8**(3), 331–348 (2016). https://doi.org/10.1007/s12403-016-0193-y

D. Marghade, D.B. Malpe, K. Duraisamy, P.D. Patil, P. Li, Hydrogeochemical evaluation, suitability, and health risk assessment of groundwater in the watershed of Godavari basin, Maharashtra, Central India. Environ. Sci. Pollut. Res., 1–24 (2020)

K. Moharir, C. Pande, S. Singh, P. Choudhari, R. Kishan, L. Jeyakumar, Spatial interpolation approach-based appraisal of groundwater quality of arid regions. J. Water Supply: Res. Technol.-AQUA (2019) https://iwaponline.com/aqua/articlepdf/doi/10.2166/aqua.2019.026/588295

D.B. Panaskar, V.M. Wagh, A.A. Muley, S.V. Mukate, R.S. Pawar, M.L. Aamalawar, Evaluating groundwater suitability for the domestic, irrigation, and industrial purposes in Nanded Tehsil, Maharashtra, India, using GIS and statistics. Arab. J. Geosci. **9**(13), 615 (2016). https://doi.org/10.1007/s12517-016-2641-1

B. Panneerselvam, S. Karuppannan, K. Muniraj, Evaluation of drinking and irrigation suitability of groundwater with special emphasizing the health risk posed by nitrate contamination using nitrate pollution index (NPI) and human health risk assessment (HHRA). Hum. Ecol. Risk Assess. Int. J. **27**(5), 1324–1348 (2020a)

B. Panneerselvam, S.K. Paramasivam, S. Karuppannan, N. Ravichandran, P. Selvaraj, A GIS-based evaluation of hydrochemical characterisation of groundwater in hard rock region, South Tamil Nadu, India. Arabian J. Geosci. **13**(17), 1–22 (2020b)

B. Panneerselvam, K. Muniraj, C. Pande, N. Ravichandran, Prediction and evaluation of groundwater characteristics using the radial basic model in semi-arid region, India. Int. J. Environ. Anal. Chem., 1–17 (2021a)

B. Panneerselvam, K. Muniraj, M. Thomas, N. Ravichandran, B. Bidorn, Identifying influencing groundwater parameter on human health associate with irrigation indices using the Automatic Linear Model (ALM) in a semi-arid region in India. Environ. Res. **202**, 111778 (2021b)

R. Ravi, S. Aravindan, K. Shankar, P. Balamurugan, Suitability of groundwater quality for irrigation in and around the main Gadilam river basin on the east coast of southern India. Arch. Agr. Environ. Sci. **5**(4), 554–562 (2020)

S. Shukla, A. Saxena, Sources and leaching of nitrate contamination in groundwater. Curr. Sci. **00113891**, 118(6) (2020)

A. Singh, A.K. Patel, M. Kumar, Mitigating the risk of arsenic and fluoride contamination of groundwater through a multi-model framework of statistical assessment and natural remediation techniques, in *Emerging Issues in the Water Environment during Anthropocene*, (Springer, Singapore, 2020), pp. 285–300

G.S. Solangi, A.A. Siyal, M.M. Babar, P. Siyal, Application of water quality index, synthetic pollution index, and geospatial tools for the assessment of drinking water quality in the Indus Delta, Pakistan. Environ. Monit. Asess. **191**(12), 731 (2019)

A. Verma, B.K. Yadav, N.B. Singh, Data on the assessment of groundwater quality in Gomti-Ganga alluvial plain of Northern India. Data Brief **105660** (2020a)

P. Verma, P.K. Singh, R.R. Sinha, A.K. Tiwari, Assessment of groundwater quality status by using water quality index (WQI) and geographic information system (GIS) approaches: A case study of the Bokaro district. India. Appl. Water Sci. **10**(1), 27 (2020b)

V.M. Wagh, D.B. Panaskar, S.V. Mukate, M.L. Aamalawar, U. Laxman Sahu, Nitrate associated health risks from groundwater of Kadava river basin Nashik, Maharashtra, India. Hum. Ecol. Risk Assess. Int. J. **26**(3), 654–672 (2020)

WHO, *Guidelines for Drinking-Water Quality*, vol 216 (World Health Organization, Geneva, 2011), pp. 303–304

P. Xu, W. Feng, H. Qian, Q. Zhang, Hydrogeochemical characterization and irrigation quality assessment of shallow groundwater in the central-western Guanzhong basin, China. Int. J. Environ. Res. Public Health **16**(9), 1492 (2019)

K.K. Yadav, V. Kumar, N. Gupta, S. Kumar, S. Rezania, N. Singh, Human health risk assessment: Study of a population exposed to fluoride through groundwater of Agra city, India. Regul. Toxicol. Pharmacol. **106**, 68–80 (2019)

Q. Yang, Z. Li, C. Xie, J. Liang, H. Ma, Risk assessment of groundwater hydrochemistry for irrigation suitability in Ordos Basin, China. Nat. Hazards, 1–17 (2020)

A. Younas, N. Mushtaq, J.A. Khattak, T. Javed, H.U. Rehman, A. Farooqi, High levels of fluoride contamination in groundwater of the semi-arid alluvial aquifers, Pakistan: Evaluating the recharge sources and geochemical identification via stable isotopes and other major elemental data. Environ. Sci. Pollut. Res. **26**(35), 35728–35741 (2019)

M.S. Zango, E.D. Sunkari, M. Abu, A. Lermi, Hydrogeochemical controls and human health risk assessment of groundwater fluoride and boron in the semi-arid North East region of Ghana. J. Geochem. Explor. **207**, 106363 (2019)

L. Zhang, L. Zhao, Q. Zeng, G. Fu, B. Feng, X. Lin, et al., Spatial distribution of fluoride in drinking water and health risk assessment of children in typical fluorosis areas in North China. Chemosphere **239**, 124811 (2020a)

Q. Zhang, P. Xu, H. Qian, Groundwater quality assessment using improved water quality index (WQI) and Human Health Risk (HHR) valuation in a semi-arid region of Northwest China. Exposure Health, 1–14 (2020b)

Y. Zhou, P. Li, L. Xue, Z. Dong, D. Li, Solute geochemistry and groundwater quality for drinking and irrigation purposes: a case study in Xinle City, North China. Geochemistry, 125609 (2020)

D. Zhu, G. Zhu, Z. Zhang, Z. Wang, X. Yan, Y. Yan, Effects of independent and combined water-deficit and high-nitrogen treatments on flag leaf proteomes during wheat grain development. Int. J. Mol. Sci. **21**(6), 2098 (2020)

Chapter 9
Seasonal Variation of Groundwater Quality for Irrigation Purposes in a Semi-Arid Region in Tamil Nadu, India

Shunmuga Priya Kaliyappan, Kirubakaran Muniraj, Gopinath Rajendran, Butsawan Bidorn, Maciej Thomas, and Balamurugan Panneerselvam

Abstract Contamination of groundwater is a serious issue in arid and semi-arid regions all over the world. This study aims to evaluate the quality of groundwater for irrigation purposes using various irrigation indices associated with a geographic information system (GIS). A total of 50 groundwater samples were collected from agricultural fields, industrial zones, and densely populated areas in the study region. The collected samples were analyzed using standard procedures recommended by the American Public Health Association (APHA). The results revealed that the concentration of sodium in all the sample locations was under the permissible limit and fit for irrigation purposes. An excess concentration of magnesium and salinity highly influenced the quality of groundwater for irrigation purposes. This study divulged that natural sources of contamination such as rock–water interaction,

S. P. Kaliyappan
Department of Civil Engineering, Nehru Institute of Technology, Coimbatore, India

K. Muniraj (✉)
Department of Civil Engineering, University VOC College of Engineering, Thoothukudi, India

G. Rajendran
Department of Civil Engineering, Hindustan Institute of Technology & Science, Chennai, India

B. Bidorn
Department of Water Resources Engineering, Chulalongkorn University, Bangkok, Thailand

M. Thomas
Cracow University of Technology, Faculty of Environmental and Power Engineering, Cracow, Poland
e-mail: maciej.thomas@pk.edu.pl

B. Panneerselvam
Department of Construction Architecture and Environmental Engineering, University of Naples Federico II, Naples, Italy

B. Panneerselvam et al. (eds.), *Climate Change Impact on Groundwater Resources*, https://doi.org/10.1007/978-3-031-04707-7_9

weathering of rocks, and the nature of an aquifer are the primary factors that cause groundwater contamination. This research helps decision- and policy-makers to take remedial action to avoid groundwater pollution in the study region.

Keywords Groundwater · Irrigation purposes · Magnesium and salinity concentration

9.1 Introduction

The nature of surface water and groundwater is controlled by the environment in which they are distributed. Increasing water requirements of a growing population and industrial and agricultural activities show the importance of water resources (Jesudhas et al. 2017). At present, freshwater supply is not sufficient to meet the requirements for industrial consumption and irrigation use. Moreover, the demand is constantly increasing (Balamurugan et al. 2021). Groundwater plays a significant role in human health, socioeconomic development, and ecology (Balamurugan et al. 2020b). Groundwater yields, chemical compositions, and temperatures change from time to time, depending on the aquifer structure, geological formation, and human impact (Kirubakaran et al. 2016). Over the past few decades, groundwater pollution has become increasingly complex and is an environmental issue around the world (Balamurugan et al. 2021).

Land use puts pressure on surface water quality by disposing domestic and other wastewater and also reduces groundwater quality through this contaminated surface–water infiltration (Jesudhas et al. 2021). Sustainable groundwater resource management and development activities face significant problems, especially in semi-arid areas with shallow water shortages, unreliable rainfall, soil and water conservation practices, reservoirs, and the need for higher water irrigation and consumption (Kirubakaran et al. 2015). By evaluating the quality of groundwater, a better strategy can be developed to prevent further contamination of groundwater. The use of groundwater for industrial purposes, in general, is determined based on its quality characteristics.

Groundwater quality is an important indicator of demand utilization (Jesudhas et al. 2017). Anthropological activities such as irrigation and industrial practices generally cause pollution and are major factors contributing to the loss of the natural quality of groundwater as a result. Groundwater pollution has been on the rise in Tamil Nadu in southern India, according to several recent studies (Murugesan et al. 2020; Nawale et al. 2021; Nejad et al. 2021; Pande et al. 2019; Panneerselvam et al. 2020a; Panneerselvam et al. 2020b; Rao et al. 2019; Ravi et al. 2020). Several recent studies (Abbasnia et al. 2019; Adimalla 2019; Ahmed et al. 2020; Aliyu et al. 2019; Asadi et al. 2020; Azhdarpoor et al. 2019; Balamurugan et al. 2020a; Balasubramani et al. 2020; Karunanidhi et al. 2020) have revealed the importance of groundwater resources, groundwater pollution, and its management. Batarseh et al. (2021) further assessed groundwater quality using the irrigation water quality index (IWQI) and GIS in the arid region of Abu Dhabi in 2021 and found that 44% of the total sample sites were most affected by high concentrations of electrical conductivity. He further noted that with the support of the IWQI, 52% of the sample sites were strictly prohibited for irrigation purposes. The integrated approach of a GIS spatial analysis and the IWQI shows the appropriate crops in the study area and the recommended zones for the red zone. IWQI's new integrated approach based on entropy weighting and critical assessment is extremely important for assessing groundwater quality for irrigation, and these results are far more valuable than those of previous methods (Zhang et al. 2021).

Panneerselvam et al. (2021a, b, c) conducted a GIS-based assessment and found that high levels of nitrate and fluoride in groundwater have affected a high percentage of sample sites due to irrigation compatibility and municipal disposal using the water quality index (WQI), pollution index, and irrigation. A geographic information system (GIS) is now known as an excellent and efficient tool for manipulating spatial data, and such spatial datasets are used to produce conclusive results for decision-making purposes in many research domains, including environmental studies. Many researchers around the world, in recent years (Kouadri et al. 2021; Kumar and Balamurugan 2018a; Kumar and Balamurugan 2019; Kumar and Balamurugan 2018b; Rosanoff 2013), have demonstrated the effectiveness of GIS technology in various environmental applications. In order to continue to expand agricultural areas, the quality of irrigation water should be maintained. Scientists have developed thematic maps describing various chemical parameters with the help of GISs to show areas with high-quality groundwater that can be used together in agriculture. Recently, many researchers (Rufino et al. 2019; Sarani et al. 2016; Senthilkumar et al. 2019; Sharma et al. 2020; Shunmugapriya et al. 2021; Wang et al. 2020; Xu et al. 2019) around the world have utilized GISs to explore the suitability of groundwater and have been able to better reveal its superiority for various purposes.

9.2 Materials and Methods

9.2.1 Study Area Description (Fig. 9.1)

9.2.2 Piper Diagram

Piper (1944) developed a diagram to represent the hydrogeochemical composition of groundwater. The diagram contains two triangles and one diamond shape to reveal the chemical composition of groundwater in a specific region. The triangles on the right side represent the concentrations of cations such as calcium, magnesium, sodium, and potassium, whereas the triangles on the left side represent the concentrations of anions such as chloride, sulphate, carbonate, and bicarbonate in groundwater. The major composition of groundwater, such as the combination of

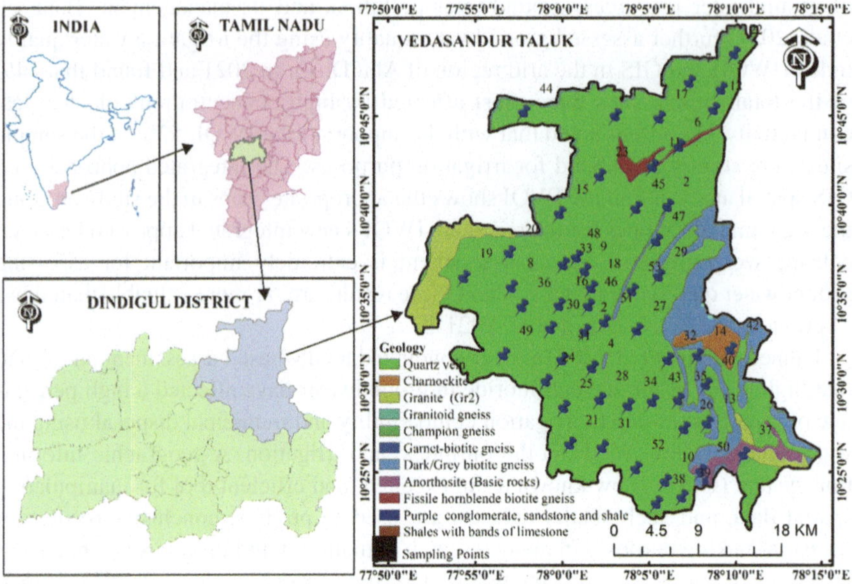

Fig. 9.1 Study area and sample location map

calcium, magnesium, sodium, chloride, and sulphate, is revealed by the diamond shape in the diagram. The Piper diagram results help identify the sources of contamination, for example, rock–water interaction, evaporation, and other anthropogenic activities such as municipal waste disposal, sewage effluent disposal, industrial waste disposal, and dumping yard leachates.

9.2.3 *The US Salinity Laboratory (USSL) Diagram*

The US Salinity Laboratory diagram is another effective method for evaluating the suitability of groundwater for irrigation purposes. It is a graphical plot of the sodium absorption ratios (SARs) on the *y*-axis and the salinity hazard values on the *x*-axis. Classification of groundwater based on the diagram is presented in Table 9.1. The results of the USSL diagram help identify the salinity and sodium concentration of groundwater.

9.2.4 *Irrigation Indices*

Various irrigation indices such as the sodium absorption ratio, percentage sodium (% Na), permeability index (PI), magnesium hazard (MH) index, residual sodium carbonate (RSC), Kelly ratio (KR), residual sodium bicarbonate (RSBC), and

Table 9.1 Classification of groundwater based on the USSL diagram (Rajendran et al. 2021)

Category	Class	Purpose of use
C1-S1	Low sodium–less salinity	Safe for drinking and irrigation
C1-S2	Low sodium–medium salinity	Need treatment for irrigation
C1-S3	Low sodium–high salinity	Can be used for irrigation with primary treatment and satisfactory for most of the crops
C1-S4	Low sodium–very high salinity	Unsatisfactory for most of the crops
C2-S1	Medium sodium–less salinity	Moderate leaching and good for irrigation
C2-S2	Medium sodium–medium salinity	Need treatment for irrigation
C2-S3	Medium sodium–high salinity	Can be used for irrigation with primary treatment and satisfactory for most of the crops
C2-S4	Medium sodium–very high salinity	Unsatisfactory for most of the crops
C3-S1	High sodium–less salinity	Doubtful for drinking and need treatment for irrigation
C3-S2	High sodium–medium salinity	Unfit for drinking and need treatment for irrigation
C3-S3	High sodium–high salinity	Can be used for irrigation with primary treatment and satisfactory for most of the crops
C3-S4	High sodium–very high salinity	Unsatisfactory for most of the crops
C4-S1	Very high sodium–less salinity	Unfit for drinking and need treatment for irrigation
C4-S2	Very high sodium–medium salinity	Unfit for drinking and need treatment for irrigation
C4-S3	Very high sodium–high salinity	Can be used for irrigation with primary treatment and satisfactory for most of the crops
C4-S4	Very high sodium–very high salinity	Unsatisfactory for most of the crops

potential salinity (PS) are important for evaluating the quality of groundwater for irrigation purposes. The following formulas are used to calculate each index value:

$$\mathrm{SAR} = \frac{Na^{+}}{\left(\frac{\mathrm{Ca}^{2+} + \mathrm{Mg}^{2+}}{2}\right)^{1/2}} \tag{9.1}$$

$$\%\mathrm{Na} = \frac{(\mathrm{Na} + \mathrm{K}) \times 100}{\mathrm{Ca} + \mathrm{Mg} + \mathrm{Na} + \mathrm{K}} \tag{9.2}$$

$$\mathrm{PI} = \frac{\sqrt{\mathrm{HCO3}} + \mathrm{Na}}{\sqrt{(\mathrm{Na} + \mathrm{Mg} + \mathrm{Ca}}} \tag{9.3}$$

$$\mathrm{MH} = \frac{\mathrm{Mg} \times 100}{\mathrm{Ca} + \mathrm{Mg}} \tag{9.4}$$

$$RSC = [HCO_3 + CO_3] - [Ca + Mg] \tag{9.5}$$

$$KR = \frac{Na^+}{Ca^{2+} + Mg^{2+}} \tag{9.6}$$

$$RBSC = HCO_3^- - Ca^{2+} \tag{9.7}$$

$$PS = Cl^- + \frac{1}{2} \times SO_4^{2-} \tag{9.8}$$

9.3 Results and Discussion

9.3.1 Piper Diagram

In this study, the Piper classification of groundwater revealed that 65% of the sample locations fell under the mixed Ca–Mg–Cl type, 20% under the Na–Cl type, 15% under the Ca–Cl type, and 5% under the mixed Ca–Na–HCO_3 type during the pre-monsoon season (Fig. 9.2). In the post-monsoon season, the Piper diagram divulged that 75% of the sample locations fell under the mixed Ca–Mg–Cl type, 15% under the Na–Cl type, 15% under the Ca–Cl type, and 5% under the mixed Ca–HCO_3 type (Fig. 9.3). The results indicate that natural processes such as the ion exchange process, rock–water interaction, and the nature of an aquifer are the major sources of contamination in the study region. Spatial analysis of the Piper diagram clearly indicates that the south-eastern and central parts of the study region are of the mixed Ca–Mg–Cl type of groundwater.

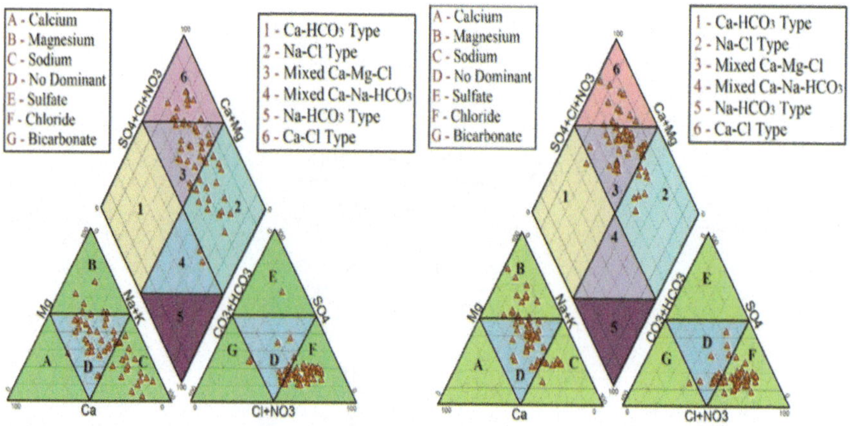

Fig. 9.2 Piper classification of groundwater in the study area

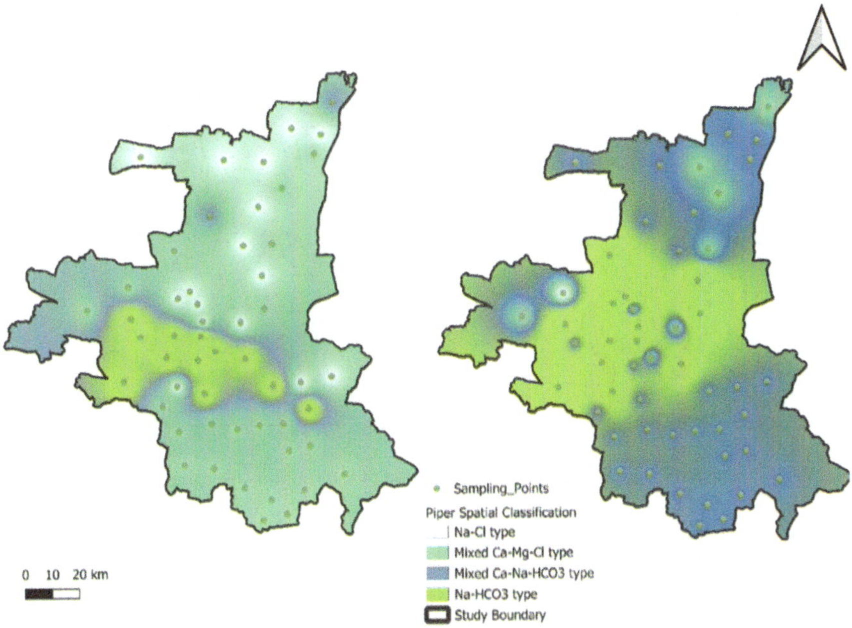

Fig. 9.3 Spatial analysis of Piper-based groundwater classification

9.3.2 SAR

The sodium absorption ratio is an effective and significant index value to evaluate the quality of groundwater for irrigation purposes. In this study, the spatial analysis of the SAR revealed that 1030.77 km^2 of groundwater is excellent, 25.57 km^2 is good, and 0.17 km^2 is permissible for irrigation purposes during the pre-monsoon period. The total area of 1056.51 km^2 is excellent during the post-monsoon period and it indicates that all the samples are excellent and fit for irrigation purposes.

Based on the SAR results, rock–water interaction, weathering, and ion exchange process are the primary factors that govern the quality of groundwater during the pre-monsoon period (Fig. 9.4).

9.3.3 Residual Sodium Carbonate

The residual sodium carbonate index helps evaluate the influence of sodium concentration on groundwater quality (Eaton 1950). In this study, spatial analysis of residual sodium carbonate in the study region covering an area of 995.6 km^2 is satisfactory, 36.95 km^2 is marginally polluted, and 23.96 km^2 is unsatisfactory during the pre-monsoon period. About 998.41 km^2 is satisfactory, 42.01 km^2 is marginally polluted, and 16.09 km^2 is unsatisfactory during the post-monsoon period (Fig. 9.5).

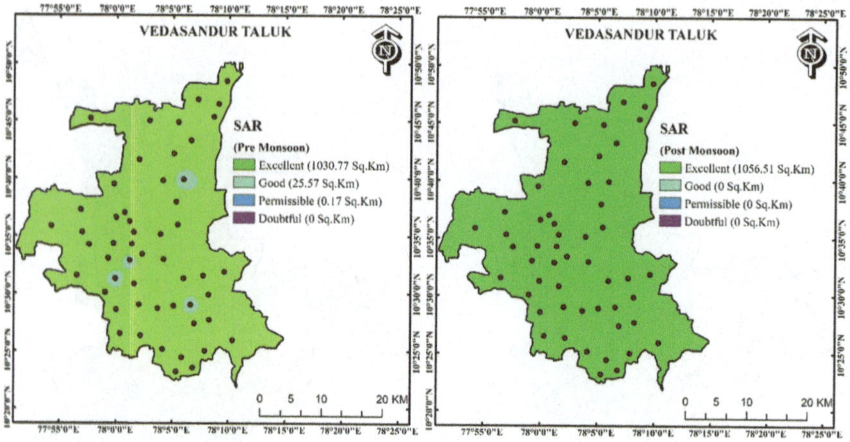

Fig. 9.4 Spatial distribution of the seasonal variation of the sodium absorption ratio

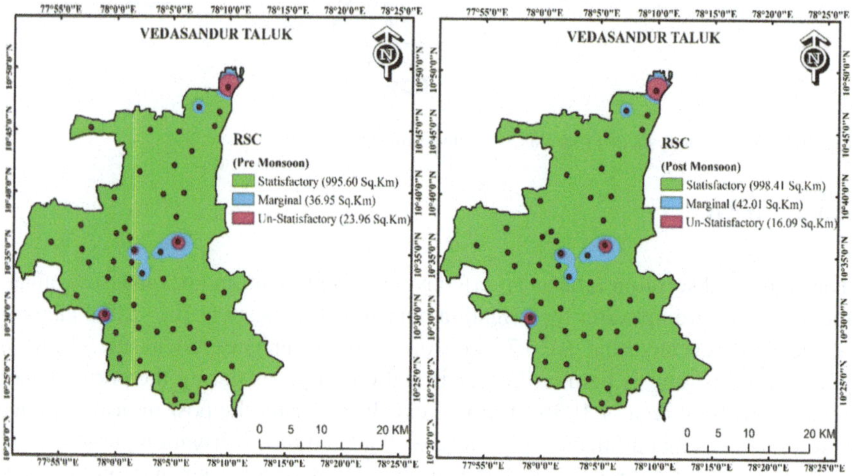

Fig. 9.5 Spatial distribution of the seasonal variation of residual sodium carbonate

The results show that concentration of sodium ions increased between the pre- and post-monsoon periods. The natural sources of contamination such as the presence of minerals in aquifers, release of ions from the rock stratum, and rock–water interaction are the major factors that affect the quality of groundwater in the study region.

9.3.4 Magnesium Hazards

Magnesium ions in groundwater play a vital role in suitability for agricultural purposes. An excess concentration of magnesium in groundwater affects the soil layer and alters the physical properties of the soil. It causes serious damage to crop patterns and yields. In this study, 28.3% of the samples were suitable and 71.7% were unsuitable for irrigation purposes during the pre-monsoon period (Table 9.2). About 16.98% and 83.02% of the samples were suitable and unsuitable for irrigation purposes during the post-monsoon period, respectively (Table 9.3). The spatial analysis of magnesium hazards covering areas of 173.36 km^2 and 56.72 km^2 is

Table 9.2 Irrigation index classification of groundwater in the pre-monsoon season

Range	No. of samples	%	Class
Sodium absorption ratio			
LT 10	49	92.45	Excellent
10–18	3	5.66	Good
18–26	1	1.89	Doubtful
GT 26	0	0.00	Unsuitable
Residual sodium ratio			
LT 1.25	48	90.57	Satisfactory
1.25–2.5	3	5.66	Marginal
GT 2.5	2	3.77	Unsatisfactory
Magnesium hazards			
LT 50	15	28.30	Suitable
GT 50	38	71.70	Unsuitable
Permeability index			
GT 75	12	22.64	Class I
75–25	41	77.36	Class II
LT 25	0	0.00	Class III
Residual sodium bicarbonate			
LT 5	46	86.79	Satisfactory
5–10	7	13.21	Marginal
GT 10	0	0.00	Unsatisfactory
Potential salinity			
LT 5	1	1.89	Satisfactory
5–10	36	67.92	Marginal
GT 10	16	30.19	Unsatisfactory
Kelly ratio			
LT 1	35	66.04	Good
1–2	11	20.75	Doubtful
GT 2	7	13.21	Unsuitable
Percentage sodium			
0–20	6	11.32	Excellent
20–40	19	35.85	Good
40–60	17	32.08	Permissible
60–80	9	16.98	Doubtful
GT 80	2	3.77	Unsuitable

Note: *GT* - Greater than; *LT* - Less than

Table 9.3 Irrigation index classification of groundwater in the post-monsoon season

Range	No. of samples	%	Class
Sodium absorption ratio			
LT 10	53	100.00	Excellent
10–18	0	0.00	Good
18–26	0	0.00	Doubtful
GT 26	0	0.00	Unsuitable
Residual sodium ratio			
LT 1.25	45	84.91	Satisfactory
1.25–2.5	3	5.66	Marginal
GT 2.5	5	9.43	Unsatisfactory
Magnesium hazards			
LT 50	9	16.98	Suitable
GT 50	44	83.02	Unsuitable
Permeability index			
GT 75	11	20.75	Class I
75–25	39	73.58	Class II
LT 25	3	5.66	Class III
Residual sodium bicarbonate			
LT 5	45	84.91	Satisfactory
5–10	7	13.21	Marginal
GT 10	1	1.89	Unsatisfactory
Potential salinity			
T 5	0	0.00	Satisfactory
5–10	32	60.38	Marginal
GT 10	21	39.62	Unsatisfactory
Kelly ratio			
LT 1	46	86.79	Good
1–2	7	13.21	Doubtful
GT 2	0	0.00	Unsuitable
Percentage sodium			
0–20	8	15.09	Excellent
20–40	29	54.72	Good
40–60	14	26.42	Permissible
60–80	2	3.77	Doubtful
GT 80	0	0.00	Unsuitable

Note: *GT* - Greater than; *LT* - Less than

suitable and 883.15 km^2 and 999.79 km^2 is unsuitable for agricultural purposes during the pre- and post-monsoon periods (Fig. 9.6). The results indicate that the concentration of magnesium affects the crop yields and highly influences the quality of groundwater in the study region.

Note: GT - Greater than and LT - Less than for both Table 9.2 and 9.3

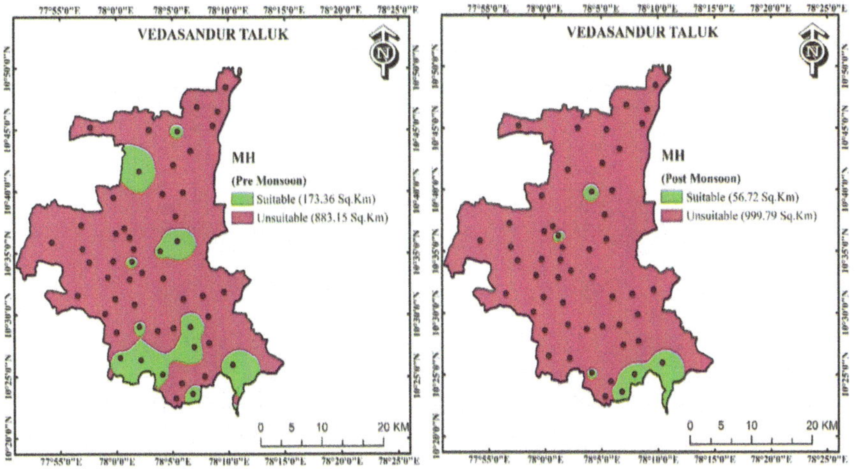

Fig. 9.6 Spatial distribution of the seasonal variation of magnesium hazards

9.3.5 Permeability Index

Permeability index is the most important index value to assess the quality of groundwater for irrigation purposes. The permeable properties of the soil affect the absorption capacity of water through plant roots (Doneen 1964). These also decrease crop yields and affect the soil properties. In this study, 22.64% of the samples belonged to class I and 77.36% belonged to class II for irrigation purposes during the pre-monsoon period (Table 9.2). About 20.75%, 73.85%, and 83.02% of the samples belonged to class I, class II, and class III for irrigation purposes during the post-monsoon period, respectively (Table 9.3). The spatial analysis of the permeability index covering areas of 1056.51 km^2 and 1048.06 km^2 is suitable and 8.45 km^2 is unsuitable for agricultural purposes (Fig.9.7). The results indicate that an excess concentration of magnesium and sodium affects crop yields and highly influences the quality of groundwater in the study region.

9.3.6 Residual Sodium Bicarbonate

Residual sodium bicarbonate is a significant index value to assess the quality of groundwater, and it also indicates the residual amount of sodium in groundwater. An excess concentration of residual sodium bicarbonate affects crop yields and also the soil properties. In this study, 86.79% of the samples were satisfactory and 13.21% were marginally polluted for irrigation purposes during the pre-monsoon period (Table 9.2). About 84.91%, 13.21%, and 1.89% of the samples were satisfactory, marginally polluted, and unsuitable for irrigation purposes during the post-monsoon period, respectively (Table 9.3). The spatial analysis of residual sodium

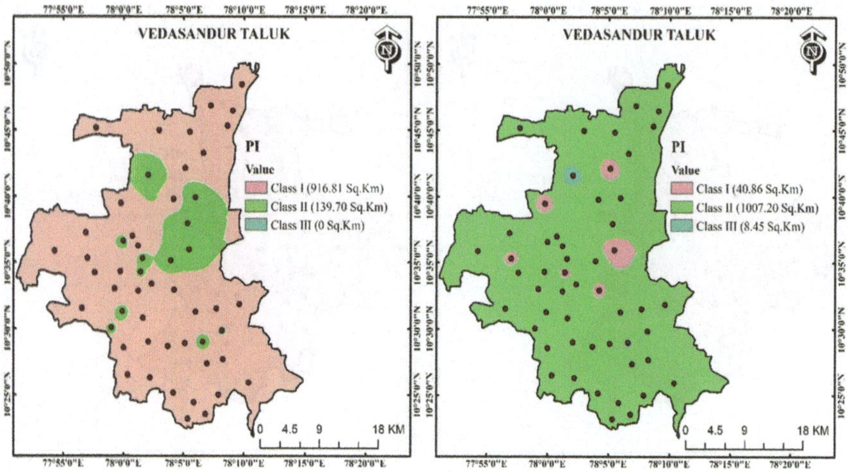

Fig. 9.7 Spatial distribution of the seasonal variation of the permeability index

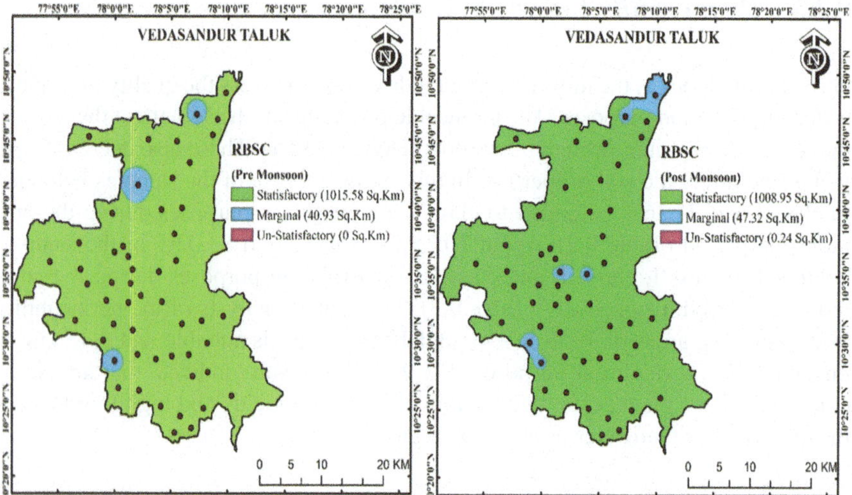

Fig. 9.8 Spatial distribution of the seasonal variation of residual sodium bicarbonate

bicarbonate covering areas of 1015.58 km^2 and 40.93 km^2 is suitable and 1008.95 km^2, 47.32 km^2, and 0.24 km^2 is suitable, moderately suitable, and unsuitable for agricultural purposes, respectively (Fig. 9.8). The results indicate that the concentration of residual sodium ions in groundwater affects the quality of groundwater for irrigation purposes and soil layers.

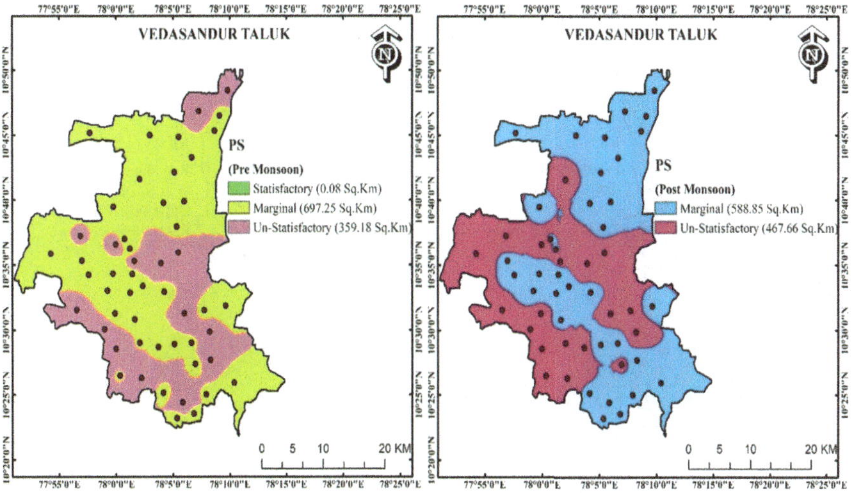

Fig. 9.9 Spatial distribution of the seasonal variation of potential salinity

9.3.7 *Potential Salinity*

Salinity plays a vital role in deciding the suitability of groundwater for agricultural purposes. An excess concentration of salinity affects the chemical composition of groundwater (Richards 1968).

In this study, 1.89% of the samples were satisfactory, 67.92% were marginally polluted, and 30.19% were unsuitable for irrigation purposes during the pre-monsoon period (Table 9.2). About 60.38% and 39.62% of the samples were marginally polluted and unsuitable for irrigation purposes during the post-monsoon period, respectively (Table 9.3). The spatial analysis of potential salinity covering areas of 0.08 km^2, 697.25 km^2, and 359.18 km^2 is suitable, moderately suitable, and unsuitable for agricultural purposes during the pre-monsoon period, respectively. About 588.85 km^2 and 467.66 km^2 of the area are suitable and unsuitable, respectively (Fig. 9.9).

9.3.8 *Kelly Ratio*

Kelly developed a numerical expression to evaluate the concentration of sodium, calcium, and magnesium for checking the suitability of groundwater for irrigation purposes (Kelley 1963). The value exceeding 1 is unsuitable for irrigation purposes. In this study, 66.04% of the samples were suitable and 34.96% were unsuitable for irrigation purposes during the pre-monsoon period (Table 9.2). About 86.79% and 13.21% of the samples were suitable and unsuitable for irrigation purposes during the post-monsoon period, respectively (Table 9.3). The spatial analysis of kelly

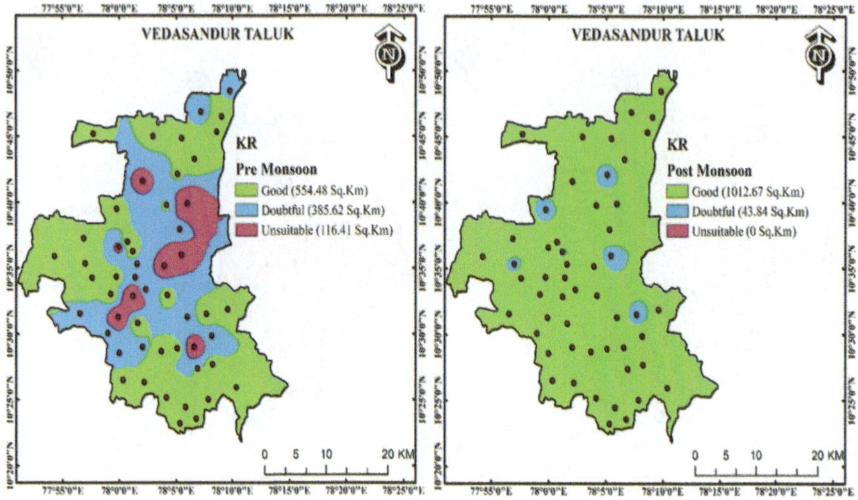

Fig. 9.10 Spatial distribution of the seasonal variation of the Kelly ratio

ratio revealed that 554.48 km^2, 385.62 km^2 and 116.41 km^2 of area are suitable, moderate, and unsuitable for irrigation purposes. The area of about 1012.67 km^2 and 43.84 km^2 is suitable and unsuitable for agricultural purposes, respectively (Fig. 9.10).

9.3.9 Percentage Sodium

In this study, 11.32%, 35.85%, 32.08%, 16.98%, and 3.77% of the samples were excellent, good, permissible, doubtful, and unsuitable for irrigation purposes during the pre-monsoon period, respectively (Table 9.2). About 15.09%, 54.72%, 26.42%, and 3.77% of the samples were excellent, good, permissible, and doubtful for irrigation purposes during the post-monsoon period, respectively (Table 9.3). The spatial analysis results of sodium percentage revealed that **5.31** km^2, 401.54 km^2, 543.85 km^2, 104.28 km^2 and 1.53 km^2 during pre-monsoon and **25.41** km^2, 729.58 km^2, 299.86 km^2 and 1.66 km^2 of area excellent, good, permissible, doubtful and unsuitable for agriculture purpose (Fig. 9.11). The results indicate that the concentration of magnesium affects crop yields and highly influences the quality of groundwater in the study region (Wilcox 1955).

9.3.10 USSL

The US Salinity Laboratory (USSL 1954) has suggested a diagram to understand the concentration of sodium ions with salinity. In this study, 65% of the samples fell under the class of C3-S1, 23% under C3-S2, 6% under C4-S1, and 6% under C3-S3.

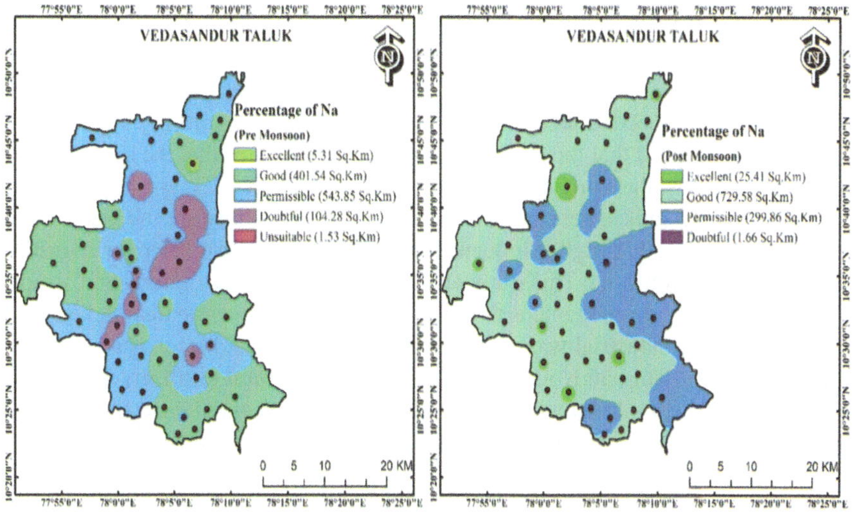

Fig. 9.11 Spatial distribution of the seasonal variation of percentage sodium

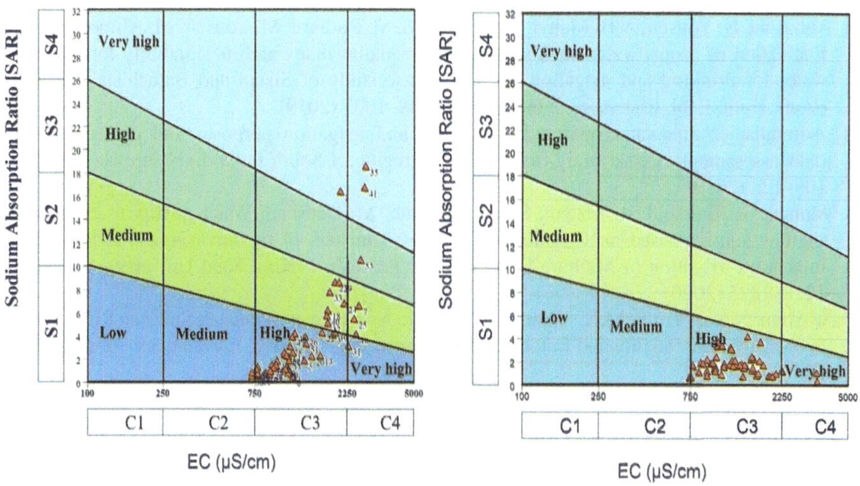

Fig. 9.12 USSL classification of groundwater in both seasons

The test results indicate high salinity with low sodium during the pre-monsoon period. The groundwater samples during the post-monsoon period fell under the category of C3-S1, and this indicates that high salinity highly influences the quality of groundwater during the post-monsoon period (Fig. 9.12). The results indicate that rock–water interaction, weathering, and ion exchange process and anthropogenic activities such as municipal waste disposal, sewage disposal, and dumping yard leachates are the major reasons for salinity concentration in groundwater.

9.4 Conclusions

Recently, assessing the suitability of groundwater for irrigation purposes has become the much needed area of research in arid and semi-arid regions in India. This study concluded that the Piper diagram clearly indicates that the south-eastern and central parts of the study region are of the mixed Ca–Mg–Cl type of groundwater. The irrigation indices revealed that magnesium and salinity are the major water quality parameters that highly influence the quality of groundwater for irrigation purposes in the study area. This study also identified rock–water interaction, evaporation, and other anthropogenic activities such as municipal waste disposal, sewage effluent disposal, industrial waste disposal, and dumping yard leachates as sources of contamination. This study recommends that, to use natural fertilizers, proper and periodic waste disposal systems and maintenance of water supply and sewage pipelines are required in the study region to maintain a sustainable practice.

References

A. Abbasnia, N. Yousefi, A.H. Mahvi, R. Nabizadeh, M. Radfard, M. Yousefi, M. Alimohammadi, Evaluation of groundwater quality using water quality index and its suitability for assessing water for drinking and irrigation purposes: Case study of Sistan and Baluchistan province (Iran). Hum. Ecol. Risk Assess. Int. J. **25**(4), 988–1005 (2019)

N. Adimalla, Groundwater quality for drinking and irrigation purposes and potential health risks assessment: A case study from semi-arid region of South India. Exposure Health **11**(2), 109–123 (2019)

S. Ahmed, S. Khurshid, R. Madan, B.A.A. Amarah, M. Naushad, Water quality assessment of shallow aquifer based on Canadian Council of Ministers of the environment index and its impact on irrigation of Mathura District, Uttar Pradesh. J. King Saud University-Sci. **32**(1), 1218–1225 (2020)

G.A. Aliyu, N.R.B. Jamil, M.B. Adam, Z. Zulkeflee, Assessment of Guinea Savanna River system to evaluate water quality and water monitoring networks. Global J. Environ. Sci. Manage. **5**(3), 345–356 (2019)

E. Asadi, M. Isazadeh, S. Samadianfard, M.F. Ramli, A. Mosavi, N. Nabipour, et al., Groundwater quality assessment for sustainable drinking and irrigation. Sustainability **12**(1), 177 (2020)

A. Azhdarpoor, M. Radfard, M. Pakdel, A. Abbasnia, A. Badeenezhad, A.A. Mohammadi, M. Yousefi, Assessing fluoride and nitrate contaminants in drinking water resources and their health risk assessment in a semiarid region of Southwest Iran. Desalin. Water Treat. **149**, 43–51 (2019)

P. Balamurugan, S. Pauline, M. Kirubakaran, M. Thomas, Assessment of inverse fluidized bed reactor on the treatment efficiency of distillery spent wash water. Int. J. Environ. Sci. Technol., 1–14 (2021)

P. Balamurugan, P.S. Kumar, K. Shankar, R. Nagavinothini, P.J. Sajil Kumar, Impact of climate and anthropogenic activities on groundwater quality for domestic and irrigation purposes in Attur region, Tamil Nadu, India. Desalin. Water Treat. **208**, 172–195 (2020a)

P. Balamurugan, P.S. Kumar, K. Shankar, R. Nagavinothini, K. Vijayasurya, Non-carcinogenic risk assessment of groundwater in southern part of Salem District in Tamil Nadu, India. J. Chil. Chem. Soc. **65**(1), 4697–4707 (2020b)

K. Balasubramani, K.R. Murthy, M. Gomathi, K. Kumaraswamy, Integrated assessment of groundwater resources in a semi-arid watershed of South India: Implications for irrigated agriculture. Geo J. **85**(6), 1701–1723 (2020)

M. Batarseh, E. Imreizeeq, S. Tilev, M. Al Alaween, W. Suleiman, A. M. Al Remeithi, M. K. Al Tamimi, M. Al Alawneh, Dataset for the physio-chemical parameters of groundwater in the Emirate of Abu Dhabi UAE. Data in Brief 38, 107353 (2021). https://doi.org/10.1016/j.dib.2021.107353

L.D. Doneen, *Notes on Water Quality in Agriculture. Department of Water Science and Engineering* (University of California, Davis, 1964)

F.M. Eaton, Significance of carbonates in irrigation waters. Soil Sci. **69**(2), 123–134 (1950)

C.J. Jesudhas, S.M. Chidambaram, K. Muniraj, Isolation of wells contaminated by tannery industries using principal component analysis. Arab. J. Geosi. **10**(14), 1–8 (2017)

C.J. Jesudhas, A. Chinnasamy, K. Muniraj, A, Sundaram, Assessment of vulnerability in the aquifers of rapidly growing sub-urban: a case study with special reference to land use. Arabian Journal of Geosciences, **14**(1) (2021). https://doi.org/10.1007/s12517-020-06439-8

D. Karunanidhi, P. Aravinthasamy, P.D. Roy, R.M. Praveenkumar, K. Prasanth, S. Selvapraveen, K. Srinivasamoorthy, Evaluation of non-carcinogenic risks due to fluoride and nitrate contaminations in a groundwater of an urban part (Coimbatore region) of South India. Environ. Monit. Assess. **192**(2), 102 (2020)

W.P. Kelley, Use of saline irrigation water. Soil Sci. **95**(6), 385–391 (1963)

M. Kirubakaran, C. Ashokraj, J. Colins Johnny, R. Anjali, Groundwater quality analysis using WQI and GIS techniques: A case study of Manavalakurichi in Kanyakumari District, Tamil Nadu, India. Int J Innov Sci Eng Technol **2**(11), 341–347 (2015)

M. Kirubakaran, J.C. Johnny, C. Ashokraj, S. Arivazhagan, A geostatistical approach for delineating the potential groundwater recharge zones in the hard rock terrain of Tirunelveli taluk, Tamil Nadu, India. Arab. J. Geosci. **9**(5), 382 (2016)

S. Kouadri, C.B. Pande, B. Panneerselvam, K.N. Moharir, A. Elbeltagi, Prediction of irrigation groundwater quality parameters using ANN, LSTM, and MLR models. Environ. Sci. Pollut. Res., 1–25 (2021)

P.S. Kumar, P. Balamurugan, Evaluation of groundwater quality for irrigation purpose in Attur Taluk, Salem, Tamil Nadu, India. Water Energy Int. **61**(4), 59–64 (2018a)

P.S. Kumar, P. Balamurugan, Evaluation of groundwater quality for irrigation purpose in attur taluk, Salem, Tamil Nadu, India. Water Ener. Int. **61**(4), 59–64 (2018b)

P.S. Kumar, P. Balamurugan, Suitability of ground water for irrigation purpose in Omalur Taluk, Salem, Tamil Nadu, India. Indian J. Ecol **46**(1), 1–6 (2019)

B. Murugesan, S. Karuppannan, A.T. Mengistie, M. Ranganathan, G. Gopalakrishnan, Distribution and trend analysis of COVID-19 in India: Geospatial approach. J. Geogr. Stud. **4**(1), 1–9 (2020)

V.P. Nawale, D.B. Malpe, D. Marghade, R. Yenkie, Non-carcinogenic health risk assessment with source identification of nitrate and fluoride polluted groundwater of Wardha sub-basin, central India. Ecotoxicol. Environ. Saf. **208**, 111548 (2021)

R.A. Nejad, S.A. Mohammed, D.F. Hamamin, Irrigation water susceptibility indexing method, using pesticide DRASTIC and water quality index, for Basara Basin; Kurdistan region-Iraq. J. Environ. Treat. Tech. **9**(2), 435–445 (2021)

C.B. Pande, K.N. Moharir, S.K. Singh, B. Dzwairo, Groundwater evaluation for drinking purposes using statistical index: Study of Akola and Buldhana districts of Maharashtra, India. Environ. Develop. Sustainabil., 1–19 (2019)

B. Panneerselvam, S. Karuppannan, K. Muniraj, Evaluation of drinking and irrigation suitability of groundwater with special emphasizing the health risk posed by nitrate contamination using nitrate pollution index (NPI) and human health risk assessment (HHRA). Hum. Ecol. Risk Assess. Int. J. **27**(5), 1324–1348 (2020a)

B. Panneerselvam, S.K. Paramasivam, S. Karuppannan, N. Ravichandran, P. Selvaraj, A GIS-based evaluation of hydrochemical characterisation of groundwater in hard rock region, South Tamil Nadu, India. Arab. J. Geosci. **13**(17), 1–22 (2020b)

B. Panneerselvam, K. Muniraj, M. Thomas, N. Ravichandran, GIS-Based Legitimatic Evaluation of Groundwater's Health Risk and Irrigation Susceptibility Using Water Quality Index, Pollution Index, and Irrigation Indexes in Semiarid Region. In: Pande, C.B., Moharir, K.N. (eds) Groundwater Resources Development and Planning in the Semi-Arid Region. Springer, Cham. 239–268 (2021)

B. Panneerselvam, K. Muniraj, C. Pande, N. Ravichandran, Prediction and evaluation of groundwater characteristics using the radial basic model in semi-arid region, India. Int. J. Environ. Anal. Chem., 1–17 (2021a)

B. Panneerselvam, K. Muniraj, M. Thomas, N. Ravichandran, B. Bidorn, Identifying influencing groundwater parameter on human health associate with irrigation indices using the automatic linear model (ALM) in a semi-arid region in India. Environ. Res. **202**, 111778 (2021b)

B. Panneerselvam, K. Muniraj, C. Pande, N. Ravichandran, M. Thomas, S. Karuppannan, Geochemical evaluation and human health risk assessment of nitrate-contaminated groundwater in an industrial area of South India. Environ. Sci. Pollut. Res., 1–18 (2021c)

A.M. Piper, A graphic procedure in the geochemical interpretation of water-analyses. Transactions, American Geophysical Union, **25**, 914–923 (1944)

G. Rajendran, J. Rooby, V. Govindan, Influence of groundwater quality on human health and its suitability for drinking and irrigation uses: A review. Int. J. Environ. Anal. Chem., 1–19 (2021)

N.S. Rao, B. Sunitha, L. Sun, B.D. Spandana, M. Chaudhary, Mechanisms controlling groundwater chemistry and assessment of potential health risk: A case study from South India. Geochemistry **125568** (2019)

R. Ravi, S. Aravindan, K. Shankar, P. Balamurugan, Suitability of groundwater quality for irrigation in and around the main Gadilam river basin on the east coast of southern India. Arch. Agri. Environ. Sci. **5**(4), 554–562 (2020). https://doi.org/10.26832/24566632.2020.0504019

A.L. Richards, Diagnosis and improvement of saline and alkali soils. Agric Handb **60**, 210–220 (1968)

A. Rosanoff, Changing crop magnesium concentrations: Impact on human health. Plant Soil **368**(1), 139–153 (2013)

F. Rufino, G. Busico, E. Cuoco, T.H. Darrah, D. Tedesco, Evaluating the suitability of urban groundwater resources for drinking water and irrigation purposes: An integrated approach in the Agro-Aversano area of Southern Italy. Environ. Monit. Assess. **191**(12), 768 (2019)

F. Sarani, A.G. Ahangar, A. Shabani, Predicting ESP and SAR by artificial neural network and regression models using soil pH and EC data (Miankangi Region, Sistan and Baluchestan Province, Iran). Arch. Agron. Soil Sci. **62**(1), 127–138 (2016)

M. Senthilkumar, D. Gnanasundar, R. Arumugam, Identifying groundwater recharge zones using remote sensing & GIS techniques in Amaravathi aquifer system, Tamil Nadu, South India. Sustain. Environ. Res. **29**(1), 15 (2019)

C.M. Sharma, S. Kang, L. Tripathee, R. Paudyal, M. Sillanpää, Major ions and irrigation water quality assessment of the Nepalese Himalayan rivers. Environ. Develop. Sustain., 1–13 (2020)

K. Shunmugapriya, B. Panneerselvam, K. Muniraj, N. Ravichandran, P. Prasath, M. Thomas, K. Duraisamy, Integration of multi criteria decision analysis and GIS for evaluating the site suitability for aquaculture in southern coastal region, India. Mar. Pollut. Bull. **172**, 112907 (2021)

U. S. Salinity Laboratory (USSL), Diagnosis and improvement of saline and alkaline soil; USSA handbook No. 60 71 (1954)

Z. Wang, M.U. Hassan, F. Nadeem, L. Wu, F. Zhang, X. Li, Magnesium fertilization improves crop yield in most production systems: A meta-analysis. Front. Plant Sci. **10**, 1727 (2020)

L.V. Wilcox, Classifcation and use of irrigation waters. USDA Circular No 969, Washington, DC, 19 (1955)

P. Xu, W. Feng, H. Qian, Q. Zhang, Hydrogeochemical characterization and irrigation quality assessment of shallow groundwater in the central-western Guanzhong basin, China. Int. J. Environ. Res. Public Health **16**(9), 1492 (2019)

Q. Zhang, H. Qian, P. Xu, K. Hou, F. Yang, Groundwater quality assessment using a new integrated-weight water quality index (IWQI) and driver analysis in the Jiaokou Irrigation District, China. Ecotoxicol. Environ. Saf. **212**, 111992 (2021)

Chapter 10
Evaluation of Groundwater Quality and Its Suitability for Drinking and Cultivation Practices in and around Around the Deltaic Regions of South India Using DWQI, IWQI, and GIS

Shankar Karuppannan, Balamurugan Panneerselvam, Muhammed Haji, Bagyaraj Murugesan, Hassen Shube, and Nafyad Serre Kawo

Abstract The main objective of this research was to assess the quality of groundwater and its appropriateness for drinking and irrigation, along the deltaic areas of the Cauvery River in South India. The physicochemical parameters of groundwater samples were determined, and the results were compared to national and international drinking water quality norms. The pH of groundwater is alkaline, and the concentrations of HCO_3^- and Cl^- in most samples exceeded the permissible drinking limit. The spatial interpolation of the drinking water quality index (DWQI) shows that 54.5% and 45.5% of the groundwater in the study region is categorized as excellent and good water for drinking, respectively. Residual sodium carbonate (RSC) levels are within the safe range for agrarian purposes throughout the year. According to the permeability index, groundwater is moderate to good for agrarian purposes. The region's groundwater quality is adequate and within safe limits. In contrast, the quality of groundwater in the region generally exceeds allowable limits and has been contaminated by anthropogenic activities and industrial effluent discharge into the river. This study has suggested that the existing norms and regulations

S. Karuppannan (✉) · M. Haji · H. Shube
Department of Applied Geology, School of Applied Natural Sciences, Adama Science and Technology University, Adama, Ethiopia

B. Panneerselvam
Department of Civil, Architectural and Environmental Engineering, University of Naples Federico II, Naples, Italy

B. Murugesan
Department of Geology, College of Natural and Computational Sciences, Debre Berhan University, Debre Berhan, Ethiopia

N. Serre Kawo
School of Natural Resources, University of Nebraska-Lincoln, Lincoln, NE, USA

B. Panneerselvam et al. (eds.), *Climate Change Impact on Groundwater Resources*, https://doi.org/10.1007/978-3-031-04707-7_10

be strictly enforced in order to prevent illegal discharge from various industries along with the catchment areas.

Keywords Groundwater quality · Drinking water quality index · Irrigation indices · GIS · South India

10.1 Introduction

Water is never found in its pure form on the earth's crust; minerals suspended in a solvent are present in all groundwater, the composition of which varies depending on the shallow subsurface environment, groundwater flow rate, and aquifer origin (Chenini et al. 2010; Aravindan et al. 2008, 2010). Precipitation is partially liberated from minerals until it comes into contact with various constituents of dirt, which are conveyed as an arrangement by the flow of water. The dissolvability of the minerals present in the groundwater, the length of time that the water is in interaction with the rocks, and the amount of dissolved carbon dioxide in the water all have an effect on the cation and anion concentrations (Haji et al. 2021a). Regardless of natural mineralization, humans can adversely alter the synthetic nature of groundwater through allowable means dictated by its compound, physical, and bacteriological properties. In reality, the hydrogeochemical processes responsible for changing the compound synthesis of groundwater differ. Groundwater quality has deteriorated as a result of various natural and anthropogenic activities that profoundly mineralize water into freshwater layers by removing creature squandering, sewage, or other modern industrial by-products (Kavitha et al. 2019a, b). One of the major concerns is the deterioration of the quality of underground water in semi-arid and arid areas (Aravinthasamy et al. 2021; Karunanidhi et al. 2021; Chandrasekar et al. 2021; Panneerselvam et al. 2021d).

10.2 Background of This Study

Groundwater quality has been surveyed and observed on a regular basis for many years, utilizing the geographic information system (GIS) procedure associated with the inverse distance weighting (IDW) introduction strategy, which has been ascertained to be a formidable tool for evaluating and dissecting spatial data of groundwater sources (Shankar et al. 2010, 2011a, b; Venkateswaran et al. 2012; Balamurugan et al. 2020b, d; Bagyaraj et al. 2020; Sarfo and Shankar 2020). In this study, the GIS technique was used to assess the spatial distribution of various groundwater quality parameters. In various parts of South India, many researchers have conducted studies on groundwater quality based on physicochemical parameters (Aravindan et al. 2011; Jecudhas et al. 2017; Johnny et al. 2018; Panneerselvam et al. 2020a, b, 2021a, b, c; Ravi et al. 2020). In this study region, due to their immense usage, pesticides and inorganic fertilizers have dissolved in soil particles, percolating into the aquifer system and affecting the groundwater quality. Numerous research findings have been incorporated into various groundwater quality indices (GWQIs) for the evaluation and monitoring of various water contamination levels (Shankar and Kawo 2019; Balamurugan et al. 2020c).

In most developing countries, subsurface water is the foremost source of freshwater for domestic consumption, agriculture, and industrial uses (Shunmugapriya et al. 2021). Groundwater is crucial because it meets all of civilization's needs for domestic use, irrigation, and industrialization. Due to the chemical change of meteoric water and the re-energizing of the spring framework, each groundwater framework in any region has inimitable chemistry (Senthil Kumar and Shankar 2014; Wagh et al. 2020; Ashabir et al. 2020; Kadam et al. 2019). Groundwater's chemical constituents change as it moves from catchments to discharge regions, reflecting changes caused by various geochemical processes (Haji et al. 2021b). Water quality and quantity are important factors in sustaining healthy and prosperous human populations. Water's hydrogeochemistry is an important factor in determining whether it should be used for domestic, agrarian, or industrial purposes.

10.3 Objective of This Study

The primary goal of this research was to evaluate groundwater quality and suitability for drinking and agrarian purposes using the drinking water quality index (DWQI) and geographic information systems (GISs).

10.4 Materials and Methods

10.4.1 Study Area

The Orathanadu Taluk was chosen for study in the Thanjavur District, situated on the eastern margin of Tamil Nadu. The research area lies between 10° 24′ to 10° 48'N and 79° 07′ to 79° 24′ E. It begins in the north at the Coleroon River. The research area covers 585 km^2 and is located on toposheets 58 N/02, N/03, N/05, N/06, and N/07 (Fig. 10.1). The research area is located at the tail end of the Cauvery River basin's new delta and has a flat topography. The primary water source for drinking and irrigation is the Grand Anicut canal, which covers a major part of the region. The mean annual precipitation is 1180 mm, and it occurs mostly during the northeast monsoon. The region's minimum and maximum temperatures range from 23 °C to 37 °C. Geologically, the research area comprises cretaceous, tertiary, and alluvial deposits, with the alluvial and tertiary deposits occupying most of the area.

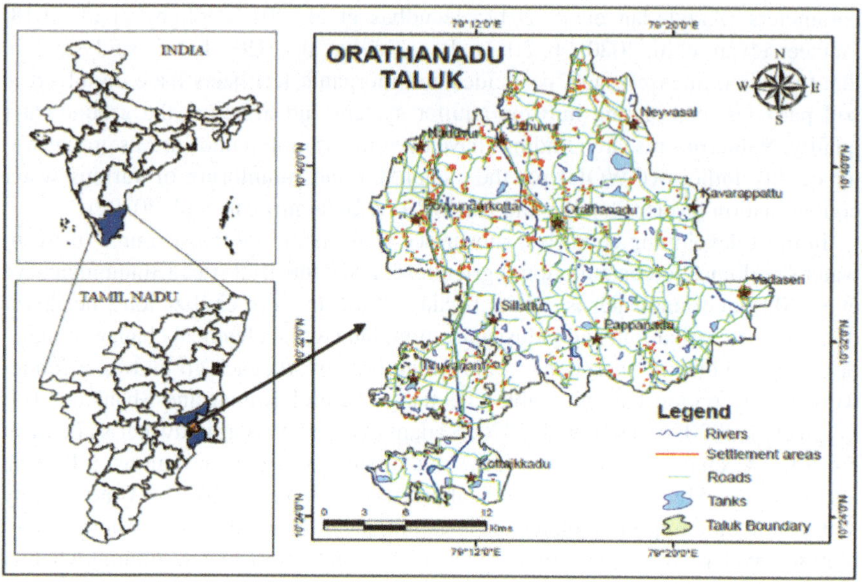

Fig. 10.1 Location of the study area

The cretaceous formations are found in small patches in the south-western and western parts of the region. Tertiary Cuddalore Sandstone is overlain on the coast. These sandstones are covered by a thin layer of wind-blown sandy clay, unconsolidated sand, bound clay sand, and speckled clay. These formations are 30–400 m thick.

10.4.2 Sampling and Analytical Techniques

Groundwater samples (n = 11) were collected from the study area during the pre-monsoon season (PRM) (June 2016) using the standard American Public Health Association (APHA, 2005, 2012) methods. Prior to sample collection, all the containers were washed with 10% nitric acid solution with double-distilled water to avoid contamination. The samples were filled in pre-cleaned high-density polyethylene (HDPE) containers. Physical parameters like pH and electrical conductivity (EC) were measured in the field itself using portable handheld meters. The collected groundwater samples were filtered in the field using a 0.45-μm membrane filter paper using a portable vacuum hand pump and preserved at <4 °C temperature as per the recommended protocols till the analysis. Anions and cations in the groundwater were analyzed using international standard analytical protocols (APHA 2012). The titrimetric method was adopted for determining calcium (Ca^{2+}), magnesium (Mg^{2+}), bicarbonate (HCO_3^-), and chloride (Cl^-). Sodium (Na^+) and potassium (K^+) were determined by a flame photometer; nitrate (NO_3^-), fluoride (F^-), sulfate (SO_4^{2-}) were analyzed using a spectrophotometer. The quality assurance and quality control (QA/QC) procedure of the data was considered during the study. The ion balance error (IBE) was examined to cross-check the data accuracy, which is within ±10% (Freeze and Cherry 1979). The ion balance was calculated using Eq. 10.1.

$$IBE = \frac{\sum \text{Cations} - \sum \text{Anions}}{\sum \text{Cations} - \sum \text{Anions}} \times 100 \tag{10.1}$$

The spatial dispersal maps of groundwater quality parameters were plotted using ArcGIS 10.8 software with (IDW) the interpolation method (Aravindan and Shankar 2011a, b; Balamurugan et al. 2020a; Sarfo and Shankar 2020; Soujanya et al. 2020). The hydrochemical facies of groundwater were evaluated using AquaChem 4.0 software to produce a Piper (1944) diagram.

10.4.3 The Drinking Water Quality Index (DWQI)

The DWQI was developed to evaluate groundwater suitability, and it provides composite information on the impact of each water quality parameter on the overall groundwater quality (Panneerselvam et al. 2022). The DWQI was estimated using pH, total dissolved solids (TDSs), major anions, and cations. The calculation of the DWQI was based on the study by Brown et al. (1972). The groundwater samples' analyzed quality was compared to the Bureau of Indian Standards (BIS 2012), which is also considered for the computation of the DWQI. After selecting and scaling the desired parameters for the DWQI, weight (w_i) was assigned to individual parameters based on their relative importance, ranging from 1 to 5. The relative weight (W_i) of variables was calculated with the help of Eq. 10.2 (Table 10.1). Following that, the quality rating scale (Q_i), SI_i index, and, ultimately, the DWQI value were evaluated using Eqs. 10.3, 10.4, and 10.5, respectively.

$$Wi = \frac{w_i}{\sum_{i=1}^{n} w_i} \tag{10.2}$$

$$Q_i = \frac{C_i}{S_i} \times 100 \tag{10.3}$$

$$SI_i = W_i \times Q_i \tag{10.4}$$

$$\text{DWQI} = \Sigma_n^{i=1} SI_I \tag{10.5}$$

where n is the number of parameters, C_i is concentration of each water sample (mg/L), and S_i is taken from the BIS (2012) guidelines. Based on DWQI scores (Sahu and Sikdar 2008), groundwater assessments were classified into five clusters:

Table 10.1 A comparative statistical summary of groundwater analysis according to WHO and BIS

	PRM				
	February 2016			WHO	BIS
Ions	Min	Max	Average	2011	2012
pH	7.9	8.4	8.1	6.5–8.5	6.5–8.5
EC	170	1370	684.1	1500	–
TDS	86	764	349.1	500	500
TH	65	315	178.1	500	500
Ca^{2+}	8	40	27.5	75	75
Mg^{2+}	4.9	52.2	26.6	50	30
Na^{+}	9	154	64.5	200	–
K^{+}	0.10	27	6.7	–	–
Cl^{-}	11	269	99.4	250	200
HCO_3^{-}	68.9	427	173.5	200	200
SO_4^{2-}	1	50	18.8	200	200
NO_3^{-}	0.05	11	2.9	50	45
F^{-}	0.05	0.8	0.3	1.5	1

Note: EC in μS/cm; all ions in mg/L
Min minimum, *Max* maximum

exemplary (<50), appropriate (50–100), unsatisfactory (100–200), extremely poor (200–300), and unacceptable (>300).

10.4.4 The Irrigation Water Quality Indices (IWQIs)

Groundwater is essential to agriculture in the study region, where there are a large number of agricultural professionals, which will have a significant impact on the rural economy and is a substantial source of income for villagers in South India. Hence, it is important to appraise the aptness of groundwater for agricultural purposes based on the agrarian water quality indices. Sodium adsorption ratio (SAR), percentage sodium (% Na), permeability index (PI), and residual sodium carbonate (RSC) can be described by the following equations to assess the appropriateness of groundwater for agrarian purposes:

$$TH = 2.497\,Ca + 4.115\,\mathrm{Mg} \tag{10.6}$$

$$SAR = \frac{Na^{+}}{\left(\frac{Ca^{2+} + \mathrm{Mg}^{2+}}{2}\right)^{1/2}} \tag{10.7}$$

$$\%Na = \frac{(Na + \mathrm{K}) \times 100}{Ca + \mathrm{Mg} + Na + \mathrm{K}} \tag{10.8}$$

$$PI = \frac{\sqrt{HCO_3} + Na}{\sqrt{(Na + Mg + Ca}} \tag{10.9}$$

$$RSC = [HCO_3 + CO_3] - [Ca + Mg] \tag{10.10}$$

10.5 Results and Discussion

Overall, the summarized statistics of all groundwater parameters with the statistical measures and their drinking water quality limits as suggested by the BIS (2012) and the World Health Organization (WHO 2011) for drinking water quality are displayed in Table 10.1.

10.5.1 Physical Characteristics of Groundwater

10.5.1.1 pH

According to the BIS standards, the pH of groundwater for domestic use should be between 6.5 and 8.5. During the Pre Monsoon Season (PRM) period, the pH values for various groundwater samples ranged from 7.9 to 8.4. Groundwater is found to be alkaline in nature. Owing to the presence of alkalis in the fluvial sediments adjacent to the coast, the study area is dominated by alkaline water (Fig. 10.2a).

10.5.1.2 Electrical Conductivity (EC)

Rainwater has electrical conductivity (EC) of 0.05 μS/cm (Hem 1991). During the PRM season, the EC of groundwater was found to be in the range of 170–1370 μS/cm (Table 10.1; Fig. 10.2b). The allowable limit for EC, according to WHO (2008), is 1000 s/cm (see Table 10.1). As a result, all of the groundwater samples were found to be safe to drink. The presence of high EC values in the research region might be attributed to the addition of some salts as a result of the prevalent agricultural activities.

10.5.1.3 Total Dissolved Solids (TDSs)

TDS levels are a critical parameter in potable water and other water quality criteria. To determine the acceptability of groundwater for any purpose, the groundwater has to be classified based on its hydrochemical characteristics as determined by TDS

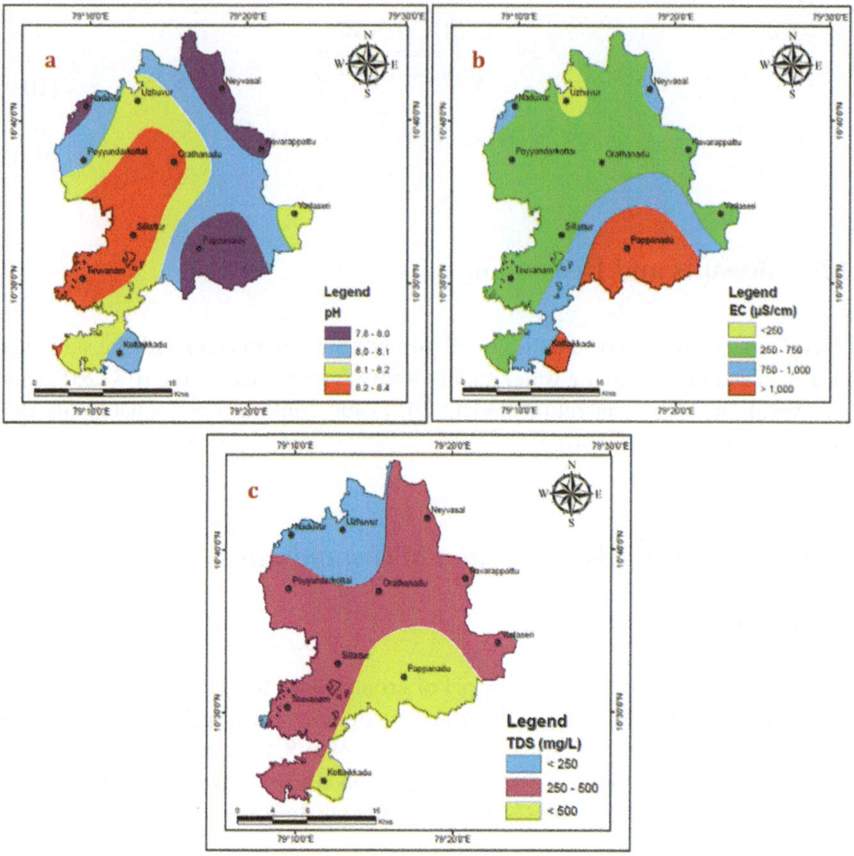

Fig. 10.2 Spatial distributions of pH (**a**), EC (**b**), and TDSs (**c**)

values (Freeze and Cherry 1979). The TDS values ranged between 86 and 764 mg/L, and the results of Freeze and Cherry's classification indicate that all of the samples were fresh (Table 10.1 and Fig. 10.2c).

10.5.2 Chemical Characteristics of Groundwater

10.5.2.1 Major Cations

Calcium (Ca^{2+})

Ca^{2+} ion concentration varied from 8 mg/L to 40 mg/L, with a mean value of 27.5 mg/L. The allowable limit of Ca^{2+} ions is 75–200 mg/L as per the BIS (2012) and WHO (2011) norms. All samples were found to be below the tolerable limit of drinking and in good groundwater quality zones (Fig. 10.3a).

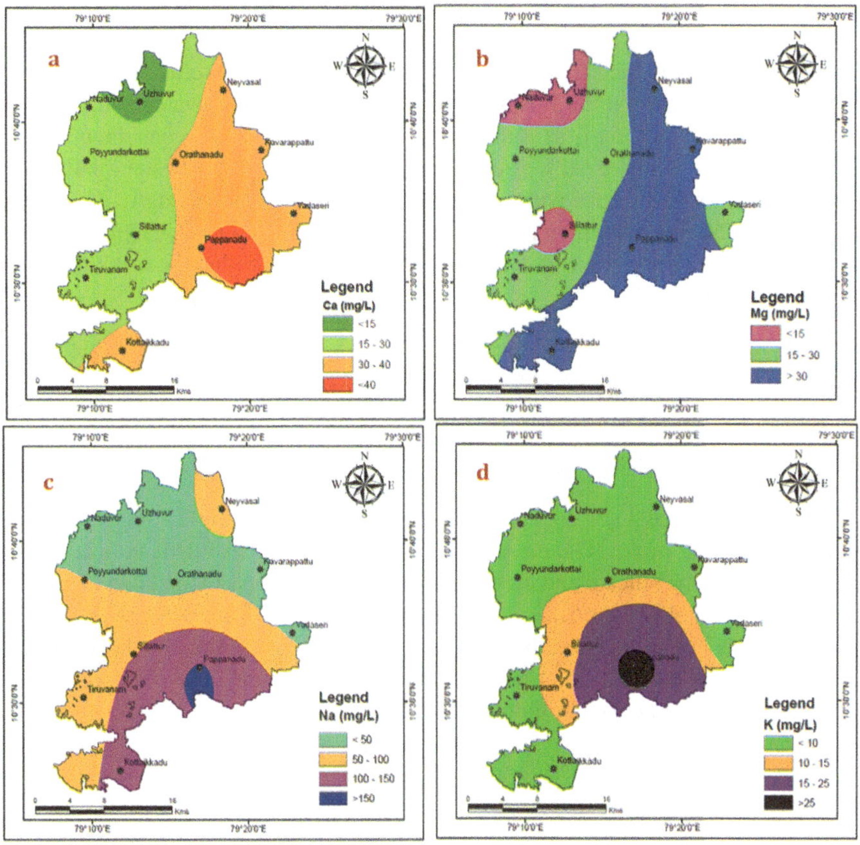

Fig. 10.3 Spatial distributions of Ca^{2+} (a), Mg^{2+} (b), Na^{+} (c), and K^{+} (d)

Magnesium (Mg^{2+})

Mg^{2+} ion contents ranged from 4.9 to 52.2 mg/L, with a mean value of 26.6 mg/L. According to the BIS (2012) and WHO (2011) standards, the permissible drinking limit for Mg^{2+} ions is 30–100 mg/L. Mg^{2+} concentrations in groundwater were found to be greater than 30 mg/L in the north-eastern (NE), south-eastern (SE), and eastern (E) parts of the research area (Fig. 10.3b). Mg^{2+} is typically derived from SiO_4 and feldspathic rocks (Hem 1970).

Sodium (Na^{+})

Na^{+} ion concentration ranged from 9 to 154 mg/L, with a mean value 64.5 mg/L (Table 10.1). All the samples were found to be below the permissible limit (200 mg/L). The spatially southern zone of the research area has a moderate groundwater quality zone (Fig. 10.3c).

Potassium (K^+)

The content of K^+ ranged from 0.10 to 27 mg/L, with a mean value of 6.7 mg/L. It was classified relying on the WHO norms of <50 mg/L as an elevated zone suitable for drinking, 50–100 mg/L as a moderate zone appropriate for consumption, and 100 mg/L as the standard set for home-grown utilization. Groundwater quality was designated as poor in zones with concentrations greater than 200 mg/L. According to WHO (2008), the maximum potassium concentration is 27 mg/L (Table 10.1). As a consequence, all groundwater tests are considered to be within the drinking water norms (Fig. 10.3d).

10.5.2.2 Major Anions

Bicarbonate (HCO_3^-)

The bicarbonate (HCO_3^-) concentration ranged from 68.9 to 427 mg/L, with an average of 173.5 mg/L. The content of HCO_3^- is higher in PRM, indicating the contribution of the silicate and carbonate weathering process (Mondal et al. 2018). A concentration of 100 mg/L, according to the WHO standards, falls under the poor zone category and is only suitable for industrial use (Fig. 10.4a). Good and moderate zones cover 85% of the commanded area (Fig. 10.4a) and the remaining portion of the area when the concentration exceeds 250 mg/L, respectively.

Chloride (Cl^-)

The chloride levels ranged from 11 to 269 mg/L, with a mean value of 99.4 mg/L. The chloride content revealed that 15% of the observations surpassed the norm (250 mg/L). The dissolution of evaporate reserves causes the presence of chloride in

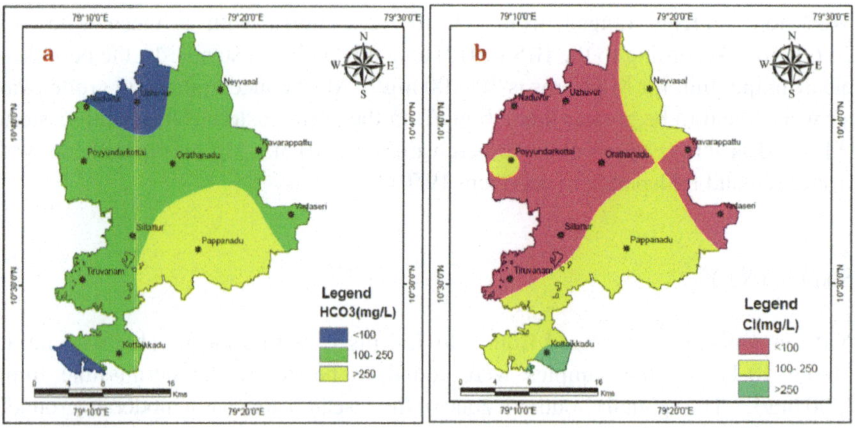

Fig. 10.4 Spatial distributions of HCO_3^- (**a**) and Cl^- (**b**)

groundwater (the dissolution of halite). The salinity of some runoff water, as well as the salinity of aquifers, is prompted by this morphology (Balasubramanian et al. 2022). Furthermore, it is induced by mineral spills as well as residential and industrial effluents into groundwater (Muthukumaravel et al. 2021). Except for a small portion (15%) of the study area in the south, the spatial variation of Cl^- was found to be below the acceptable threshold (250 mg/L) for drinking water use (Fig. 10.4b).

Sulfate (SO_4^{2-}), Nitrate (NO_3^-), and Fluoride (F^-)

The elevated values for SO_4^{2-}, NO_3^-, and F^- ions were 50 mg/l, 11 mg/l, and 0.8 mg/l, respectively (Table 10.1). The consequences revealed that all of the samples were below the allowable limits for SO_4^{2-}, NO_3^-, and F^-, which are 200, 45, and 1.5 mg/l (WHO 2011), respectively. The sources of nitrate and sulfate in groundwater were primarily anthropogenic, which could be due to sewage leachates, excessive use of inorganic fertilizers, or poor manure farming practices (Ravi et al. 2020).

10.5.3 Hydrogeochemical Facies

The Piper trilinear plot is a graph that shows the relationship between variables. The nature of the major cations and anions and their relations can be explained by hydrochemical facies. Based on the hydrochemical facies, water types (Piper 1944) can be classified into six types: Ca–HCO_3 type, Na–Cl–SO_4 type, mixed Ca–Na–HCO_3 type, mixed Ca–Mg–Cl type, Ca–Cl type, and Na–HCO_3 type (Fig. 10.5). The groundwater in the study area is of various types, indicating mixing with different aquifer layers and groundwater evolution processes. In the hydrochemical facies of the research area, there was a predominance of the Ca–HCO_3 type, mixed Ca–Mg–Cl type, and mixed Ca–Na–HCO_3 type. The Ca–Mg–HCO_3-type combination present in the region is due to agrarian return flows and human activities (Shankar et al. 2011b, c; Balasubramanian et al. 2022). Ca–Mg–HCO_3 water is likely to originate from the recharging of precipitation containing saline water interacting with recently recharged water.

10.5.4 The Drinking Water Quality Index (DWQI)

The DWQI is an important hydrochemical indicator for ingesting water (Table 10.2). Table 10.3 shows the calculated DWQI values for each groundwater sample. According to the DWQI results (Table 10.2), 54.5% and 45.5% of groundwater samples had excellent and good water quality, respectively. The north-western part

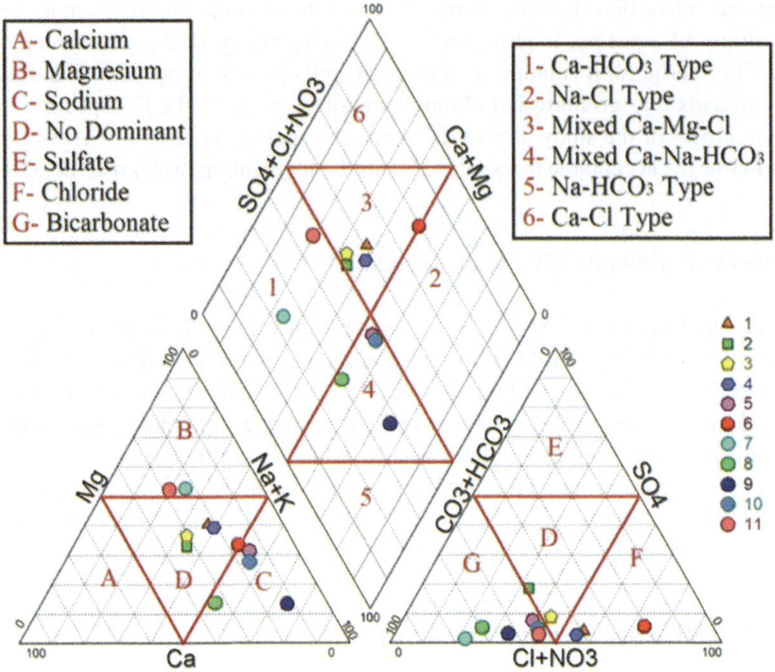

Fig. 10.5 Hill–Piper plot for groundwater

Table 10.2 DWQIs and their status of groundwater

DWQI range	Type of water	Percentage of the samples
Less than 50	Excellent	54.5
50–100	Good	45.5
100–200	Poor	0
200–300	Very poor	0
Greater than 300	Unsuitable	0

Table 10.3 Calculated DWQI values for each groundwater sample

Well no.	Village	DWQI	Classification of water
1	Neyvasal	57	Good
2	Vadaseri	49	Excellent
3	Orathanadu	40	Excellent
4	Poyyundarkottai	42	Excellent
5	Pappanadu	82	Good
6	Kottaikadu	56	Good
7	Uzhuvur	23	Excellent
8	Naduvur	30	Excellent
9	Silattur	50	Good
10	Thiruvanam	48	Excellent
11	Kavarapattu	51	Good

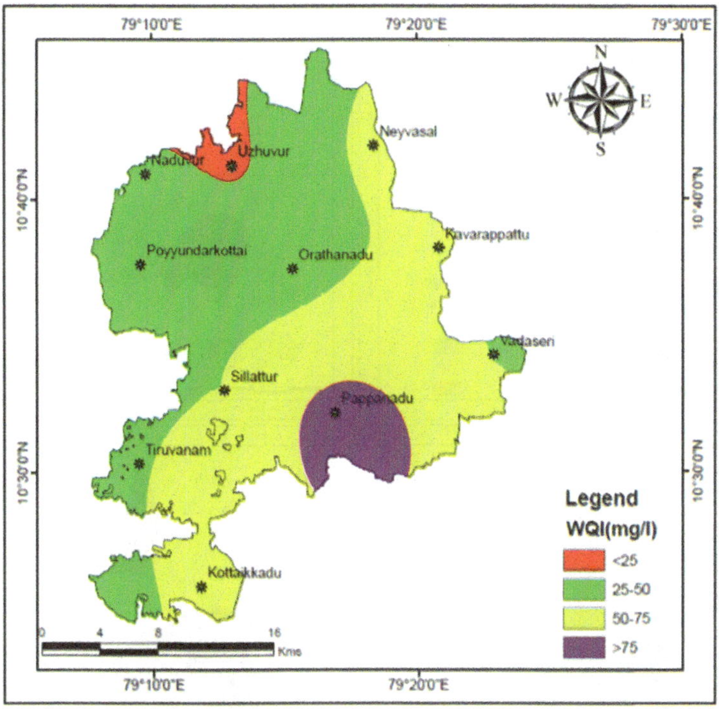

Fig. 10.6 Spatial distribution of the DWQI

of the research area had excellent water quality, whereas the rest of the area had good water quality (Fig. 10.6). This indicates that chemical fertilizers are being used excessively, domestic and animal wastes are being dumped, and groundwater is being overexploited for agrarian purposes, resulting in a decline in groundwater quality.

10.5.5 Irrigation Suitability

10.5.5.1 Total Hardness (TH)

The average hardness was 179.9 mg/L, with a range of 65–315 mg/L. According to the WHO norms, the groundwater in the study site was hard (Table 10.1), and all samples were within the tolerable limit (500 mg/L) and could be used for domestic purposes. According to Sawyer and McCarty (1967), the subsurface water in the research area was moderately hard to hard water (Fig. 10.7a).

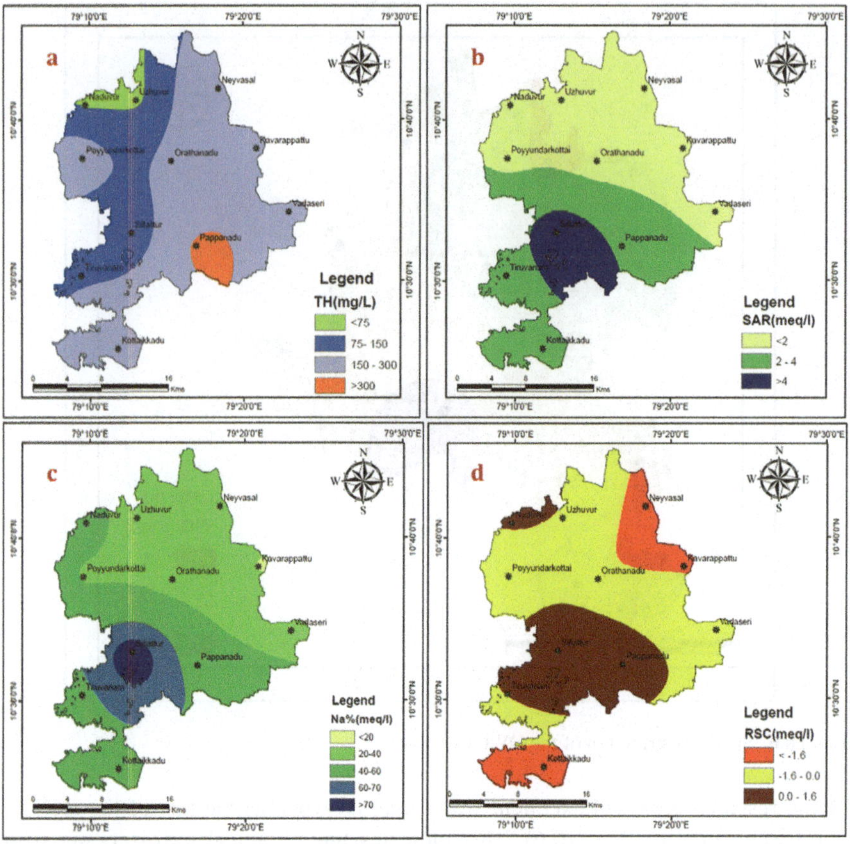

Fig. 10.7 Spatial distributions of TH (**a**), SAR (**b**), % Na (**c**), and RSC (**d**)

10.5.5.2 Sodium Adsorption Ratio (SAR)

The SAR is generally estimated as an index of sodium hazard. It is the relative ratio between the Na^+ ions and the Ca^{2+} and Mg^{2+} ions in water. The physical bonds of soil break down due to the continued utility of high SAR containing water. Sodium ions get absorbed and combine with the soil particles. Then, the soil starts becoming compact and hard, which leads to a decrease in the porosity of the soil and makes it difficult for water to percolate. As per the Richards (1954) classification, all of the samples fell under the excellent category (Table 10.4; Fig.10.7b), indicating that the samples were suitable for irrigation. As shown in Fig. 10.8b, the SAR and EC values were then plotted using the US Salinity Laboratory (USSL) diagram (USSL 1954). This figure demonstrates that about 64% of the samples fell under the C2-S1 category, 27% under the C3-S1 category, and 9% under the C1-S1 category. These graphical data give us an indication that the majority of the samples had a medium to high salinity hazard and a low sodium hazard.

10.5.5.3 Percentage Sodium (% Na)

Sodium is the most important element in agrarian purposes. Sodium binds to the soil, reducing its permeability, and, when it reacts with carbonate and chlorine, it forms basic and brine soils, which inhibit plant growth (Ayers and Westcot 1985). Furthermore, a base-exchange reaction substitutes high sodium levels in water with Ca^{2+} and Mg^{2+} ions, influencing soil structures, oxygenation, and permeation (Ravi et al. 2020; Kawo and Karuppannan 2018). Using the Wilcox classification (Wilcox 1955), 73%, 18%, and 9% of the groundwater samples showed excellent, good, and permissible percentage sodium for irrigation, respectively (Table 10.4; Fig. 10.8a). The % Na for groundwater in the PRM seasons varied in the range of 19.8–74.4. The % Na indicates that around 91% of samples fell under the category of excellent to the permissible limit. It was revealed from the analysis that the groundwater of doubtful quality was found in less than 9% of the area (Fig. 10.7c).

10.5.5.4 Residual Sodium Carbonate (RSC)

When CO_3^- and HCO_3^- ions exceed Ca^{2+} and Mg^{2+} ions, it indicates Ca^{2+} ion loss via an exchange reaction. A sodium build-up results in a negative RSC value because adequate calcium and magnesium are present over the carbonates. A positive RSC value indicates the likelihood of sodium accumulation, whereas a high RSC value (>2.5 meq/L) indicates that sodium accumulation in the soil is imminent. According to Richards (1954), the values of RSC are classified as follows: <1.25, good category; 1.25–2.50, medium category; and > 2.50, bad category. Based on RSC values,

Table 10.4 Appropriateness of groundwater for agrarian purposes

Irrigation indices	Range	No. of samples	Percentage	Class
Sodium adsorption ratio (SAR)	Less than 10	11	100	Excellent
	10–18	0	0	Good
	18–26	0	0	Doubtful
	Greater than 26	0	0	Unsuitable
Percentage sodium (% Na)	Less than 20	1	9.1	Excellent
	20–40	5	45.45	Good
	40–60	4	36.36	Permissible
	60–80	1	9.1	Doubtful
	Greater than 80	0	0	Bad
Residual sodium carbonate (RSC)	Less than 1.25	10	90.91	Satisfactory
	1.25–2.5	1	9.09	Marginal
	Greater than 2.5	0	0	Unsatisfactory
Permeability index (PI)	Greater than 75	4	36.36	Class I
	75–25	7	63.64	Class II
	Less than 25	0	0	Class III

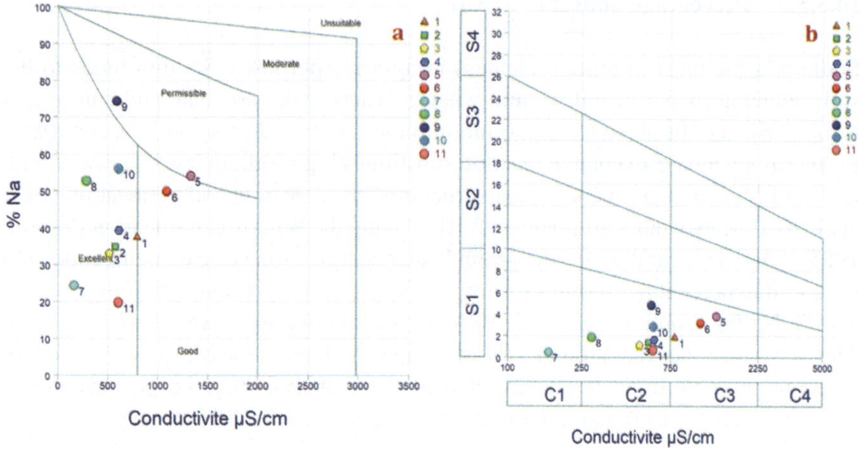

Fig. 10.8 Wilcox (**a**) and USSL (**b**) diagrams for classifying groundwater for irrigation

the majority of the samples (Table 10.1; Fig. 10.7d) were deemed suitable for irrigation (RSC 1.25 meq/L). The spatial distribution of RSC values during PRM indicates that 95% are of good quality, whereas the remaining 5% are of doubtful to unsuitable quality (Fig. 10.7d), particularly in the south-western part of the region in and around Silattur, due to the presence of carbonate residues in the soil.

10.5.5.5 Permeability Index (PI)

The parameter PI is crucial in water irrigation potential classification (Kawo and Karuppannan 2018). It determines the levels of Ca^{2+}, Mg^{2+}, Na^{+}, and HCO_3^{-} in underground water, which can reduce soil aeration and hinder seedling growth if used for irrigation for a prolonged period of time (Khalid 2019). Doneen (1964) classified the PI into three classes based on values of >75%, 25–75%, and 25%, which correspond to suitable, good, and unsuitable groundwater for irrigation, respectively. The calculated PI values for irrigation were 4 (36%) for class I (suitable) and 7 (64%) for class II (good) (Table 10.4).

10.6 Conclusions

Groundwater quality assessment is critical in the Orathanadu Taluk's deltaic areas where surface water is scarce and there is a decrease in mean precipitation, which has been exacerbated by climate change.

- The groundwater is found to be alkaline in nature. The EC and TDS levels were observed to be within the allowable threshold for potable water.

- The contents of Ca^{2+}, Mg^{2+}, Na^{+}, and K^{+} are reported to be within the tolerable limit of consumption. Bicarbonate and chloride are high in the southern part of Pappanadu and Kottaikadu with poor and very poor water.
- In the hydrochemical facies of the research area, there was a predominance of the Ca–HCO_3 type, mixed Ca–Mg–Cl type, and mixed Ca–Na–HCO_3 type.
- The DWQI results showed that 54.5% of the groundwater samples had admirable water quality and 45.5% had virtuous water quality. The water quality in the north-western part of the area was admirable, whereas that of the remaining portion of the zone was virtuous.
- It has been observed that the groundwater in the study area is moderate to hard. According to the Wilcox diagram, the majority of the samples were suitable for agrarian purposes, and as the SAR values were < 10, all of the samples were suitable for agrarian purposes. Other indicators, including RSC and PI, exemplify the appropriateness of irrigation purposes.

References

APHA, *Standard Methods for the Examination of Water and Wastewater*, 21st edn. (American Public Health Association, New York, 2005)

APHA (American Public Health Association), *Standard Methods for the Examination of Water and Wastewater*, 22nd edn. (American Water Works Association, Denver, 2012)

S. Aravindan, K. Shankar, Ground water quality maps of Paravanar River Sub Basin, Cuddalore District, Tamil Nadu, India. J. Indian Soc. Remote Sensing **39**(4), 565–558 (2011a). https://doi.org/10.1007/s12524-011-0152-9

S. Aravindan, K. Shankar, S.S. Mini, Integrated geohydrological studies in the sedimentary part of Gadilam River Basin, Cuddalore District, Tamil Nadu. Asian J. Earth Sci. **4**, 183–192 (2011). https://doi.org/10.3923/ajes.2011.183.192

S. Aravindan, K. Shankar, Groundwater quality in Paravanar River sub-basin, Cuddalore District, Tamilnadu, India. Gondwana Geol. Magazine **26**(2), 139–146 (2011b)

S. Aravindan, M. Manivel, S. Rajendran, Bhuvaneshwari and Shankar, K., Groundwater investigations in the hard rock region of gadilam river basin, Tamil Nadu. J. Eco-chronicle **3**(1), 21–30 (2008)

S. Aravindan, K. Shankar, B. Poovalinga Ganesh, K. Dharani Rajan, Groundwater geochemical mapping of in the hard rock area of Gadilam River basin, using GIS technique, Tamil Nadu. Indian J. Appl. Geochem. **12**(2), 209–216 (2010)

P. Aravinthasamy, D. Karunanidhi, K. Shankar, T. Subramani, R. Setia, P. Bhattacharya, S. Das, COVID-19 lockdown impacts on heavy metals and microbes in shallow groundwater and expected health risks in an industrial city of South India. Environ. Nanotechnol. Monit. Manage. **16** (2021). https://doi.org/10.1016/j.enmm.2021.100472

G. Ashabir, N.S. Kawo, S. Karuppannan, A.T. Hordofa, P. Paron, Numerical modeling of groundwater flow system in the Modjo River catchment, Central Ethiopia. Modeling Earth Syst. Environ. (2020). https://doi.org/10.1007/s40808-020-01040-0

R.S. Ayers, D.W. Westcot, *Water Quality for Agriculture. FAO Irrigation and drainage paper 29 Rev. 1*, vol 1 (Food and Agricultural Organization, Rome, 1985), p. 74

M. Bagyaraj, K. Shankar, T. Alemayehu, R. Muthukumarasamy, G. Gnanachandrasamy, Distribution and trend analysis of COVID-19 in India: Geospatial approach. J. Geographical Stud. **4**(1), 1–9 (2020). https://doi.org/10.21523/gcj5.20040101

P. Balamurugan, Kirubakaran, Muniraj Karunanidhi, Duraisamy Chaitanya, Pande Shankar, Karuppannan Maciej, Thomas An integrated approach to explore the suitability of nitrate-contaminated groundwater for drinking purposes in a semiarid region of India. Environmental Geochemistry and Health. https://doi.org/10.1007/s10653-022-01237-5

P. Balamurugan, P.S. Kumar, K. Shankar, Dataset on the suitability of groundwater for drinking and irrigation purposes in the Sarabanga River region, Tamil Nadu, India. Data Brief **29**, 105255 (2020a)

P. Balamurugan, P.S. Kumar, K. Shankar, R. Nagavinothini, K. Vijayasurya, Non-carcinogenic risk assessment of groundwater in southern part of Salem District in Tamilnadu, India. J. Chil. Chem. Soc. **65**(1), 4697–4707 (2020b)

P. Balamurugan, K. Shunmugapriya, R. Vanitha, GIS based assessment of ground water for domestic and irrigation purpose in Vazhapadi Taluk, Salem, Tamil Nadu, India. Taiwan Water Conservancy **68**(2), 1–10 (2020c)

P. Balamurugan, P.S. Kumar, K. Shankar, R. Nagavinothini, P.J. Sajil Kumar, Impact of climate and anthropogenic activities on groundwater quality for domestic and irrigation purposes in Attur region, Tamilnadu, India. Desalin. Water Treat. **208**, 172–195 (2020d)

M. Balasubramanian, S.G.D. Sridhar, R. Ayyamperumal, S. Karuppannan, G. Gopalakrishnan, M. Chakraborty, X. Huang, Isotopic signatures, hydrochemical and multivariate statistical analysis of seawater intrusion in the coastal aquifers of Chennai and Tiruvallur District, Tamil Nadu, India. Mar. Pollut. Bull. **174**, 113232 (2022). https://doi.org/10.1016/j.marpolbul.2021.113232

BIS (Bureau of Indian Standards), *Indian Standard Drinking Water Specification, (2nd Revision IS 10500:2012), Drinking Water Sectional Committee, FAD25* (New Delhi, 2012)

R.M. Brown, N.I. McClelland, R.A. Deininger, M.F. O'Connor, A water quality index—Crashing the psychological barrier, in *Indicators of Environmental Quality*, (Springer, Boston, 1972), pp. 173–182

T. Chandrasekar, T. Keesari, G. Gopalakrishnan, et al., Occurrence of heavy metals in groundwater along the lithological Interface of K/T boundary, Peninsular India: A special focus on source, geochemical mobility and health risk. Arch. Environ. Contam. Toxicol. **80**, 183–207 (2021). https://doi.org/10.1007/s00244-020-00803-1

I. Chenini, B. Farhat, A. Ben Mammou, Identification of major sources controlling groundwater chemistry from a multilayered aquifer system. Chem. Spec. Bioavailab. **22**(3), 183–189 (2010). https://doi.org/10.3184/095422910x12829228276711

L.D. Doneen, *Notes on Water Quality in Agriculture* (Department of Water Science and Engineering, University of California, Davis, 1964)

R.A. Freeze, J.A. Cherry, *Groundwater* (Englewood Cliffs, 1979), p. 604

M. Haji, S. Karuppannan, D. Qin, H. Shube, N.S. Kawo, Potential human health risks due to groundwater fluoride contamination: A case study using multi-techniques approaches (GWQI, FPI, GIS, HHRA) in Bilate River Basin of Southern Main Ethiopian Rift, Ethiopia. Arch. Environ. Contam. Toxicol. **80**, 277–293 (2021a). https://doi.org/10.1007/s00244-020-00802-2

M. Haji, D. Qin, Y. Guo, L. Li, D. Wang, S. Karuppannan, H. Shube, Origin and geochemical evolution of groundwater in the Abaya Chamo basin of the Main Ethiopian Rift: Application of multitracer approaches. Hydrogeol. J., 1–20 (2021b). https://doi.org/10.1007/s10040-020-02291-y

J.D. Hem, Study and interpretation of the chemical characteristics of natural water, USGS., Water supply paper 1473 (1970)

J.D. Hem, Study and interpretation of the chemical characteristics of natural water, 3rd edn. Scientific Publishers, Jodhpur, p 2254 (1991)

C.J. Jesudhas, S.M. Chidambaram, K. Muniraj, Isolation of wells contaminated by tannery industries using principal component analysis. Arab. J. Geosci. **10**(14), 1–8 (2017)

J.C. Johnny, M.C. Sashikkumar, M. Kirubakaran, L.M. Mathi, GIS-based assessment of groundwater quality and its suitability for drinking and irrigation purpose in a hard rock terrain: A case study in the upper Kodaganar basin, Dindigul district, Tamil Nadu, India. Desalin. Water Treat. **102**, 49–60 (2018)

A. Kadam, V. Wagh, B. Umrikar, R. Sankhua, An implication of boron and fluoride contamination and its exposure risk in groundwater resources in semi-arid region, Western India. Environ. Develop. Sustainabil., 1–24 (2019)

D. Karunanidhi, P. Aravinthasamy, M. Deepali, T. Subramani, K. Shankar, Groundwater pollution and human health risks in an industrialized region of southern India: Impacts of the COVID-19 lockdown and the monsoon seasonal cycles. Arch. Environ. Contam. Toxicol. **80**, 259–276 (2021). https://doi.org/10.1007/s00244-020-00797-w

M.T. Kavitha, K. Shankar, R. Divahar, T. Meenambal, R. Saravanan, Impact of industrial wastewater disposal on surface water bodies in Kalingarayan canal, Erode district, Tamil Nadu, India. Arch. Agri. Environ. Sci. **4**(4), 379–387 (2019a). https://doi.org/10.26832/24566632.2019.040403

M.T. Kavitha, R. Divahar, T. Meenambal, K. Shankar, R. VijaySingh, T.D. Haile, C. Gadafa, Dataset on the assessment of water quality of surface water in Kalingarayan Canal for heavy metal pollution, Tamil Nadu. Data Brief **22**, 878–884 (2019b). https://doi.org/10.1016/j.dib.2019.01.010

N.S. Kawo, S. Karuppannan, Groundwater quality assessment using water quality index and GIS technique in Modjo River Basin, Central Ethiopia. J. Afr. Earth Sci. **147**, 300–311 (2018). https://doi.org/10.1016/j.jafrearsci.2018.06.034

S. Khalid, An assessment of groundwater quality for irrigation and drinking purposes around brick kilns in three districts of Balochistan province, Pakistan, through water quality index and multivariate statistical approaches. J. Geochemical Explor. **197**, 14–26 (2019). https://doi.org/10.1016/j.gexplo.2018.11.007

N.C. Mondal, Hydrochemical approach for irrigation suitability of groundwater in Krishkindapalem of Krishna delta, Andhra Pradesh. J. Geosci. Res. **4**(1), 1–10 (2018)

K. Muthukumaravel, K.P. Pradhoshini, N. Vasanthi, T. Raja, M.A. Jaleel, K.D. Arunachalam, M.S. Musthafa, R. Ayyamperumal, S. Karuppannan, R. Rajagopal, A. Alfarhan, Assessment of seasonal variation in distribution and abundance of plankton and ichthyofaunal diversity in relation to environmental indices of Karankadu Mangrove, South East Coast of India. Mar. Pollut. Bull. **173**, 113142 (2021). https://doi.org/10.1016/j.marpolbul.2021.113142

B. Panneerselvam, S.K. Paramasivam, S. Karuppannan, et al., A GIS-based evaluation of hydrochemical characterisation of groundwater in hard rock region, South Tamil Nadu, India. Arab. J. Geosci. **13**(17), 1–22 (2020a). https://doi.org/10.1007/s12517-020-05813-w

B. Panneerselvam, S. Karuppannan, K. Muniraj, Evaluation of drinking and irrigation suitability of groundwater with special emphasizing the health risk posed by nitrate contamination using nitrate pollution index (NPI) and human health risk assessment (HHRA). Human Ecol. Risk Assess.: Int. J., 1–25 (2020b). https://doi.org/10.1080/10807039.2020.1833300

B. Panneerselvam, K. Muniraj, C. Pande, N. Ravichandran, Prediction and evaluation of groundwater characteristics using the radial basic model in Semi-arid region, India. Int. J. Environ Anal. Chem., 1–17 (2021c)

B. Panneerselvam, K. Muniraj, C. Pande, N. Ravichandran, M. Thomas, S. Karuppannan, Geochemical evaluation and human health risk assessment of nitrate-contaminated groundwater in an industrial area of South India. Environ. Sci. Pollut. Res. (2021d). https://doi.org/10.1007/s11356-021-17281-0

B. Panneerselvam, K. Muniraj, M. Thomas, N. Ravichandran, GIS-based legitimatic evaluation of groundwater's health risk and irrigation susceptibility using water quality index, pollution index, and irrigation indexes in semiarid region, in *Groundwater Resources Development and Planning in the Semi-Arid Region*, (Springer, Cham, 2021a), pp. 239–268

B. Panneerselvam, K. Muniraj, M. Thomas, N. Ravichandran, B. Bidorn, Identifying influencing groundwater parameter on human health associate with irrigation indices using the Automatic Linear Model (ALM) in a semi-arid region in India. Environ. Res. **202**, 111778 (2021b)

B. Panneerselvam, K. Muniraj, K. Duraisamy, C. Pande, S. Karuppannan, M. Thomas, An integrated approach to explore the suitability of nitrate-contaminated groundwater for drinking purposes in a semiarid region of India. Environ. Geochem. Health. 1–17 (2022). https://doi.org/10.1007/s10653-022-01237-5

A.M. Piper, A graphic procedure in the geochemical interpretation of water analyses. Trans. Am. Geophys. Union **25**, 914–928 (1944)

R. Ravi, S. Aravindan, K. Shankar, P. Balamurugan, Suitability of groundwater quality for irrigation in and around the main Gadilam river basin on the east coast of southern India. Arch. Agri. Environ. Sci. **5**(4), 554–562 (2020). https://doi.org/10.26832/24566632.2020.0504019

L.A. Richards, Diagnosis and improvement of saline and alkali soils. Soil Sci. **78**(2), 154 (1954). LWW

P. Sahu, P.K. Sikdar, Hydrochemical framework of the aquifer in and around East Kolkata wetlands, West Bengal, India. Environ. Geol. **55**, 823–835 (2008)

A.K. Sarfo, K. Shankar, Application of geospatial technologies in the COVID-19 fight of Ghana. Transact. Indian Nat. Acad. Eng. **5**, 193–204 (2020). https://doi.org/10.1007/s41403-020-00145-3

C.N. Sawyer, P.L. McCarty, *Chemistry for Sanitary Engineers*, 2nd edn. (McGraw Hill, New York, 1967), p. 518

G.R. Senthil Kumar, K. Shankar, Assessment of groundwater potential zones using GIS. Front. Geosci. **2**, 1–10 (2014)

K. Shankar, S. Aravindan, S. Rajendran, GIS based groundwater quality mapping in Paravanar River Sub-Basin, Tamil Nadu, India. Int. J. Geomat. Geosci. **1**(3), 282–296 (2010)

K. Shankar, S. Aravindan, S. Rajendran, Hydrogeochemistry of the Paravanar river sub-basin, Cuddalore District, Tamilnadu, India. E-J Chem. **8**(2), 835–845 (2011a). https://doi.org/10.1155/2011/107261

K. Shankar, S. Aravindan, S. Rajendran, Hydrochemical profile for assessing the groundwater quality of Paravanar River Sub-Basin, Cuddalore district, Tamil Nadu, India. Curr. World Environ. **6**(1), 45–52 (2011b). https://doi.org/10.12944/CWE.6.1.05

K. Shankar, S. Aravindan, S. Rajendran, Assessment of groundwater quality in Paravanar River Sub-Basin, Cuddalore district, Tamil Nadu, India. Adv. Appl. Sci. Res. **2**(5), 92–103 (2011c)

K. Shankar, N.S. Kawo, Groundwater quality assessment using geospatial techniques and WQI in North East of Adama Town, Oromia Region, Ethiopia. Hydrospatial Anal **3**(1), 22–36 (2019). https://doi.org/10.21523/gcj3.19030103

K. Shunmugapriya, B. Panneerselvam, K. Muniraj, N. Ravichandran, P. Prasath, M. Thomas, K. Duraisamy, Integration of multi criteria decision analysis and GIS for evaluating the site suitability for aquaculture in southern coastal region, India. Mar. Pollut. Bull. **172**, 112907 (2021)

B. Soujanya Kamble, P.R. Saxena, R.M. Kurakalva, K. Shankar, Evaluation of seasonal and temporal variations of groundwater quality around Jawaharnagar municipal solid waste dumpsite of Hyderabad city, India. SN Appl. Sci. **2**(3), 498 (2020). https://doi.org/10.1007/s42452-020-2199-0

U.S. Salinity Laboratory, (1954). Diagnosis and improvement of saline and alkaline soils. U.S.U.S Dept. of Agriculture. Hand Book. 60, p. 160.le

S. Venkateswaran, S. Karuppannan, K. Shankar, Groundwater quality in Pambar Sub-Basin, Tamil Nadu, India using GIS. Int. J. Recent Sci. Res. **3**(10), 782–787 (2012)

V. Wagh, S. Mukate, A. Muley, A. Kadam, D. Panaskar, A. Varade, Study of groundwater contamination and drinking suitability in basaltic terrain of Maharashtra, India through PIG and multivariate statistical techniques. J. Water Supply Res. Technol. AQUA **69**(4), 398–414 (2020)

WHO, World Health Organization guidelines for drinking-water quality, in *Incorporating the First and Second Addendum, Vol. 1, Recommendations*, 3rd edn., (World Health Organization, Geneva, 2008), p. 668

WHO, World Health Organisation guidelines for drinking water quality, in *Incorporating the First and Second Addenda, 1Recommendation*, 4rd edn., (Geneva, 2011)

L. Wilcox, Classification and uses of irrigation waters. USDA Circular No. 969, Washington, DC (1955)

Chapter 11
Groundwater Quality Assessment Using GIS-Based Multi-Criteria Decision Analysis for Pattukkottai Taluk, Tamil Nadu

Kalidhas Muthu, Karuppasamy Sudalaimuthu, and Kirubakaran Muniraj

Abstract Groundwater (GW) has become an inevitable source of drinking water in many parts of India due to the insufficient availability of surface water (SW). Therefore, it is necessary to examine whether GW, which is considered essential, is fit for human consumption. The objective of this study was to assess the water quality of Pattukkottai Taluk, Tamil Nadu, South India. Emphasis was placed on agricultural land areas in the sample point detection event for this study. A total of 78 samples were collected and the physicochemical parameters were analyzed including pH, fluoride, sulfate, total dissolved solids (TDSs), alkalinity, nitrate, hardness, and chloride. With the help of the derived physicochemical characteristics, a geographic information system (GIS)-based multi-criteria decision analysis (MCDA) was carried out to categorize the water quality as low, moderate, and good as per the Bureau of Indian Standards (BIS). The integrated GIS-MCDA approach revealed that approximately 368.65 km^2 belongs to the good groundwater quality, 102.43 km^2 moderate groundwater quality, and nearly 224 km^2 poor water quality.

Keywords MCDA · GIS · Groundwater Quality

K. Muthu
Department of Civil Engineering, Anna University Regional Campus, Tirunelveli, Tamil Nadu, India

K. Sudalaimuthu (✉)
Department of Civil Engineering, SRMIST, Kattankulathur, Tamil Nadu, India
e-mail: karuppas@srmist.edu.in

K. Muniraj
Department of Civil Engineering, University VOC College of Engineering, Thoothukudi, Tamil Nadu, India

B. Panneerselvam et al. (eds.), *Climate Change Impact on Groundwater Resources*, https://doi.org/10.1007/978-3-031-04707-7_11

11.1 Introduction

Agriculture is the backbone of a country like India; more likely, the Indian economy relies on agricultural activities, which contribute to 46% of the national gross domestic product (GDP) (Subba Rao 2006). Although surface water (SW) is the main source for irrigation activities, due to the failure of monsoons, most of the SW tanks have dried up, and, therefore, SW is not sufficient for arid and semi-arid regions' irrigation activities. Considering these SW scarcity issues, exploitation of groundwater (GW) has become normal for domestic and irrigation purposes. One-third of the population uses GW for drinking purposes (Nickson et al. 2005). Approximately 6.5 million hectares were cultivated with GW resources in 1950–1951 (CGWB 1992); however, GW use increased by 46.5 million ha (Sivanappan 2002) in 2000–2001, an increase of 70% of the total irrigation requirement for the year. This shows the increasing pressure on GW sources. Besides, soluble chemicals produced by urbanization and industrial and modern farming methods pose a threat to GW resources (Insaf et al. 2007).

It is an assumption that GW is safe for drinking because it is beneath the earth's surface and also not directly in contact with external environmental conditions (Quist et al. 1988). However, land use changes and anthropogenic activities have led to groundwater degradation (Narany et al. 2017). In addition to these, overpumping and low recharge rate (Abbasnia et al. 2018), bedrock geology (Sujatha and Rajeswara 2003), landfill (Christensen et al. 1994), poor sanitation, inadequate disposal (Salifu et al. 2013), improper land use management (Narany et al. 2017), groundwater and SW relationships (Zanotti et al. 2019), quality of recharged water, atmospheric precipitation, and subsurface geochemical process are some of the parameters that influence groundwater quality (Vasanthavigar et al. 2010).

11.2 Background of This Study

In general, GW quality constitutes physical, chemical, and biological qualities (Thomas 2003). However, many literature works focus on the chemical parameters rather than the physical and biological parameters. GW quality should consider is same as quantity of the water. Therefore, it was initiated by the Central Ground Water Board in 1969 in view of the need to monitor GW quality in planning for safe drinking water (CGWB 2016) because poor GW quality not only deteriorates human health but also rebates agricultural yields (Wilcox 1948; WHO 1984; Karanth 1997). Thus, it is mandatory to have good-quality GW for human health and the best ecosystem-based functioning (Humphreys 2009).

Various studies have been carried out for GW quality assessment across India (NiranjanBabu et al. 1997; Subba Rao et al. 1999; Sreedevi 2004; Johnny et al. 2018; Panneerselvam et al. 2020, 2021; Kirubakaran et al. 2015). They followed the traditional approach to determine GW quality by comparing the derived parameters with the standard guideline values, and the samples that exceeded the standard guideline values were considered as detrimental to the users (Dede et al. 2013).

In the traditional approach, interpretation of the result is ambiguous, and, thus, an index-based study was developed for identifying GW contamination and quality, such as the groundwater contamination index (Backman et al. 1998). The water quality index (WQI) was introduced for determining GW quality (Mohamad Najib 2019). The WQI is a mathematical approach to transform large water quality data into numerical values based on common categorizations such as unsuitable, very poor, poor, moderate, good, very good, and excellent, based on the conditions used to represent the overall water quality level. According to the study, a novel WQI was proposed (Horton 1965) for the purpose of drinking water supply assessments. Furthermore, the WQI was developed to compare the water quality of different water bodies (Brown et al. 1970). Worldwide, several studies were conducted (Ramakrishnaiah et al. 2009; Ashok Lumb et al. 2011; Abbasi and Abbasi 2012; Akkaraboyina and Raju 2012; Sutadian et al. 2016) using the WQI to compute groundwater and SW quality by different users. A geographic information system (GIS) is used as a tool in various GW-related studies, such as those on vulnerability, mapping (Jesudhas et al. 2021), flow and quality (Engel and Navulur 1999), predicting nitrate leaching to GW (Rodda et al. 1999), GW contamination (Ahn and Chon 1999), GW pollution, and hazard maps (Ducci 1999), and is also used to derive the relationship between water quality parameters and other various parameters (Dhanasekarapandian et al. 2016). Besides, GIS is capacious to spatially represent WQI-derived values with graphical visualization.

GIS- and multi-criteria decision analysis (MCDA)-integrated approaches offer robust spatial analysis functions (Yu et al. 2009; Shunmugapriya et al. 2021). In the current scenario, GIS was integrated with MCDA for GW quality assessment and mapping. Various studies have been carried out with the help of integrated GIS and MCDA for GW potential mapping (Pietersen 2006; Althuwaynee et al. 2014; Kaliraj et al. 2014; Champak and Dinesh 2018; Muniraj et al. 2020; Arivazhagan and Kirubakaran 2016; Kirubakaran et al. 2016). MCDA is used in deriving the weightages of multi-variables based on their priority of individual variables. The MCDA-based analytical hierarchy process (AHP) is used for deriving the weightages to identify the GW quality index (Jhariyaa et al. 2017). Moreover, Fuzzy-based techniques are used to derive the weightages for assessing GW quality (Srinivasa et al. 2015).

11.3 Objective of This Study

In this study, GW quality assessment was carried out by using a GIS-based weighted index overlay analysis (WIOA). Many studies have been carried out using WIOA during different time intervals (Brindha and Elango 2012; Tapasya 2015; Jeyavel Raja Kumar et al. 2019). In this context, 78 samples were collected and processed to derive various physicochemical parameters such as nitrate > pH > total dissolved solids > fluoride > sulphate > alkalinity > hardness > chloride throughout the study area. The prioritization solely depended on the study area conditions and the parameters influencing GW quality in the study area. The derived weightages from WIOA for different parameters were assigned a ranking based on their influences using GIS tools to identify GW quality in different parts of the study area. Although, it adopts the predecessor methodology, none of the studies has reported in the study area also the parameters are selected for this study have differences.

11.4 Study Area

Pattukkottai is located between the northern latitudes 10°17′30″ to 10°32′38″ and the eastern longitudes 79° 7′58″ to 79°27′58″ and belongs to the Thanjavur district of Tamil Nadu as shown in Fig. 11.1. According to the 2011 census, the population was about 396,476 over an area of 695 km^2 of the study area. It is located in the Cauvery delta region, in which agriculture and fishery are the main sources of income as per the study area statistics. It has a tropical climate, with an average temperature of around 37.3 degree Celsius, and the rainfall received per annum is approximately 1106 mm. The land use pattern of the study area is crop land, plantation, fallow land, water bodies, salt-affected land, built-up area, and mangroves. The geological formations of the study area are sedimentary and alluvial, mainly because it is situated in the region of Eastern Ghats, and seashore deltaic. Geomorphological

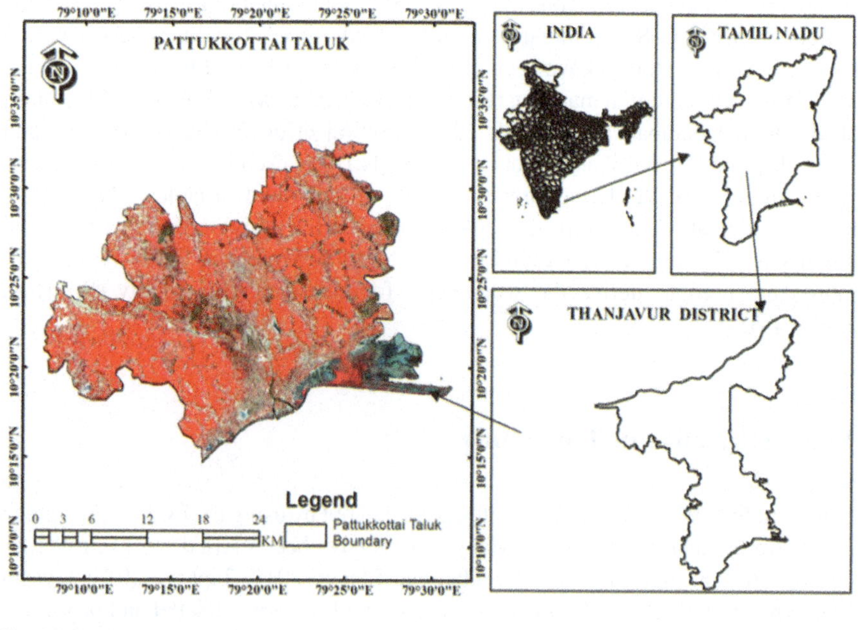

Fig. 11.1 Study area map

features, by nature, in the study area are alluvial plain, coastal plain, old coastal plain, salt flat, shallow flood plain, pediplain, and mud flat.

The study area uses both SW and GW for agricultural and drinking purposes, mainly due to insufficient rainfall. SW declination has led to the demand for GW for irrigation activities. Since the study area is famous for agricultural production, most of the economy relies on the same. Therefore abstraction of GW, excessive fertilizer used in the agricultural land, and the GW quality is suspicious to this region.

11.5 Materials and Methods

In order to estimate GW quality of the Pattukkottai Taluk, field samples were collected. In this study, 78 GW samples were collected from different parts of the taluk. GW samples were spatially well distributed as shown in Fig. 11.2.

The water level of the bore well varies from approximately 100 to 160 feet. In particular, the GW samples were collected from nearby agricultural areas. It is considered essential, since this study focused on GW quality in agricultural areas. The collected samples were properly sealed and then subjected to chemical analysis. The physicochemical parameters determined from the laboratory are pH, nitrate, total dissolved solids (TDSs), chloride, fluoride, alkalinity, sulfate, and hardness. The standardized procedure for determining GW quality parameters was followed

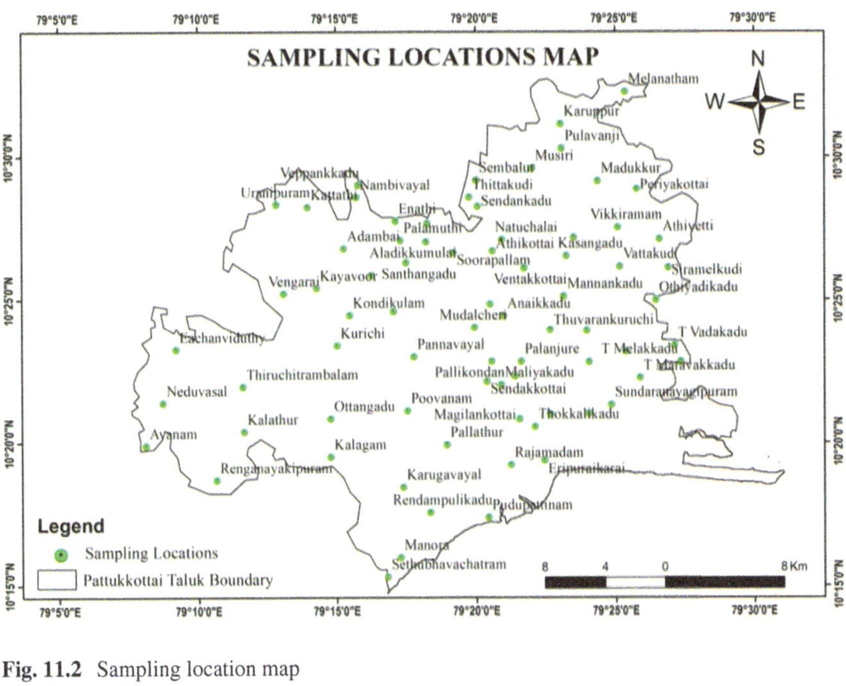

Fig. 11.2 Sampling location map

Table 11.1 Water quality parameters against the BIS standards

Parameters	Values from the collected samples		BIS standards	
	Minimum	Maximum	Acceptable	Maximum
pH	6.7	8.6	7	8.5
Nitrate	2.07	46.92	45	100
TDS	786	1511	500	1500
Chloride	202	364	250	1000
Fluoride	0.2	0.9	1.0	1.5
Alkalinity	206	383.96	200	600
Sulphate	9	61.89	200	400
Hardness	204	361	300	600

(APHA, 1998). The water quality parameters and their BIS minimum and maximum acceptable standards are depicted in Table 11.1.

11.5.1 Estimation of the Physicochemical Parameters

Water quality determination is a primary requirement for drinking and agricultural practices prior to utilization. Traditional laboratory-based water quality measurement methods used for water quality monitoring are time-consuming and also lack sensitivity and selectivity (Jyotirmoy Bhardwaj et al. 2015). Therefore, in recent times, advanced optical sensors have been utilized for water quality detection, mainly because of their rapidness, high sensitivity, and selectivity. According to the study, the physicochemical parameters were estimated using different approaches, like pH was measured using a pH scan meter, nitrate, chloride, sulphate, and fluoride were estimated using a spectrophotometer, TDSs were derived using a TDS meter, and alkalinity and hardness estimations were carried out using titration.

11.5.2 Spatial Interpolation

The estimated water quality physicochemical parameters were used as input sources for the generation of spatial variation maps of water quality parameters using inverse distance weighting (IDW). IDW is a basic assumption that attributes information of unknown sample points based on the weighted averages of neighbourhood cell values; also, the weights are inversely proportional to the distances between the predicted and observed values (George and David 2008). IDW weights are calculated based on Eq. 11.1.

$$\lambda = \frac{Di^{-a}}{\sum_{i=1}^{n} Di^{-a}} \tag{11.1}$$

λ = the weight of the point, Di = the distance between point i and the unknown point, and a = the power ten of the weight.

IDW is one of the popular methods used by many researchers, like geoscientists, geographers, etc., in their studies because it can be used to graphically visualize the water quality parameters for detailed interpretation. Besides, IDW is integrated with many GIS software as a plug-in. Therefore, it was adopted in this study to prepare spatial variation maps for pH, nitrate, total dissolved solids (TDSs), chloride, fluoride, alkalinity, sulphate, and hardness.

11.5.3 Weighted Index Overlay Analysis (WIOA)

According to the literature stated in the Introduction section, many studies have been carried out using WIOA for water quality assessment. In WIOA, weights are assigned based on their influences on water quality, topographical conditions, land use types of the area, etc. According to the study, nitrate was given the highest weightage of 20% among all other parameters, pH and TDSs were given 15%, and the rest of the parameters such as fluoride, sulfate, alkalinity, hardness, and chloride were given 10% weightage. Rank 1 was given for the least influential parameter and rank 3 was given for the most influential parameter. This was followed for all the physicochemical parameters used in this study. The weightages and ranking assigned to the different parameters are given in Table 11.1. Weightage-incorporated thematic layers were combined to prepare the water quality map. As per the vulnerability, it was categorized as low, moderate, and good GW quality.

11.6 Results and Discussion

11.6.1 Physicochemical Parameters

The estimated physicochemical parameter values from the laboratory analysis were compared with the BIS standards. Table 11.1 shows the minimum and maximum values of the estimated parameters and the acceptable, maximum values as per the BIS standards. It is understood from Table 11.1 that the maximum values of pH, nitrate, and TDSs are higher than those of the BIS standards; however, the remaining chemical parameters such as chloride, fluoride, alkalinity, sulphate, and hardness are within the acceptable limit.

11.6.1.1 pH

The pH of water is an essential determinant parameter because many of the geochemical-based reactions within GW rely on pH. Hence, it is mandatory to control it within an acceptable limit, especially for chemical processes such as

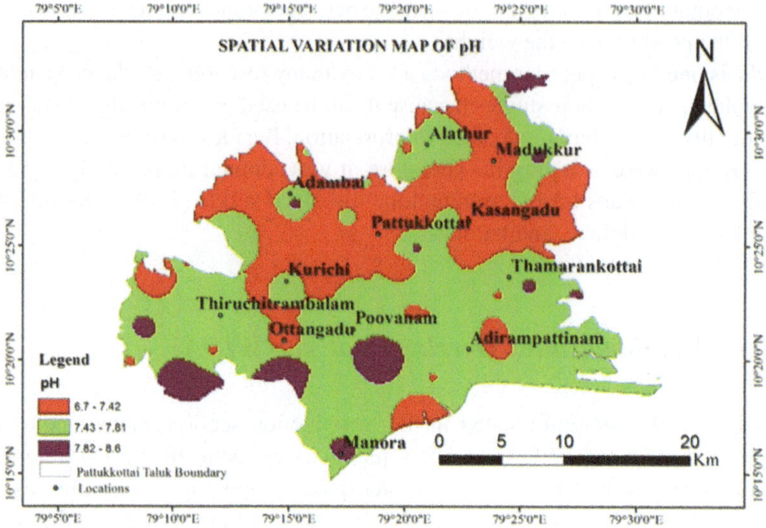

Fig. 11.3 Spatial variation map of pH of the study area

disinfection, coagulation, softening, and corrosion control. In the study area, pH values of the analyzed samples ranged from 6.7 to 8.6. Mohamad Najib Ibrahim (Mohamad Najib 2019) had revealed that pH values ranged from 7.15 to 8.71, which is an indication of the minimum level of alkaline nature present in the GW (Mohamad Najib 2019). It was understood from the study conducted by Praharaj et al. that tube well pH ranged from 7.8 to 8.6, with an average of about 8.2. However, in the open well, it varied from 7.7 to 8.4, with an average of 7.9, but it was within the United States Environmental Protection Agency (USEPA) permissible limit (Praharaj et al. 2002). According to the study, pH samples were moderate within the prescribed standards for drinking water. Ranges are classified in the spatial variation map shown in Fig. 11.3. Based on the figure, pH values ranged from 6.7 to 7.42 in Madukkur, Pattukkottai area, 7.43 to 7.81 in Thiruchitrambalam, Ottangadu area, and 7.82 to 8.6 in the Manora area.

11.6.1.2 Chloride

Chloride is generally distributed as salts in nature. Although it is relevant in seawater, it is also permitted in a freshwater system. It is another imperative parameter for assessing water quality. It is known that a higher level of chloride concentration leads to organic pollution. According to the BIS standards, the chloride permissible level in drinking water is 250 mg/l. Chloride is suitable for irrigation purposes when it is below 70 parts per million. According to the literature, higher levels of chloride concentration in freshwater are due to the discharge of municipal and industrial wastes (Kant and Raina 1990), inorganic fertilizers, septic tank effluents, and

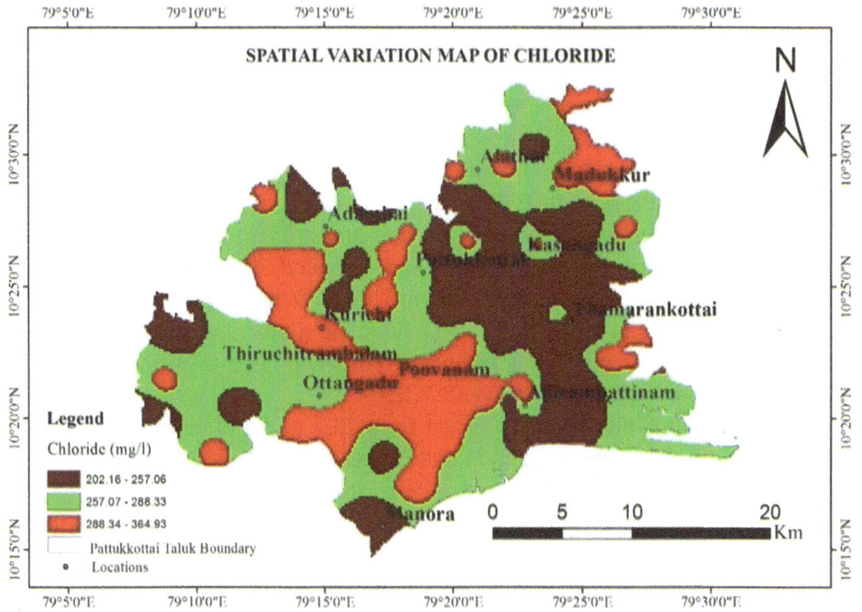

Fig. 11.4 Spatial variation map of chloride of the study area

irrigation drainage (Sayyed Juned and Bhosle Arjun 2011). In the study area concerned, chloride values ranged from 202 to 364 mg/l, and the chloride contents in the study area led to laxative effects. The spatial variation map of chloride is shown in Fig. 11.4. It was revealed from the figure that the chloride concentration is higher than the permissible limit in more than 50% of the study area due to irrigation drainage, exclusively in Pallathur, Kalagam, Kayavoor, and Melanatham, which had a higher-level concentration of chloride above 340 mg/l. Another study from the Jind district of Haryana revealed that chloride concentration varied from 1.5 to 2045 mg/l; it was above 80% of the study area and exceeded the World Health Organization (WHO) limits (Meenakshi Garg et al. 2004).

11.6.1.3 Nitrate

Nitrate is the final product from Nitrogen aerobic oxidation, there are other criteria also reason to form nitrates such as nitrogenous fertilizers, decomposition of organic matter in the soil, and etc. In addition to that, anthropogenic activities are the primary reason for the consistent release of reactive nitrogen (N) into the environment (Prakasa Rao and Puttanna 2000). In general, nitrate existence in GW may be associated with excessive usage of organic and synthetic fertilizers. As per the BIS standards, the permissible nitrate level in drinking water is 45 mg/l. The nitrate level increases in groundwater and SW due to various reasons, especially due to the use of Nitrate fertilizers in agricultural areas. The study revealed that nitrate concentration in water may be decreased with increasing water table; however, significant levels are observed in 4- to 11-m-deep water tables (Bijay and Sekhon 1976). Nitrate leaching is common in agricultural fields; nevertheless, it is based on the land use types, i.e. forest areas have less leaching than do agricultural areas (Biswas et al. 2010). According to the study, nitrate concentrations in the shallow well have increased by almost 2 mg/l due to the use of Nitrate fertilizers from the year 1982 to 1988 (Bijay et al. 1991). Moreover, a study from Punjab revealed that nitrate levels are higher in vegetable-growing areas (17%) compared to those in areas growing potato and wheat (6%) and rice and wheat (3%) (Bajwa and BijaySingh 1993). Another study from China showed that nitrate concentration is higher in the central and northern regions than in the southern region because of the higher usage of fertilizers for vegetable cultivation and faster movement of GW in the central and northern regions (Yu et al. 2020).

In the study area, nitrate levels varied from 2.07 to 46.92 mg/l, especially the Kalagam, Renganayaki Puram, and Pallathur areas had 46 mg/l, which is approximately 4% of the study area and above the permissible limit; it is depicted in the spatial variation map shown in Fig. 11.5. The study was carried out in the Thanjavur district where nitrate concentration above its permissible limit, which is approximately 8% of the study area, is due to the use of chemical fertilizers in agricultural fields and also other non-agricultural activities such as seepage from septic tanks.

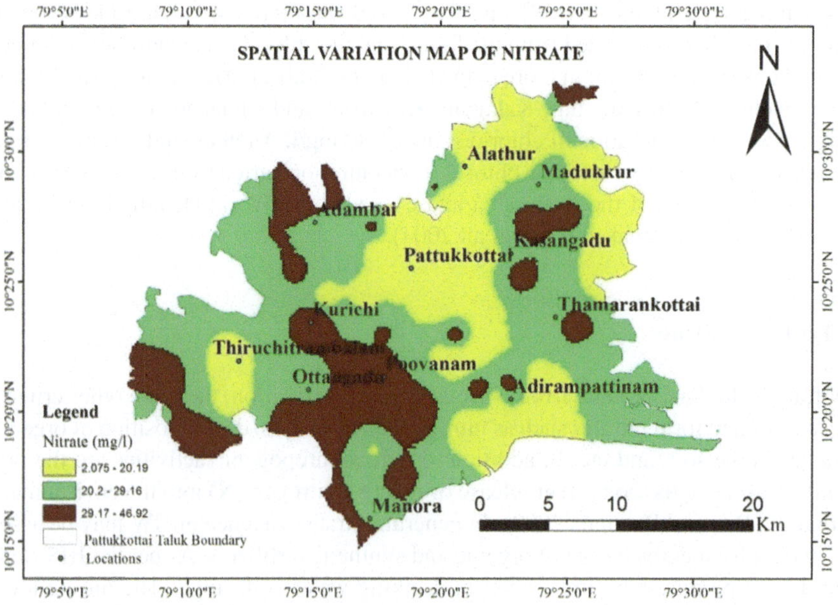

Fig. 11.5 Spatial variation map of nitrate of the study area

11.6.1.4 Total Dissolved Solids

Total dissolved solids (TDSs) is a term that constitutes inorganic and organic matter and a variety of salts (Miranda and Krishnakumar 2015); this measure is also used to determine the saline behaviour of water (Heydari and Bidgoli 2012). Miranda and Krishnakumar (2015) found that the highest TDS values of 25,266.67 mg/L were recorded in the month of April from one of their study areas in Mangalore. The study by Madan Kumar Jha et al. (Madan Kumar et al. 1990) revealed that TDSs in GW range from 157 to 3202 mg/L and exceed the permissible limit as per the standards, occupying around 5% of the study area. According to the study of Ahmed M. El-Zeiny and Salwa F. Elbeih (2019) in Bangladesh, most of the GW wells, nearly 92% of the study area, have TDSs lower than 2000 mg/l, which is suitable for irrigation. In another study by Syed Muhammad et al. (2014) from Karachi city, TDS levels varied from 350 to 600 mg/l; thus, some parts of the study area were above the WHO permissible limits. It is understood from the various literature works in India and other parts of the world that TDS levels are above the permissible limits for drinking and irrigation.

However, the TDS level for irrigation is considered just below 2000 mg/l as per their terrain conditions. In this study, TDS varied from 786 to 1411 mg/l in, as per the study area concern, all the sampled locations are above the permissible limit according to BIS standards, in which 45% of the study area TDS ranges are above 1000 mg/l. It is found that Melanatham had the highest value of 1412 mg/l. In general, water with more than 1000 mg/l TDS level has an unlikeable taste and is not suitable for drinking. The spatial variation map of the study area is prepared and depicted in Fig. 11.6. It is identified from Fig. 11.6 that Ottangadu, Thamarankottai, and Narasingapuram are some of the areas that had TDSs less than 1000 mg/l, whereas Pallathur, Renganayaki Puram, and Nainankollai areas had higher than 1000 mg/l.

11.6.1.5 Total Hardness (TH)

Water hardness is determined by the inhabitation of soap action in the water due to the condensation of magnesium, and calcium salts such as sulphates, carbonates, and chlorides (Vasanthavigar et al. 2010). In general, the hardness of natural water also depends upon the geological structure of the soils, the lithological nature of the geological formation, and the amount of calcium and magnesium contained in the water (Enyegue et al. 2020). Although hardness has no known deleterious effects, there exists evidence indicating its influences in heart-related problems (Schroeder 1960). Besides, excessive hardness in water restricts industrial purposes like scaling of boilers, closure of irrigation pipes, etc.

A study carried out in the Ranga Reddy District in Andhra Pradesh revealed that the total hardness ranged from 90 to 680 mg/l. TH above 80 mg/l cannot be utilized for domestic purposes (Sujatha, Rajeswara Reddy (Sujatha and Rajeswara 2003). Another study from the Guntur district, Andhra Pradesh, revealed that the TH mean

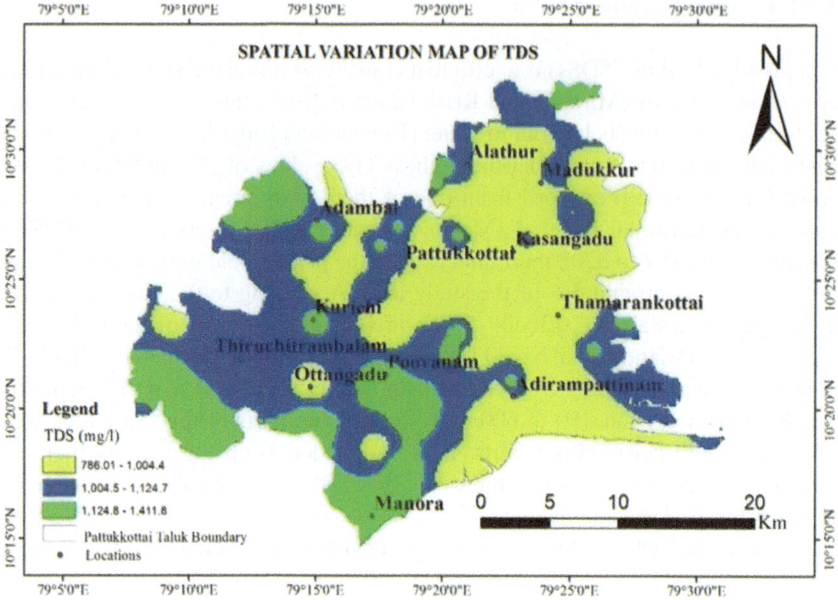

Fig. 11.6 Spatial variation map of total dissolved solids of the study area

values are 442 and 449 mg/l in the pre- and post-monsoon periods, respectively. In the post-monsoon period, the TH concentration is higher than that in the pre-monsoon period, and it is also higher than the BIS permissible limit (Subba Rao 2006).

According to the study area, the TH varied from 204 to 361 mg/l, and also 20% of the study area had hardness above the prescribed limit. Especially, Pallathur, Neduvasal, and Kalagam areas had higher than 340 mg/l levels in the study area. The spatial variation map for the TH is depicted in Fig. 11.7. It is revealed from the figure that Alathur, Madukkur, and Thiruchitrambalam areas had TH between 204 mg/l and 255 mg/l, Adirampattinam and Thamarankottai areas had TH above 250 and below 305 mg/l, and the Kalagam area had TH above 362 mg/l.

11.6.1.6 Sulfate

Sulfate is found in small quantities in GW. The presence of sulfate in GW may be due to industrial or anthropogenic additions in the form of sulphate fertilizers. As per the World Health Organization (WHO), freshwater's average sulphate value is 20 mg/L; also, sulphate values range from 0 to 630 mg/L in rivers and from 2 to 250 mg/L in lakes, whereas sulfate levels in GW range from 0 to 230 mg/L (WHO 2004). In the study area concerned, the maximum sulfate level of 62 mg/L was found in the Pallathur area and the lowest level of 9 mg/L was found in the Pudupattinam area as shown in Fig. 11.8.

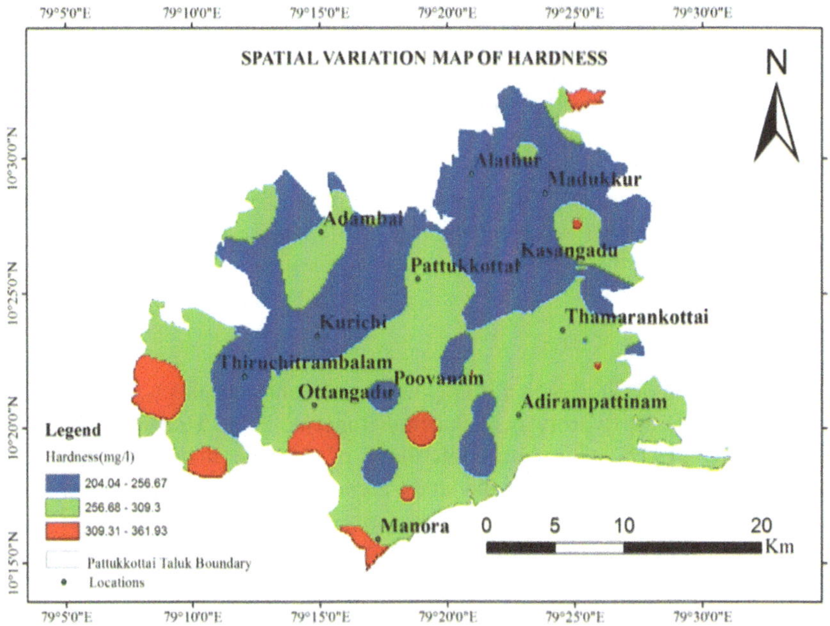

Fig. 11.7 Spatial variation map of hardness of the study area

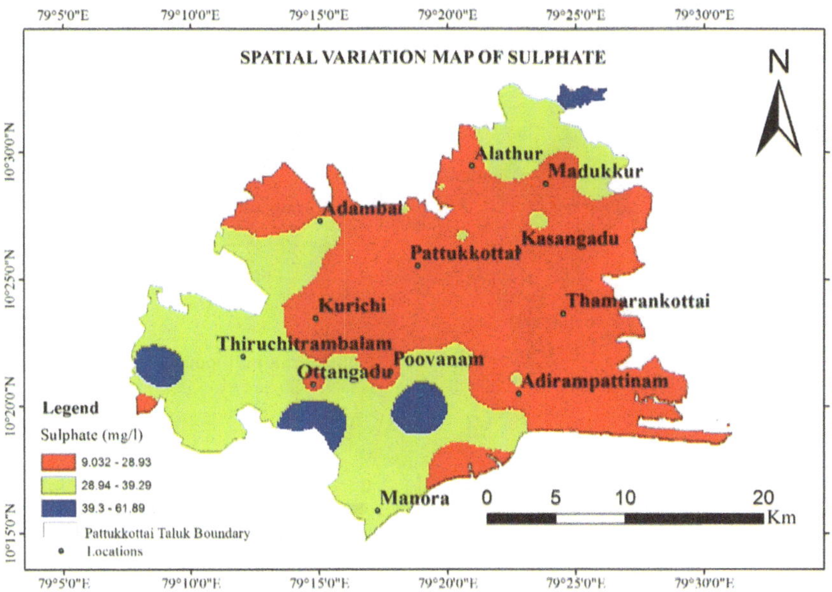

Fig. 11.8 Spatial variation map of sulfate of the study area

According to a study from the Jind District, Haryana, India, sulfate ranged from 32 to 3100 mg/L, with only 9 locations out of 60 within the acceptable limit (Meenakshi Garg et al. 2004). In another study from Shekhawati, North India, Singhal et al. (2020) collected samples in the pre- and post-monsoon periods, in which the sulfate levels ranged from 0 to 140 mg/L and from 0 to 170 mg/L, respectively; it was within the permissible limit as per the BIS standards. Considering this, the study area's sulfate level was within the permissible limit in all parts of the study area.

11.6.1.7 Fluoride

The concentration of fluoride in a small number of water samples was high, which may be related to the hydrolysis of fluorite (Banerjee 2015; Yousefi et al. 2018). As per WHO (1971), the permissible limit of fluoride in drinking water is 1.0 mg/L. The fluoride levels in the study area ranged from 0.2 to 0.9 mg/L as shown in Fig. 11.9, which is below the prescribed limit, in which Pallathur had a maximum concentration of 0.9 mg/L and Pudupattinam had a minimum concentration of 0.2 mg/L.

Traces of fluoride were found in the study area because of the effluents from the glass and aluminium manufacturing industries. It was observed in the Jind District,

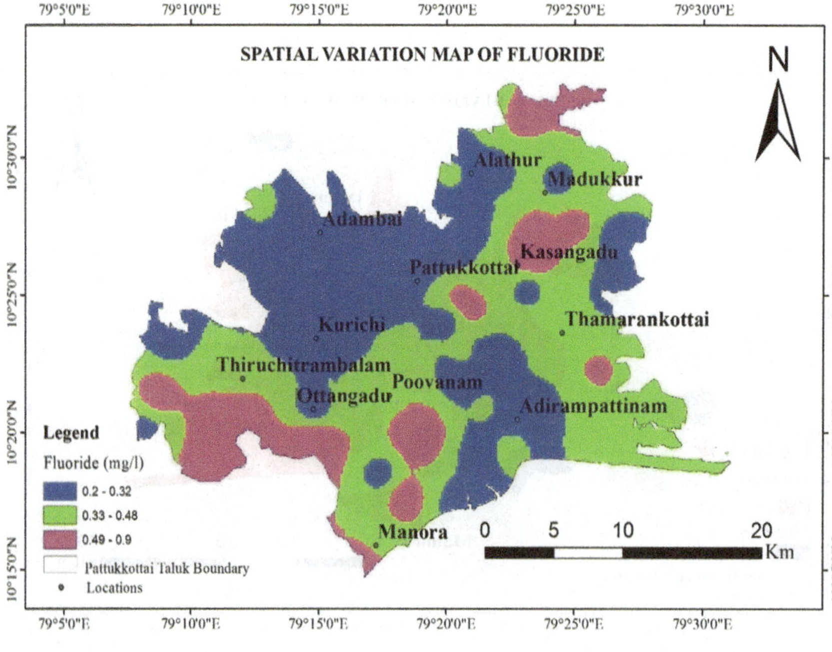

Fig. 11.9 Spatial variation map of fluoride of the study area

Haryana, that fluoride levels ranged from 6 to 30 mg/L, which is extremely high and above the permissible limit (Meenakshi Garg et al. 2004). According to the study from the Ranga Reddy district, Hyderabad, fluoride levels in GW ranged from 0.5 to 4.5 mg/L (Sujatha and Rajeswara 2003). The high levels of fluoride in this area is mainly because of the high utilization of phosphatic fertilizers, which are leached down to irrigation flows by saturation (Handa 1975).

11.6.1.8 Alkalinity

The standard desirable limit of the alkalinity of portable water is 200 mg/l. The maximum permissible level is 600 mg/l. Figure 11.10 shows that the spatial distribution of alkalinity in the GW of the study area varied from 206 to 383 mg/l. A large amount of alkalinity imparts a bitter taste to water and makes it unpalatable, and, thus, consumer acceptance decreases with this water. The spatial variation map of alkalinity is shown in Fig. 11.10. A study carried out in the Guntur district, Andhra Pradesh, revealed that alkalinity ranged from 340 to 361 mg/L, which is not significant as per the standards (Subba Rao 2006). As per the study area, it ranged from 206 to 384 mg/L, which is within the prescribed limit as shown in Fig. 11.10.

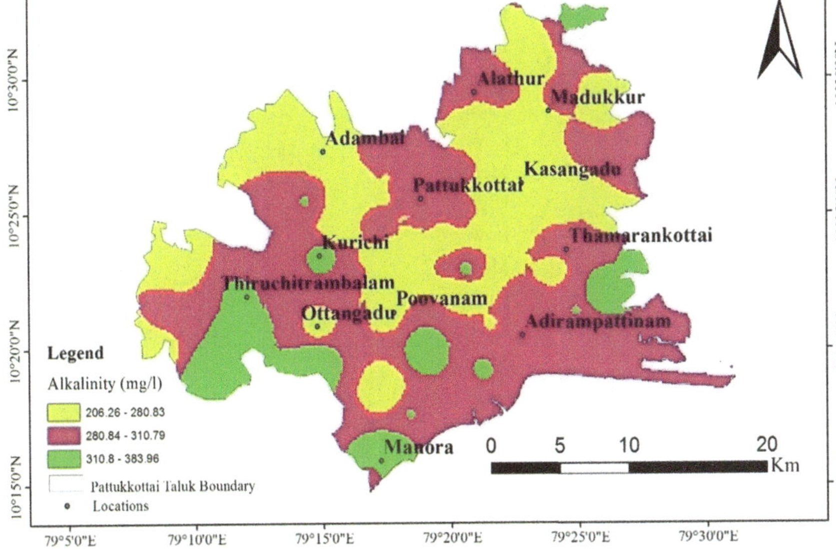

Fig. 11.10 Spatial variation map of alkalinity of the study area

11.7 Overlay Analysis

In the weighted index overlay analysis (WIOA) method, weights are assigned to various classes of different themes like pH, nitrate, total dissolved solids, hardness, sulphate, fluoride, alkalinity, and chloride according to their influences. A weight is defined as a relative percentage, and the summation of the percentages of influence weights should add up to 100. It is known that each cell value of the raster is multiplied by its % influences and then added to create the resultant raster. In this study, weights for the different themes were assigned based on the expertise inputs from various studies (Elbeih 2007) and nitrate was given the highest weight of 20%, TDSs and pH a weight of 15%, and the rest of the parameters a weight of 10%. The assigned weights for the different physicochemical parameters are given in Table 11.2. WIOA was used to classify the results as good, poor, and moderate water quality as shown in Fig. 11.11. According to the results, 368.65 km^2 of the study area came under the good GW quality class, 102.43 km^2 of the area under the moderate water quality class, and 224 km^2 of the area under the poor quality class.

Table 11.2 Physicochemical parameters: weights and ranks

Physicochemical parameters	Parameter range	Rank	Weight
pH	Low	1	15
	Medium	2	
	High	3	
Fluoride	Low	1	10
	Medium	2	
	High	3	
Sulphate	Low	1	10
	Medium	2	
	High	3	
TDS	Low	1	15
	Medium	2	
	High	3	
Alkalinity	Low	1	10
	Medium	2	
	High	3	
Nitrate	Low	1	20
	Medium	2	
	High	3	
Hardness	Low	1	10
	Medium	2	
	High	3	
Chloride	Low	1	10
	Medium	2	
	High	3	

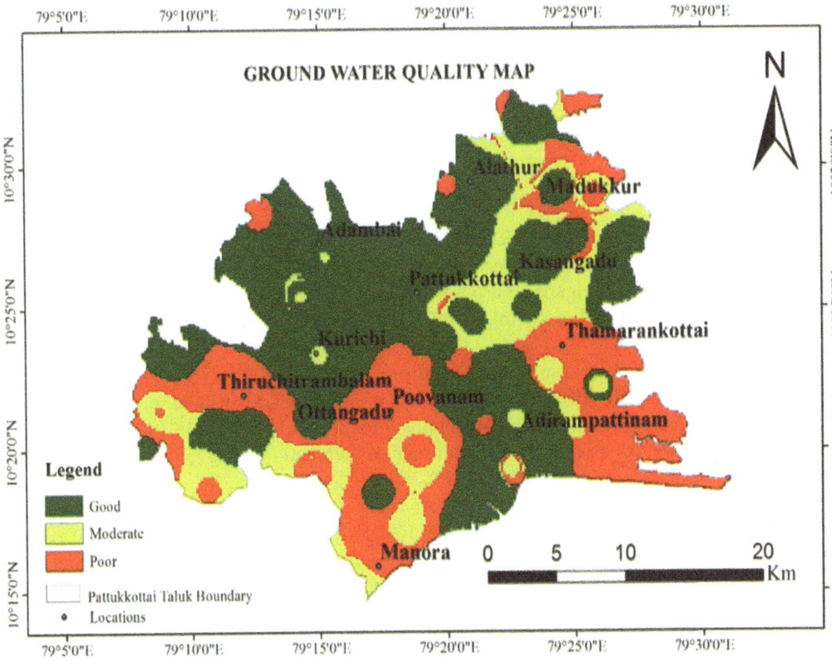

Fig. 11.11 Groundwater quality map of the study area derived using the weighted index overlay analysis method

11.8 Conclusions

This study intended to analyze the GW quality of the Pattukkottai Taluk in Tamil Nadu, India, where prominent agricultural activities take place. This assessment is considered essential because GW is the major source of water for drinking and agricultural activities. Due to its excessive and exhaustive usage, the quality and quantity of GW have drastically deteriorated over the decades in the study area. This has reduced agricultural yield, thus leading to demands for advent technology to visualize the geospatial distribution of GW quality. In order to fulfil the requirements of this study, a GIS-integrated MCDA was used to assess the GW quality. In MCDA, expert's weights and ranks were assigned as inputs to identify the influential factors for GW quality degradation. According to the study, 33% of the area fell under the poor water quality category, whereas the rest of them fell under the good and moderate categories. It is an indication that the GW quality of the study area is worsening due to various activities, in which anthropogenic activities and excessive usage are the primary factors. Considering the above-stated reasons, this study is imperative for decision- and policy-making authorities to restrict excessive usages of GW and impose regulations on civilians to prevent GW quality from deteriorating and preserve it for future generations. Although studies have used integrated modern technology for identifying the GW quality vulnerable zones, it is recommended to adopt prediction models for dynamic evaluation of GW quality for better planning for the future generations.

References

A. Abbasnia, N. Yousefi, A.H. Mahvi, R. Nabizadeh, M. Radfard, M. Yousefi, M. Alimohammadi, Evaluation of groundwater quality using water quality index and its suitability for assessing water for drinking and irrigation purposes: Case study of Sistan and Baluchistan province (Iran). Hum. Ecol. Risk Assess. Int. J. **0**(0), 1–18 (2018). https://doi.org/10.1080/10807039.2018.1458596

T. Abbasi, S. A. Abbasi, Water Quality Indices. Elsevier, (2012). ISBN:978-0-444-54304-2

M.E.-Z. Ahmed, S.F. Elbeih, GIS-based evaluation of groundwater quality and suitability in Dakhla oases, Egypt. Earth Syst. Environ. (2019). https://doi.org/10.1007/s41748-019-00112-1

H.-I. Ahn, H.-T. Chon, Assessment of groundwater contamination using geographic information systems. Environ. Geochem. Health **21**, 273–289 (1999)

M.K. Akkaraboyina, B.S.N. Raju, Assessment of water quality index of river Godavari at rajahmundry. Univers J Environ Res Technol **2**(3), 161–167 (2012)

P.B. Althuwaynee, H.J. Park, J.H. Lee, A Novel ensemble bivariate statistical evidential belief function with knowledge-based analytical hierarchy process and multivariate statistical logistic regression for landslide susceptibility mapping. Catena **114**, 21–36 (2014)

APHA, Standard Methods for the Examination of Water and Wastewater. 20th Edition, American Public Health Association, American Water Works Association and Water Environmental Federation, Washington DC (1998)

S. Arivazhagan, M. Kirubakaran, Detection of urban changes and statistical site suitability analysis in Tiruchencode taluk using geoinformatic techniques. Int. J. Earth Sci. Eng. **9**(4), 1465–1473 (2016)

B. Backman, D. Bodis, P. Lahermo, S. Rapant, T. Tarvainen, Application of a groundwater contamination index in Finland and Slovakia. Environ. Geol. **36**, 55–64 (1998)

M.S. Bajwa, B. Singh, P. Singh, *Nitrate Pollution of Groundwater under Different Systems of Land Management in the Punjab. 1st Agriculture Science Congress., 1992 Proceedings of National Academy of Agriculture Science, New Delhi* (1993), pp. 223–230

A. Banerjee, Groundwater fluoride contamination: A reappraisal. Geosci. Front. **6**(2), 277–284 (2015). https://doi.org/10.1016/j.gsf.2014.03.003

S. Bijay, G.S. Sekhon, Nitrate pollution of ground water from nitrogen fertilizers and animal wastes in the Punjab. Indian Agri. Environ. **3**, 57–67 (1976)

S. Bijay, Y. Singh, C.S. Khind, O.P. Meelu, Leaching losses of urea-N applied to permeable soils under wetland rice. Fertilizer Res. **28**, 179–184 (1991)

A.K. Biswas, S. Sanjay, S.A. Rao, G.V. Lakshmi, B. Mandal, B. Singh, R.N. Katkar, B. Murugesa, *Nitrate Contamination of Groundwater in Some Heavily Fertilized and Intensively Cultivated Districts of India* (IISS Research Bulletin No.1, Indian Institute of Soil Science (ICAR), Nabibagh, Bhopal, 2010), pp. 1–89

K. Brindha, L. Elango, Groundwater quality zonation in a shallow weathered rock aquifer using GIS. Geo-spatial Inform. Sci. **15-2**, 95–104 (2012). https://doi.org/10.1080/10095020.2012.714655

R.M. Brown, N.I. McClelland, R.A. Deininger, R.G. Tozer, A water quality index: Do we dare? Water Sewage Works **117**(10), 339–343 (1970)

CGWB, Groundwater. In: Proceedings of 3rd National Water Commission, Nagpur, India. Central ground water board (CGWB), government of India, India (1992)

CGWB, Ground Water Scenario in India, January 2016 (2016)

B.S. Champak, P. Dinesh, Review on practices and state of the art methods on delineation of ground water potential using GIS and remote sensing, bulletin of Department of Geology, Tribhuvan university, Kathmandu, Nepal. **20**(21), 7–20 (2018)

I. Chenini, A.B. Mammou, M.M. El, Groundwater recharge zone mapping using GIS-based multi-criteria analysis: A case study in Central Tunisia Maknassy Basin. Water Resour. Manag. **245**, 921–939 (2009). https://doi.org/10.1007/s11269-009-9479-1

T.H. Christensen, P. Kjeldsen, H.-J. Albrechtsen, G. Heron, P.H. Nielsen, P.L. Bjerg, P.E. Holm, Attenuation of landfill leachate pollutants in aquifers. Crit. Rev. Environ. Sci. Technol. **24**, 119–202 (1994)

O.T. Dede, I.T. Telci, M. Aral, The use of water quality index models for the evaluation of surface water quality: A case study for Kirmir Basin, Ankara, Turkey. Water Qual Expo Health **5**, 41–56 (2013). https://doi.org/10.1007/s12403-013-0085-3

M. Dhanasekarapandian, S. Chandran, D. Saranya Devi, V. Kumar, Spatial and temporal variation of groundwater quality and its suitability for irrigation and drinking purpose using GIS and WQI in an urban fringe. J. Afr. Earth Sci. **124**, 270–288 (2016)

D. Ducci, GIS techniques for mapping groundwater Contamina-tion risk. Nat. Hazards **20**(2–3), 279–294 (1999)

S. Elbeih, Impact of groundwater recharge on the surrounding environment PhD thesis, Faculty of Engineering, Ain shams University (2007)

B.A. Engel, K.C.S. Navulur, The role of geographical information systems in groundwater engineering, in *The Handbook of Groundwater Engineering, Chap 21*, ed. by J. W. Delleur, (CRC Press LLC, Boca Raton, 1999), pp. 1–16

F.M. Enyegue, A. Nyam, A. Eyike Yomba, A. Ngouokouo Tchikangoua, C.P. Bounoung, R. Nouayou, Assessment and characterization of groundwater quality under domestic distribution using hydrochemical and multivariate statistical methods in Bafia, Cameroon (2020)

B.A. Farzaneh, M. Ali Akbar, H.D. Mohammad, Y. Mahmood, Data on assessment of groundwater quality with application of ArcGIS in Zanjan, Iran. Data in Brief **18**, 375–379 (2018)

Y.L. George, W.W. David, An adaptive inverse-distance weighting spatial interpolation technique. Comp. Geosc. **39**, 1044–1055 (2008)

Groundwater for Sustainable Development, Vol. 10, pp. 100347

B.K. Handa, Geochemistry and genesis of fluoride containing groundwaters in India. Groundwater **13**(3), 275–281 (1975)

M.M. Heydari, H.N. Bidgoli, Chemical analysis of drinking water of Kashan District, Central Iran. World Appl. Sci. J. **16**(6), 799–805 (2012)

R.K. Horton, An index number system for rating water quality. J. Water Pollut. Contr. Federat. **37**(3), 300–306 (1965)

W.F. Humphreys, Hydrogeology and groundwater ecology: Does each inform the other? Hydrogeol. J. **17–1**, 5–21 (2009)

S.B. Insaf, A.A. Mohamed, T. Hiyama, Assessing groundwater quality using GIS. Water Res. Manag. **21**, 699–715 (2007). https://doi.org/10.1007/s11269-006-9059-6

C.J. Jesudhas, A. Chinnasamy, K. Muniraj, A. Sundaram, Assessment of vulnerability in the aquifers of rapidly growing sub-urban: A case study with special reference to land use. Arab. J. Geosci. **14**(1), 1–20 (2021)

T. Jeyavel Raja Kumar, S. Vasanth Raja, P. Saravanan, C. Dushiyanthan, B. Thirunelakandan, R. Suresh, K. Deivanayagi Karthikeyan, Assessment of groundwater quality using weighted index overlay analysis and water quality index technique in chittar sub basin of tamiraparani river basin, Tamil Nadu, India, eco-chronicle. **14**(1), 43–52 (2019)

D.C. Jhariyaa, T. Kumara, R. Dewangan, D. Pal, P.K. Dewangan, Assessment of groundwater quality index for drinking purpose in the Durg District, Chhattisgarh using geographical information system (GIS) and multi-criteria decision analysis (MCDA) techniques. J. Geol. Soc. India **89**, 453–459 (2017)

J.C. Johnny, M.C. Sashikkumar, M. Kirubakaran, L.M. Mathi, GIS-based assessment of groundwater quality and its suitability for drinking and irrigation purpose in a hard rock terrain: A case study in the upper Kodaganar basin, Dindigul district, Tamil Nadu, India. Desalin. Water Treat. **102**, 49–60 (2018)

Jyotirmoy Bhardwaj, Karunesh K Gupta, Rajiv Gupta, 2015, A Review of Emerging Trends on Water Quality Measurement Sensors, International Conference on Technologies for Sustainable Development, Feb. 04–06, 2015, Mumbai, India

S. Kaliraj, N. Chandrasekar, N.S. Magesh, Identification of potential groundwater recharge zones in Vaigai upper basin, Tamil Nadu, using GIS-based analytical hierarchical process (AHP) technique. Arab. J. Geosci. **7**, 1385–1401 (2014). https://doi.org/10.1007/s12517-013-0849-x

S. Kaliraj, N. Chandrasekar, N.S. Magesh, Identification of potential groundwater recharge zones in Vaigai upper basin, Tamil Nadu, using GIS-based analytical hierarchical process AHP technique. Arabian J. Geosci. **74**, 1385–1401 (2013). https://doi.org/10.1007/s12517-013-0849-x

S. Kant, A.K. Raina, Limnological studies of two ponds in Jammu II physiological parameters. J. Environ. Biol. **11**(2), 137 (1990)

K.R. Karanth, *Groundwater Assessment, Development and Management* (Tata McGraw-Hill Publishing Company Limited, New Delhi, India, 1997)

M. Kirubakaran, C. Ashokraj, J. Colins Johnny, R. Anjali, Groundwater quality analysis using WQI and GIS techniques: A case study of Manavalakurichi in Kanyakumari District, Tamilnadu, India. Int. J. Innov. Sci. Eng. Technol. **2**(11), 341–347 (2015)

M. Kirubakaran, J.C. Johnny, C. Ashokraj, S. Arivazhagan, A geostatistical approach for delineating the potential groundwater recharge zones in the hard rock terrain of Tirunelveli taluk, Tamil Nadu. India. Arabian J. Geosci. **9**(5), 382 (2016)

A. Lumb, T.C. Sharma, J.-F. Bibeault, A Review of Genesis and Evolution of Water Quality Index (WQI) and Some Future Directions. Water Qual. Expo. Health. **3**(1), 11–24 (2011). https://doi.org/10.1007/s12403-011-0040-0

J. Madan Kumar, A.J. AnkitShekhar, M. Mass, in *Crop Salt Tolerance. Agricultural Salinity Assessment and Management Manual*, ed. by K. K. Tanji, (ASCE, New York, 1990), pp. 262–304

V.K. Meenakshi Garg, K. Renuka, A. Malik, Groundwater quality in some villages of Haryana, India: Focus on fluoride and fluorosis. J. Hazard. Mater. **106B**, 85–97 (2004)

J. Miranda, G. Krishnakumar, Microalgal diversity in relation to the physicochemical parameters of some industrial sites in Mangalore. South India. Environ Monit. Assess **187**(11) (2015)

I. Mohamad Najib, Assessing groundwater quality for drinking purpose in Jordan: Application of water quality index. J. Ecol. Eng. **20**(3), 101–111 (2019). https://doi.org/10.12911/22998993/99740

K. Muniraj, C.J. Jesudhas, A. Chinnasamy, Delineating the groundwater potential zone in Tirunelveli taluk, South Tamil Nadu, India, using remote sensing, geographical information system (GIS) and analytic hierarchy process (AHP) techniques. Proc. Natl. Acad. Sci. India Section A: Phys. Sci. **90**(4), 661–676 (2020)

T.S. Narany, A.Z. Aris, A. Sefe, S. Keesstra, Detecting and pre-dicting the impact of land use changes on groundwater quality, a case study in Northern Kelantan, Malaysia. Sci. Total Environ. **599–600**, 844–853 (2017)

R.T. Nickson, J.M. McArthur, B. Shrestha, T.O. Kyaw-Nyint, D. Lowrt, Arsenic and other drinking water quality issues, Muzaf-fargarh District, Pakistan. Appl. Geochem. **20**, 55–68 (2005)

P. NiranjanBabu, N. Subba Rao, P. Chandra Rao, J. Prakasa Rao, Groundwater quality and its importance in the land developmental programmes. Indian J. Geol. **69**(4), 305–312 (1997)

B. Panneerselvam, S. Karuppannan, K. Muniraj, Evaluation of drinking and irrigation suitability of groundwater with special emphasizing the health risk posed by nitrate contamination using nitrate pollution index (NPI) and human health risk assessment (HHRA). Hum. Ecol. Risk Assess. Int. J. **27**(5), 1324–1348 (2020)

B. Panneerselvam, K. Muniraj, M. Thomas, N. Ravichandran, GIS-based legitimatic evaluation of groundwater's health risk and irrigation susceptibility using water quality index, pollution index, and irrigation indexes in semiarid region, in *Groundwater Resources Development and Planning in the Semi-Arid Region*, (Springer, Cham, 2021), pp. 239–268

F. Pedrero, A. Albuquerque, M.H. Marecosdo, V. Cavaleiro, J.J. Alarcon, Application of GIS-based multi-criteria analysis for site selection of aquifer recharge with reclaimed water. Resour. Conserv. Recycl. **561**, 105–116 (2011). https://doi.org/10.1016/j.resconrec.2011.08.003

K. Pietersen, Multiple criteria decision analysis (MCDA): A tool to support sustainable management of groundwater resources in South Africa. Water SA **32**(2), 119–128 (2006)

T. Praharaj, S.P. Swain, M.A. Powell, B.R. Hart, S. Tripathy, Delineation of groundwater contamination around an ash pond geochemical and GIS approach. Environ. Int. **27**, 631–638 (2002)

E.V.S. Prakasa Rao, K. Puttanna, Nitrates, agriculture and environment. Curr. Sci. **79**(9), 1163–1168 (2000)

L.G. Quist, R.R. Bannerman, S. Owusu, *Groundwater in Rural Water Supply in Ghana. Ground Water in Rural Water Supply, Report of the West African Sub-regional Workshop Held in Accra, Ghana, 20–24 October 1986* (UNESCO Tech. Doc. Hydrol., Paris, 1988), pp. 101–126

C.R. Ramakrishnaiah, C. Sadashivaiah, G. Ranganna, Assessment of water quality index for groundwater in TumkurTaluk, Karna- taka state. India E-J Chem. **6**(2), 523–530 (2009)

H.J.E. Rodda, S. Demuth, U. Shankar, The application of a GIS-based decision support system to predict nitrate leaching to groundwater in southern Germany. Hydrol. Sci. **44**, 221–235 (1999)

M. Salifu, S.M. Yidana, S. Osae, Y.S. Armah, The infuence of the unsaturated zone on the high fuoride contents in ground- water in the Middle Voltaian aquifers—The Gushegu District, Northern Region of Ghana. J. Hydrogeol. Hydrol. Eng. (2013). https://doi.org/10.4172/2325-9647.1000107

A. Sayyed Juned, B. Bhosle Arjun, Analysis of chloride, sodium and potassium in groundwater samples of Nanded City in Mahabharata, India. Eur. J. Exp. Biol. **1**, 74–82 (2011)

H.A. Schroeder, Relations between hardness of water and death rates from certain chronic and degenerative diseases in the United States. J. Chronic Dis. **12**, 586–591 (1960)

K. Shunmugapriya, B. Panneerselvam, K. Muniraj, N. Ravichandran, P. Prasath, M. Thomas, K. Duraisamy, Integration of multi criteria decision analysis and GIS for evaluating the site suitability for aquaculture in southern coastal region, India. Marine Pollut. Bull. **172**(112), 907 (2021)

A. Singhal, R. Gupta, A.N. Singh, A. Shrinivas, Assessment and monitoring of groundwater quality in semi –arid region. Groundw. Sustain. Dev. **11**, 1–9 (2020)

R.K. Sivanappan, Expansion of storage capacity, critical, in *Survey of Indian Agriculture*, ed. by N. Ravi, (The Hindu, India, 2002)

P.D. Sreedevi, Groundwater quality of Pageru river basin, Cuddapah district, Andhra Pradesh. J. Geol. Soc. India **64**(5), 619–636 (2004)

R. Srinivasa, P. Bhakara, A.P. Singh, Groundwater quality assessment in some selected area of Rajasthan, India using fuzzy multi-criteria decision making tool. Aquatic Procedia **4**, 1023–1030 (2015)

N. Subba Rao, G. Srinivasa Rao, S. Venkateswara Rao, P. Madhusudhana Reddy, D. John Devadas, Environmental control of groundwater quality in a tribal region of Andhra Pradesh, India. Indian J. Geol. **71**(4), 299–304 (1999)

N. Subba Rao, Seasonal variation of groundwater quality in a part of Guntur District, Andhra Pradesh, India. Environ. Geol. **49**, 413–429 (2006). https://doi.org/10.1007/s00254-005-0089-9

D. Sujatha, R.B. Rajeswara, Quality characterization of groundwater in the south-eastern part of the Ranga Reddy district, Andhra Pradesh, India. Environ. Geol. **44**, 579–586 (2003). https://doi.org/10.1007/s00254-003-0794-1

A.D. Sutadian, N. Muttil, A.G. Yilmaz, B.J.C. Perera, Development of river water quality indices—a review. Environ. Monit. Assess. **188**(1), (2016). https://doi.org/10.1007/s10661-015-5050-0

M. Syed, R. Saqibnadeem, Determination of water quality parameters of water supply in different areas of Karachi City. Eur. Acad. Res. **I**(12), 6030–6050 (2014)

T. Tapasya, Water quality indices used for groundwater quality assessment. Int. J. Res. Environ. Sci. Technol. **5–3**, 76–80 (2015)

H. Thomas, Groundwater quality and groundwater pollution. Farm Water Qual. Plann., 1–3 (2003)

D.K. Todd, *Ground Water Hydrology* (Wiley, New York, 1980), p. 535

United States Salinity Laboratory Staff, *Diagnosis and Improvement of Saline and Alkaline Soils* (US Department of Agriculture, Hand Book, Washington, DC, 1954)

M. Vasanthavigar, K. Srinivasamoorthy, K. Vijayaragavan, R. Rajiv Ganthi, S. Chidambaram, P. Anandhan, R. Manivannan, S. Vasudevan, Application of water quality index for groundwater quality assessment: Thirumanimuttar sub-basin, Tamilnadu, India. Environ. Monit. Assess. **171**, 595–609 (2010). https://doi.org/10.1007/s10661-009-1302-1

WHO, *International Standards for Drinking Water*, 3rd edn. (WHO, Geneva, 1971)

WHO, *Guidelines for Drinking Water Quality* (World Health Organization, Geneva, 1984)

L.V. Wilcox, The quality of water for irrigation use. US Department of Agricultural Technical Bulletin 1962, Washington (1948)

World Health Organisation (2004). Sulphate in Drinking-Water. Background Document for Development of WHO Guidelines for Drinking-Water Quality

M. Yousefi, M. Ghoochani, A.H. Mahvi, Health risk assessment to fluoride in drinking water of rural residents living in the Poldasht city, Northwest of Iran. Ecotoxicol. Environ. Saf. **148**, 426–430 (2018). https://doi.org/10.1016/j.ecoenv.2017.10.057

J. Yu, Y. Chean, J.P. Wu, *Cellular Automata and GIS Based Land Use Suitability Simulation for Irrigated Agriculture. 18th world IMACS/MODSIM Congress July, Cairns* (2009), pp. 13–17

L. Yu, T. Zheng, X. Zheng, Y. Hao, R. Yuan, Nitrate source apportionment in groundwater using Bayesian isotope mixing model based on nitrogen isotope fractionation. Sci. Total Environ. **137**, 242 (2020). https://doi.org/10.1016/j.scitotenv.2020.137242

C. Zanotti, M. Rotiroti, L. Fumagalli, G.A. Stefania, F. Canonaco, G. Stefenelli, A.S.H. Prev, B. Leoni, T. Bonomi, Groundwater and surface water quality characterization through positive matrix factorization combined with GIS approach. Water Res. **159**, 122–134 (2019)

Chapter 12
Study of Heavy Metal Contamination at Durgapur Barrage Site on the Damodar River

Susmita Pandit, Sumit Kumar Khan, and Bahnisikha Das

Abstract A study was conducted at Durgapur Barrage site on the Damodar River to assess the dynamism and bioaccumulation of heavy metals. The presence of heavy metals in the environment is considered to have negative impacts because their toxic characteristics cause various health issues. The Damodar River is adversely affected by industrial and mining activities, which have brought about an ecological disaster to the study zone. To determine the contamination of metals, a total of eight different samples were collected from the study area. Evaluation of the heavy metal contamination was performed using the enrichment factor. From the study it was found that cadmium is present in the sediment in the range between non concentration to high concentration. The enrichment factor put on view a man-made enrichment in both side of the barrage for zinc, copper, lead, and cadmium which is mainly originated from industrial activates and moderately from mining activates. For geoaccumulation, the values were categorized ranging from low-contaminated to non-contaminated values. The Damodar River showed a moderate to strong contamination of lead and cadmium. This study clearly indicates that heavy metals have contaminated the river water and hence have a dangerous impact on the biota because of high metal bioavailability.

Keywords Durgapur Barrage · Damodar River · Heavy metal · Contamination · Bioaccumulation · Enrichment factor · Geoaccumulation

S. Pandit (✉) · S. K. Khan · B. Das
Department of Civil Engineering, School of Engineering & Technology, Adamas University, Kolkata, India
e-mail: bahnisikha1.das@adamasuniversity.ac.in

B. Panneerselvam et al. (eds.), *Climate Change Impact on Groundwater Resources*, https://doi.org/10.1007/978-3-031-04707-7_12

12.1 Introduction

Nowadays, environmental toxicity is rapidly increasing in water bodies; this is due to water and sediment pollution, which has an adverse impact on the environment. Natural as well as anthropogenic activities like huge amounts of solid waste disposal, agricultural activities, etc. (Adigiah et al. 2013), are observed to be some of the reasons behind this aquatic pollution. In the maximum number of cases, human activities are responsible for polluting the aquatic environment. This results in an increase in the amount of heavy metals in the aquatic environment and also has a significant effect on the nearby flora and fauna. The increase in the concentrations of heavy metals like zinc (Zn), copper (Cu), lead (Pb), and cadmium (Cd) results in the pollution of water and sediments on which the activities of the aquatic ecosystem are completely dependent (AI Nayeem et al. 2017). When these heavy metals are absorbed by the soil and contaminates the water, the food chain of organisms is disrupted.

12.2 Background of this Study

As fish are situated at the bottom of the food chain in an aquatic environment, they easily consume heavy metals from the sediments, which are then passed on to human beings who consume the fish, thus leading to several health issues. This condition occurs due to the direct discharge of wastes from industries and agricultural runoffs, which get quickly deposited into the river (Azevedo et al. 2009; Bhardwaj & Verma 2017; Braatne et al. 2008; Burger & Gochfeld 2004; Dallinger et al. 1987; de Pinho et al. 2002; Chatterjee et al. 2009). Fish absorb heavy metals through their organs and tissues and also from sediments; these heavy metals accumulate in their brain and, therefore, consumption of fish on a regular basis will cause serious health issues in humans (Leon et al. 2018).

Treated water is not available in every part of India; people in remote areas use river water directly to meet their daily needs. River water is contaminated with heavy metals like Zn, Cu, Pb, and Cd, which cause many health issues; for example, Zn causes diarrhoea and nausea, Cu damages the liver, kidney, heart, and brain, Cd causes hepatic dysfunction, testicular damage, and osteomalacia, and Pb causes anaemia and brain damage (Dudgeon 2000; Dus et al. 2005; Fu et al. 2020; Gorde & Jadhav 2013; Hanna et al. 2018; Hofman et al. 2004; Holt 2000; Hooper & Austen 2013; Hossain and Patra 2020b; Genchi et al. 2020).

12.3 Objective of this Study

This work consists of selecting a study area for determining the heavy metals present in water and sediments, the geoaccumulation index, and heavy metals present in fish (in the muscles, brain, liver, heart). Four metals, namely, Zn, Cd, Pb, and Cu were chosen, which are considered as essential elements for survival if they are present under permissible limits; otherwise, they are hazardous to the aquatic environment and humans. In this study, the heavy metal concentration level is checked and compared to that prescribed by the World Health Organization (WHO) to determine the toxicity limits of water, sediments, and fish.

12.4 Materials and Methods

12.4.1 Study Area

This study was conducted at the site of Durgapur Barrage on the Damodar River near the Bansuri District (Durgapur) and partly in Paschim Bardhaman in the Indian state of West Bengal. The river covers nearly 547 km with a longitude and latitude of 23° 28′ 31.44″ N and 87° 18′ 8.28″ E, respectively. The study area is an industrial area, and the pollution rate is extremely high. Therefore, the main aim of this study was to determine the effects on an underwater ecosystem in both the upstream and downstream areas and its subsequent effects on the nearby flora and fauna. If the focus is made on the climatic condition of this area, then the month of October and November lies in the monsoon period when the flow of water is high. Therefore, this study was conducted in the month of January–February 2019 (Fig. 12.1).

Fig. 12.1 Map of the study area (source from Google Maps) (Table 12.1)

Table 12.1 Average climatic conditions and rainfall in the Durgapur region (Regional Meteorological Centre Kolkata)

Month	Temperature (°C)	Humidity (%)	Precipitation (mm)
January	17.7	64	13
February	21.5	67	22
March	20.4	48	23
April	30.2	50	38
May	30.8	63	117
June	29.8	75	256
July	28	84	300
August	27.7	85	268
September	27.3	86	238
October	25.8	70	99
November	22.4	67	10
December	18.8	65	10

12.4.2 Sampling

Surface water samples were collected from the upstream and downstream of Durgapur Barrage on the Damodar River (from different upstream and downstream sites). A total of eight samples were collected from the river. The gap between each sampling point was 100 m, and water samples were collected from a depth of 0.5 m.

Sample collection started in the month of January 2019 at some selected points upstream and downstream of the site. Water samples were collected from a depth of 0.5 m below the surface.

Polyethylene bottles were collected for storing the samples after cleaning them with a solution of diluted nitric acid (1 litre of distilled water and 3 ml of diluted nitric acid). Cleaning the apparatus before sampling is extremely important so as to ensure that the equipment is not a source of foreign substances that could affect the concentration of the target analysts in the samples or affect the chemical properties. After this, the collected samples were kept in a refrigerator at 4 °C to avoid loss from evaporation.

Not only the water samples but also some fish samples (*Labeo rohita*) were collected for bioaccumulation.

12.4.3 Sediment Collection

For collecting the sediments, a stainless steel scoop was used, which was cleaned before sampling and kept in an air-tight polyethylene zipper pouch.

12.5 Analysis Methods

12.5.1 Procedure

For the water sample heavy metal analysis, 100 ml of the water sample was taken in a clean beaker and then 5 ml of concentrated nitric acid (HNO_3) was added and mixed well. A hot plate was used to heat the mixture that was digested at 100 °C in the minimum volume possible (10–15 ml) before precipitation. The completion of the digestion process can be known if the sample changes into light colour. Therefore, the sample was heated and concentrated HNO_3 was added till the colour changed. The sample was cooled and then filtered using a filter paper (Whatman Grade No. 42), and, after filtration, it was diluted using 30 ml distilled water.

For the sediment sample heavy metal analysis, the samples of the sediments were dried in a hot air oven. In all, 1 mg of the sample was collected in a solution of 10 ml of HNO_3 and 10 ml of distilled water. A hot plate was used to heat the sample at 90–100 °C for 25 minutes. The process of heating and addition of concentrated HNO_3 was continued till the complete digestion of the mixed solution had taken place. The mixture was then allowed to cool. In the cooled sample, 20 ml of concentrated hydrochloric acid (HCl) was added and again heated using the hot plate at 90 °C–100 °C for 30 min. Then, the digestion was completed and the sample was filtered using a filter paper (Whatman Grade No. 41), and, after filtration, it was diluted using 30 ml distilled water.

For the fish sample heavy metal analysis, the fish sample was taken and cut into four separate sections (brain, scale, mass, and skeleton). The samples of fish were dried in a hot air oven at 40 °C for 1 hour, and then the temperature was increased to 140 °C for the next 3 hours. After that, it was ground using mortar and pestle. In all, 1 mg of the sample was collected in a solution of 10 ml of concentrated HNO_3 and 10 ml of distilled water. The sample was then kept in a digester at 90–100 °C for 20 minutes. The process of heating and addition of concentrated HNO_3 was continued till the end of the process. The mixture was then allowed to cool. Into the cooled sample,10 ml of concentrated HCl was added and heated using the hot plate at 90–100 °C for 30 minutes. The digestion was completed, and the sample was filtered using the filter paper (Whatman Grade No. 41), and after filtration, it was diluted using 30 ml distilled water.

12.5.2 Heavy Metal Analysis

The digested filter samples of water, sediments, and fish were quantified to determine the amount of heavy metals (Cd, Zn, Cu, Pb) to be analyzed by anodic stripping voltammetry (Model - VA 797, Metrohm, Switzerland) with a voltametric cell consisting of a hanging mercury drop electrode (HMDE) as a working electrode (WE), a platinum rod electrode as an auxiliary electrode (AE), and an Ag/AgCl electrode (3M-KCL) as a reference electrode (RE) for differential pulse analysis. The voltametric analyzer was controlled by VA Computrace version 2.797.00X0, SR1 software (8.797.8013, 797 VA Computrace, Metrohm).

Based on the known concentration of the heavy metal solution, the instrument was calibrated. Then, analysis of the sample was started based on the known concentration of the metal.

12.5.3 Estimation of the Enrichment Factor and the Geoaccumulation Index

The enrichment factor helps to determine the status of sediment in it naturally or anthropogenically and it is used to conclude anthropogenic influence commonly. The enrichment factor is defined as the concentration of a heavy metal with respect to the reference metal. It is calculated (Olayinka et al. 2021) using the following formula:

$$\text{Enrichment factor}\left(\text{EF}\right) = \frac{\left(Cs\right)\text{sample}}{\left(Cb\right)\text{background}} \tag{12.1}$$

where the *Cs* sample and *Cb* background refer to the heavy metal in the sediment's sample ratio and the heavy metal concentration in the background, respectively. Enrichment factors divided into four categories and are as follows (Tables 12.2 and 12.3):

The geoaccumulation index is used to determine the heavy metal contamination levels in sediments. It is calculated by Eq. 12.2 (Olayinka et al. 2021).

Table 12.2 Enrichment factor classes (Hossain and Patra 2020a)

Value	Quality
1 > EF	Low EF
3 > EF > 1	Moderate EF
6 > EF > 3	Considerably high EF
6 = EF	Very high EF

Table 12.3 Shale values of international geochemical background (Olayinka et al. 2021)

Heavy metal	Shale limit (mg/kg)
Zn	55.7
Pb	22.4
Cd	0.28
Cu	45

Table 12.4 Geoaccumulation classes (Hossain et al. 2021)

Class	Value	Quality of water and sediment
0	$0 > I_{geo}$	No contamination
1	$1 > I_{geo} > 0$	No contamination to moderate contamination
2	$2 > I_{geo} > 1$	Moderate contamination
3	$3 > I_{geo} > 2$	Moderate to heavy contamination
4	$4 > I_{geo} > 3$	Heavy contamination
5	$5 > I_{geo} > 4$	Heavy to extreme contamination

Table 12.5 Heavy metal concentration in water

Location						
Upstream	pH	Temp.	Heavy metal (mg/l)			
	6.8	19	Zn	Pb	Cd	Cu
			0.667	0.6025	0.1057	0.475
Bank3’	6.9	17	0.669	0.6030	0.1062	0.470
Bank4	6.7	11	0.672	0.6028	0.1067	0.454
Bank4’	6.4	20	0.677	0.6017	0.1052	0.460
Downstream						
Bank1	6.1	26	0.682	0.6045	0.1077	0.455
Bank1’	6.3	24	0.687	0.6023	0.1044	0.474
Bank2	6.8	14	0.612	0.6047	0.1087	0.445
Bank2’	6.9	19	0.697	0.6060	0.1092	0.432

$$\text{Geoaccumulation index}\left(I\text{geo}\right) = \ln\left(\frac{Cn}{1.5Bn}\right) \tag{12.2}$$

where *Cn* is the sample element’s concentration and *Zn* is the same baseline sample element’s concentration (Table 12.4).

12.6 Results and Discussion

12.6.1 Concentration of Metals in River Water

The concentration of heavy metals in the water samples from the Damodar River is given in Table 12.5.

Table 12.5 and Fig. 12.2 show that the pH of the river water lies in the range 6.9–6.1, which is marginally acidic in nature, and for this, the bioavailability of heavy metal will increase. The temperature varies from 11 °C to 26 °C. This study was conducted on both sides of the river, upstream and downstream. Table 12.5 and Fig. 12.3 show that the concentration of Zn ranges from 0.612 mg/l to 0.697 mg/l.

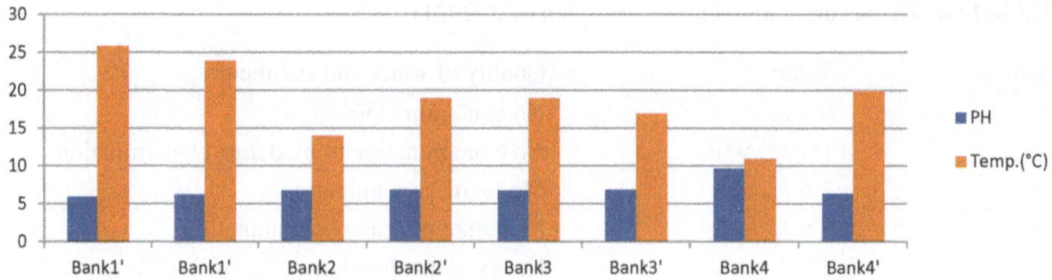

Fig. 12.2 Physical characteristics of water

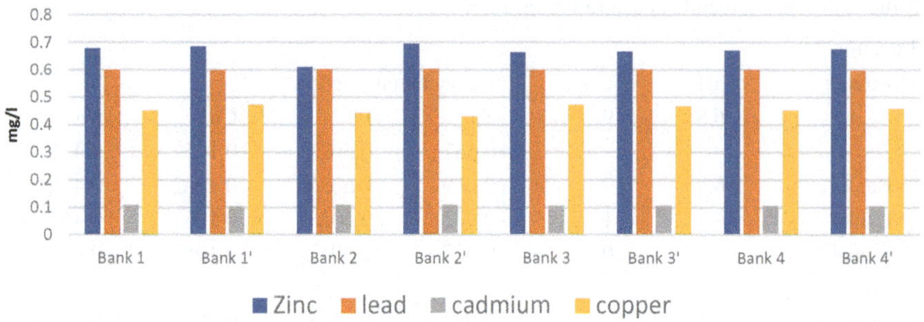

Fig. 12.3 Heavy metal concentration in water

The results conclude that all the concentrations are within the permissible limit of 5.00 mg/l as recommended by the WHO.

The concentration of Pb varies from one bank to the other and ranges from 0.6017 mg/l to 0.6060 mg/l, which is above the permissible limit of 0.1 mg/l as recommended by the WHO. A higher concentration of Pb is dangerous for aquatic life as well as for human health. The concentration of Cd ranges from 0.1044 mg/l to 0.1092 mg/l, which is also higher than the permissible limit of 0.001 mg/l–0.005 mg/l. Metals like Pb and Cd help in bioaccumulation and biomagnification, which is harmful to aquatic life and the environment. The concentration of Cu ranges from 0.432 mg/l to 0.475 mg/l, which is below the permissible limit of 3.00 mg/l (Table 12.6).

12.6.2 Heavy Metal Concentrations in Sediments

Heavy metal concentrations examined in the sediments from the Damodar River near the Durgapur Barrage region are shown (Table 12.7 and Fig. 12.4).

The Zn concentration in the river sediments was found to be high at the downstream (Bank1) location and low at the upstream location (Bank3). Zn is an important element; many aquatic organisms regulate it and it does not take part in biomagnification for aquatic life. Aquatic organisms absorb Zn majorly from the sediments, but they do not consume it. The sources of Zn in sediments are batteries, fungicides, pesticides, and various industrial effluents (Kaur 2007; Keup 1968;

Table 12.6 Permissible limit for heavy metals in water and sediments

Elements	Water (WHO) (mg/l)	Sediments (WHO) (mg/l)
Zinc	5.00	<1
Lead	0.1	5.00
Cadmium	0.001–0.005	0.1
Copper	3.00	0.05–0.15

Table 12.7 Heavy metals in sediments (mg/l)

Location	Zn	Pb	Cd	Cu
Bank1	2.86	2.227	1.52	2.838
Bank1'	2.79	2.231	1.5	2.734
Bank2	2.58	2.038	1.47	2.534
Bank2'	2.59	2.052	1.49	2.423
Bank3	1.95	1.680	1.75	1.744
Bank3'	2.01	1.674	1.81	1.679
Bank4	2.1	1.451	1.62	1.525
Bank4'	2.08	1.523	1.55	1.685

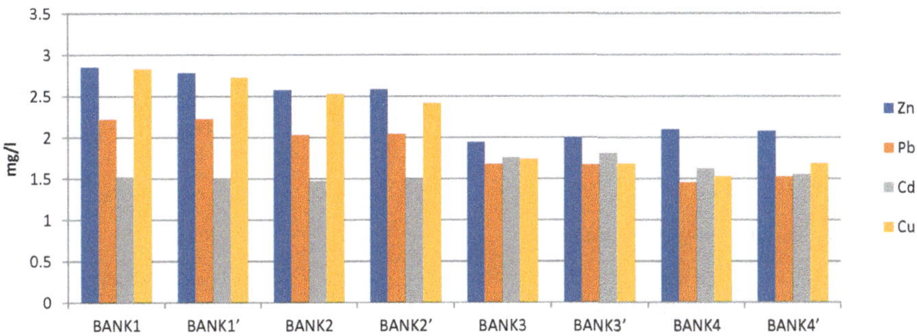

Fig. 12.4 Heavy metal concentration in sediments

McCain 2013; Mebane et al. 2020; Nendza et al. 2018; Ojija 2015; Ouyang et al. 2018; Palmer et al. 2014; Paul 2017; Ratte 1999).

Pb is highly toxic at a low concentration and is also harmful to human health. In other words, Pb toxicity is a long-term issue, and it causes damage to the central nervous system and also to sensitive organs like the brain, liver and kidney. The amount of Pb is high at Bank1 and Bank1'. Industries located near the river site release wastewater directly into the river, which can be the cause of the increasing concentration of Pb in river water.

Cd and Cu are highly toxic metals. In this study, a high concentration of Cd and Cu was found in sediments, which is more than its permissible limit of 0.1 mg/l and 0.15 mg/l, respectively. The high concentration of Cd and Cu might be due to the presence of industries near the river site. If the concentration of Cd is more than the permissible limit, then it may cause serious health issues like liver damage, brain damage, etc. A high concentration of Cu is also extremely harmful as it has a high ability of bioaccumulation and can cause vomiting, headache, and low blood pressure in humans. The metals were arranged according to their concentration in the sediments: Zn > Cu > Pb > Cd (Sattar & Raslan 2014; Sehgal et al. 2012; Sumpter 2009; Zhuang & Gao 2014; Žižek et al. 2007).

12.6.3 Heavy Metal Contamination in Fish Organs

Table 12.8 shows that the collected fish samples from both downstream and upstream of the Damodar River belong to the cyprinid family, the *Labeo rohita* species. The samples were collected from the pelagic zone and the weights of the two fish samples were 572 g and 484 g, respectively. Heavy metal concentration was found in fish organs (brain, scale, bone, muscle). Table 12.9 and Fig. 12.5 show that the concentration of Cd is above the permissible limit of 0.1 mg/l. A high concentration of Cd is extremely dangerous to the kidney and central nervous system. The result shows that the fish brain had the highest concentration of Cu with 0.6553 mg/l, whereas the permissible limit is 3 mg/l, so the Cu concentration was below the permissible limit. Cu is an important metal for the animal as it helps in the enzyme activity. Pb concentration in the fish organs was slightly higher than the permissible limit of 0.05 mg/l. Zn concentration in the fish organs was lower than the permissible limit of 5 mg/l. From the study, it was found that the concentration of Cu and Zn was below the permissible limit but that of Cd and Pb was above the permissible limit.

Table 12.8 Family and species of fish

Location	Family	Species	Zone Total weight (g)
Downstream	Cyprinidae	*Labeo rohita*	Pelagic 572
Upstream	Cyprinidae	*Labeo rohita*	Pelagic 484

Table 12.9 Heavy metal concentration in fish organs

Downstream	Heavy metals	Organs			
	Zn	Bones	Muscle	Scale	Brain
		0.6428	0.6444	0.6345	0.7475
	Pb	0.0534	0.0628	0.0554	0.0396
	Cd	0.1026	0.1141	0.1104	0.0956
	Cu	0.4454	0.3934	0.3997	0.6553
Upstream	Zn	0.6428	0.6423	0.6365	0.7455
	Pb	0.0544	0.0648	0.0564	0.0386
	Cd	0.1076	0.1161	0.1127	0.0946
	Cu	0.4554	0.3974	0.4017	0.6543

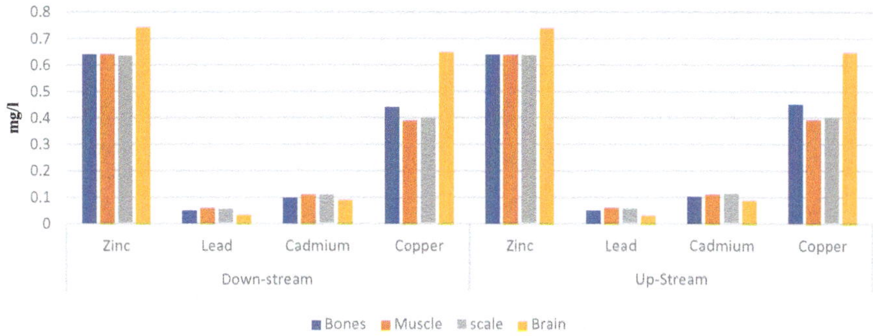

Fig. 12.5 Presence of heavy metals in fish organs

According to Table 12.9, the concentrations of Zn and Cu metals in the fish bones were below the permissible limit, whereas the concentrations of Pb and Cd were above the permissible limit.

In the fish muscles, the Zn concentration was 0.6444 mg/l, which was below the permissible limit; Pb concentration was 0.0648 mg/l and was slightly high; Cd concentration was higher than the permissible limit; and Cu concentration was 0.4554 mg/l, which was below the permissible limit.

In the fish scales, the Zn concentration was 0.6345 mg/l, Pb concentration was 0.0554 mg/l, Cd concentration was 0.1104 mg/l, and Cu concentration was 0.4017 mg/l, all of which were below the permissible limit.

In the fish brain, the Zn concentration was 0.7475 mg/l, Pb concentration was 0.0396 mg/l, Cd concentration was 0.0956 mg/l, and Cu concentration was 0.6553 mg/l, all of which were below the permissible limit.

12.7 The Contamination Index

12.7.1 The Enrichment Factor in Water

The enrichment factor is defined as the concentration of a heavy metal with respect to the reference metal. According to Table 12.10 and Fig. 12.6, the enrichment values for Pb, Cd, Zn, and Cu from all the eight banks in the observed study were low because the calculated values are less than 1.

12.7.2 The Geoaccumulation Index in Water

The degree of contamination is calculated by the formula

$$I_{\text{geo}} = \log_2\left(Cn / 1.5Bn\right)$$

Table 12.10 Enrichment factor and the I_{geo} factor in water

Location	*Enrichment factor*			
	Zn	*Pb*	*Cd*	*Cu*
Bank1	0.0122	0.0269	0.384	0.0101
Bank1'	0.0123	0.0268	0.372	0.0105
Bank2	0.0109	0.0269	0.388	0.0098
Bank2'	0.0125	0.0266	0.39	0.0096
Bank3	0.0117	0.0268	0.377	0.0105
Bank3'	0.0120	0.0269	0.379	0.0104
Bank4	0.0120	0.0269	0.381	0.01
Bank4'	0.0121	0.0278	0.375	0.0102
Location	*Geoaccumulation factor*			
	Zn	*Pb*	*Cd*	*Cu*
Bank1	−4.808	−4.018	−1.360	−4.999
Bank1'	−4.8	−4.021	−1.392	−4.958
Bank2	−4.916	−4.017	−1.351	−5.021
Bank2'	−4.786	−4.015	−1.347	−5.051
Bank3	−4.830	−4.021	−1.379	−4.956
Bank3'	−4.827	−4.020	−1.374	−4.967
Bank4	−4.22	−4.021	−1.370	−5.001
Bank4'	−4.815	−4.022	−1.384	−4.988

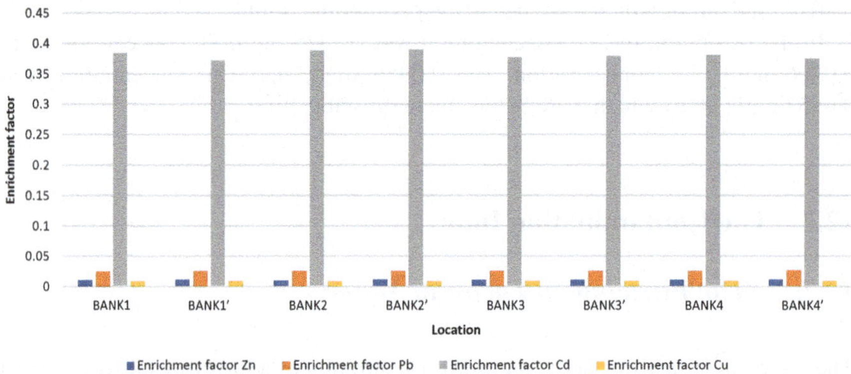

Fig. 12.6 Enrichment factor in water

where *Cn* is the measured concentration of the metal and *Bn* is the geochemical background concentration of the average continental shale. Table 12.10 and Fig. 12.7 show the calculated values of Zn, Pb, Cd, and Cu, which were less than their permissible limit of 1, so these are categorized as low contaminated.

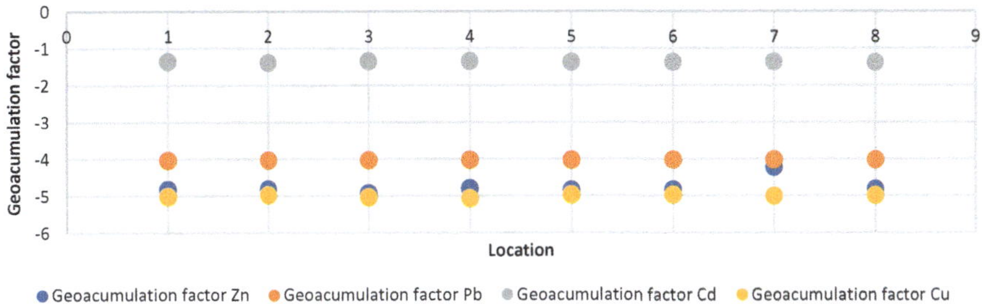

Fig. 12.7 I_{geo} factor in water

Table 12.11 ER and I_{geo} factor in sediments

Location	*Enrichment factor*			
	Zn	*Pb*	*Cd*	*Cu*
Bank1	0.051	0.099	5.429	0.063
Bank1’	0.05	0.1	5.357	0.061
Bank2	0.046	0.091	5.25	0.056
Bank2’	0.046	0.092	5.321	0.054
Bank3	0.035	0.075	6.251	0.039
Bank3’	0.036	0.075	6.464	0.037
Bank4	0.038	0.065	5.786	0.034
Bank4’	0.037	0.068	5.536	0.037
Location	*Geoaccumulation factor*			
	Zn	*Pb*	*Cd*	*Cu*
Bank1	−3.375	−2.714	1.286	−3.169
Bank1’	−3.399	−2.712	1.273	−3.206
Bank2	−3.474	−2.806	1.253	−3.282
Bank2’	−3.727	−2.796	1.266	−3.271
Bank3	−3.726	−2.996	1.427	−3.656
Bank3’	−3.727	−2.999	1.460	−3.693
Bank4	−3.684	−3.142	1.350	−3.790
Bank4’	−3.693	−3.094	1.306	−3.690

12.7.3 The Enrichment Factor in Sediments

Table 12.11 and Fig. 12.8 show the enrichment factor in sediments where Zn, Pb, and Cu concentrations were less than 1, which concludes that these have low-contaminating capacity, and Cd has the highest enrichment factor of 6.464, which shows its high contaminating capacity.

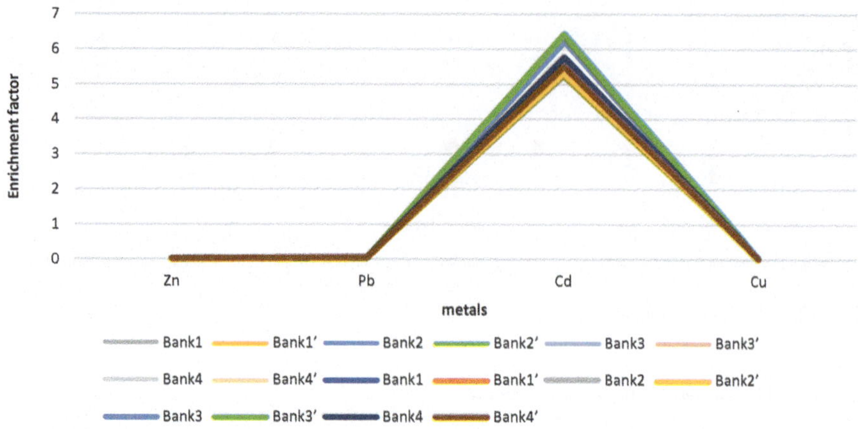

Fig. 12.8 Enrichment factor in sediments

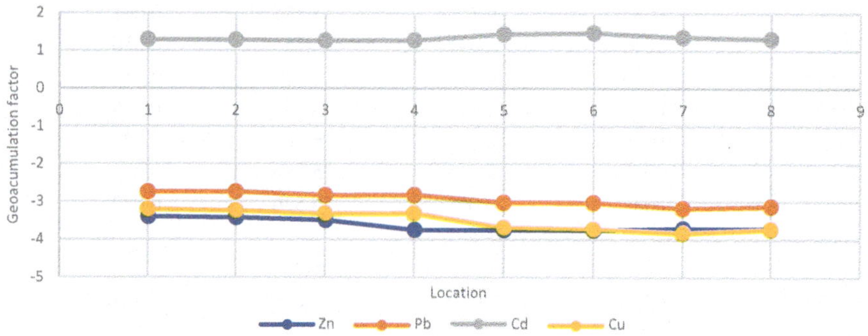

Fig. 12.9 I_{geo} factor in sediments

12.7.4 The Geoaccumulation Index in Sediments

Table 12.11 and Fig. 12.9 show that the values of Zn, Pb, and Cu are less than 1, which indicates that the area is uncontaminated. However, the value of Cd is 1, which means that it is low-contaminated. If the concentration of Cd increases, then it will have a great impact on human health (Chatterjee et al. 2010).

12.8 Conclusions

This study was conducted to measure the heavy metal bioavailability in the sediments of the Damodar River near Durgapur Barrage. The results obtained from the samples of water and fish organs collected from the site show that the concentrations of Pb and Cd were higher than the recommended WHO permissible limit for water and living wood. The concentrations of Pb, Cd, and Cu in sediments were found to be higher than the WHO permissible limit, which may be due to the presence of the paper mill and paint factory near the study area. It is important to note that if the level of contaminants deposited into the river is not reduced, then the chances of health hazards will increase. The result obtained from the enrichment factor index of sediments was higher for Cd. The reason for this might be the industrial activities, the effluents of which are discharged into the Damodar River through the drainage system. The result of the enrichment factor index for water and sediments was found to be within the range. The result for the geoaccumulation index of Cd was found to be higher compared to those of other metals but low-contaminated. On the other hand, the geoaccumulation index for water was found to be within the range. Therefore, the concentration of heavy metals in the samples of sediments, fish, and water of the Damodar River was found to be in the range of moderate-to-high. So, factories should take precautionary measures regarding the treatment of their effluents before dumping them directly into the river.

References

J. Adiyiah, H. Aboagye-Larbi, M.A. Acheampong, Comparative assessment of the upstream and downstream water qualities of River Tano in Ghana. J. Environ. Sci. Eng. A **2**(5A), 283 (2013)

A. Al Nayeem, A.K. Majumder, M.S. Hossain, Assessment of water quality parameters in Karnafuli River: A case study in Karnafuli Paper Mill Area. J. Water Resour. Pollut. Stud. **2**(3) (2017)

J.S. Azevedo, A. Serafim, E.S. Braga, D.I. Fávaro, M.J. Bebianno, Biomarkers of exposure to metal contamination and lipid peroxidation in the benthic fish Cathoropsspixii from two estuaries in South America, Brazil. Ecotoxicology **18**(8), 1001–1010 (2009)

D. Bhardwaj, N. Verma, Research paper on analysing impact of various parameters on water quality index. Int. J. Adv. Res. Comput. Sci. **8**(5) (2017)

J.H. Braatne, S.B. Rood, L.A. Goater, C.L. Blair, Analyzing the impacts of dams on riparian ecosystems: A review of research strategies and their relevance to the Snake River through hells canyon. Environ. Manag. **41**(2), 267–281 (2008)

J. Burger, M. Gochfeld, Metal levels in eggs of common terns (Sterna hirundo) in New Jersey: Temporal trends from 1971 to 2002. Environ. Res. **94**(3), 336–343 (2004)

S.K. Chatterjee, I. Bhattacharjee, G. Chandra, Water quality assessment near an industrial site of Damodar River, India. Environ. Monit. Assess. **161**(1–4), 177–189 (2010)

Chatterjee, M., Massolo, S., Sarkar, S. K., Bhattacharya, A. K., Bhattacharya, B. D., Satpathy, K. K., & Saha, S. (2009). An assessment of trace element contamination in intertidal sediment cores of Sunderban mangrove wetland, India for evaluating sediment quality guidelines. Environmental monitoring and assessment, 150(1), 307–322.

R. Dallinger, F. Prosi, H. Segner, H. Back, Contaminated food and uptake of heavy metals by fish: A review and a proposal for further research. Oecologia **73**(1), 91–98 (1987)

A.P. de Pinho, J.R.D. Guimarães, A.S. Martins, P.A.S. Costa, G. Olavo, J. Valentin, Total mercury in muscle tissue of five shark species from Brazilian offshore waters: Effects of feeding habit, sex, and length. Environ. Res. **89**(3), 250–258 (2002)

D. Dudgeon, The ecology of tropical Asian rivers and streams in relation to biodiversity conservation. Annu. Rev. Ecol. Syst. **31**(1), 239–263 (2000)

L. Dus, Z. Svobodová, D. Janous, B. Vykusová, J. Jarkovský, R. Šmíd, P. Pavlis, Bioaccumulation of mercury in muscle tissue of fish in the Elbe River (Czech Republic): Multispecies monitoring study 1991–1996. Ecotoxicol. Environ. Saf. **61**(2), 256–267 (2005)

Q. Fu, D. Fedrizzi, V. Kosfeld, C. Schlechtriem, V. Ganz, S. Derrer, J. Hollender, Biotransformation changes bioaccumulation and toxicity of diclofenac in aquatic organisms. Environ. Sci. Technol. **54**(7), 4400–4408 (2020)

G. Genchi, M.S. Sinicropi, G. Lauria, A. Carocci, A. Catalano, The effects of cadmium toxicity. Int. J. Environ. Res. Public Health **17**(11), 3782 (2020)

S.P. Gorde, M.V. Jadhav, Assessment of water quality parameters: A review. J. Eng. Res. Appl. **3**(6), 2029-2035 (2013)

D.E. Hanna, S.A. Tomscha, C.O. Dallaire, E.M. Bennett, A review of riverine ecosystem service quantification: Research gaps and recommendations. J. Appl. Ecol. **55**(3), 1299–1311 (2018)

J. Hofman, L. Dušek, J. Klánová, J. Bezchlebová, I. Holoubek, Monitoring microbial biomass and respiration in different soils from the Czech Republic—A summary of results. Environ. Int. **30**(1), 19–30 (2004)

M.S. Holt, Sources of chemical contaminants and routes into the freshwater environment. Food Chem. Toxicol. **38**, S21–S27 (2000)

T. Hooper, M. Austen, Tidal barrages in the UK: Ecological and social impacts, potential mitigation, and tools to support barrage planning. Renew. Sustain. Energy Rev. **23**, 289–298 (2013)

M. Hossain, P.K. Patra, Contamination zoning and health risk assessment of trace elements in groundwater through geostatistical modelling. Ecotoxicol. Environ. Saf. **189**, 110038 (2020a)

M. Hossain, P.K. Patra, Water pollution index–a new integrated approach to rank water quality. Ecol. Indic. **117**, 106668 (2020b)

M. Hossain, P.K. Patra, B. Ghosh, A. Khatun, & S. Nayek (2021). Sensitive assessment of groundwater-associated, multi-exposure health hazards in a fluoride-enriched region of West Bengal, India. Environmental Geochemistry and Health, 1–18

R. Kaur (2007). Regional resource characterization through remote sensing & GIS for effective decision making—A case study of NCR.(a note prepared for an indo-US workshop on innovative E-technologies for distance education, extension/outreach in efficient water management, march 2007, ICRISAT, Patancheru, AP, India)

L.E. Keup, Phosphorus in flowing waters. Water Res. **2**(5), 373–386 (1968)

G.J. León-García, D.M. Meza-Figueroa, J.L. Valenzuela-García, M.A. Encinas-Romero, A.I. Villalba-Atondo, K.K. Encinas-Soto, & A. Gómez-Álvarez. Study of heavy metal pollution in arid and semi-arid regions due to mining activity: Sonora and Bacanuchi Rivers. Int. J. Environ. Sci. Nat. Resour. https://doi.org/10.19080/IJESNR, 555804 (2018)

K.N. McCain, Moving large river ecology from past theories to future actions: A review. Rev. Fish. Sci. **21**(1), 39–48 (2013)

C.A. Mebane, T.S. Schmidt, J.L. Miller, L.S. Balistrieri, Bioaccumulation and toxicity of cadmium, copper, nickel, and zinc and their mixtures to aquatic insect communities. Environ. Toxicol. Chem. **39**(4), 812–833 (2020)

M. Nendza, R. Kühne, A. Lombardo, S. Strempel, G. Schüürmann, PBT assessment under REACH: Screening for low aquatic bioaccumulation with QSAR classifications based on physicochemical properties to replace BCF in vivo testing on fish. Sci. Total Environ. **616**, 97–106 (2018)

F. Ojija, Analysis of water quality parameters and ecosystem services of Nzovwe stream. Int. J. Biol. Sci. Technol. **7**(1), 1 (2015)

J.O. Olayinka-Olagunju, A.A. Dosumu, A.M. Olatunji-Ojo, Bioaccumulation of heavy metals in pelagic and benthic fishes of Ogbese River, Ondo state, South-Western Nigeria. Water Air Soil Pollut. **232**(2), 1–19 (2021)

W. Ouyang, Y. Wang, C. Lin, M. He, F. Hao, H. Liu, W. Zhu, Heavy metal loss from agricultural watershed to aquatic system: A scientometrics review. Sci. Total Environ. **637**, 208–220 (2018)

M.A. Palmer, K.L. Hondula, B.J. Koch, Ecological restoration of streams and rivers: Shifting strategies and shifting goals. Annu. Rev. Ecol. Evol. Syst. **45**, 247–269 (2014)

D. Paul, Research on heavy metal pollution of river ganga: A review. Ann. Agrarian Sci. **15**(2), 278–286 (2017)

H.T. Ratte, Bioaccumulation and toxicity of silver compounds: A review. Environ. Toxicol. Chem. Int. J. **18**(1), 89–108 (1999)

A.M. Sattar, Y.M. Raslan, Predicting morphological changes ds new Naga-Hammadi barrage for extreme Nile flood flows: A Monte Carlo analysis. J. Adv. Res. **5**(1), 97–107 (2014)

M. Sehgal, A. Garg, R. Suresh, P. Dagar, Heavy metal contamination in the Delhi segment of Yamuna basin. Environ. Monit. Assess. **184**(2), 1181–1196 (2012)

J.P. Sumpter, Protecting aquatic organisms from chemicals: The harsh realities. Philosophical Transact. Royal Soc. A Mathematical. Phys. Eng. Sci **367**(1904), 3877–3894 (2009)

W. Zhuang, X. Gao, Integrated assessment of heavy metal pollution in the surface sediments of the Laizhou Bay and the coastal waters of the Zhangzi Island, China: Comparison among typical marine sediment quality indices. PLoS One **9**(4), e94145 (2014)

S. Žižek, M. Horvat, D. Gibičar, V. Fajon, M.J. Toman, Bioaccumulation of mercury in benthic communities of a river ecosystem affected by mercury mining. Sci. Total Environ. **377**(2–3), 407–415 (2007)

Chapter 13
Heavy Metal Assessment of Groundwater on the Periphery of a Semi-Urban Solid Waste Dumpsite and Mitigation Strategy

Prasanna Kuchelar and Jayaraman Sethuraman Sudarsan

Abstract Increase in population and improper urbanization have led to a huge quantity of waste generation and an increase in resource utilization in the form of energy, water, etc. This study focuses on groundwater contaminations near dump yards and their impacts. The study area was selected based on the existing issues raised by the public, and also it was observed that no proper geosynthetic clay liners were installed before starting the dumping process in the location because the dumping was initially started as minimal disposing along the bund of the lake by the nearby residents of the area but was later slowly extended by the municipality into a dumping yard. The subsurface water (SSW) sampling points were identified based on the shallow water table sources with a minimum depth of 10–20 m. Similarly, the low lying spots in the lake for surface water sampling were identified and marked for repeated sampling during successive periods. Leachate sampling points and soil sampling points were identified based on the suitability of depressed locations within and on the periphery of the site. A total of 27 convenient sampling points for groundwater were collected from different shallow bore wells and open wells, which are at distances ranging from 9 m to 1800 m radially from the boundary of the dumping site. The depth of the water table varies from 10 m to 15 m during the pre-monsoon (Pre-MN) and from 7.5 m to 11 m during the post-monsoon (Post-MN) seasons. Heavy metal concentrations were assessed, and it was reported that lead levels in both Pre-MN and Post-MN were beyond the acceptable limit of 0.01 mg/L. Chromium concentrations in all the samples in both the Pre-MN and Post-MN periods were beyond the acceptable value of 0.05 mg/L because of the dumping of discarded wastes from nearby tanneries at the site.

P. Kuchelar
Department of Civil Engineering, College of Engineering and Technology, SRM Institute of Science and Technology, Kattankulathur, India

J. S. Sudarsan (✉)
School of Construction management (SOCM), National Institute of Construction Management and Research (NICMAR), Pune, India
e-mail: ssudarsan@nicmar.ac.in

B. Panneerselvam et al. (eds.), *Climate Change Impact on Groundwater Resources*, https://doi.org/10.1007/978-3-031-04707-7_13

Keywords Leachate · Groundwater · Heavy metals · Dumping site · Contamination

13.1 Introduction

Proper handling and safe disposal of all aspects of solid waste without affecting the environment have become a challenge. Municipal solid waste (MSW) is generally referred to as garbage and other discarded materials, which is said to be unwanted thrown outs.(Tripathi and Misra 2012; Raman and Sathiyanarayanan 2011) Across the globe, stakeholders should play a major role in taking care of these issues, and appropriate measures have to be taken to sustain the environment and its cycle. (Sallam et al. 2015; Maiti et al. 2016) In cities, municipal solid waste constitutes wastes produced from households, commercial institutions, and industries. The ground water source is one of the essential resources and also one of the limited resource that has been depleted in modern era due to the illegimate management and disposal of solid waste which results in substandard governance at all levels.(Ramesh and Thirumangai 2014; Ali et al. 2014) Landfills and dumping sites are the major two methods of disposal used widely across the globe and mostly dumping in low-lying areas and nearby bunds of water bodies has become a regular practice especially in developing countries.(Nanda et al. 2011; Dharmarathne and Gunatilake 2013; Khalil et al. 2018; Phillips and Mondal 2014) As an outcome, contamination of surface and subsurface sources of water occurs due to improper solid waste management practices.(Swati et al. 2019; Foufou et al. 2017; Neshat et al. 2015; Salem et al. 2008)

Dumping of municipal soild waste is the general practice adopted in open land without any prior segregation which occupies vast area of open land where the garbage is exposed to the surrounding areas and subjected to decay on its own. (Jeevanandam et al. 2012; Mor et al. 2006; Liu et al. 2018) In most of the developing nations, which includes India, open dumping is the prime method of disposing the solid waste. This is due to less financial support from the government peratainig to solid waste management and lack of trained labour. Normally, in most of the cities and towns, unauthorized dumping is prevalent and leads to big threats to the environment and the ecosystem.(Negi et al. 2018; Rawat et al. 2013; Kaushal et al. 2012; Ravindra and Mor 2019; Kumar et al. 2017) The solid wastes dumped in low lying areas are subjected to various weathering conditions and results in generating coloured liquid which is known as leachate. The movement of leachate generated in dumping yards into subsurface water sources endangering the survival and health of living beings residing close to the landfills and dumping sites.(Mor et al. 2006; Liu et al. 2018) Due to leachate percolations and prolonged contact with the subsurface water, the existing quality of the groundwater starts deteriorating, which results in a sizeable rise in heavy metal concentration. The composition of the leachate

Fig. 13.1 Bore well near the dumpsite area

reflects the composition of the solid waste that is dumped.(Yukalang et al. 2017; Yang et al. 2018; Ferronato and Torretta 2019; Gupta et al. 2017) In many cases, even community bore wells are located extremely close to dumping sites as presented in Fig. 13.1.

Generally, the wastes dumped in landfills undergo degradation and the degraded wastes release a liquid containing heavy metals like chromium (Cr), copper (Cu), cadmium (Cd), lead (Pb), nickel (Ni), arsenic (As), and zinc (Zn). Heavy metals originate from the crust of the earth, and due to the natural phenomenon of erosion and being eroded, they get dissolved in the surface water and eventually in the subsurface water.(Ahluwalia 2016; Elangovan et al. 2018; Parvin and Tareq 2021) Heavy metals, when present in excess in water sources due to various human-made activities, results in serious health issues in almost all living beings who consume the same sources without any treatment. The runoff from the leachate affects the soil properties surrounding the dumping sites.(Abd El-Salam and Abu-Zuid 2015; Joshi and Ahmed 2016; Mozhiarasi et al. 2020) Recent studies have shown the presence of high concentrations of heavy metals such as arsenic, cadmium, copper, lead, and zinc in soils in many nations around the world. Countries like India, Bangladesh, and China have recorded rising cases of arsenic poisoning.(Babaei et al. 2019; Rana et al. 2018; Pujara et al. 2019; Xaypanya et al. 2018; Arunbabu et al. 2017) Similarly, cadmium has been observed to be bioaccumulated in the bodies of cattle and other animals in New Zealand and Australia. The liquid generated which influence the characteristics of the soil and it gets degraded which is found on the dumping site. (Karpagam and Ramesh 2015; Ikem et al. 2002; Jose et al. 2019) This contamination of land and soil may or may not be reversible, and, thus, the resale value of the land will get reduced and soil fertility is mainly affected (Adelopo et al. 2018; Paluch et al. 2018; Moody and Townsend 2017).

The leachate movement along the lateral direction on surface water and in a vertical downward direction on groundwater leads to contamination of the sources of water. Groundwater is the major source of freshwater that is comparatively less polluted than surface water sources. However, leachates have the tendency to percolate into the groundwater table and end up contaminating it, thus causing major health risks (Ajaz et al. 2019). The level of this threat normally depends on the concentration level of toxicity present as the contaminant-based level of the prevailing water table and the direction of groundwater movement. The role of microbial activities in the leachate and soil under biosorption and bioaccumulation has some significance in resisting the heavy metals in both the leachate and soil. The contamination of surface and subsurface water quality during successive pre- and post-monsoon variations and its effect on consumption will have significant impacts (Abd El-Salam and Abu-Zuid 2015; Alhassan 2012; Ivanov and Chu 2008; Pandey and Tiwari 2017).

13.2 Review of Previous Studies

Groundwater and surface water quality, affected by the wastes in landfills and dumping sites, has been studied across the globe. Few literature studies that coincide with this study are enumerated below.

Rakesh Kumar Pandey et al. (2017) characterized the municipal solid waste and evaluated the quality of subsurface water near a dumping ground in Madhya Pradesh. The waste composition, moisture content, specific gravity, the Atterberg limits, and compressibility were determined. The physicochemical characteristics of the subsurface water were also analysed. The groundwater at the time of study was not threatened by contamination of leachates from the dumping site (Pandey and Tiwari 2017).

Adelopo A.O. et al. (2018) compared heavy metal pollution in closed and active dumpsites to comprehend the risk levels with regard to parameters like age, particle size, and depth of the sample. Heavy metals were higher in a closed landfill site. The size of the landfill influenced the distribution of metals more than the depth factor. Heavy metal pollution index was calculated and was beyond the acceptable limit, thus hinting at a major hazard to the environment (Adelopo et al. 2018).

Arunbabu V. et al. (2017) investigated how a leachate affects the extent of germination, chlorophyll content, and growth and enzyme activities of the plant species *Vigna unguiculata*. The collected leachate showed a leachate pollution index (LPI) of 31.99. The phytotoxicity study conducted suggested that lower concentrations of the leachate encouraged the germination. A higher concentration of the leachate stopped the enzyme activities, resulted in membrane lipid peroxidation, and had an ill effect on the growth of plants. A relationship was established between the LPI value of the leachate and its effect on plant growth. If the LPI value is within 10, then a better growth of plants is seen. An LPI value above 10 causes adverse effects on the growth (Arunbabu et al. 2017). **Maiti S.K. et al. (2016)** characterized the leachate and how it affects the surrounding water resources of a closed dumping site located at Kolkata, India. Samples were collected from three points in and around the dumping site. The samples were tested for their physicochemical characteristics and how they varied with seasonal changes. It was observed that total dissolved solids (TDSs), Chemical Oxygen Demand (COD), ammonia, chlorine, and nitrogen concentrations were higher than the recommendations given by the Municipal Solid Wastes (Management and Handling) Rules, 2000. The concentrations of lead and mercury in surface water were also alarmingly high when compared to the permissible limits as per the recommendations made by the Bureau of Indian Standards (BIS). A correlation between TDSs, electrical conductivity (EC), and pH was established. Biochemical Oxygen Demand (BOD) and COD were correlated with lead. Chlorine was well correlated with EC and TDSs (Maiti et al. 2016).

C. M. Moody et al. (2017) collected leachates from 14 different landfills from Florida, United States, which predominantly contained municipal solid waste. The assessment of samples was performed for trace metals, anions, and nutrients to understand the mass balance of the total dissolved solid content. The solid waste incineration ash leachate constituted 98% of metallic salts and had an acidic pH. The results were treated as an indicator of the differences between the laboratory and field conditions of the leachate in terms of quality. The organic matter was characterized both qualitatively and quantitatively, and it was hence found that COD alone was not a sufficiently accurate indicator of the organic content present in the leachate that contains solid waste incineration ash (Moody and Townsend 2017).

BarjinderBhalla et al. (2012) assessed the landfill leachate treatment efficiency by adopting the leachate pollution index (LPI) for selected characteristics. Leachate samples were collected from the Jamalpur landfill site located in Ludhiana city, Punjab, and analysed for pH, TDSs, BOD, COD, and chloride to estimate their pollution potential. The LPI value for the disposal of leachate to the inland surface water was 7.378 (Bhalla et al. 2012).

13.3 Study Area

Pallavaram, formerly known as Pallava Puram located in the suburbs of Chennai, is one of the oldest inhabited places in southern India. The total area is about 18 square kilometres. The present-day town of Pallavaram is a large special-grade municipality that is divided into Zamin Pallavaram, Old Pallavaram, and Cantonment Pallavaram (Raman and Sathiyanarayanan 2011). The municipality is divided into 42 wards and has a total population of approximately 2.34 lakhs. Pallavaram is home to many small- and large-scale industries and is also closely situated near Chromepet, which itself is home to many leather industries. The open dumpsite is located adjacent to the radial road with a daily dumping quantum of 140 MT/day. The dumping of waste has been stopped since 2017. The location coordinates are from 12°57′18.50"N to12°57′23.56"N and from 80° 8′58.74"E to 80°9'00.24"E. The total area of dumping site is about 3.44 hectares and has a measured circumference of 1.03 km. The satellite imaginary of the dumpsite is presented in Fig. 13.2. It has been highlighted with locations and key landmarks near the dumpsite. The overall geology spreads from south to north and west to east. The environment of the space

Fig. 13.2 Aerial view of the dumpsite

is with low stickiness and high temperature, and the temperature is around 18 °C–25 °C during winter and is 35 °C–42 °C during summer, and, thus, the area is by and large warm. Temperature begins ascending towards the end of February. The region gets greatest precipitation from the northwest rainstorm and the yearly precipitation is 1200 mm. The progression of groundwater is from west to east. The unloading site is arranged straightforwardly inverse to the Big Lake. The lake and the unloading site are isolated by a 200 feet street.

13.4 Methodology

A total of 27 convenient sampling points were collected for groundwater from different shallow bore wells, open wells, and hand pumps in residences, streets, roads, open yards, and parks, which are at distances ranging from 9 m to 1800 m radially from the boundary of the dumping site. The depth of the water table varies from 10 m to 15 m during the pre-monsoon (Pre-MN) and from 7.5 m to 11 m during the post-monsoon (Post-MN) seasons.

13.4.1 Water Sampling Locations

Two successive years were taken for sampling, i.e. 2016–2018. The subsurface water samples and leachate samples were collected from the sample locations for successive periods. Suitable sign boards and pegs were used to mark the locations

Table 13.1 Coordinates of the sampling locations of subsurface water

Samples	Latitude	Longitude
SSW1	12°57′21"N	80° 8′57″E
SSW2	12°57′16"N	80° 8′52″E
SSW3	12°57′22"N	80° 8′53″E
SSW4	12°57′18"N	80° 8′50″E
SSW5	12°57′15"N	80° 9′8″E
SSW6	12°57′9"N	80° 9′3″E
SSW7	12°57′36"N	80° 8′59″E
SSW8	12°57′27"N	80° 8′49″E
SSW9	12°57′6"N	80° 8′53″E
SSW10	12°57′10"N	80° 8′34″E
SSW11	12°57′22"N	80° 8′48″E
SSW12	12°57′17"N	80° 8′50″E
SSW13	12°57′24"N	80° 8′55″E
SSW14	12°57′16"N	80° 8′58″E
SSW15	12°57′14"N	80° 9′9″E
SSW16	12°57′10"N	80° 9′6″E
SSW17	12°57′34"N	80° 8′56″E
SSW18	12°57′28"N	80° 8′48″E
SSW19	12°57′8"N	80° 8′54″E
SSW20	12°57′12"N	80° 8′36″E
SSW21	12°57′8"N	80° 9′5″E
SSW22	12°57′34"N	80° 8′56″E
SSW23	12°57′29"N	80° 8′49″E
SSW24	12°57′7"N	80° 8′52″E
SSW25	12°57′12"N	80° 8′36″E
SSW26	12°57′24"N	80° 8′47″E
SSW27	12°57′19"N	80° 8′51″E

for leachate samples. For surface water, random samples were collected from the Periya Lake. The pre-monsoon season in 2016 was August and September and the post-monsoon season in 2017 was January and February. Similarly, the pre-monsoon season in 2017 and the post-monsoon season in 2018 were in the same months. The periods are with respect to the northeast monsoon (NEM), in which Tamil Nadu receives the maximum rainfall. A total 27 subsurface water (SSW) samples were collected in and around the lake, and the sampling locations denoted as SSW1–27 and the corresponding latitudes and longitudes are presented in Table 13.1.

13.5 Results and Discussion

Heavy metals, namely, Cr, Cd, Cu, Fe, Zn, and Pb, were assessed for the subsurface samples collected in two successive pre- and post-monsoon periods during 2016–2018.

Out of the major heavy metal concentrations assessed, it was reported that Pb levels in both the Pre-MN and Post-MN periods were beyond the permissible limit of 0.01 mg/L. Cr levels in all the samples in both the Pre-MN and Post-MN periods were beyond the permissible limit of 0.05 mg/L because of the dumping of discarded wastes from nearby tanneries at the site. Fe levels in all the samples were under the permissible limit of 0.3 mg/L, except for three samples in the Pre-MN period of 2016 and four samples in the Pre-MN period of 2017. Zinc and cadmium levels in all the subsurface water samples were not beyond the permissible limit of 5 mg/L. However, in the case of copper, almost all the samples were beyond the acceptable limit of 0.05 mg/L. The variations of heavy metal concentrations in subsurface water during the Pre-MN and Post-MN periods of 2016–18 are shown in Figs. 13.3, 13.4, and 13.5.

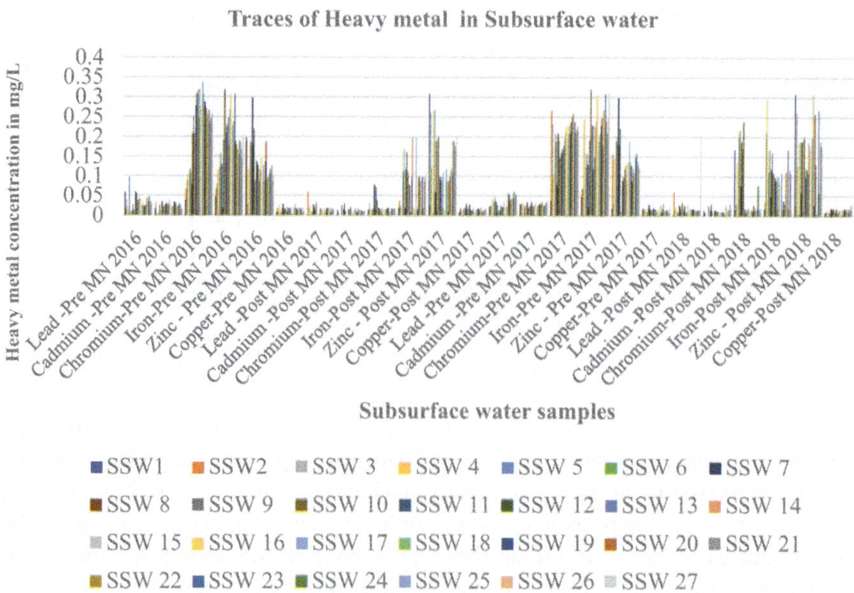

Fig. 13.3 Traces of heavy metals in surface water of 27 sample locations

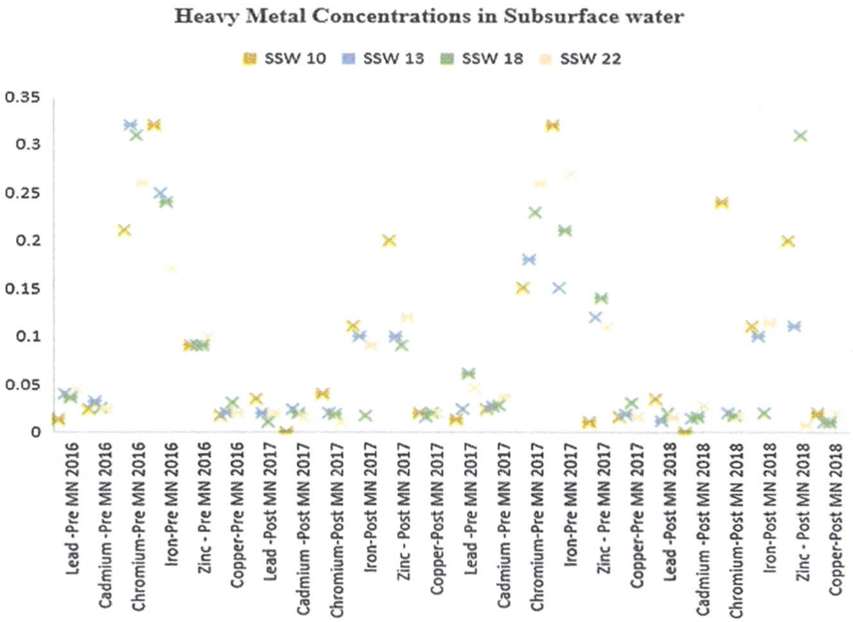

Fig. 13.4 Higher traces of heavy metal concentration

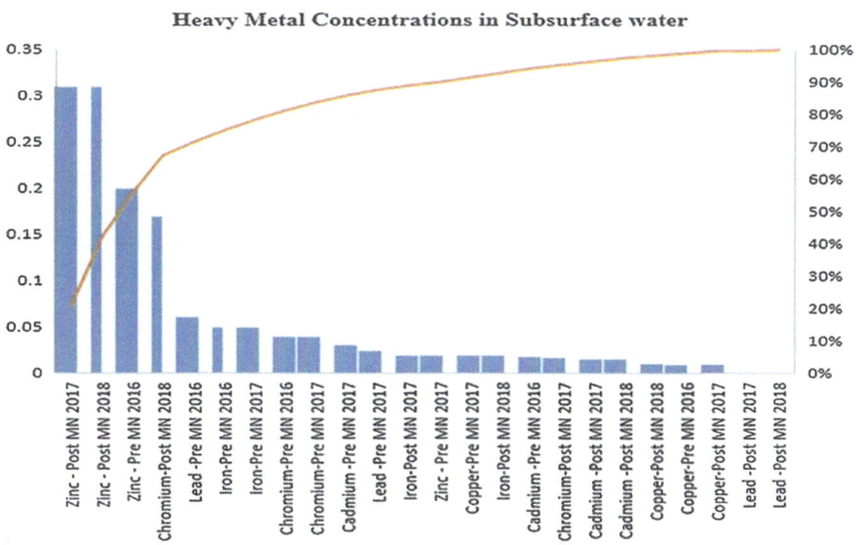

Fig. 13.5 Heavy metal concentration in subsurface water, 2016–2018

13.6 Outcomes

Heavy metal analysis was conducted as per the guidelines of the American Public Health Association (APHA) and the Central Pollution Control Board (CPCB). It is clear from Figs. 13.3, 13.4, and 13.5 that several heavy metals are found in the subsurface water samples, and most of the heavy metals are within the prescribed limit but few are slightly higher than the limit. It was found that Zn alone seems to be within the standards of 5 mg/L in almost all the water samples. However, the concentrations of heavy metals Cd, Cr, Cu, and Pb of all the samples collected in the Pre-MN and Post-MN seasons are slightly higher than those prescribed by the BIS. During assessment, with the exception of two samples in the Pre-MN and four samples in the Post-MN periods, all others were observed to be less than the standard of 0.3 mg/L or beyond the standards.

It was observed that, there is a sizeable impact on the subsurface water, which is to the samples taken along the periphery, and the same variation was observed in their respective heavy metals and successive monsoon periods. Hence, the heavy metals in the leachate samples play an important role in increasing the concentration levels of subsurface water and lake water.

Most of the urban clusters have open dumps. These dumps are unhygienic and usually spoil the entire environment. These also give room to several animals like rats, insects, pigs, and other pests. As clearly evident from the study, surface water seeps through many contaminants and can leach dangerous chemical substances, which might be then carried away from the dumpsites in surface or subsurface runoff. It is also clear from this research's findings that heavy metals are particularly insidious in surface water and can cause bioaccumulation and biomagnification, leading to severe health risks in humans. Most of the health risks are chronic, mainly affecting children, women, and the elderly. If heavy metals enter the human system or biotic component system through food, they may initially cause anaemia and weakness and then slowly affect human body parts like the kidney causing diseases like brain damage, etc.; children are more susceptible than are adults. Cd is a highly toxic carcinogen, and it is harmful to the human body. In the long term, this kind of impact leads to serious health risks like cancer.

13.7 Mitigation

It is high time to take some serious measures, and a policy framework must be formulated to streamline the system of waste management mainly in the interior parts and outskirts of cities, towns, and villages; there is a need for adjusting elements by developing better and strong waste management techniques. This will help in reducing the leachate contamination of groundwater in addition to continuous tracking of groundwater quality by means of periodical surveys and monitoring.

Some of the mitigation measures to reduce groundwater contamination and to manage leachates are presented in the following points.

1. It is mandatory to adopt source reduction technique which is the prime attribute in solid waste management to be implemented at the beginning stage of waste technology.
2. Waste must be disposed of by developing engineered landfills with proper gas vents and leachate recirculation system.
3. Geosynthetic clay liners (GCLs) can also be installed to decrease the seepage of leachates into the soil to reach the water table. This will shield the groundwater from contaminants.
4. The entire waste dumping yard must be monitored using continuous central surveillance systems.
5. It is necessary to conduct periodical sampling of groundwater in and around the waste dumping yards to keep vigil on the quality of groundwater.

The above-mentioned measures are based on the research study's output and further, these can be effectively followed and implemented by strict policies in accordance with the legislations set by the national, regional, and local bodies.

13.8 Conclusions

This study was carried out at a dumpsite in the Pallavaram Municipality, and various aspects have been identified with respect to the quality of subsurface water. The impact of the leachate movement along the lateral direction on the surface water and that in a vertical downward direction on groundwater leads to contamination of the sources of water. Groundwater is the major source of freshwater that is comparatively less polluted than surface water sources. However, leachates have the tendency to percolate into the groundwater table and end up contaminating it. Higher values of heavy metal concentrations of Pb, Cr, Cu, Zn, Cd, and Fe were reported for the Pre-MN period of 2016 than the Pre-MN period of 2017. Similarly, the heavy metal concentrations in the Pre-MN period of 2017 were higher than those in the Post-MN of 2018.

This is due to the intensity of rainfall and dilution factor. It was assessed that the effects of heavy metals have some significant impact on the subsurface water samples due to the leachate contact along the periphery, and, also, the same variation was observed in their respective heavy metals and successive monsoon periods. Heavy metal analysis found that Zn alone seems to be within the standards of 5 mg/L in almost all the water samples. However, the heavy metals Cd, Cr, Cu, and Pb reported that all the samples collected in the Pre-MN and the Post-MN seasons were slightly higher than those prescribed by the BIS; in particular, Cr seemed to have a higher concentration due the chrome factories in the nearby areas. While assessing Fe, two ground water samples in Pre-MN and four samples in Post-MN were observed within the standard limit of 0.3 mg/L and rest of the samples are beyond the permissible limits. It was observed that there is a significant effect on the subsurface water samples, which is near the leachate samples collected along the periphery, and the same variation was observed in their respective heavy metals and successive monsoon periods. Hence, the heavy metals in the leachate samples play a major role in increasing the concentration levels of subsurface water.

The major sources of lead are batteries and minor e-wastes along with other wastes dumped in the site. Similarly, copper and iron are from the dumped wastes containing corroded metals generated from various residential, commercial, and industrial sources. If the solid waste is segregated at the source itself, the waste, which contains more heavy metals, can be arrested while dumping. However, recycling and reusing options would be the best option for contamination and contact with the groundwater. The majority of heavy metals are from inorganic waste, which is combined with municipal solid waste. Hence, the leachate contact with the subsurface water has to be arrested by providing suitable collecting channels and collecting tanks so that the percolation in both the lateral and longitudinal movements of the leachate in the surface can be controlled to the maximum extent.

13.9 Recommendations

This study recommends bio-mining process of recovering the dumped waste in the site and retriving the lake portion which has been occupied by the waste dumped in the site. This recommendation is very much essential, since residential buildings surround the dumpsite and no dumping of waste is allowed anymore. The local bodies have taken appropriate steps to remove the existing wastes dumped and have completely shut down the dumping of wastes. It has become a nuisance and significant issue for the humans residing around the dumpsite. The existing dumpsite should stop further disposal as followed by the municipality currently. Until complete removal of the solid waste dumped in the site, a temporary measure needs to be taken by erecting steel fibre sheets along the boundary of the site so that the lateral movement of the leachate can be arrested, and at the same time, to control percolation, the leachate can be collected safely through possible pathways and existing depressed points.

References

M.M. Abd El-Salam, G.I. Abu-Zuid, Impact of landfill leachate on the groundwater quality: A case study in Egypt. J. Adv. Res. **6**(4), 579–586 (2015)

A.O. Adelopo, P.I. Haris, B.I. Alo, K. Huddersman, R.O. Jenkins, Multivariate analysis of the effects of age, particle size and landfill depth on heavy metals pollution content of closed and active landfill precursors. Waste Manag. **78**, 227–237 (2018)

I.J. Ahluwalia, Challenges of urbanisation in India. Contemp. Issues Develop. Econ., 163–177 (2016)

M. Ajaz, M.Y. Zargar, M. Asif, Development of microbial consortia of cold tolerant microbes for solid waste management. Int. J. Chem. Stud. **7**(4), 148–155 (2019)

M. Alhassan, Effect of municipal solid waste on geotechnical properties of soils. Int. J. Environ. Sci. Manag. Eng. Res. **1**(5), 204–210 (2012)

S.M. Ali, A. Pervaiz, B. Afzal, N. Hamid, A. Yasmin, Open dumping of municipal solid waste and its hazardous impacts on soil and vegetation diversity at waste dumping sites of Islamabad City. J. King Saud Univ. Sci. **26**(1), 59–65 (2014)

V. Arunbabu, K.S. Indu, E.V. Ramasamy, Leachate pollution index as an effective tool in determining the phytotoxicity of municipal solid waste leachate. Waste Manag. **68**, 329–336 (2017)

A. Babaei, I. Parseh, D. Jalili, M. Hashemi, H. Mohammadi, Municipal solid waste characterization, quantification, and management : A case study in Shadegan international wetland region, Iran. J. Adv. Environ. Health Res. **7**, 145–154 (2019)

B. Bhalla, M. Saini, M. Jha, Characterization of leachate from Municipal Solid Waste (MSW) landfilling sites of Ludhiana, India: A comparative study. Int. J. Eng. Res. Appl. **2**(6), 732–745 (2012)

N. Dharmarathne, J. Gunatilake, Leachate characterization and surface groundwater. Int. J. Sci. Res. **3**(11), 1–7 (2013)

N.S. Elangovan, V. Lavanya, S. Arunthathi, Assessment of groundwater contamination in a sub-urban area of Chennai, Tamil Nadu, India. Environ. Dev. Sustain. **20**(6), 2609–2621 (2018)

N. Ferronato, V. Torretta, Waste mismanagement in developing countries: A review of global issues. Int. J. Environ. Res. Public Health **16**(6), 1060 (2019)

A. Foufou, S. Djorfi, N. Haied, R. Kechiched, M. Azlaoui, A. Hani, Water pollution diagnosis and risk assessment of Wadi Zied plain aquifer caused by the leachates of Annaba landfill (N-E Algeria). Energy Procedia **119**, 393–406 (2017)

S. Gupta, R. Rameshwar, S.N. Gupta, N. Gupta, Nation challenges for solid waste management. J. Soc. Welf. Manag. **9**(2), 75–83 (2017)

A. Ikem, O. Osibanjo, M.K.C.S. And, A. Sobande, Evaluationof groundwater quality characteristics near two waste sites in Ibadan and Lagos, Nigeria a case study. Water Air Soil Pollut. **140**, 307–332 (2002)

V. Ivanov, J. Chu, Applications of microorganisms to geotechnical engineering for bioclogging and biocementation of soil in situ. Rev. Environ. Sci. Biotechnol. **7**(2), 139–153 (2008)

M. Jeevanandam, R. Nagarajan, M. Manikandan, M. Senthilkumar, S. Srinivasalu, M.V. Prasanna, Hydrogeochemistry and microbial contamination of groundwater from lower Ponnaiyar Basin, Cuddalore District, Tamil Nadu, India. Environ. Earth Sci. **67**(3), 867–887 (2012)

J.M. Jose, A. Vats, S. Rajendran, M.C. John Milton, Biomonitoring effluents from Perungudi MSW/STP Facility and its impact on surface waters of Pallikaranai wetland, Chennai, Tamil Nadu, India. Nat. Environ. Pollut. Technol. **18**(2), 369–378 (2019)

R. Joshi, S. Ahmed, Status and challenges of municipal solid waste management in India: A review. Cogent Environ. Sci. **2**(1), 1–18 (2016)

V. Karpagam, K. Ramesh, Assessment of groundwater quality of Chrompet industrial area by water quality index method. Int. J. Eng. Technol. Manag. Appl. Sci. **3**(7), 123–132 (2015)

R.K. Kaushal, G.K. Varghese, M. Chabukdhara, Municipal solid waste management in India-current state and future challenges: A review. Int. J. Eng. Sci. Technol. **4**(4), 1473–1489 (2012)

C. Khalil, C. Al Hageh, S. Korfali, R.S. Khnayzer, Municipal leachates health risks: Chemical and cytotoxicity assessment from regulated and unregulated municipal dumpsites in Lebanon. Chemosphere **208**, 1–13 (2018)

S. Kumar, S.R. Smith, G. Fowler, C. Velis, S.J. Kumar, S. Arya, K.R. Rena, C. Cheeseman, Challenges and opportunities associated with waste management in India. R. Soc. Open Sci. **4**(3), 160764 (2017)

Y. Liu, F. Kong, E.D.R. Santibanez Gonzalez, Dumping, waste management and ecological security: Evidence from England. J. Clean. Prod. **167**, 1425–1437 (2018)

S.K. Maiti, S. De, T. Hazra, A. Debsarkar, A. Dutta, Characterization of leachate and its impact on surface and groundwater quality of a closed dumpsite – A case study at Dhapa, Kolkata, India. Procedia Environ. Sci. **35**, 391–399 (2016)

C.M. Moody, T.G. Townsend, A comparison of landfill leachates based on waste composition. Waste Manag. **63**, 267–274 (2017)

S. Mor, K. Ravindra, R.P. Dahiya, A. Chandra, Leachate charecterization and assesement of groundwater pollution nearmunicipal soild waste landfill site. Environ. Monit. Assess. **118**, 435–456 (2006)

V. Mozhiarasi, R. Raghul, C.J. Speier, P.M.B. Rose, D.W. And, S.V. Srinivasan, *Sustainable waste management: Policies and case studies* (Springer Nature Singapore Pte Ltd, Springer Singapore, 2020), pp. 143–151

H.S. Nanda, R. Shivaraju, C. Ramakrishnegowda, Impact of Municipal Solid Waste Disposal on Geotechnical Properties of Soil. *Proceedings of Indian Geotechnical Conference*, pp. 715–716 (2011)

P. Negi, S. Mor, K. Ravindra, Impact of landfill leachate on the groundwater quality in three cities of North India and health risk assessment. Environ. Dev. Sustain. **22**, 1455–1474 (2018)

A. Neshat, B. Pradhan, S. Javadi, Risk assessment of groundwater pollution using Monte Carlo approach in an agricultural region: An example from Kerman plain, Iran. Comput. Environ. Urban. Syst. **50**, 66–73 (2015)

J. Paluch, R.B.R. Mesquita, V. Cerdà, J. Kozak, M. Wieczorek, A.O.S.S. Rangel, Sequential injection system with in-line solid phase extraction and soil mini-column for determination of zinc and copper in soil leachates. Talanta **185**(March), 316–323 (2018)

R.K. Pandey, K.P.M. Tiwari, Characterization of municipal solid waste and evaluation of subsurface water quality in the vicinity of dumping sites. Int. J. Res. Appl. Sci. Eng. Technol. **V**(X), 425–431 (2017)

F. Parvin, S.M. Tareq, Impact of landfill leachate contamination on surface and groundwater of Bangladesh: A systematic review and possible public health risks assessment. Appl Water Sci **11**(100), 15–17 (2021)

J. Phillips, M.K. Mondal, Determining the sustainability of options for municipal solid waste disposal in Varanasi, India. Sustain. Cities Soc. **10**, 11–21 (2014)

Y. Pujara, P. Pathak, A. Sharma, J. Govani, Review on Indian municipal solid waste management practices for reduction of environmental impacts to achieve sustainable development goals. J. Environ. Manag. **248**(June), 109–238 (2019)

N. Raman, D. Sathiyanarayanan, Quality assessment of ground water in Pallavapuram municipal solid waste dumpsite area nearer to Pallavaram in Chennai, Tamil Nadu. Rasayan J. Chem. **4**(2), 481–487 (2011)

K. Ramesh, V. Thirumangai, Impacts of tanneries on quality of groundwater in Pallavaram, Chennai Metropolitan City. Int. J. Eng. Res. Appl. **4**(1), 63–70 (2014)

R. Rana, R. Ganguly, A.K. Gupta, Physico-chemical characterization of municipal solid waste from Tricity region of northern India: A case study. J. Mater. Cycles Waste Manag. **20**(1), 678–689 (2018)

K. Ravindra, S. Mor, Distribution and health risk assessment of arsenic and selected heavy metals in groundwater of Chandigarh, India. Environ. Pollut. **250**, 820–830 (2019)

M. Rawat, A. Ramanathan, T. Kuriakose, Characterisation of Municipal Solid Waste Compost (MSWC) from selected Indian cities—A case study for its sustainable utilisation. J. Environ. Protect. (Irvine,. Calif). **04**(02), 163–171 (2013)

Z. Salem, K. Hamouri, R. Djemaa, K. Allia, Evaluation of landfill leachate pollution and treatment. Desalination **220**(1–3), 108–114 (2008)

A.S. Sallam, A.R.A. Usman, H.A. Al-Makrami, M.I. Al-Wabel, A. Al-Omran, Environmental assessment of tannery wastes in relation to dumpsite soil: A case study from Riyadh, Saudi Arabia. Arab. J. Geosci. **8**(12), 11019–11029 (2015)

T.I.S. Swati, V.K. Vijay, P. Ghosh, Scenario of landfilling in India: Problems, challenges, and recommendations, in *Handbook of Environmental Materials Management*, (Springer International Publishing, Cham, 2019), pp. 321–336

A. Tripathi, D.R. Misra, A study of physico-chemical properties and heavy metals in contaminated soils of municipal waste dumpsites at Allahabad, India. Agris On-line Paper Econ. Inform. **2**(4), 2043–2052 (2012)

P. Xaypanya, J. Takemura, C. Chiemchaisri, H. Seingheng, M. Tanchuling, Characterization of landfill leachates and sediments in major cities of Indochina peninsular countries—Heavy metal partitioning in municipal solid waste leachate. Environments **5**(6), 65 (2018)

Q. Yang, L. Fu, X. Liu, M. Cheng, Evaluating the efficiency of municipal solid waste management in China. Int. J. Environ. Res. Public Health **15**(11) (2018)

N. Yukalang, B. Clarke, K. Ross, Barriers to effective municipal solid waste management in a rapidly urbanizing area in Thailand. Int. J. Environ. Res. Public Health **14**(9), 9–14 (2017)

Chapter 14
Climate Change Impacts of Nitrate Contamination on Human Health

Jeyaragash Danaraj, Uthirakrishnan Ushani, Sybiya Vasantha Packiavathy, Jeba Sweetly Dharmadhas, Tamilarasan Karuppiah, S. Anandha Kumar, and E. S. Aooj

Abstract Global climate change has become a major concern in recent years as it threatens human health by altering the hydrobiological systems and their water resource quality with increased nitrate concentration, which is considered to be a paradigmatic pollutant in groundwater, based on the climatic forecasts of temperature and precipitation patterns for the years 2021 and 2050. The consequences of decreased water availability and increased nitrate concentration are premature death and serious health issues such as colorectal cancer, thyroid disease, and neural tube defects in human beings. Therefore, nitrate pollution in groundwater is a universal problem and has been the major focus of researchers' agenda for several decades. Human populations have doubled the natural rate at which nitrogen is left behind on the land via the production and application of nitrogen fertilizers and fossil fuel combustion. It is exclusively vulnerable to ambient waters as toxic pollutants that in turn affects the human health and natural vegetations which will be controlled by planting nitrogen fixing crops. Domesticated wastewater stands as a steadfast local resource, yet its amount will only fulfil a fraction of the loss of available resources

J. Danaraj (✉) · S. V. Packiavathy · J. S. Dharmadhas
Centre for Ocean Research (DST-FIST Sponsored Centre), Col. Dr. Jeppiaar Research Park, Sathyabama Institute of Science and Technology (Deemed to be University), Chennai, Tamil Nadu, India

U. Ushani
Department of Biotechnology, Karpaga Vinayaga College of Engineering and Technology, Chinnakolambakkam, Chengalpattu, Tamil Nadu, India

T. Karuppiah
Department of Civil Engineering, Vel Tech Rangarajan Dr. Sagunthala R&D Institute of Science and Technology, Avadi, Chennai, Tamil Nadu, India

S. Anandha Kumar
Department of Civil Engineering, Aditya Engineering College, Surampalem, East Godavari (D.T), Andhra Pradesh, India

E. S. Aooj
Xcellogen Biotech India Pvt., Ltd., Thiruvananthapuram, India

B. Panneerselvam et al. (eds.), *Climate Change Impact on Groundwater Resources*, https://doi.org/10.1007/978-3-031-04707-7_14

due to climate change. This chapter documents how the nitrate exposure of groundwater affects human health and a consolidative management outlook that evade an unplanned actions from private sectors since it threatens the accomplishment of sustainable water resources manipulation under different hydrobiological scenarios.

Keywords Climate change · Nitrate · Groundwater · Pollution · Human health

14.1 Introduction

Groundwater is considered as the most significant natural resource with high economic and social impacts as it meets the demand of nearly half of all drinking water needs in the world (WWAP 2009). It also plays an imperative role in food production with about 40% of the consumptive use in agricultural irrigation globally (Siebert et al. 2010). In addition, it also provides a direct support to lakes, rivers, and wetlands. However, reports have documented that groundwater resources face increased pressure globally, and in recent years, many cases have been reported that increased pressure has induced generalizations beyond sustainable stages and increased levels of pollution (Groundwater Governance 2014). Climate change, population growth, and the use of lands pose an array of threats to groundwater resources globally, thereby influencing water quality and quantity. Haller et al. (2013) reported that, on a global scale, agricultural use of lands represents the largest pollution for groundwater quality, as the use of fertilizers and the spreading of animal manure on the ground significantly increase the nutrient concentrations that have been reported in both municipal and private well systems (Hallberg and Keeney 1993). Among the nutrients, nitrate (NO_3^-) is considered to be one of the main groundwater pollutants. Nitrate pollution in groundwater has been found to be ubiquitous and poses a threat to public health and environmental pollution worldwide and controlling this has been researchers' agenda for several decades (Rosenstock et al. 2014). Consumption of nitrate-containing water leads to immediate adverse effects on human health (acute toxicity) and could cause the blue baby syndrome (methemoglobinemia), in which the blood cells in a child's body lack sufficient amounts of oxygen (Uhlman and Artiola 2011). Long-term exposure to nitrate contamination in drinking water induces cancers such as gastric, colorectal, urothelial, and bladder and brain tumours (CDPH 2013). Hence, it is of fundamental importance to control nitrate concentration in groundwater to minimize the public health risks. Therefore, it is essential to mitigate the significance of groundwater nitrate pollution and understand its global impacts on human health, and this chapter focuses on the global climate change impacts of nitrate contamination on human health. Numerous efforts have been taken to advance agricultural practices in order to reduce nitrogen leaching owing to fertilization (Cameron et al. 2013), to analyze the biogeochemistry so that denitrification can be identified, to use aquifers to clean up the nitrate from groundwater (Bölhke 2002; Rivett et al. 2008), to follow its movement via unsaturated zones (Baram et al. 2017), and, finally, to treat polluted groundwater obtained from aquifers so that the content of nitrate might be reduced before entering the water supply systems (Seidel et al. 2011). Furthermore, legislation has been issued to manage these problems and those who fail to adhere are liable to face legal prosecution (i.e. the Nitrates Directive, 1991/676/ EEC; and the Water Framework Directive, 2000/60/EEC).

14.2 Sources of Nitrate and Its Main Groups

Various processes and sources play an essential role in nitrate contamination in groundwater systems. Identification of these sources and their underlying mechanisms is significant in order to address the quality of groundwater. It was documented that the sources of nitrate pollution are divided into two groups, viz., point source and non-point source pollution. Areas dumped with livestock confinement or chemicals or manure storage are examples of point source pollution, which result in a higher accumulation of nitrate concentration in localized areas (Haller et al. 2013), whereas the use of fertilizes in agricultural systems in turn affects the groundwater quality in the localized areas, and these are considered as examples of non-point source pollution (Viers et al. 2012).

14.2.1 Point Source Pollution

This type of pollution occurs due to the accidental fall of nitrogen rich compounds in the soil, lack of slurry storage and manure tank facility in the rural areas and lack of slurry storage facility. Furthermore, abandoned wells and boreholes, partially constructed wells, and sinkholes might lead to a wide range of groundwater contaminations on the surface area (Hallberg and Keeney 1993). It causes the most harmful levels of nitrate accumulation, and many studies have reported that poorly constructed manure storage pits, storage tanks, and lagoons can contribute to noteworthy nitrate contamination in groundwater systems. Moreover, nitrate leaching in barnyards is found to be mostly higher in sandy soils or thin soils over creviced bedrock. At this point, the yards remain porous and aerobic, converting nitrogen into nitrite, and, in turn, leaching occurs rapidly into the groundwater (Bundy et al. 1980). In addition, septic systems are considered as examples of anthropogenic sources of groundwater nitrogen contamination (Haller et al. 2013). In this process, wastes are discharged to tanks and then to the underground fields during which the nitrogen remains as ammonium and other organic forms until it reaches the aerobic zone of the disposable field. Further oxidation leads to the conversion of ammonia nitrate, which then enters the groundwater system.

14.2.2 Non-Point Source Pollution

This type of pollution is mainly due to the widespread use of organic and synthetic nitrogen fertilizers. The Haber–Bosch process is the principle procedure for the production of synthetic fertilizers (Smil, 2001). During this process, the catalytic activity of atmospheric nitrogen with hydrogen forms ammonia, and further oxidation produces nitrate. Erisman et al. (2011) documented that the Haber–Bosch process is most commonly employed at the industrial scale, and, thus, agricultural productivity increased in most regions globally. Nearly two billion people are estimated to be affected by synthetic fertilizers produced by the Haber–Bosch process (Schmoll et al. 2006). It was reported that plants do not completely use the nitrate present in synthetic fertilizers, and, hence, the remaining nitrate accumulates in the soil with increased potential of nitrogen leaching into the groundwater system. Figure 14.1 shows the complete nitrogen transformation process in the soil. Nitrogen can be exhibited in various forms as shown in Table 14.1. Plants cannot use organic nitrogen directly for their growth; nevertheless, decomposition of organic matter results in nitrogen release in the form of nitrate that can be utilized by plants. Mahler et al. (2007) reported that nitrogen leaching depends on the soil types and water in the form of precipitation or irrigation. The transformation of nitrogen to nitrate and

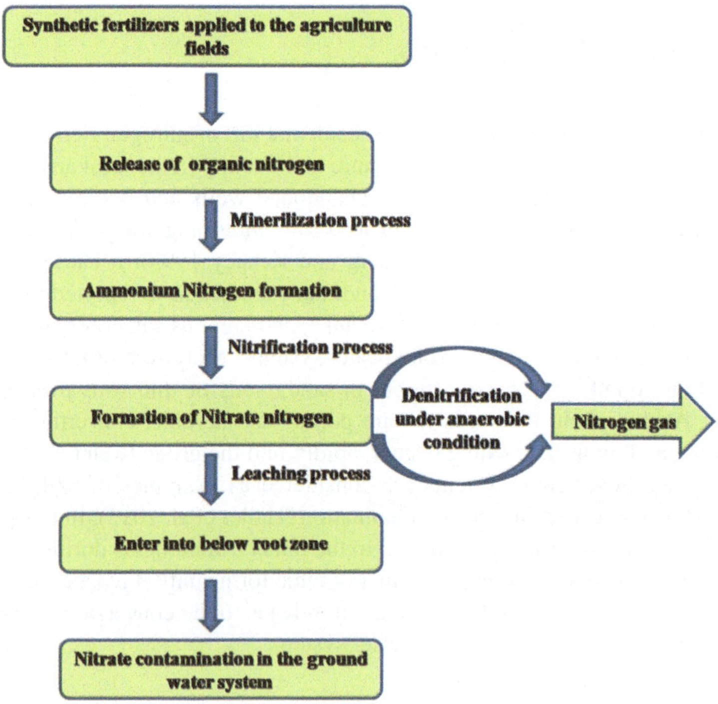

Fig. 14.1 Transformation of nitrogen in the soil for groundwater contamination

Table 14.1 Nitrogen forms in the soil

Nitrogen forms	Process	Crops availability
Organic nitrogen	Mineralization	Non-plant available nitrogen
Ammonium nitrogen	Nitrification	Plant available nitrogen
Nitrate nitrogen	Leaching into the root zone	Plant available nitrogen

Source: Frate (2007)

its accumulation in the soil are mostly not harmful to the environment. However, an extreme use of fertilizers and their inappropriate management promote nitrogen leaching that leads to groundwater contamination. Organic nitrogen fertilizers used in agricultural systems include human wastes, animal manures, composts, sewage sludges, and green manure crops.

14.3 Climate Change and Its Impacts on Nitrate Contamination

In recent years, climate change has become a major threat due to increased greenhouse gas emissions. As a result, there is increased change in precipitation and more intense weather, which directly affect the availability and reliance on groundwater. There is a higher risk of aquifer depletion in longer drought periods, specifically in the case of small and shallow aquifers. However, populations in water-scarce areas will mostly depend on groundwater owing to its buffer capacity. Subsequently, indirect climate change impacts the escalation of human activities and increases the groundwater demand due to increased land use. Strategies to sustain ground water quality globally and comprehending the hydrobiological cycle during climate change are more important for water resources for future. In recent years, numerous studies have been attempted to incorporate the results from site and case studies due to the change in the complexity of potential environmental responses.

Among the water nutrients, nitrate has been found to be ubiquitous in groundwater, often at higher concentrations and beyond the drinking water standards globally (Rivett et al. 2008), and nowadays, it is considered to be the most widespread problem of groundwater quality. It has been predicted that nitrate concentration will continue to increase globally over the next few decades (Stuart et al. 2011). If this to be continued, many groundwater resources may lose the drinking water standard, which lead to the accumulation of hazardous nutrients in near future. Holman and Loveland (2001) developed a cross-sectorial regional assessment of climate change (RegIS) and used the drivers-pressure-state-impact-response approach in order to link the climate change, water resources, agricultural sectors, and receptors like biodiversity, water quality, and societal response. Similarly, Stuart et al. (2011) applied the source-pathway-receptor model to the nitrogen cycle to understand the climate change impacts on groundwater nitrate concentrations, as shown in Fig. 14.1. This model facilitates a conceptual framework to visualize various factors to be measured individually, and it will interpret a simplification process excluding the communication of this cycle from other element cycles and water cycles. It includes direct flow from the soil to surface water, evading groundwater.

It has been very well explained that the nitrate input to groundwater is derived from sewage sludges, manures, fertilizers, and crop residues from agricultural areas (DEFRA 2006), in addition to other smaller inputs from aerial deposition and urban point sources (Wakida and Lerner 2005). Nitrogen fertilizers can be used as ammonia or urea, as well as nitrate; however, the non-nitrate forms are usually converted into nitrate in the soil. Destouni and Darracq (2009) reported that a small percentage of nitrate applied in agricultural fields will be lost to the atmosphere as NO, NH_3, or N_2O. Furthermore, the leaching process converts a small proportion of soil nitrogen into the nitrate form in the soil base, which can be either stored or transmitted via the unsaturated zones to groundwater. In recent years, various models have been developed to study groundwater problems; however, exemplifying the ambiguity in climate change modelling studies has recently been the focus of widespread work. Prudhomme and Davies (2009) reported three sources of climate model impact studies on river flows in the UK, viz., downscaling techniques, hydrological modelling techniques, and general circulation models (GCMs). Among these models, GCM model was found appropriate for climate change and ground water quality prediction, than other nitrate contamination GCMs models. In addition, Jackson et al. (2011) used 13 different GCMs to understand the regional distribution of groundwater of the Chalk aquifer of Berkshire and found about a 5% reduction in the annual groundwater recharge across the study area. Few additional literature works were documented, although Fowler et al. (2007) studied the broader framework of hydrological modelling and compared the different methods for recharge potential.

14.4 Climate Change and Nitrate Leaching in Agricultural Lands

As stated, the primary source of nitrate is derived from agricultural lands, and the NEAP-N model predicted the concentration of nitrate in the base of the agricultural land soil zone at 180 mg L^{-1} NO_3 (Silgram et al. 2001). Rozemeijer et al. (2009) recorded a concentration higher than that reported above within the potential recharge exiting the soil zone. From the reports, it is clear that climate change has direct impacts on the geographical distribution and crop yields; however, it is critical to forecast what these will be and relate these directly to changes in soil nitrate availability. De Jong et al. (2008) reported the long-term impacts of climate change on crop production in agricultural lands and found that changes occurred in nitrogen leaching from the soil base. The flow diagram in which green house gas concentration impact agricultural lands by nitrate leaching and affect the groundwater system

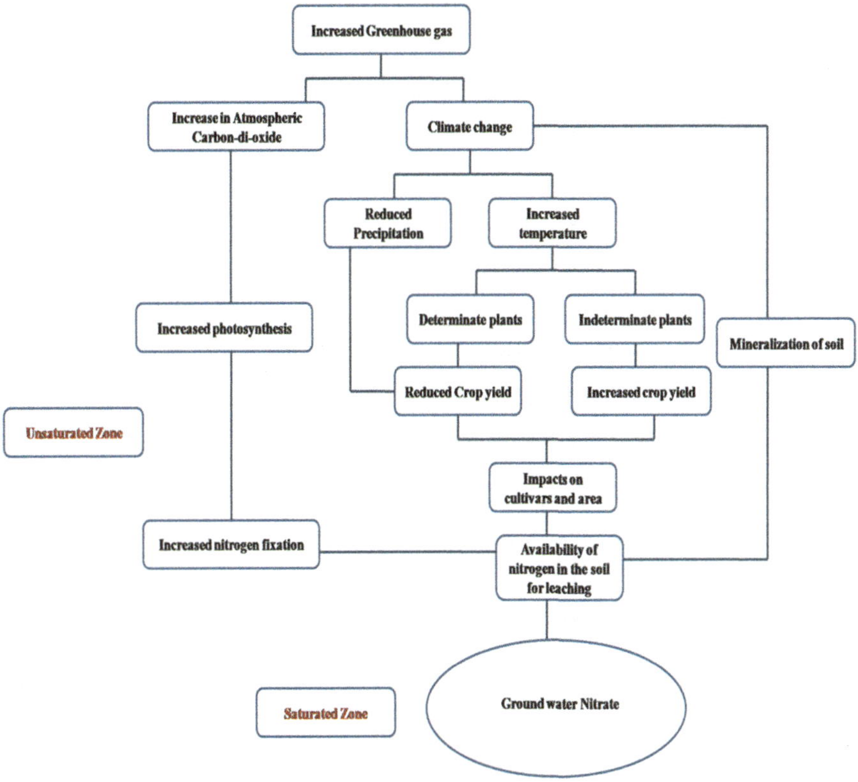

Fig. 14.2 Impacts of climate change on soil nitrate available for leaching from agricultural areas

was shown in the Fig. 14.2. The literature has documented that climate change includes both temperature and CO_2 rise and may also impact leaching from forests. If N is sufficient, the primary production will increase with further N leaching. Nadelhoffer et al. (1984) reported that organic matter mineralization in forests exhibited seasonal variation more than nitrification, and this mineralization is promoted by the higher soil temperature releasing nitrogen that could be taken up by the plants, followed by leaching into groundwater. Wright et al. (1998) attributed the rise in the N flux in runoff of approximately 5–12 mmol m^{-2} $yr.^{-1}$ with 5% rise in the decomposition rate of organic matter with respect to higher temperatures. Furthermore, for contaminants to attain a receptor, there must be a probable route to induce nitrate pollution in the groundwater system. The pathway through which nitrates move from the soil surface and the factors that control their migration include soil infiltration, migration via the unsaturated zones and saturated zones and inter-linkage of groundwater and receptors (Fig. 14.3).

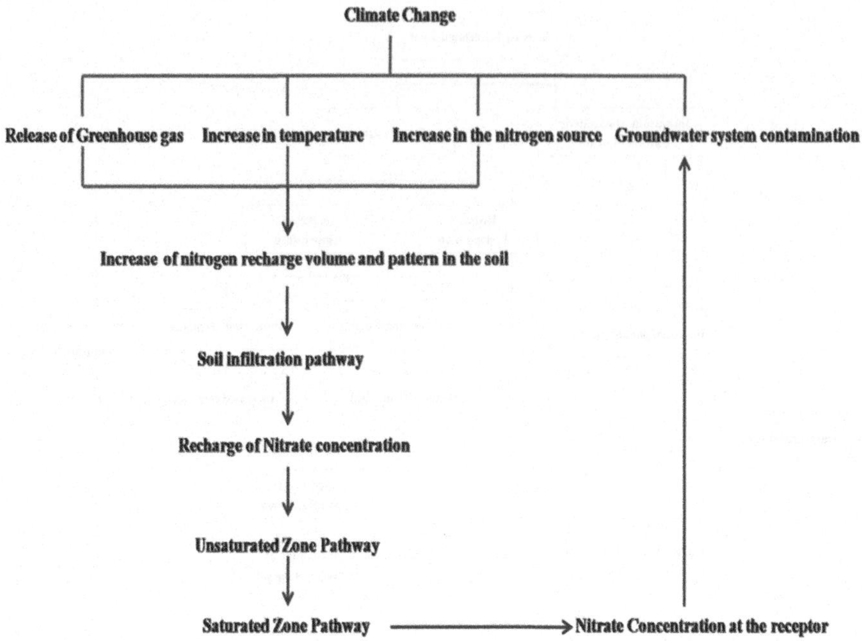

Fig. 14.3 Impact of climate change on nitrate leaching pathways

14.5 Sources of Human Exposure to Nitrate Contamination

Nitrate contamination in surface water and groundwater has been reported not only in India but also in most rural and urban areas of the world. It was reported that cities such as Tehran, Semnan, Rasht, Mashhad, Sari, Arak, Hamedan, and Isfahan with surface water and groundwater resources have the same problem of nitrate contamination, with further rise in the concentration due to industrialization (Mesdaghi et al. 2011). The literature surveys on nitrate pollution of underground water in the rural areas in India (Sankararamakrishnan et al. 2008), Newyork (Gelberg et al. 1999), Lowa (Kross et al. 1993), Argentina (Coasta et al. 2002) and China demonstrated that this problem is not unique in general; however, human activities such as agricultural activities (Mahvi et al. 2013), purification and sanitation (Rotaru and Raileanu, 2008), diffusion from industrial progression, and removal of solid wastes (Wakida and Lerner 2005) are the common sources of nitrate pollution. The literature has documented that human beings are usually exposed to nitrate contamination via drinking water, food, air, and the soil. The most probable way of nitrate contamination to the general population is via food ingestion and drinking water, and about 5–8% of nitrate ingested is converted into nitrite by many bacteria in the mouth (Walker 1996). Apart from these exposures, the remaining nitrate concentration enters the body from other external sources (Mensinga et al. 2003). In drinking water, nitrate contamination is mainly due to disposal of waste pile leachates and agricultural, human, and animal wastes, leading to surface water and groundwater system pollution (Liebscher et al. 1992). Foods such as processed meats, vegetables, and food additives act as the major sources of nitrate contamination that cause adverse effects on human beings. Food preservatives contain nitrate and nitrite that prevent the growth of *Clostridium botulinum*, and fruits contain nitrate concentration in the range of 200–250 mg/kg (Duijvenbooden et al. 1989). In case of atmospheric air, the annual nitrate concentration was measured to be in the range between 0.11 and 0.36 mg/cubic metres in Iceland (Prospero and Savoie 1989) and, in the Netherlands, it was recorded between 1.5 and 9.1 mg/cubic metres (Janssen et al. 1989). Among these sources, the public exposure to nitrate in the soil has not been reported. Since nitrate is highly soluble in the soil, it has a way to leak into the groundwater system. However, Lefebvre et al. (2010) documented the evaluation of nitrogen concentration in the soil, and the findings reported that the nitrate concentration in drainage water was higher at 10 mg/L. Various methods have been optimized by researchers to reduce the nitrate contamination including horizontal roughing filtration (Khazaei et al. 2016), ion exchange resins (Shahbazi et al. 2010), and nanofiltration (Yousefi et al. 2016). The World Health Organization (WHO) has set the maximum permissible concentration of nitrate to prevent methaemoglobinaemia, and, further, this WHO guideline could be followed only for the public water systems. Nevertheless, the studies on the acute and chronic outcomes of nitrate contamination in the groundwater in different parts of the world have demonstrated that different concentrations of nitrate in water lead to numerous diseases in different groups of the human population (Balamurugan et al. 2020b; Panneerselvam et al. 2020). So far, broad research on the positive and negative impacts of nitrate contamination in drinking water has been accomplished in terms of human health.

14.6 Endogenic Formation of Lethal N-Nitroso Compounds in Humans

Precursors are the initial compounds to be involved in the biosynthetic pathway in order to attain the target product, and, here, nitrate acts as the precursor for the synthesis of N-nitroso compounds. These compounds are considered to be carcinogenic to humans and animals, as they produce genotoxic substances. Nitrate that enters the human body will be inert and further can be broken down by the enzymes involved in the metabolic process. Nevertheless, the process by which the bacteria reduce nitrate to nitrite and other nitrogenous compounds can impact the human physiology and health. It was documented that 25% of nitrate is excreted by saliva of which 20% is converted into nitrite by bacteria (Spiegelhalder et al. 1976). Nitrite in the form of nitrous acid (HNO_2) is further converted into nitric oxide, nitrogen dioxide, and nitrogen trioxide under acidic conditions. However, nitric oxide plays a major role in the defensive mechanisms against oral bacteria (Leaf et al. 1989).

Nitrogen trioxide is responsible for nitrosamine compounds that are carcinogenic in nature. Hence, these mechanisms are considered to be a source for the human exposure to nitric oxide of about 40–75%, and the other sources include occupational exposure, unprocessed meat and fish, and tobacco products (Tricker et al. 1996).

14.7 Adverse Effects of Nitrate Concentration on Human Health

Higher nitrate accumulation in drinking water can pose negative threats to human health and specifically infants are more susceptible to serious health issues. Nitrate poisoning in infants can result in "methemoglobinemia" or "blue baby syndrome", which was first reported in 1945. The disease is characterized by the transformation of the normal skin colour to blue or lavender colour, accompanied by the inability of blood cells to carry oxygen from the lungs to other parts of the body. Disease effects caused by nitrate concentration can be categorized into two forms: chronic and acute effects (Balamurugan et al. 2020a; Panneerselvam et al. 2020a).

14.7.1 Chronic Effects

14.7.1.1 Carcinogenicity

Long-term exposure to nitrate concentration results in the formation of nitrosamine compounds that are carcinogenic in nature. The literature has documented epidemiological studies on humans in association with nitrate uses, and the results showed that there was no significant correlation between nitrate concentration and the risk of disease from colon cancer. Statistical reports from 1999 to 2010 revealed that the death rate due to colon cancer was directly proportional to nitrate concentration from 0.84 to 1.14 with a 98% confidence interval (Yang et al. 2007). van Loon et al. (1998) found no relationship between nitrate concentration and stomach cancer, and further the cohort study showed that there was no positive correlation between cancer non-appearance and vitamin C use as an effective factor. In addition, a cohort study by Knekt et al. (1999), conducted for 24 years, reported no correlation between tumour incidence and nitrate uptake. Hence, it was considered that antioxidants play a major role in cancer formation. In contrast to the above reports, another case–control study was reported in Italy in 2001, which found that there was a huge risk of gastric cancer among people who consumed higher levels of nitrate and lower antioxidants. Palli et al. (2001) documented a positive correlation between the high level of nitrate uptake and gastric cancer. A study was conducted among 258 citizens in Spain that analyzed the relationship between different nitrate concentrations in drinking water and other cancers such as stomach, bladder, prostate, and colon. The results showed that the mortality rates due to gastric cancer had increased with increase in the concentration of nitrates in males and females. Among the different forms of cancers, prostate cancer in males has been reported between the age group of 55 and 75 years. A similar study conducted in Slovakia reported the formation of colorectal cancer and Hodgkin's lymphoma among males and females who were exposed to higher nitrate concentrations in the range from 4.5 to 11.3 mg nitrate nitrogen per L (Gulis et al. 2002; Panneerselvam et al. 2021a, b). However, a negative incidence was recorded in America where no association between nitrate concentration and Hodgkin's lymphoma was found (Law et al., 1999). A cohort study conducted by Weyer et al. (2001) among the females in America about the Hodgkin's lymphoma occurrence and the results exhibited a weak converse relationship between nitrate uptake concentration of 2.46 mg per litre of nitrate nitrogen in drinking water and Hodgkin's lymphoma. Ward et al. (2003) documented case–control studies in America for both males and females and found that the average nitrate concentration in drinking water was greater than 4 mg of nitrate nitrogen per litre. A dose–response relationship exhibited that the higher the concentration of nitrate, the greater is the risk of Hodgkin's lymphoma and its association remains with the contemplation of confounding factors such as vitamin C and carotenes. It was concluded that exposure to high nitrate concentration in the long term could lead to the risk of Hodgkin's lymphoma. In general, most of the studies have shown that nitrate concentrations have no direct link to Hodgkin's lymphoma; however, the risk is due to the interference of vitamin C and carotenes that weaken the relationship between nitrate and its incidence of Hodgkin's lymphoma (Ward et al. 1996). A nitrate concentration of 2.86 mg per litre of nitrate nitrogen has been found to be associated with bladder cancers (Chui et al. 2007). Volkmer et al. (2005) evaluated the effects of nitrate concentrations in drinking water on the occurrence of urinary tract cancer in Germany, which were exposed to different nitrate concentrations ranging from 10 to 60 mg per litre during 2005.

Volkmer et al. (2005) reported that they found a positive correlation between urinary tract infections and nitrate concentration, which led to the formation of penile and prostate cancer. Ward et al. (2003) conducted a case study and reported that they found no association between urinary tract tumour formation in both males and females who consumed nitrates from plant and animal sources. The literature has documented that the higher concentration of nitrate from plant sources was connected with the typical risk of infection among males, and so far there have been no processes that changed the infection rate at the same time as the concentration increased. During 1986 to 1989, a population-based case study was conducted on how nitrate concentration impact human health using the kidney cancer cell lines and the results exhibited that the number of sources for groundwater contamination was found increased, thereby contaminate drinking water lead to numerous health issues. It was concluded that, for years, the standard nitrate levels derived from public source drinking water and their use had no correlation with the risk of renal cell carcinoma. However, a higher exposure to nitrate has been reported to cause infections among those who consumed higher amounts of red meat and lower amounts of vitamin C in their diets (Ward et al. 2007). It was hypothesized that a higher exposure to nitrosamine compounds increased the risk of brain tumour, and a cohort study investigated and reported that the consumption of meat with higher nitrate and nitrite levels results in the formation of *N*-nitrosopyrrolidine and nitrous dimethyl amine, which in turn leads to brain tumour formation (Michaud et al. 2009). Studies also reported that there was no relationship between nitrite and nitrate and pancreatic cancer in both males and females, with little evidence about that the consumption of processed meat leading to the formation of pancreatic cancer (Aschebrook-Kilfoy et al. 2011).

14.8 Acute Effects

14.8.1 Methemoglobinemia

Greer et al. (2005) have described the entire physiological process that leads to the blue baby syndrome or methemoglobinemia in infants, and the ingested nitrate is further reduced to nitrite by the bacteria present in the mouth and stomach of infants, which have been found to be less in the case of adults. Nitrite thus formed is then bound to haemoglobin to form methemoglobin that obstructs the oxygen transport capacity of the blood. Methemoglobinemia is a life-threatening form by which the concentration of methemoglobin exceeds 10% and the risks include water containing a higher concentration of nitrate, including food and medications (Sanchez-Echaniz et al. 2001), leading to enteric infections (Charmandari et al. 2001; Balamurugan et al. 2020; Panneerselvam et al. 2021). In 1945, the first case of methemoglobinemia was reported, caused due to a higher level of nitrates in drinking water (Comly 1987). Walton (1951) reported that there were no observed cases during the prediction of the US Environmental Protection Agency (EPA) nitrate limit set at 10 mg/L NO_3. The World Health and Organization (WHO) report on water and health documented that there were 41 cases in Hungary annually and 2913 cases in Romania (1985–1996) and 46 cases in Albania during 1996, and the results from numerous epidemiological studies conducted prior to 2005 exhibited a relationship between nitrate concentration in drinking water and methemoglobin levels in infants. In brief, nitrate concentration > 10 mg/L N-NO_3 is connected with an increased level of methemoglobin; however, clinical methemoglobinemia does not always exist.

Abu Naser et al. (2007) conducted a case study in Gaza and found that the higher levels of methemoglobin in infants were due to the highest mean nitrate concentration in well water. Sadeq et al. (2008) conducted a cross-sectional study in Morocco with a 22% increased risk of methemoglobinemia in infants who consumed drinking water with a nitrate concentration of >50 mg/L as compared to the threshold nitrate level. Zinman et al. (2011) performed a retrospective cohort study in Lowa in individuals in the age group of 1–60 years who consumed well water with nitrate levels less than 10 mg/L N-NO_3 and found a positive relationship between nitrate levels and methemoglobin levels upon ingestion. However, nitrate concentration less than 3 mg/L N-NO_3 showed no relationship between nitrate intake and methemoglobin levels in pregnant women around 36 weeks of gestation (Manassarem et al. 2010). It was reported that infants who were breastfed may have been exposed to concentrations of nitrate and nitrite and hence methemoglobinemia occurs. It may also be due to the preparation of baby food and dried milk powder using water contaminated with nitrate. Children in the age group of 1–8 years get infected on consuming nitrate with a concentration of 97–491 mg per litre. Contradictory reports were documented that 38 infants who consumed water from the well contained higher nitrate concentration which in turn cause methemoglobinemia. Furthermore, the study conducted by the WHO found no relationship with nitrate concentration in drinking water and the incidence of methemoglobinemia. It was reported that methemoglobinemia could also be the result of exposure to other chemicals such as sulphate, chlorite, chloramines, and chlorate and drugs such as lidocaine, benzocaine, sulphonamides, and also nitroglycerin (Sanchez-Echaniz et al. 2001).

14.8.2 Adverse Pregnancy Outcomes

Consumption of drinking water containing nitrate during pregnancy has been reported to exhibit several risk factors including pre-maturity, abortions, foetal deaths, growth retardation, low birth weight, malformations, and, sometimes, deaths. Numerous studies have been reported on the correlation between nitrate concentration in drinking water and congenital malformations in offspring, most likely due to the accessibility of a birth defect observation system around the world. Migeot et al. (2013) conducted a cohort study in the Deux-Sèvres district (France) and connected the maternal data from birth records to nitrate, atrazine, and other pesticide concentrations in the community water systems. Exposure to nitrate concentrations in the range from 14 to 27 mg/L NO_3 without atrazine exhibited gestational-age births and there was no association with atrazine detected in the drinking water samples. Table 14.2 describes the published reports with respect to nitrate in drinking water samples and pregnancy outcomes. In contrast, Albouy-Llaty et al. (2016) did not observe any relation between higher nitrate concentration and preterm birth.

14.8.3 Thyroid Disease

The literature has documented that a higher concentration of nitrate ingestion competitively slows down the uptake of iodine, thereby inducing hypertrophy of the thyroid gland (De Groef et al. 2006). Earlier studies in the Netherland proved that women who consumed water with nitrate contamination were susceptible to thyroid

Table 14.2 Studies on nitrate in drinking water and pregnancy outcomes

Description of exposure	Pregnancy outcomes	Findings summary	Reference
Abdominal wall failure connected to nitrate and atrazine concentrations on a monthly interval	Abdominal wall birth defects	No relationship was observed between nitrate concentration in drinking water and abdominal wall defects; however, a positive correlation was observed with atrazine	Mattix et al. (2007)
Evaluated the rate of combined and specific birth defects with nitrate, atrazine, and other pesticide concentrations	Defective birth categorized into 22 groups	Defective birth category as "other congenital anomalies"	Winchester et al. (2009)
Distances were calculated between the maternal habitation and the closest monitoring site with nitrate or atrazine concentration in the surface waters	Formation of gastroschisis	Gastroschisis was not connected with the maternal habitation to higher nitrate or nitrite concentration in the surface waters	Waller et al. (2010)
Connection between atrazine and nitrate concentrations to the births	Small gestational-age births	No association with nitrate; however, atrazine has a role in births	Migeot et al. (2013)
Addressed Maternal delivery associated with nitrate concentrations in municipal water supplies	Congenital malformations	No association with nitrate concentrations	Holtby et al. (2014)
Atrazine and nitrate concentrations in the community water systems were measured and connected with births	Preterm births	No relationship for >26.99 mg/L vs. <14.13 mg/L nitrate concentrations in the community water systems	Albouy-Llaty et al. (2016)

hypertrophy (Van Maanen et al. 1996). Several cohort studies have correlated nitrate consumption from drinking water with an occurrence of thyroid diseases. A study conducted in school-age children in Slovakia found an increase in the occurrence of sub-clinical hypothyroidism with a nitrate concentration of 51–274 mg/L NO_3. It was reported that high nitrate concentration (75 mg/NO_3) and low concentration of (8 mg/L) was recorded in the drinking water of Bulgarian village which exhibit the clinical symptoms of thyroids in pregnant women and school aged children respectively (Radikova et al. 2008) and further the concentration of iodine was also found increased. A cohort study conducted in post-menopausal women in Lowa reported that hypothyroidism and hyperthyroidism were found to be associated with nitrate concentrations. Nevertheless, the predominant intake of dietary nitrate is associated with the higher risks of hypothyroidism but not hyperthyroidism (Tajtakova et al. 2006).

14.8.4 *Diabetes Mellitus*

Kostraba et al. (1992) conducted ecological studies with nitrate contamination and reported a significant relationship between the occurrence of type 1 insulin-dependent diabetes mellitus and nitrate concentration in drinking water. Dahlquist et al. (1990) also showed a positive correlation between the nitrates and nitrosamine compounds in the food reflected with disease outbreak among the public who had a diet with high nitrate concentrations in Sweden. In contrast to the above studies, no significant risk was observed in diabetic children who were exposed to nitrate concentrations in 594 drinking water samples in Scotland and in some parts of England. Similar studies were also reported in the Netherlands (van Maanen and Van Rijn 2008) and Italy (Casu et al. 2000), with no correlation between nitrate concentration and diabetes mellitus. The results obtained from the above studies exhibited that the data were found limited and are in disagreement to achieve a broad and ultimate diagnosis of nitrate, nitrite and nitrosamine in the body and diabetes mellitus. Hence, a proper estimation should be done in future with these parameters via various food and beverage items.

14.9 Methods to Mitigate Nitrate Risks

Groundwater systems are affected by nitrate contamination and hence the drinking water supply has become a serious global problem, and as such many studies have highlighted the need to develop efficient management processes to control and reduce nitrate pollution.

14.10 Preventive Measures

The most important step to reduce nitrate contamination is nitrate leaching, and it is enough to use slow-release nitrogen sources or low rates of soluble nitrogen. Furthermore, farmers should be more careful about nitrogen use during periods in which the ground is not yet frozen but the grass has not started to grow. In addition, overirrigation should be avoided by the farmers so that there will be a chance to increase nitrogen leaching. By following these measures, agricultural lands will highly reduce the chance of nitrogen leaching into the groundwater.

14.11 Protection Policies and Regulations of Environments

Nitrate pollution has become the most commonly addressed environmental issue that is mostly related to increased atmospheric pollution due to climate change, and government policy measures have been implemented to reduce the atmospheric pollution, thereby limiting nitrate contamination in surface water and groundwater resources. This includes several European directives, such as the European Water Framework Directive, the Nitrates Directive, the Groundwater Directive, the National Emission Ceilings Directive, the Urban Waste Water Treatment directive, and ambient air quality (Sutton et al. 2011).

14.12 Monitoring of Drinking Water Standards

In order to regulate nitrate concentrations and minimize health risks, the World Health Organization (WHO 2004) has provided guidelines that nitrate concentration in drinking water should not exceed 50 mg/L. Moreover, the local environmental protection agency of various countries has set the standard for the maximum nitrate contamination level in drinking water supply. In addition, the US Environmental Protection Agency has established the standards and limits for nitrate levels in public water supplies. The maximum containment level for drinking water should not exceed 10 mg/L of nitrate and 1 mg/L of nitrite.

14.13 Management of Sewage Systems

It is important to treat wastewater before it is discharged in order to minimize environmental pollution, and, in many developed countries, money and resources have been invested in wastewater treatment aspects and their disposal technology (WHO 2004). The ultimate aim is to purify sewage effluents by removing harmful and hazardous contaminants and plant nutrients present in the sewage as solids and as dissolved matters. The United Nations Environmental Program (UNEP) has published national policies and measures to make sure that all disposed wastes conform with the national and international quality guidelines by the year 2025. Furthermore, awareness programmes about the formulations and implementations should be made public in order to achieve general recognition of suitable and environment-friendly sewage facilities.

14.14 Conclusions

In summary, nitrate contamination in groundwater systems is a widespread problem all over the world. Climate change is considered to be one of the major causes due to atmospheric gas emissions, which are above the preindustrial levels. Due to climate change, groundwater quality is getting affected resulting in public health problems and environmental ecosystem degradation. The most adverse health effects of nitrates related to drinking water are due to the mixture of high nitrate consumption and factors that increase the formation of endogenous nitrosamine compounds. Recent reports on cancer and birth defects have suggested that it would be possible to recognize those subgroups of populations likely to have greater potential for endogenous nitrosamine formations. Nevertheless, direct methods are needed for the assessment of these individuals. New methodologies need to be developed for the quantification of oral microbiomes that play a vital role in nitrate reduction, and characterizing their genetic variation in Nopaline Catabolism gene (NOC) metabolism requires large amounts of data to identify the higher-risk groups in epidemiological studies. To date, numerous studies have been documented about the risks of nitrate ingestion; however, additional studies need to be conducted with respect to the improved exposure assessment for population and the factors affecting nitrogenation need to be studied. Studies related to chronic and acute diseases that showed the most consistent results in connection to nitrate ingestion will be specifically useful for clarifying these risks. Furthermore, more limited evidences are available on the other health effects that should also be in focus. All these negative impacts can be studied and minimized by developing strategies of groundwater management and good governance to reduce nitrate contamination risks. Nevertheless, to better control groundwater nitrate contamination, researchers, water resources specialists, and policy-makers need to concentrate and gather more information on the scope, distribution, and severity of groundwater nitrate contamination due to climate change.

Acknowledgement The authors thank Karpagam Academy of Higher Education for their support in completing this chapter.

Conflict of Interests The authors disclose no conflicts of interest.

References

A.A. Abu Naser, N. Ghbn, R. Khoudary, Relation of nitrate contamination of groundwater with methaemoglobin level among infants in Gaza. East Mediterr. Health J. **13**, 994–1004 (2007)

M. Albouy-Llaty, F. Limousi, C. Carles, A. Dupuis, S. Rabouan, V. Migeot, Association between exposure to endocrine disruptors in drinking water and preterm birth, taking neighborhood deprivation into account: A Historic Cohort Study. Int. J. Environ. Res. Public Health **13**, 796 (2016)

B. Aschebrook-Kilfoy, A.J. Cross, R.Z. Stolzenberg-Solomon, A. Schatzkin, A.R. Hollenbeck, R. Sinha, et al., Pancreatic cancer and exposure to dietary nitrate and nitrite in the NIH-AARP diet and health study. Am. J. Epidemiol. **174**(3), 305–315 (2011)

P. Balamurugan, P.S. Kumar, K. Shankar, Dataset on the suitability of groundwater for drinking and irrigation purposes in the Sarabanga River region, Tamil Nadu, India. Data Brief **29**, 105255 (2020)

P. Balamurugan, P.S. Kumar, K. Shankar, R. Nagavinothini, K. Vijayasurya, Non-carcinogenic risk assessment of groundwater in southern part of Salem District in Tamilnadu, India. J. Chil. Chem. Soc. **65**(1), 4697–4707 (2020a)

P. Balamurugan, P.S. Kumar, K. Shankar, R. Nagavinothini, P.J. Sajil Kumar, Impact of climate and anthropogenic activities on groundwater quality for domestic and irrigation purposes in Attur region, Tamilnadu, India. Desalin. Water Treat. **208**, 172–195 (2020b)

S. Baram, V. Couvreur, T. Harter, M. Read, P.H. Brown, M. Kandelous, D.R. Smart, Hopmans, Estimating nitrate leaching to groundwater from orchards: Comparing crop nitrogen excess, deep vadose zone data-driven estimates, and HYDRUS modeling. Vadose Zone J. (2017). https://doi.org/10.2136/vzj2016.07.0061

J.K. Bölhke, Groundwater recharge and agricultural contamination. Hydrogeol. J. **10**, 153–179 (2002)

L.G. Bundy, L. Knobeloch, B. Webendorfer, G.W. Jackson, B.H. Shaw, *Nitrate in Wisconsin Groundwater: Sources and Concerns* (U.S Department of agricultrue, university of wisconsin-extension, Wisconsin, 1980)

K.C. Cameron, H.J. Di, J.L. Moir, Nitrogen losses from the soil/plant system: A review. Ann. Appl. Biol. **162**, 145–173 (2013)

A. Casu, M. Carlini, A. Contu, G.F. Bottazzo, M. Songini, Type 1 diabetes in sardinia is not linked to nitrate levels in drinking water. Diabetes Care **23**(7), 1043–1044 (2000)

CDPH, *California Department of Public Health*. Retrieved from Drinking water contaminants: Nitrate: http://www.ehib.org/page.jsp?page_key=14 (2013)

E. Charmandari, N. Meadows, M. Patel, A. Johnston, N. Benjamin, Plasma nitrate concentrations in children with infectious and noninfectious diarrhea. J. Pediatr. Gastroenterol. Nutr. **32**, 423–427 (2001)

H.F. Chiu, S.S. Tsai, C.Y. Yang, Nitrate in drinking water and risk of death from bladder cancer: An ecological case-control study in Taiwan. J. Toxicol. Environ. Health A **70**(12), 1000–1004 (2007)

H.H. Comly, Landmark article 8 September 1945: Cyanosis in infants caused by nitrates in well-water. By Hunter H. Comly. JAMA **257**, 2788–2792 (1987)

J.L. Costa, H. Massone, D. Martinez, E.E. Suero, C.M. Vidal, F. Bedmar, Nitrate contamination of a rural aquifer and accumulation in the unsaturated zone. Agric. Water Manag. **57**(1), 33–47 (2002)

G.G. Dahlquist, L.G. Blom, L.A. Persson, A.I. Sandstrom, S.G. Wall, Dietary factors and the risk of developing insulin dependent diabetes in childhood. BMJ **300**(6735), 1302–1306 (1990)

B. De Groef, B.R. Decallonne, S. Van der Geyten, V.M. Darras, R. Bouillon, Perchlorate versus other environmental sodium/iodide symporter inhibitors: Potential thyroid-related health effects. Eur. J. Endocrinol. **155**, 17–25 (2006)

R. De Jong, B. Qian, J.Y. Yang, Modelling nitrogen leaching in Prince Edward Island under climate change scenarios. Can. J. Soil Sci. **88**, 61–78 (2008)

DEFRA, *Procuring the Future—The Sustainable Procurement Task Force National Action Plan* (DEFRA, London, 2006)

G. Destouni, A. Darracq, Nutrient cycling and N2O emissions in a changing climate: The subsurface water system role. Environ. Res. Lett. **4**, 035008 (2009) (7 pp)

W.V. Duijvenbooden, A.J.C.M. Matthijsen, G.J.A. Speijers, C.F.V. Went, M.E.V. Apeldoorn, Integrated Criteria Document Nitrate. National Institute of Public Health and Environmental Protection Report. 1989; (758473012)

J.W. Erisman, H. Grinsven, B. Grizzetti, B. Fi, D. Powlson, *The European Nitrogen Problem in a Global Perspective* (Cambridge University Press, 2011)
H.J. Fowler, G. Stephen, C. Tebaldi, Linking climate change modelling to impacts studies: Recent advances in downscaling techniques for hydrological modeling. Int. J. Climatol. **27**(12), 1547–1578 (2007)
C. Frate, Nitrate Transformation in Soil. Tulare County: Universtiy of California Cooperative Extension. (2007)
K.H. Gelberg, L. Church, G. Casey, M. London, D.S. Roerig, J. Boyd, et al., Nitrate levels in drinking water in rural New York State. Environ. Res **80**(1), 34–40 (1999)
Groundwater governance. (2014, August 25). Retrieved from http://www.groundwatergovernance.org SCO and London: Earthscan.
F. R. Greer, M. Shannon., C. O. Nutrtion, C. O. E. Health. Infant Methanoglobinemia: The role of dietary nitrate in food and water. Paediatrics. **119**, 784–786 (2005)
G. Gulis, M. Czompolyova, J.R. Cerhan, An ecologic study of nitrate in municipal drinking water and cancer incidence in Trnava District, Slovakia. Environ. Res. **88**(3), 182–187 (2002)
G.A. Hallberg, D.R. Keeney, Sources of nitrate to groundwater. 300-301 (1993)
L. Haller, P. McCarthy, T. O'Brien, J. Riehle, T. Stuhldreher, Nitrate pollution of groundwater. 2014: Alpha Water Systems INC. (2013)
I. P. Holman, P. J. Loveland, (eds.), Regional Climate Change Impacts in East Anglia and the North West (the RegIS project), Final report of MAFF Project No. CC0337 (2001)
C.E. Holtby, J.R. Guernsey, A.C. Allen, J.A. VanLeeuwen, V.M. Allen, R.J. Gordon, A population-based case-control study of drinking-water nitrate and congenital anomalies using geographic information systems (GIS) to develop individual-level exposure estimates. Int. J. Environ. Res. Public Health **11**, 1803–1823 (2014)
C.R. Jackson, R. Meister, C. Prudhomme, and, Modelling the effects of climate change and its uncertainty on UK Chalk groundwater resources from an ensemble of global climate model projections. J. Hydrol. **399**(1–2), 12–28 (2011)
D.B. Janssen, F. Pries, J. Van Der Ploeg, B. Kazemier, P. Terpstra, B. Witholt, Cloning of 1,2-dichloroethane degrading genes of Xanthobacter autotrophicus GJ10 and expression and sequencing of the dhlA Gene. J. Bacteriol. **171**(12), 6791–6799 (1989)
M. Khazaei, R. Nabizadeh, A.H. Mahvi, H. Izanloo, R. Ansari Tadi, F. Gharagazloo, Nitrogen and phosphorous removal from aerated lagoon effluent using horizontal roughing filter (HRF). Desalin. Water Treat. **57**(12), 5425–5434 (2016)
P. Knekt, R. Jarvinen, J. Dich, T. Hakulinen, Risk of colorectal and other gastro-intestinal cancers after exposure to nitrate, nitrite and N-nitroso compounds: A follow-up study. Int. J. Cancer **80**(6), 852–856 (1999)
J.N. Kostraba, E.C. Gay, M. Rewers, R.F. Hamman, Nitrate levels in community drinking waters and risk of IDDM. An ecological analysis. Diabetes Care **15**(11), 1505–1508 (1992)
B.C. Kross, G.R. Hallberg, D.R. Bruner, K. Cherryholmes, J.K. Johnson, The nitrate contamination of private well water in Iowa. Am. J. Public Health. (AJPH), **83**(2), 270–272 (1993)
G. Law, R. Parslow, P. McKinney, R. Cartwright, Non-Hodgkin's lymphoma and nitrate in drinking water: A study in Yorkshire, United Kingdom. J Epidemiol. Commun. Health **53**(6), 383–384 (1999)
C.D. Leaf, J.S. Wishnok, S.R. Tannenbaum, L-arginine is a precursor for nitrate biosynthesis in humans. Biochem. Biophys. Res. Commun. **163**(2), 1032–1037 (1989)
V. Lefebvre, S. Thomas, F. Louis, A. Olivier, Did a Katian large igneous province trigger the Late Ordovician glaciation?: A hypothesis tested with a carbon cycle model. Palaeogeogr. Palaeoclimatol. Palaeoecol. **296**(3–4), 310–319 (2010)
B. Lefebvre, T. Timmers, M. Mbengue, S. Moreau, C. Hervé, K. Tóth, J. Bittencourt-Silvestre, D. Klaus, L. Deslandes, L. Godiard, J.D. Murray, M.K. Udvardi, S. Raffaele, S. Mongrand, J. Cullimore, P. Gamas, A. Niebel, T. Ott, A remorin protein interacts with symbiotic receptors and regulates bacterial infection. Proc. Natl. Acad. Sci. U S A. **107**(5), 2343–8 (2010)

D.-E, Liebscher W., Priester J., Hoell, Lymanα forests and the evolution of the universe. Astronomische Nachrichten **313**(5) 265–273 (1992). https://doi.org/10.1002/asna.2113130502

R.L. Mahler, A. Colter, R. Hirnyck, *Quality Water for Idaho Nitrate and Groundwater* (University of Idaho College of Agricultural and Life Sciences, Idaho, 2007)

A.H. Mahvi, J. Nouri, A.A. Babaei, R. Nabizadeh, Agricultural activities impact on groundwater nitrate pollution. Int. J. Environ. Sci. Technol. **2**(1), 41–47 (2013)

D.M. Manassaram, L.C. Backer, R. Messing, L.E. Fleming, B. Luke, C.P. Monteilh, Nitrates in drinking water and methemoglobin levels in pregnancy: A longitudinal study. Environ. Health **9**, 60 (2010)

K.D. Mattix, P.D. Winchester, L.R. Scherer, Incidence of abdominal wait defects is related to surface water atrazine and nitrate levels. J. Pediatr. Surg. **42**, 947–949 (2007)

T.T. Mensinga, G.J.A. Speijers, J. Meulenbelt, Health implications of exposure to environmental nitrogenous compounds. Toxicol. Rev. **22**, 41–51 (2003)

A.R. Mesdaghinia, A.H. Mahvi, S. Naseri, A.A. Mohamadi, M. Shekarriz, M. Alimohamadi, Study of nitrate reduction from water using nanosized iron [in Persian]. Iran J Health Environ. **4**(3), 313–320 (2011)

D.S. Michaud, C.N. Holick, T.T. Batchelor, E. Giovannucci, D.J. Hunter, Prospective study of meat intake and dietary nitrates, nitrites, and nitrosamines and risk of adult glioma. Am. J. Clin. Nutr. **90**(3), 570–577 (2009)

V. Migeot, M. Albouy-Llaty, C. Carles, F. Limousi, S. Strezlec, A. Dupuis, S. Rabouan, Drinking-water exposure to a mixture of nitrate and low-dose atrazine metabolites and small-for-gestational age (SGA) babies: A historic cohort study. Environ. Res. **122**, 58–64 (2013)

K. Nadelhoffer, J. Aber, J. Melillo, Seasonal patterns of ammonium and nitrate uptake in nine temperate forest ecosystems. Plant Soil **80**, 321–335 (1984)

D. Palli, C. Saieva, C. Coppi, G. Del Giudice, C. Magagnotti, G. Nesi, et al., O6-alkylguanines, dietary N-nitroso compounds, and their precursors in gastric cancer. Nutr. Cancer **39**(1), 42–49 (2001)

B. Panneerselvam, S. Karuppannan, K. Muniraj, Evaluation of drinking and irrigation suitability of groundwater with special emphasizing the health risk posed by nitrate contamination using nitrate pollution index (NPI) and human health risk assessment (HHRA). Hum. Ecol. Risk Assess. Int. J. **27**(5), 1324–1348 (2020)

B. Panneerselvam, K. Muniraj, C. Pande, N. Ravichandran, Prediction and evaluation of groundwater characteristics using the radial basic model in semi-arid region, India. Int. J. Environ. Anal. Chem. 1–17 (2021)

B. Panneerselvam, K. Muniraj, M. Thomas, N. Ravichandran, GIS-based legitimatic evaluation of groundwater's health risk and irrigation susceptibility using water quality index, pollution index, and irrigation indexes in semiarid region, in *Groundwater Resources Development and Planning in the Semi-Arid Region*, (Springer, Cham, 2021a), pp. 239–268

B. Panneerselvam, K. Muniraj, M. Thomas, N. Ravichandran, B. Bidorn, Identifying influencing groundwater parameter on human health associate with irrigation indices using the Automatic Linear Model (ALM) in a semi-arid region in India. Environ. Res. **111778** (2021b)

B. Panneerselvam, S.K. Paramasivam, S. Karuppannan, N. Ravichandran, P. Selvaraj, A GIS-based evaluation of hydrochemical characterisation of groundwater in hard rock region, South Tamil Nadu, India. Arab. J. Geosci. **13**(17), 1–22 (2020a)

C. Prudhomme, H. Davies, Assessing uncertainties in climate change impact analyses on the river flow regimes in the UK. Part 1: Baseline climate. Clim. Change **93**, 177–195 (2009)

J. Prospero, D. Savoie, *Effect of Continental Sources We Nitrate Concentrations over the Pacific*. (Springer, 1989) **339**, pp. 687–689

Z. Radikova, M. Tajtakova, A. Kocan, T. Trnovec, E. Sebokova, I. Klimes, P. Langer, Possible effects of environmental nitrates and toxic organochlorines on human thyroid in highly polluted areas in Slovakia. Thyroid Off. J. Am. Thyroid Assoc. **18**, 353–362 (2008)

M.O. Rivett, S.R. Buss, P. Morgan, J.W.N. Smith, C.D. Bemment, Nitrate attenuation in groundwater: A review of biogeochemical controlling processes. Water Res. **42**, 4215–4232 (2008)

T.S. Rosenstock, D. Liptzin, K. Dzurella, A. Fryjoff-Hung, A. Hollander, V. Jensen, A. King, G. Kourakos, A. McNally, G. Stuart Pettygrove, J. Quinn, J.H. Viers, T.P. Tomich, T. Harter, Agriculture's contribution to nitrate contamination of Californian groundwater (1945–2005). J. Environ. Qual. **43**(3), 895–907 (2014)

A. Rotaru, P. Răileanu, Groundwater contamination from waste storage works. Environ. Eng. Manag. J. **7**(6), 731–735 (2008)

J.C. Rozemeijer, H.P. Broers, F.C.V. Geer, B. MFP, Weather-induced temporal variations in nitrate concentrations in shallow groundwater. J. Hydrol. **378**, 119–127 (2009)

M. Sadeq, C.L. Moe, B. Attarassi, I. Cherkaoui, R. ElAouad, L. Idrissi, Drinking water nitrate and prevalence of methemoglobinemia among infants and children aged 1–7 years in Moroccan areas. Int. J. Hyg. Environ. Health **211**, 546–554 (2008)

J. Sanchez-Echaniz, J. Benito-Fernandez, S. Mintegui-Raso, Methemoglobinemia and consumption of vegetables in infants. Pediatrics **107**(5), 1024–1028 (2001)

N. Sankararamakrishnan, A.K. Sharma, L. Iyengar, Contamination of nitrate and fluoride in ground water along the Ganges alluvial plain of Kanpur district, Uttar Pradesh. India. Environ. Moni.t Assess. **146**(1–3), 375–382 (2008)

O. Schmoll, G. Howard, J. Chilton, I. Chorus, *Protecting Groundwater for Health: Managing the Quality of Drinking Water Sources*. (UK: World Health Organization and IWA, 2006)

C. Seidel, C. Gorman, J.L. Darby, V.B. Jensen, *An Assessment of the State of Nitrate Treatment Alternatives—Final Report* (The American Water Works Association, Inorganic Contaminant Research and Inorganic Water Quality Joint Project Committees, 2011). p. 161.

P. Shahbazi, F. Vaezi, A.H. Mahvi, K. Naddaffi, A.R. Rahmani, Nitrate removal from drinking water by point of use ion exchange. J Res Health Sci. **10**(2), 91–97 (2010) [PubMed]

S. Siebert, J. Burke, J.M. Faures, K. Frenken, J. Hoogeveen, P. Döll, F.T. Portmann, Groundwater use for irrigation – A global inventory. Hydrol. Earth Syst. Sci. Disc., 3977–3402 (2010)

M. Silgram, R. Waring, S. Anthony, J. Webb, Intercomparison of national & IPCC methods for estimating N loss from agricultural land. Nutr. Cycl. Agroecosyst. **60**, 189–195 (2001)

V. Smil, *Cycles of Life: Civilization and the Biosphere* (Scientific American Library, New York, 2001)

B. Spiegelhalder, G. Eisenbrand, R. Preussmann, Influence of dietary nitrate on nitrite content of human saliva: Possible relevance to in vivo formation of N-nitroso compounds. Food Cosmet. Toxicol. **14**(6), 545–548 (1976)

M.E. Stuart, D.C. Goody, J.P. Bloomfield, A.T. Williams, A review of the potential impact of climate change on future nitrate concentrations in groundwater of the UK. Sci. Total Environ. **409**, 2859–2873 (2011)

M.A. Sutton, G. Billen, et al. *Technical Summary-European Nitrogen Assessment*, Cambridge University Press, Europe (2011)

M. Tajtakova, Z. Semanova, Z. Tomkova, E. Szokeova, J. Majoros, Z. Radikova, E. Sebokova, I. Klimes, P. Langer, Increased thyroid volume and frequency of thyroid disorders signs in schoolchildren from nitrate polluted area. Chemosphere **62**, 559–564 (2006)

A.R. Tricker, N-nitroso compounds and man: Sources of exposure, endogenous formation and occurrence in body fluids. Eur. J. Cancer Prev. **6**(3), 226–268 (1997)

R. Tricker, R. Casaburi, T.W. Storer, B. Clevenger, N. Berman, A. Shirazi, S. Bhasin, S, The effects of supraphysiological doses of testosterone on angry behavior in healthy eugonadal men—a clinical research center study. J. Clin. Endocr. Metab. **81**, 3754–3758 (1996)

K. Uhlman, J. Artiola, *Nitrate Contamination Potential in Arizona Groundwater: Implications Ofr Drinking Water Wells* (The University of Arizona Cooperative Extension, Arizona, 2011)

A.J. van Loon, A.A. Botterweck, R.A. Goldbohm, H.A. Brants, J.D. van Klaveren, P.A. van den Brandt, Intake of nitrate and nitrite and the risk of gastric cancer: A prospective cohort study. Br. J. Cancer **78**(1), 129–135 (1998)

J.M. Van Maanen, I.J. Welle, G. Hageman, J.W. Dallinga, P.L. Mertens, J.C. Kleinjans, Nitrate contamination of drinking water: Relationship with HPRT variant frequency in lymphocyte DNA and urinary excretion of N-nitrosamines. Environ. Health Perspect. **104**, 522–528 (1996)

J.M. van Maanen, H.J. Albering, T.M. de Kok, S.G. van Breda, D.M. Curfs, I.T. Vermeer, et al., Does the risk of childhood diabetes mellitus require revision of the guideline values for nitrate in drinking water? Environ. Health Perspect. **108**(5), 457–461 (2000)

L. Van Maanen, H. Van Rijn, An accumulator model of semantic interference. Cognitive Systems Research, **8**, 174–181 (2008))

J.H. Viers, D. Liptzin, T.S. Rosenstock, V.B. Jensen, A.D. Hollander, *Nitrogen Sources and Loading to Groundwater* (California State Water Resources Control Board, California, 2012)

B.G. Volkmer, B. Ernst, J. Simon, R. Kuefer, G.J. Bartsch, D. Bach, et al., Influence of nitrate levels in drinking water on urological malignancies: A community-based cohort study. BJU Int. **95**(7), 972–976 (2005)

F.T. Wakida, D.N. Lerner, Non-agricultural sources of groundwater nitrate: A review and case study. Water Res. **39**, 3–16 (2005)

S.A. Waller, K. Paul, S.E. Peterson, J.E. Hitti, Agricultural-related chemical exposures, season of conception, and risk of gastroschisis in Washington State. Am. J. Obstet. Gynecol. **202**, e241–e246 (2010)

R. Walker, The metabolism of dietary nitrites and nitrates. Biochem. Soc. Trans. **24**(3), 780–5 (1996)

G. Walton, Survey of literature relating to infant methemoglobinemia due to nitrate-contaminated water. Am. J. Public Health Nations Health **41**, 986–996 (1951)

M.H. Ward, K.P. Cantor, D. Riley, S. Merkle, C.F. Lynch, Nitrate in public water supplies and risk of bladder cancer. Epidemiology **14**(2), 183–190 (2003)

M.H. Ward, S.D. Mark, K.P. Cantor, D.D. Weisenburger, A. Correa-Villasenor, S.H. Zahm, Drinking water nitrate and the risk of non-Hodgkin's lymphoma. Epidemiology **7**(5), 465–471 (1996)

M.H. Ward, J.A. Rusiecki, C.F. Lynch, K.P. Cantor, Nitrate in public water supplies and the risk of renal cell carcinoma. Cancer Causes Control **18**(10), 1141–1151 (2007)

P.J. Weyer, J.R. Cerhan, B.C. Kross, G.R. Hallberg, J. Kantamneni, G. Breuer, et al., Municipal drinking water nitrate level and cancer risk in older women: The Iowa Women's health study. Epidemiology **12**(3), 327–338 (2001)

WHO, *Rolling Revision of the WHO Guidelines for Drinking-Water Quality* (World Health Organization, Geneva, 2004)

P.D. Winchester, J. Huskins, J. Ying, Agrichemicals in surface water and birth defects in the United States. Acta Paediatr. **98**, 664–669 (2009)

R.F. Wright, C. Beier, B.J. Cosby, Effects of nitrogen deposition and climate change on nitrogen runoff at Norwegian boreal forest catchments: The MERLIN model applied to Risdalsheia (RAIN and CLIMEX projects). Hydrol. Earth Syst. Sci. **2**, 99–414 (1998)

WWAP- World Water Assessment Program. (2009). The United Nations World Water Development Report 3: Water in a Changing World. 2009: UNE

C.Y. Yang, D.C. Wu, C.C. Chang, Nitrate in drinking water and risk of death from colon cancer in Taiwan. Environ. Int. **33**(5), 649–653 (2007)

N. Yousefi, A. Fatehizedeh, K. Ghadiri, N. Mirzaei, S.D. Ashrafi, A.H. Mahvi, Application of nanofilter in removal of phosphate, fluoride and nitrite from groundwater. Desalin. Water Treat. **57**(25), 11782–11788 (2016)

C. Zeman, L. Beltz, M. Linda, J. Maddux, D. Depken, J. Orr, P. Theran, New questions and insights into nitrate/nitrite and human health effects: A retrospective cohort study of private well users' immunological and wellness status. J. Environ. Health **74**, 8–18 (2011)

G. E. Zinman, S. Zhong, Z. Bar-Joseph, Biological interaction networks are conserved at the module level. BMC Syst. Biol. **5**, 134 (2011)

Chapter 15
Catastrophic Effects of Climate Change on Child Health

S. Suriya, V. Lavanya, M. Madhan Kumar, A. Mohan, and S. Anandha Kumar

Abstract Climate change is the foremost challenge faced by humans in this century. The threat of flooding, as a repercussion of climate change and growing impervious surfaces, is increasing day by day. Global climate change and hydrological extremes with high intensity and frequency have resulted in an upsurge in health risks. It is essential to integrate climate change, flooding, and health risks to develop a systematic adaptation strategy. Kannagi Nagar, located on the outskirts of Chennai city, Tamil Nadu, India, was chosen as the study area to assess the health risks due to climate change and flooding. During monsoons, the study area would transform into an island as it is located on the banks of the Buckingham Canal. Trash is littered throughout the area because there is little regard for environmental hygiene and sanitation. Local socioeconomic, demographic, and ecological conditions prevailing in the study area are also the drivers of the outbreak of diseases. In this study, the severities of floods due to climate change are assessed to identify the health risks. The health impacts on different age groups are analysed. This study shows the essential role of a health promotion agency and emphasizes that there is an urgent need for ongoing support and development work to reduce the health hazards in that area.

Keywords Climate change · Flooding · Health risk assessment

S. Suriya (✉)
Department of Civil Engineering, Jerusalem College of Engineering, Chennai, India

V. Lavanya
Department of Science & Humanities, Tagore Engineering College, Chennai, India

M. Madhan Kumar
Department of Civil Engineering, Saveetha Engineering College, Chennai, India

A. Mohan
Department of Civil Engineering, Easwari Engineering College, Chennai, India

S. Anandha Kumar
Department of Civil Engineering, Aditya Engineering College,
Surampalem, East Godavari (D.T), Andhra Pradesh, India

B. Panneerselvam et al. (eds.), *Climate Change Impact on Groundwater Resources*, https://doi.org/10.1007/978-3-031-04707-7_15

15.1 Introduction

Climate change is one of the most important challenges that humanity has been facing in recent years. Climatic changes include temperature variation, sea-level rise, and high precipitation events. These changes are due to natural and human-influenced anthropogenic activities. Climate change has a significant impact on water resources like extreme precipitation events, increase in sea levels, and storm surges, that is, flooding is expected to become more severe and frequent due to climate change effects (Nandini and Suriya 2020; Balamurugan et al. 2020c; Panneerselvam et al. 2021a). On average, urban watersheds lose 90% of storm rainfall as runoff, whereas rural watersheds retain 25% of the rainfall. Significant increases in peak discharge have been recorded in urban watersheds, which may result in a reduction of flood duration (Sheng and Wilson 2009). Due to the conversion of water bodies into agricultural fields and impermeable areas like roads, parking lots, and buildings, floods are exacerbated in metropolitan areas. The influence of climate change on flooding necessitates harmonization among various countries, policy-makers, and stakeholders (Broto & Bulkeley 2013; Balamurugan and Balakumaran 2015; Balamurugan and Kumar 2016).

Here, flood risk is defined in terms of the risk to humans and the human society and is seen as a product of the severity and probability of occurrence of flood hazards and the vulnerability of the population/system (Brooks 2003). Factors influencing the severity of the hazards are depth and duration of inundation, velocity, rate of rising, frequency of occurrence, and season (Kumar and Balamurugan 2018; Panneerselvam et al. 2021b). Apart from the immediate risk of flooding like drowning, floods tend to flush out harmful chemicals and contagious microorganisms from the soil and carry these to wherever the water flows as per gravity (Hunter 2003; Kistemann et al. 2002; Rose et al. 2001). A flood, one of the natural hazards, poses multiple risks to public health in flood plain areas as well as in residential areas along its pathway (Mirza 2011; Kumar and Balamurugan 2019; Muniraj et al. 2021). The pathway of health risk due to flood hazards as a result of climate change impact is shown in Fig. 15.1.

In terms of occurrence, health effects can be categorized as:

- Those happening during or immediately after the flooding.
- Those developing in the days or early weeks following the flooding.
- Longer-term effects, which may appear after and/or last for months or years.

All these categories can be subdivided into direct and indirect health effects (Copenhagen and Rome 2002). According to the Intergovernmental Panel on Climate Change (IPCC 2014), the health hazards linked to flooding are of serious concern, and additional research is needed on the flood-related non-fatal health consequences.

Floods, as stated in many studies, play a significant role in aggravating public health issues, such as the development of waterborne infectious diseases like

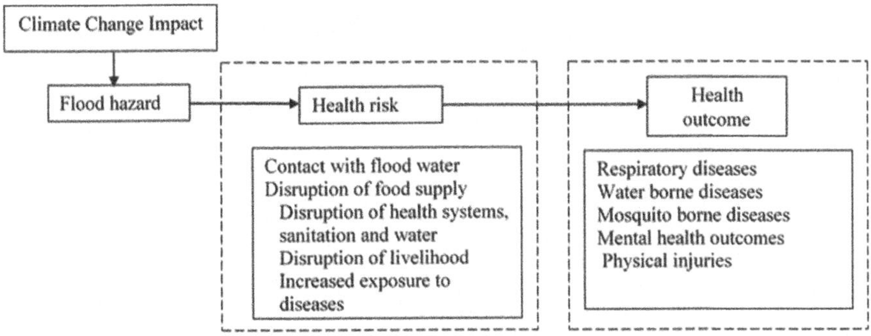

Fig. 15.1 Pathway of health risk due to flood hazards

diarrhoea (Watson et al. 2007). Only a few studies have been identified for analysing the causal relationship between floods and diarrhoea (Liao et al. 2019; Balamurugan et al. 2020a; Panneerselvam et al. 2020a). Given the prospect that flood hazards will increase because of climate change, it is necessary to make a strategic assessment of the existing knowledge base on flood risks, health, and the factors associated with diarrhoea. This study tries to describe and analyse the health consequences and responses to flooding in Kannagi Nagar, Chennai, and attempts to study the prevalence of diarrhoea in children of the 0–14-year age group.

15.2 Description of the Study Area

Kannagi Nagar is located between the latitude of 12⁰55′35” N and the longitude of 80⁰14′ 23″ E. The Google image of the study area is shown in Fig. 15.2. Kannagi Nagar is a waterlogged area located near the Buckingham Canal in Chennai, India. Since it is a flood-prone area, the Tamil Nadu Slum Clearance Board raised the ground level by 7 feet and constructed 15,000 houses. This area is allotted for people who had earlier encroached on the riverbank, canal bank, and other congested areas in Chennai. There is no proper maintenance of infrastructure in this area due to a lack of awareness and the violent and vandalizing tendencies of the people. Lack of knowledge about the sanitation and hygienic conditions among the people makes the area difficult to live in. Moreover, during monsoons, this area is cut off from all communication facilities due to floods. These details are collected based on observational techniques and preliminary investigation. People face the problem of health risks during heavy rains (Balamurugan et al. 2020b; Panneerselvam et al. 2020).

Flood-related injuries may occur in the pre-onset, onset, and post-onset phases. In the pre-onset and onset phases, injuries may be sustained when individuals are attempting to remove themselves, their family, or valued possessions from the

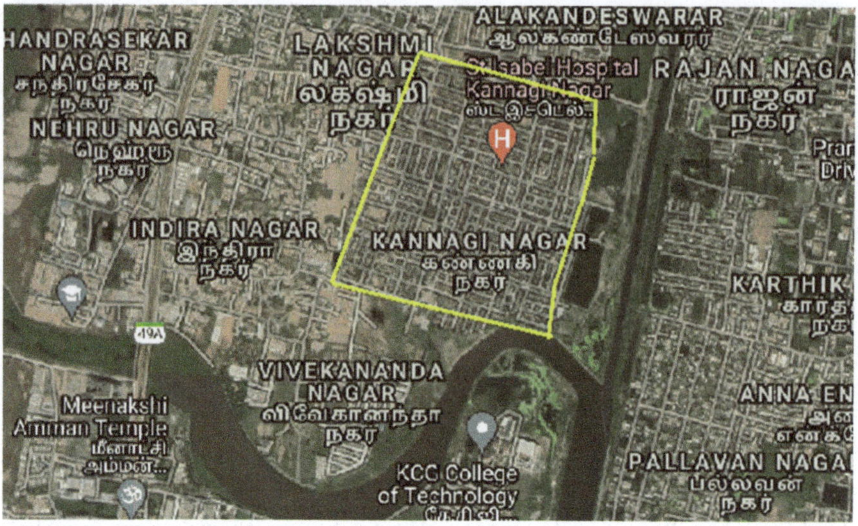

Fig. 15.2 Google image of the study area. (Source: Google Earth)

approaching waters. These injuries are relatively minor. There is more potential for injuries in the post-onset phase when people return to their homes and businesses and begin the clean-up operation. In this phase, particular care should be taken with buildings, which may have become unstable, and electrical power cables present a potential hazard. Vector-borne diseases are transmitted by mosquitoes, which breed in, or close to, stagnant or slow-moving water in the study area. There is no proper drainage system in Kannagi Nagar, and, thus, the rainwater stagnates, which increases the transmission of diseases. This research reveals that a greater complex relationship exists between the socioeconomic condition, flooding, and diarrhoeal prevalence.

15.3 Methodology

A reconnaissance survey was carried out in the study area during the 2015 floods in Chennai. Primary data such as population details and age of the residents and data regarding health during floods were collected from the primary health centre. The study instrument was designed based on literature reviews (Lavanya 2013; Balamurugan et al. 2020; Panneerselvam et al. 2021). The principal actors involved in the study were individual people, community-based organizations, local health-care providers (at the primary health centre), local water and sanitation providers, and non-governmental organizations (NGOs). Individuals included in this study were randomly selected. After the rigorous field visit and discussions with the above-mentioned key informants and panchayat board officials, the questions and variables for a questionnaire survey (QS) were modified. The main purpose of conducting a QS was to identify the major issues prevailing in the study area. The QS consists of three sections, and the data collected were analysed using SPSS software.

1. The first section covers socioeconomic data (age, the total number of family members, income, mother's education, mother's occupation, father's occupation, etc.).
2. The second section contains details on the flood inundation level, material damage, flooding depth, period of stagnation, effects of the flood on human health, etc.
3. The third section comprises water accessibility, perception of water quality and quantity, household drinking water storage days, and symptoms of diarrhoeal prevalence (confirmed by verifying medical prescriptions and the questionnaire survey).

A sample size of 250 households from Kannagi Nagar was chosen based on the availability of children aged 0–14 years using stratified random sampling. The sample also included people affected by floods, in proximity to floods, or subjected to flood risks. The same questionnaire was administered for the two phases orally to the respondents, mostly housewives during the floods, and the results were recorded. The mode of questionnaire administration was a face-to-face interview with their mothers. Relevant information about the child in the family was obtained from the respondent using a pretested structured questionnaire in the local language. If there was more than one child in the family, one of them was randomly chosen. The data on diarrhoeal episodes were obtained during the interview with a recall period of 2 weeks, as recommended by the World Health Organization (WHO) (1984). Then, the water samples used for the drinking and domestic purposes of those households were collected and analysed for an indicator organism (*Escherichia coli*).

15.4 Results and Discussion

Based on the views expressed by the residents during stakeholder meetings and the questionnaire survey, it was found that basic amenities like drinking water facility, a proper road network, drainage facility, ration shops (public distribution), and diaries are inadequate. During the monsoon, the polluted water gets mixed with the drinking water stored in the underground sump and thus is not fit for drinking. The area is squalid due to garbage dumping. There is only one primary health centre, and the availability of the doctor is minimal. Due to inadequate infrastructure and poor sanitation facilities, the number of people availing education for their children is minimal. Children have to go for about 3 km to get their education. Although the Tamil Nadu Slum Clearance Board has raised the ground level by about 7 feet as mentioned earlier, there is a problem of flooding in that area as shown in Fig. 15.3.

The effects of flooding on human health are tabulated in Table 15.1.

Diarrhoeal disease is a major disease found in Kannagi Nagar. The diarrhoeal disease can be caused by both viral and bacterial pathogens. This disease is

Fig. 15.3 Photographs taken during the 2015 Chennai floods in the study area. (**a**) Flood extent in the 25th main road (**b**) Flood extent in the seventh main road. (**c**) Flood extent in the 63rd cross street. (**d**) Flood inundation in the rear side of the 23rd cross street

associated with seasonal rains and flooding. People in this area drink water contaminated with pathogens (water mixed with polluted stormwater). Vector-borne diseases are transmitted by mosquitoes, which breed in, or close to stagnant or slow-moving water. There is no proper drainage in the Kannagi Nagar, the rainwater stagnates which increases the transmission of diseases.

In Kannagi Nagar, almost the same kind of socioeconomic group exists. The source of water is the Metro water supply, which is supplied only once in 5 days, so people store water for 5 days and perform household treatments on the water, such as boiling and filtering using a cloth or netted filter. The target group was children aged 0–14 years, and the respondents were their mothers. Samples of water used for drinking and domestic purposes were collected from the households. Diarrhoea is a symptom of intestinal infection which is caused by a host of bacterial, viral, and parasitic pathogens (Stanly et al. 2009). Diarrhoeal episodes in young children cause malnutrition. The WHO and the United Nations Children's Fund (UNICEF) (WHO and UNICEF 2000) have reported that four billion cases of diarrhoea occur every year. The health risks due to Acute Diarrhoeral Disease (ADD) likely depend on the socioeconomic condition of a household. Socioeconomic stratification is

Table 15.1 Effects of floods on human health

Direct effects		Indirect effects	
Causes	Implications	Causes	Implications
Contact with water	Respiratory diseases	Damage to water supply systems; damage to sewerage and sewage disposal systems; insufficient supply of drinking water; insufficient supply of water for washing	Possible waterborne infections; dermatitis; conjunctivitis
Contact with polluted water	Wound infections; gastrointestinal illnesses; ear, nose, and throat infections; possible serious waterborne diseases	Disruption of transport systems	Food shortages; disruption of emergency response
Clean-up activities following flooding	Electrocution; injuries	Disruption of social networks; loss of property, jobs, and family members and friends	Possible psychosocial disturbances
		Increase in physical and emotional stress	Increased susceptibility to psychosocial disturbances

relevant in several aspects of public health research. Variations in normal preventive health behaviour tend to be more highly correlated with income, education, and occupation rather than with other attributes (Green 1968).

Hygiene measures or behaviours substantially reduce the infection, which was proved by several studies published from 1980 to 2000. As such, a linkage may exist between the socioeconomic conditions, *E. coli* contamination in drinking water, and ADD prevalence in suburban communities, which are exposed to the dangers of development pressure. This chapter aims to assess the risk of ADD due to the 2015 flooding in Chennai. The focus of this study is to examine the impact of the flood on the health of the communities affected and to explore factors associated with episodes of diarrhoea. We conducted structured interviews with 250 individuals, and water samples were collected from all the interviewed households and analysed for an indicator organism (*E. coli*). *E. coli* was enumerated by placing an aliquot of 1 mL onto Petri films for *E. coli*/coliform colony count. The plates were then incubated for 48 hours at 37 °C. Blue-coloured colonies with gas entrapment on Eosin Methylene Blue agar (EMB) plates were presumed to be *E. coli*. The diarrhoeal prevalence of different intensities on exposure to flooding is shown in Table 15.2. In Fig. 15.4, the mean concentration of *E. coli* in the water samples shows that the significant relationship between the intensity of flood exposure and the contamination of *E. coli* ranges from 0 to 60 cfu/ml. Figure 15.5 depicts the mean concentration of *E. coli* contamination and diarrhoea.

In total, 133 children had diarrhoeal symptoms, and, in that, 78 children were residing on the ground floor. It was also found from the results that the occurrence of diarrhoea is less in low-level flood exposure.

Table 15.2 Diarrhoeal prevalence of different intensities on exposure to flooding

Intensity	Low	Moderate	High
With diarrhoea	20	35	78
Without diarrhoea	34	58	25
Total no. of houses	54	93	103

N = 250

Note: High-exposure houses on the ground floor, moderate-exposure houses on the second floor, and low-exposure houses on the third floor

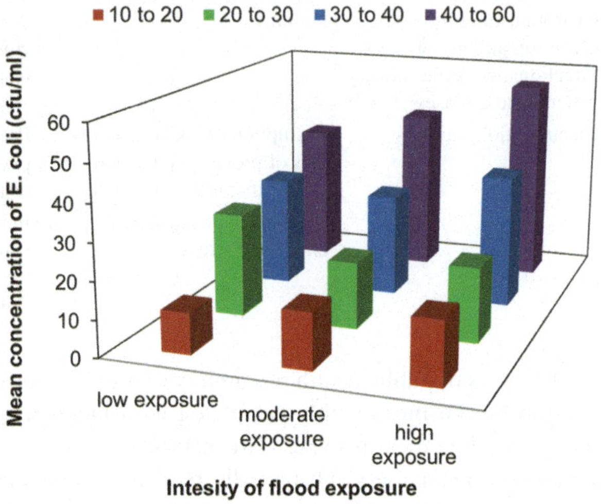

Fig. 15.4 Comparison of the mean concentration of *E. coli* with the intensity of flood exposure

From Fig. 15.4, it is understood that contamination is high in the highest intensity of flood exposure. This may be due to water stagnation and storage of water inside the house.

From Fig. 15.5, it can be seen that the occurrence of diarrhoea ranged from 0 to 60 cfu/mL. In all, 70 households lie in a mean *E. coli* concentration of 40–60 cfu/ml where diarrhoeal cases are also high during floods. Surface water has a higher level of contamination in the study area due to the supply of surface water from the Chennai Metropolitan Water Supply and sewerage Board, which augments the diarrhoeal prevalence.

In the study area, Metro water is the main source of water for domestic purposes. During the rainy and flood season, people can either boil or filter the water before giving it to the child. Table 15.3 illustrates the different treatment methods adopted by the community during floods. The data are analysed for frequency distribution using SPSS software. The results reveal that diarrhoeal prevalence in children was observed post floods, even after adopting the boiling and filtration technique. The main reason for the diarrhoeal incidence is improper storage of water after boiling and filtration, which leads to recontamination. In all, 108 households did not use

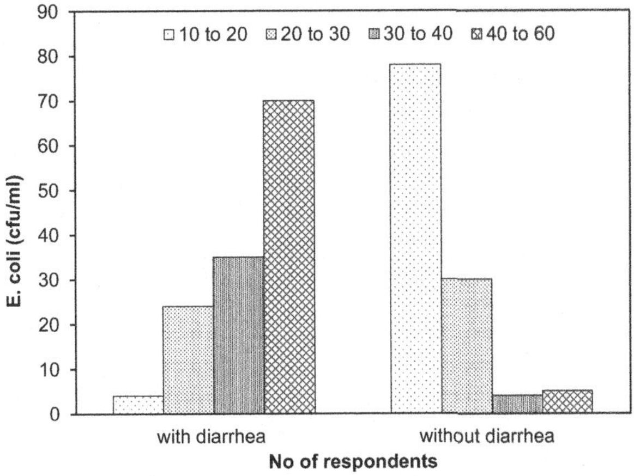

Fig. 15.5 Mean concentration of *E. coli* contamination and diarrhoea

Table 15.3 Comparison of flood intensity and different methods of water treatment with diarrhoeal incidence (with diarrhoea and without diarrhoea)

The intensity of exposure to flood	With diarrhoea				Without diarrhoea			
	Treatment of water before giving it to the child				Treatment of water before giving it to the child			
	No treatment	Boiling	Filtering	Purchase of cane water	No treatment	Boiling	Filtering	Purchase of cane water
Low	13 (65%)	2 (10%)	5 (25%)	0	4 (11.7%)	16 (47%)	7 (20.5%)	7 (20.5%)
Moderate	28 (80%)	3 (8.5%)	3 (8.5%)	1(8%)	8 (14%)	18 (31%)	12 (20.6%)	20 (4.5%)
High	53 (68%)	9 (15%)	16 (20.5%)	0 (0%)	2 (8%)	20 (80%)	3 (12%)	0.00%

any treatment method, and diarrhoeal cases were recorded in 94 households. Only one case was observed among the people who consumed cane water. In all, 14 households with diarrhoea had adopted the boiling technique, but it was not effective due to the inappropriateness in handling water.

15.5 Conclusion

Flooding is one of the most common climatic hazards, which poses a variety of health risks. Given that the possibility of flooding will increase because of climate change, this study assessed the health risks in Kannagi Nagar using primary data collection, QSs, and key informant meetings. It was found from the results that diarrhoea (53%) was the major cause of illness in children of the age group 0–14 years. Other health risks were fever (28%) and respiratory infections (19%). Efforts need to be directed towards reducing the impact of flooding on the health of infants and preschool children. Hence, solid attention should be paid to these vulnerable groups of people to improve their basic amenities, and this is the most important preventive measure to be implemented following flooding, in order to reduce the risk of outbreaks of waterborne diseases. Chlorination of water is an effective method for disinfecting drinking water, and there is a need to set up more health centres to reduce health hazards in this area.

References

P. Balamurugan, S. Balakumaran, Soil quality assessment around magnesite mines and Salem township using GIS techniques. Int. J. Adv. Eng. Technol. **8**(1), 1997 (2015)

P. Balamurugan, P.S. Kumar, Quality of ground water assessment in Salem District using GIS techniques. Adv. Nat. Appl. Sci **10**(3), 22–32 (2016)

P. Balamurugan, P.S. Kumar, K. Shankar, Dataset on the suitability of groundwater for drinking and irrigation purposes in the Sarabanga River region, Tamil Nadu, India. Data Brief **29**, 105255 (2020)

P. Balamurugan, P.S. Kumar, K. Shankar, R. Nagavinothini, K. Vijayasurya, Non-carcinogenic risk assessment of groundwater in southern part of Salem District in Tamilnadu, India. J. Chil. Chem. Soc. **65**(1), 4697–4707 (2020a)

P. Balamurugan, P.S. Kumar, K. Shankar, R. Nagavinothini, P.J. Sajil Kumar, Impact of climate and anthropogenic activities on groundwater quality for domestic and irrigation purposes in Attur region, Tamilnadu, India. Desalin. Water Treat. **208**, 172–195 (2020b)

P. Balamurugan, G. Shunmugapriya, R. Vanitha, GIS based assessment of ground water for domestic and irrigation purpose in Vazhapadi Taluk, Salem, Tamil Nadu, India. Taiwan Water Conservancy **68**(2), 1–10 (2020c)

Brooks N., Vulnerability, risk and adaptation: A conceptual framework. Tyndall centre for climate change research, Working paper 38, 1–20 (2003)

V.C. Broto, H. Bulkeley, A survey on urban climate change experiments in 100 cities. Glob. Environ. Chang. **23**, 92–102 (2013)

Copenhagen and Rome, Flooding: Health effects and preventive measures. Fact Sheet **05**, 1–7 (2002)

L.W. Green, *Status Inconsistency, Reference Group Theory and Preventive Health Behavior*. Doctoral dissertation, School of Public Health, University of California, Berkeley (1968)

P.R. Hunter, Climate change and waterborne and vector-borne disease. J. Appl. Microbiol. **94**(Suppl), 37S–46S (2003). [PubMed: 12675935]

IPCC, Climate change 2014: Synthesis report. Contribution of working groups I, II and III to the fifth assessment report of the intergovernmental panel on climate change [Core Writing Team, R.K. Pachauri and L.A. Meyer (eds.)]. IPCC, Geneva, Switzerland (2014)

T. Kistemann, T. Classen, C. Koch, F. Dangendorf, R. Fischeder, J. Gebel, V. Vacata, M. Exner, Microbial load of drinking water reservoir tributaries during extreme rainfall and runoff. Appl. Environ. Microbiol. **68**(5), 2188–2197 (2002). [PubMed: 11976088]

P.S. Kumar, P. Balamurugan, Evaluation of groundwater quality for irrigation purpose in attur taluk, Salem, Tamilnadu, India. Water Energy Int. **61**(4), 59–64 (2018)

P.S. Kumar, P. Balamurugan, Suitability of ground water for irrigation purpose in Omalur Taluk, Salem, Tamil Nadu, India. Indian J. Ecol. **46**(1), 1–6 (2019)

V. Lavanya, Environmental determinants of ADD health risk for children in peri-urban areas. Unpublished PhD. thesis submitted to Anna University, Chennai (2013)
W. Liao, L. Yang, C. Huang, The causal effects of flooding on infectious diarrheal diseases during and after flood and the related social modifiers in Anhui province, China. Environ. Epidemiol. **3**, 239 (2019)
M.M.Q. Mirza, Climate change, flooding in South Asia and implications. Reg. Environ. Chang. **11**, 95–107 (2011). https://doi.org/10.1007/s10113-010-0184-7
K. Muniraj, B. Panneerselvam, S. Devaraj, C.J. Jesudhas, K. Sudalaimuthu, Evaluating the effectiveness of emissions reduction measures and ambient air quality variability through ground-based and sentinel-5P observations under the auspices of COVID pandemic lockdown in Tamil Nadu, India. Int. J. Environ. Anal. Chem, 1–12 (2021)
K. Nandini, S. Suriya, A review of potential impact of climate change on global water resources. J. Seybold Rep. **15**(7), 495–501 (2020)
B. Panneerselvam, S. Karuppannan, K. Muniraj, Evaluation of drinking and irrigation suitability of groundwater with special emphasizing the health risk posed by nitrate contamination using nitrate pollution index (NPI) and human health risk assessment (HHRA). Hum. Ecol. Risk Assess. Int. J. **27**(5), 1324–1348 (2020)
B. Panneerselvam, K. Muniraj, C. Pande, & N. Ravichandran. Prediction and evaluation of groundwater characteristics using the radial basic model in semi-arid region, India. Int. J. Environ. Anal. Chem., 1–17 (2021)
B. Panneerselvam, K. Muniraj, M. Thomas, N. Ravichandran, GIS-based legitimatic evaluation of groundwater's health risk and irrigation susceptibility using water quality index, pollution index, and irrigation indexes in semiarid region, in *Groundwater Resources Development and Planning in the Semi-Arid Region*, (Springer, Cham, 2021a), pp. 239–268
B. Panneerselvam, K. Muniraj, M. Thomas, N. Ravichandran, B. Bidorn, Identifying influencing groundwater parameter on human health associate with irrigation indices using the Automatic Linear Model (ALM) in a semi-arid region in India. Environ. Res. **202**, 111778 (2021b)
B. Panneerselvam, S.K. Paramasivam, S. Karuppannan, N. Ravichandran, P. Selvaraj, A GIS-based evaluation of hydrochemical characterisation of groundwater in hard rock region, South Tamil Nadu, India. Arab. J. Geosci. **13**(17), 1–22 (2020a)
J.B. Rose, P.R. Epstein, E.K. Lipp, B.H. Sherman, S.M. Bernard, J.A. Patz, Climate variability and change in the United States: Potential impacts on water- and foodborne diseases caused by microbiologic agents. Environ. Health Perspect. **109**(Suppl 2), 211–221 (2001). [PubMed: 11359688]
J. Sheng, J.P. Wilson, Watershed urbanization and changing flood behaviour across the Los Angeles metropolitan region. Nat. Hazards **48**, 41–57 (2009)
M.A. Stanly, B.W.C. Sathiyasekaran, G. Palani, A Population Based Study of Acute Diarrhoea Among Children Under 5 Years in a Rural Community in South India. Sri Ramachandra Journal of Medicine, vol.1H. Poor, An Introduction to Signal Detection and Estimation; New York: Springer-Verlag, 1985, ch. 4, (2009)
J.T. Watson, M. Gayer, M.A. Connolly, Epidemics after natural disasters. Emerg. Infect. Dis., **13**, 1–5 (2007). Available from: http://www.cdc.gov/eid
WHO and UNICEF, Global Water Supply and Sanitation Assessment 2000 Report, United Nations, New York (2000)
WHO Guidelines for a sample survey of diarrhoeal disease morbidity, mortality and treatment rates. World Health Organisation Diarrheal Control Program Document, WHO/CDD/SER, Geneva. pp 84–86 (1984)

Chapter 16
Mitigation Measures of Fluoride- and Nitrate-Contaminated Regions

Ushani Uthirakrishnan, Anitha Karthik, Jayachandran Krishna, Tamilarasan Karuppiah, Pratheba Sukumar, and S. Anandha Kumar

Abstract Groundwater contamination, which has become one of the increasing problems globally, can be caused by both natural and man-made factors. Over the years, these contaminations have accumulated as a result of numerous geological processes, ion exchange reactions, nitrates, nitrate- and fluoride-bearing aquifers, and leaching of toxic industrial wastes from household chemicals, dyes, and pesticides. Fluoride and nitrate consumption through untreated groundwater by children and adults causes short- to long-term health problems ranging from dental (weak teeth) to skeletal (bone and joint damage) fluorosis to blood pressure to heart issues. This review chapter will investigate the critical climate change factors compounding the toxin levels, conventional treatment methods, concepts, and potentialities of the current measures in place to mitigate these contaminations. Based on the continuous monitoring, the mitigation measure which is obsolete can be updated with new technologies that mitigate the current toxin levels which in turn minimise the

---

Ushani Uthirakrishnan and Anitha Karthik contributed equally with all other contributors.

U. Uthirakrishnan (✉)
Department of Biotechnology, Karpaga Vinayaga College of Engineering and Technology, Chinnakolambakkam, Chengalpattu, Tamil Nadu, India

A. Karthik
Edinburgh Napier University, Edinburgh, UK

J. Krishna
Centre for Biotechnology, Anna University, Chennai, India

T. Karuppiah
Department of Civil Engineering, Vel Tech Rangarajan Dr. Sagunthala R & D Institute of Science and Technology, Avadi, Chennai, Tamil Nadu, India

P. Sukumar
Department of Civil Engineering, Karpagam Academy of Higher Education, Eachanari, Coimbatore, Tamil Nadu, India

S. Anandha Kumar
Department of Civil Engineering, Aditya Engineering College, Surampalem, East Godavari (D.T), Andhra Pradesh, India

B. Panneerselvam et al. (eds.), *Climate Change Impact on Groundwater Resources*, https://doi.org/10.1007/978-3-031-04707-7_16

direct impacts such as health risks from fluoride and nitrate contaminated regions. The findings and the suggested measures are anticipated to benefit the state and local governments and regulating bodies that ensure the health and safety of the residents.

Keywords Fluoride and nitrate contamination · Groundwater pollution · Mitigation measures · Health risks

16.1 Introduction

One-third of the global population is widely dependent on groundwater for drinking and other purposes (International Association of Hydrogeologists 2020). Being one of the renewable sources, groundwater is found in the surface soil, which retains rainwater filling lakes and streams to become a vital water source for public and private supplies, farm animals, irrigation, manufacturing, thermoelectricity generation, and various other purposes (National Ground Water Association 2018). About 6.8 billion people from the global population, utilise the basic services of improved drinking-water which is mainly from the groundwater resource. In 2017, a study conducted by the World Health Organization (WHO) revealed that 435 million people were consuming water from unprotected wells and 144 million were consuming untreated surface water primarily from ponds, lakes, streams, and rivers. Apparently, natural processes such as global warming and climate change and anthropogenic factors like urbanization, usage of fertilizers, and inadequate waste management are largely altering the quality of groundwater, and, as a result, future generations may have to live in water-stressed areas (World Health Organization 2019). Many countries have already started to use wastewater for agricultural and irrigation purposes. This means that all the chemical substances used in the form of pesticides, insecticides, and fertilizers may end up on the surface and subsurface, thus polluting the groundwater. Issues related to groundwater contamination from nitrates and fluorides have alarmingly become a global concern, especially in arid and semi-arid regions (Balamurugan et al. 2020, 2020a, Srivatsava 2020). Fatalities caused by nitrogen (N) from various sources of origin to surface water and groundwater streams have received excessive consideration in recent years. Noteworthy losses of N from farming have long been documented as the chief basis of pollution, and excess N diffusion to watercourses such as estuaries, lakes, and rivers has damaging environmental impacts as a result of eutrophication (Riemann et al. 2016). Overall, human beings are generating the combative nitrogen that impact the ground water source lead to a change in climate and ecological unit, considering the nitrogen as one of the most vital pollution that affects the humankind with numerous diseases. The European Nitrogen Assessment has recognized five main parameters of nitrogen pollution: air quality, ecosystems, biodiversity, greenhouse gas balance, and water quality.

16.1.1 Groundwater and Well Water Contamination

In the course of its flow, groundwater tends to pick up various naturally found minerals such as calcium, chloride, manganese, arsenic, fluoride, nitrate, iron, and sulphate between the rocks and soil. These minerals are absorbed by the groundwater, resulting in groundwater contamination. Figure 16.1 shows the five common ways in which groundwater or groundwater filling lakes and streams can be contaminated.

About 70% of the world's groundwater is utilized for agriculture and 38% for irrigation (National Ground Water Association 2018). Agricultural chemical usage, septic wastes, landfills, hazardous wastes, and atmospheric pollutants are further categorized as man-made contaminations. Lack of effective urban management and industrial and agricultural wastes may result in serious chemical contamination of groundwater (World Health Organization 2019). Wells that depend on artificial recharge from industries, storm runoff, and treated sewage tend to have the presence of high concentrations of nitrates and other synthetic chemical components (US Environmental Protection Agency 2015).

About 321 billion gallons of surface water is used per day (National Ground Water Association 2007). Intense agriculture has amplified the rise of riverine nitrogen and oxygen depletion, in general, in the majority of coastal zones (Balamurugan et al. 2020b; Cheng et al. 2020). Natural disasters such as earthquakes, tsunamis, storm surges, and inland flooding can also result in groundwater contamination through oil spills and sewer overflow, resulting in surface contamination. Saltwater intrusion with freshwater can even offset the fluctuating groundwater levels, resulting in adverse impacts on human, plant, and animal lives. Figure 16.2 is a pictorial representation of groundwater contamination from varied sources.

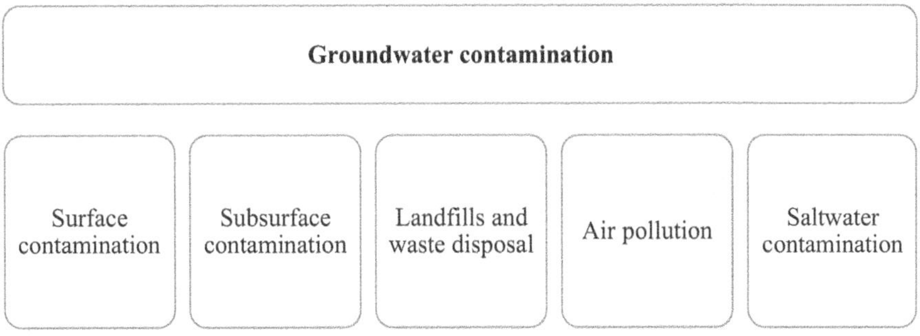

Fig. 16.1 Different types of contamination of groundwater

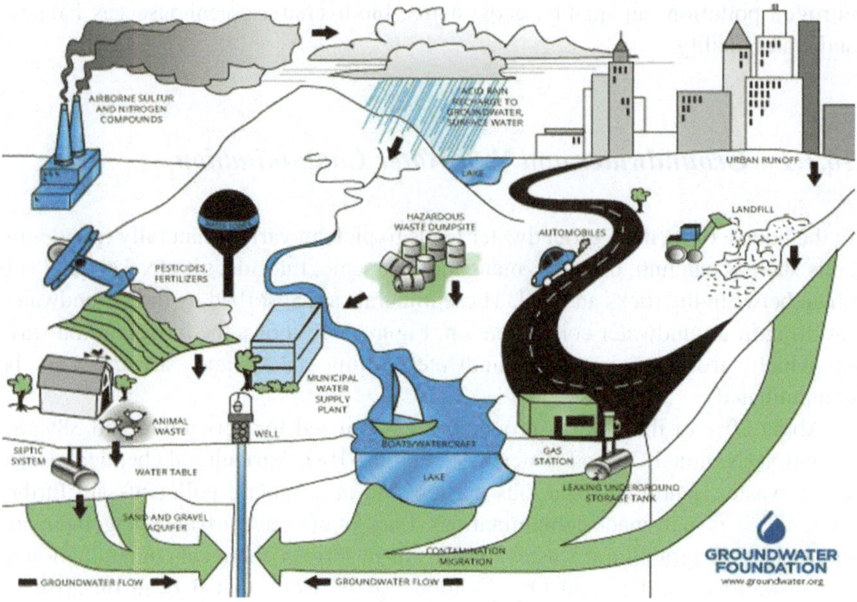

Fig. 16.2 Pictorial representation of groundwater contamination from varied sources (Groundwater Foundation 2021)

Hazardous waste disposals mixing with drinking water can impose serious health effects. Contamination from landfills and waste disposal sites can cause diseases like dysentery, diarrhoea, cholera, and even polio. Around two billion people utilize faecal-contaminated drinking water, causing 485,000 diarrhoeal deaths from contaminations (World Health Organization 2019). Improper disposal of animal and human wastes, disposal of chemical wastes and landfills from industries, septic tank leakages mixing with lakes and wells, and disposal of processed foods high in nitrates and fluorides can all cause groundwater contamination (Centers for Disease Control and Prevention 2015).

The surface soil absorbs the rainwater reaching the ground like a "sponge" and stores the water. Groundwater is generally contained in aquifers, which are permeable components that store larger quantities of water. Not all groundwater seeps through to reach the deep surface because it depends on various other factors such as geographical region, atmospheric factors, and the rate at which surface exploitation has taken place (WorldAtlas 2018). Well water is a common source of water for individual consumption, animals, and farmlands.

16.1.2 Groundwater Vulnerability of Nitrate Affecting Households, Industries, and Agricultural Systems

Nitrate contamination in groundwater has received global attention over the last decade, due to the threat of groundwater vulnerability to nitrates (Spalding and Exner 1993). There are various ways in which nitrates can blend with surface water and groundwater, which is a naturally occurring process at low concentrations. Nitrate (NO_3^-), a compound formed when nitrogen (N) combines with oxygen (O_2) or ozone (O_3), has the tendency to easily blend in water; however, it is tough to remove, particularly using household treatments such as boiling, filtering, and water softening methods (Mahler et al. 2007; Centers for Disease Control and Prevention 2015).

The end of the twentieth century paved the way for the introduction of the Haber–Bosch process during which artificial nitrogen gained significant attention, leading to a global chemical revolution. Many industries adopted this method, enabling large-scale production of ammonium nitrate fertilizers and pesticides for agricultural purposes (Smith et al. 2020). Synthetic fertilizers, pesticides, and manures act as growth nutrients for plants. To gain crop yields, surplus amounts of these chemical growth fertilizers and manures are utilized, which can result in nitrates being absorbed by the soil, eventually contaminating the groundwater (Kwon et al. 2021a, 2021b). Figure 16.3 shows crop dusting, which is one of the techniques in which pesticides and insecticides are sprayed on crops from air. This technique can be seen to be widely used in North Carolina (US Geological Survey n.d.).

Related problems have emerged such as the processing of the Haber–Bosch method, which demands usage of fossil fuels claiming responsibility for 1.2% of the global carbon emissions as more than 90% of ammonia nitrate is produced using this route (Smith et al. 2020). Nitrate contamination also occurs from the treated

Fig. 16.3 Crop dusting technique used to spray pesticides on agricultural lands of North Carolina (US Geological Survey n.d.)

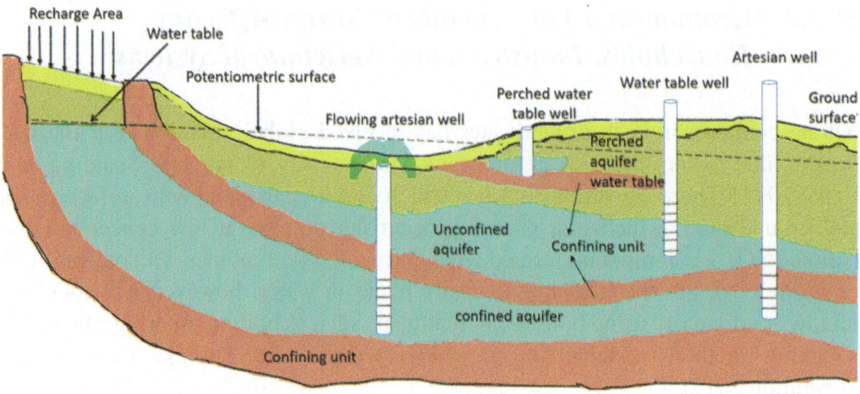

Fig. 16.4 Illustration of confined and unconfined aquifers

Table 16.1 Nitrate contamination levels and their health impacts (US Environmental Protection Agency, n.d.)

Contaminant	Maximum contaminant level goal (mg/L)	Maximum contaminant level or treatment technique (mg/L)	Health effects from prolonged exposure above the maximum contaminant level	Contamination source
Nitrate	10	10	Blue baby syndrome in infants below 6 months, cancer, thyroid, and impacts on livestock and aquatic health	Fertilizer waste; septic leakages
Nitrite	1	1	Blue baby syndrome in infants below 6 months	Fertilizer waste; septic leakages

wastewater from industries, septic tanks, or sewer leakages (Nejatijahromi et al. 2019).

Figure 16.4 Illustrates how recharged water reaches confined and unconfined aquifers. Aquifers are surface layers that aid in groundwater storing and transmission from surface water bodies to larger landscape areas (Burkartaus and Stoner 2008)

Even the most confined aquifers can be contaminated by various factors such as wellbore leakage or through direct transportation of nitrates from up/downgradient areas to confined aquifers (Panneerselvam 2020, 2021a, Kwon et al. 2021a, 2021b). Burkartaus and Stoner (2008) highlighted how shallow unconfined aquifers have high susceptibility when exposed to nitrate contamination and how, once nitrate-contaminated, shallow aquifers can take decades to improve their groundwater quality. They further underlined how well owners or farmland owners can be oblivious to the trend of nitrate contamination in wells. Richards et al. (1996) highlighted how understanding whether the contamination source is from aquifer vulnerability or from intercepted contaminated adjacent wells is critical. Findings from international research and evaluations offer key information on human-health related concerns; however, there is only limited information on aquifers that are the primary source of groundwater (Richards et al.1996).

Table 16.1 shows the nitrate and nitrite contamination levels and the direct health impacts from the intake of the contaminants and their possible sources. According to the US Public Health Service, 10 ppm of NO_3-N is a standard nitrate concentration level in unpolluted groundwater, and a level exceeding this standard value is a concerning factor. A recent study carried out by the US Environmental Protection Agency has identified that 2.7% of rural wells exceeded the standard nitrate levels. Many of these wells, which surpassed the standard levels, were identified to be in agricultural fields where high levels of nitrogen fertilizers were used (Mahler et al. 2007). These high levels may also be the resultant of surplus usage of nitrogen-rich fertilizers during farming, which causes fertilizer runoff into adjacent water bodies, lakes, and aquifers (Water Pollution 2020).

The US Geological Survey (USGS) has highlighted the consideration of the time lag between application of the chemical sediment infestations from pesticides and fertilizers and the typical time taken for them to be identified in freshwater or well water. They further advise on how the quality of deep groundwater will be affected over a period compared to the shallow groundwater, which gets immediate chemical residues from pesticide and fertilizer usage (US Geological Survey, n.d.). Nitrogen can also blend in air in the form of nitric acid, nitrate radicals, and inorganic aerosols through industrial discharges and transport vehicles (WHO 2011).

16.1.2.1 Global Nitrate Contamination Levels

According to Spalding and Exner (1993), nitrate is a "ubiquitous" contaminant identified in aquifers globally, and they emphasized on how nitrate contamination is becoming a growing concern in countries like Africa, the Middle East, Australia, New Zealand, and the USA. Nitrogen (N) is absorbed by plants during the synthesis process of nitrogen compounds, which seep through groundwater (WHO 2011). Nitrogen surplus is an indicator used to manage and control the nitrogen levels and their pollution in the soil and groundwater involved in the process of total agricultural balance.

The European countries have identified a steady rise and some of the areas have been identified to have twice the standard levels in the 1990s (WHO 2011). Between 1992 and 2012, nitrate levels in Europe were identified to be 17.5 mg/l, with the highest recorded in Malta (58.1 mg/l between 2002 and 2012). Countries like Finland and Denmark have shown decreased nitrate levels in drinking water; however, private wells and groundwater continue to have traces (Ward et al. 2018).

In the USA, it was estimated that more than 13 million households rely on private wells for drinking water, which is not controlled or monitored by the US Environmental Protection Agency (USEPA). Wells owners do not maintain any standards nor do they incorporate procedures to remove contamination if any from these wells and directly indulge in supplying these well waters as drinking supplies to rural areas (USEPA, n.d.). Private well owners have also not been mandated by the USEPA. In the USA, an assessment conducted in 1986 identified nitrate concentration levels of 44 mg/l above the standard level across 568 groundwater supplies (WHO 2011). In India, the nitrate levels were particularly high in the rural and agricultural areas with average levels being observed at 45.7 mg/l. On a global range, The Gaza Strip was observed to have nitrate levels at 500 mg/l in public wells (Ward et al. 2018).

16.1.3 Groundwater Vulnerability of Fluoride Affecting Households, Industries, and Agricultural Systems

Fluoride (F), which belongs to the halogen family, is comparatively the most reactive element because its oxidation level is −1 and it tends to occur in the form of both organic and inorganic compounds (Giwa et al. 2021). Fluorides are one of the vital nutrients for the human system, in particular, and 95% of the fluorides in humans are observed to be present in their teeth and bones. The physical properties of fluoride being colourless, odourless, and tasteless enable it to blend with water readily. This means that fluoride-contaminated water cannot be identified by its physical properties, except when going through a chemical analysis (Brindha and Elango 2011). Fluorides can reach the human system through water, food, toothpastes high in fluorides, and drugs (Kumar et al. 2019). Exposure to fluorides over a prolonged period can cause a disease called fluorosis, which is extremely dangerous to human health. Fluorosis can occur as dental fluorosis and skeletal fluorosis (Panneerselvam et al. 2021b; Yadav et al. 2019).

Table 16.2 shows the fluoride contamination levels and the direct health impacts from the intake of the contaminants and their possible sources. As we know, synthetic fluoride fertilizers are widely utilized in agriculture and the accumulative risks and exposure over a prolonged period may impose bone damage rather than supporting bone development.

Groundwater contamination of fluorides is primarily from rocks and may occur in the form of fluorite, fluorapatite, apatite, cryolite, fluormica, biotite, and other forms of rocks. Countries where agriculture is the primary source of income tend to use chemical fertilizers, which are high in fluorides and nitrates and fertilizer runoff

Table 16.2 Fluoride contamination levels and their impacts on health (US Environmental Protection Agency, n.d.)

Contaminant	Maximum contaminant level goal (mg/L)	Maximum contaminant level or treatment technique (mg/L)	Health effects from prolonged exposure above the maximum contaminant level	Contamination source
Fluoride	4.0	4.0	Pain and tenderness of the bones and mottled teeth in children	Fertilizer wastes and aluminium factory wastes

mixing with unconfined aquifers may also result in groundwater contamination. Fluoride contamination in groundwater may also occur due to the incineration of coal with volcanic ashes. Industries mainly dealing with cement production, brick and ceramic manufacturing, and improper waste disposal may also result in fluoride contamination in groundwater (Brindha and Elango 2011).

16.1.3.1 Global Fluoride Contamination Levels

In countries like India, where endemic fluorosis has been widespread since 1937, 19 different states, in which evapotranspiration of groundwater had occurred (Raj and Shaji 2017), were commonly identified. Fluoride concentrations in the arid-to-semi-arid states of Gujarat, Rajasthan, Andhra Pradesh, Karnataka, and Tamil Nadu were observed to be above the WHO's specified level (1.5 mg/l) (Indu et al. 2007); (Raj and Shaji 2017). In some parts of India, fluoride levels up to 2.88 mg/L beyond the WHO's suggested level were observed. Countries including India, China, Bangladesh, and Pakistan whose terrains were identified as hard rocky terrains were found to have been severely affected by arsenic and fluoride contamination in groundwater (Jha and Tripathi 2021; Yadav et al. 2019).

In African countries, mainly Sudan, Ethiopia, Uganda, Kenya, Tanzania, and Malawi, high fluoride concentrations beyond the suggested level were commonly observed. Particularly in Kenya, the observed levels were between 5 mg/l and 8 mg/l. Middle Eastern regions, mainly Iraq, Iran, Syria, Turkey, Algeria, and Morocco up to the Jordan Valley, were identified with high fluoride levels (Thole 2013). The National Water-Quality Assessment (NAWQA) and the USGS carried out a decadal assessment between 1992 and 2001, in which US streams and groundwater were identified to have at least one pesticide (US Geological Survey 2006). A majority of the people consuming fluoride-contaminated water are those living in tropical areas and zones particularly vulnerable to fluorides. The risk of exposure to fluorosis is observed to be higher in these tropical areas as their fluoride consumption is threefold higher than the average levels suggested by the WHO (Brindha and Elango 2011). Handa (1975) conducted a study to identify fluoride in groundwater. This chapter also revealed how these tropical zones and their groundwater contaminated to have both fluorides and nitrates at high concentration levels.

16.1.4 Impacts of Nitrates and Fluorides on Infants, Children, and Adults

16.1.4.1 The Blue Baby Syndrome

Food and water exposed to high levels of nitrates can cause various adverse effects, but the effects of long-term consumption of nitrates are still uncertain. Generally, the human digestive tract absorbs the nitrates consumed and excretes them through urination. This process is done rapidly before the body absorbs nitrates. However, infants below 6 months, who have less stomach acid to digest their food, could absorb more nitrates and eventually can be affected by serious nitrate poisoning. Infants primarily rely on gut bacteria to support their digestive process. These bacteria in the infant's digestive tract will further convert these nitrates into nitrites (NO_2). The toxic nitrites when entering the infant's blood stream interfere with the blood's ability to carry oxygen, leaving the infant in a condition called "methemoglobinemia" turning the skin around the eyes and mouth blue. This condition is otherwise called the "blue baby syndrome". This is considered as a fatal, when the infant's haemoglobin (Hb) converts into 70% methaemoglobin (metHb) (Mahler et al. 2007; Majumdar 2003).

16.1.4.2 Risk of Cancer in Adults

Although consumption of water with fluoride levels of 1.0 mg/l is claimed to have some beneficial effects, such as improvement in dental health, drinking this water beyond the World Health Organization's specified level of 1.5 mg/l could inflict negative effects on human health (Thole 2013). Nitrate infusion in drinking water has been identified to have elevated risks of different cancer types. A study conducted in Yogyakarta, Indonesia, discusses the link between nitrate-contaminated drinking water and colorectal cancer (CRC) in 75 CRC patients and identifies a significant correlation between nitrate and CRC development (Fathmawati et al. 2017). Nitrates have also been associated with thyroid, and the conversion of haemoglobin (Hb) to methaemoglobin (metHb) can also cause asphyxia and cyanosis in adults (Taneja et al. 2017). Cohn (1992) identified that the correlation between fluorides in drinking water has increased osteosarcoma (caused when fluorides are deposited in the human bones) in younger males.

16.1.4.3 Impact on Human Organs/System

Many countries whose primary source of drinking water is groundwater are clearly continuously exposed to both nitrates and fluorides. Untreated contaminated water imposes serious health impacts on babies, children, and adults. Fluoride concentration less than 0.5 mg/L can amplify the risk of tooth decay as it includes dental caries (Kumar et al. 2019). A study conducted by Yadav et al. (2019) identified how Asian countries like India, Pakistan, and China have been severely impacted by both dental and skeletal fluoroses. In addition, fluorosis also has potential impacts on human organs, mainly the heart, brain, and reproductive organs, and affects the gastrointestinal system (Yadav et al. 2019). Clinical studies suggest that exposure to excessive fluoride levels may even result in psychological dysfunction of mutagenesis, carcinogenesis, growth retardation, and even suppression of the human immune system (Kumar et al. 2019). Consuming fluoride-contaminated food and water may also influence the thyroid levels, resulting in hypothyroidism, as the thyroid-simulating hormones are the most sensitive tissues (Kumar et al. 2019). Fluorides reaching the placenta can permanently harm the foetus or can lead to infertility issues and underweighing neonates (Gupta et al. 2007). According to a study conducted by Choi et al. (2012), younger children exposed to excessive fluorides continuously over a period of time were identified to have lower intelligence quotient (IQ) levels compared to those children who were not exposed or living in low fluoride zones.

16.1.5 Prevention and Mitigation Programmes and Measures

According to the United Nation's Sustainable Development Goals (SDGs), clean accessible water for every living being on the planet is essential. To complement the UN's SDG, it is important to prevent and mitigate groundwater contaminations and make clean water accessible to everyone (United Nations, 2021). The Safe Drinking Water Act (SDWA), which was created in 1974 in the USA, monitors the data of public's drinking water (Ward et al. 2018); however, many low-income countries below poverty are still drinking contaminated water due to socioeconomic disparities. Institutional barriers could also impede executing adequate wastewater infrastructure and disposal management, particularly in low-income countries (Schaider et al. 2019). Many studies highlight how the quality of groundwater can be impacted from the past and the present land use factors such as agriculture, industries, households, etc., yet their use cannot be limited. Therefore, taking appropriate precautions to minimize the accumulation of these minerals from varied sources should be monitored and controlled at regular intervals. Identifying the significance – and restrictions – of nitrogen to our farming and ecological structures is vital for the victory of our agriculturalists and to battle climate change. Transformation is extremely time-consuming, but science, determination, and, finally, reliance on policies can lead to genuine transformation. On the other hand, it does take time.

16.1.5.1 Intense Agricultural Zones

Nitrate-infested drinking water cannot be removed by simply boiling the water; rather, there is a higher chance of increasing the nitrate content as some of the water gets evaporated during the process of boiling. Even the most common method of chlorination, which is used to remove bacteria, viruses, and microbes, becomes obsolete when it comes to removing nitrates from contaminated water. Therefore, to remove nitrates, any of the processes such as distillation, reverse osmosis, and ion exchange (Centers for Disease Control and Prevention 2015) can be used. Farmers of intensively farmed regions that are particularly vulnerable zones should consider implementing immediate measures to reduce the utilization of nitrogen and fluoride fertilizers and pesticides to minimize the nitrogen surplus and excess fluoride residues as runoff in groundwater (Howden et al. 2011). Similarly, fluoride-infested drinking water must also go through processes such as reverse osmosis, electrocoagulation, nanofiltration, ion exchange, precipitation, distillation, and adsorption. It is to be noted that, although there are existing technologies to remove fluoride contamination from water, new technologies to remove fluoride contamination are yet to be reviewed (Yadav et al. 2019; Thole 2013). It is important to identify all the potential sources that could cause contamination, for instance, farmers who use chemical fertilizers for their lands, septic tanks that connect or run adjacent to wells, unconfined aquifers, construction, or industrial process, and try to understand their waste disposal strategies, which might support sustainable community living.

16.1.5.2 Measures to Reduce Well Water Contamination

Private well owners who are supplying well water to rural areas for drinking purposes must take the required steps to test their wells at regular intervals. Measures should be taken to control the potential fertilizer runoff from adjacent agricultural lands or areas, which might tend to use synthetic fertilizers and pesticides, resulting in groundwater contamination seeping through wells and water systems. If any well is identified to have been contaminated, then the well owners should take appropriate measures to close the existing well and seek professional expertise in drilling new wells, as contaminated wells if left unattended or abandoned may also decrease the quality of the soil and groundwater. Proper well maintenance such as covering the well using a well- cap, sealing an unwanted outlets or inlets and other adaptive practices such as regular inspection should be followed based on the EPA regulation and being in touch with the corresponding environmental agencies ensuring their proper maintanace.

16.1.5.3 Minimizing Groundwater Pollution from Natural Disasters

Saltwater intrusion could occur, contaminating the freshwater sources, especially after the occurrence of natural disasters such as coastal flooding, torrential rainfall, or storm surges. In 2008, Cyclone Nargis triggered a funnell of about 16 ft. (4.9 m) through most of the creeks and channels which resulted in saltwater intrusion as seen in Fig. 16.5 encroaching around 43% of the freshwater bodies (Mohammed 2009).

Natural disasters like cyclones, tornadoes, flash floods, and torrential rainfall can contaminate groundwater and surface water by various means. A drainage system controlling surface water, especially during inundations from inland and coastal flooding, can become critical. This can also invoke outbreaks of infectious diseases, health issues, and epidemics, post the disaster. Prior action towards flood proofing wastewater treatment facilities should be taken in order to cater the flood and storm surge whin in turn prevent contamination of water due to flood overflow.

Fig. 16.5 Saltwater intrusion from storm surge flooding during Cyclone Nargis (2008). (Image source: Evan Schneider)

16.1.5.4 Regulating Industrial Contamination

Water treatment plants should also be included within the elevated infrastructure, especially to avoid saltwater intrusion and damage to treatment plants during storm surge flooding. Furthermore, the drainage system and underground sewage system should also consider their corresponding resilience measures to mitigate future flooding and contamination scenarios. Some studies claim that nearly nine million barrels of oil spills occur in the oceans, of which extreme oil spills occur along the coastline. Man-made disasters such as oil spills and industrial accidents cause chemical seepage in the nearby grounds. As oils cannot dissolve in water, they coagulate to become sludge, causing severe environmental damage affecting aquatic and marine lives. Extended damage also results in the decrease of groundwater quality of the adjacent lands to a great extent (Cohen 2013).

Figure 16.6 marks the 10-year anniversary of the Deepwater Horizon, which exploded in the Gulf of Mexico polluting 57,000 sq. miles and has been recorded as the worst oil spill (The Maritime Executive 2020). Measures should be taken to regulate and avoid these types of accidents in the future not only to save the environment but also to increase groundwater quality in such vulnerable areas.

16.1.5.5 Preventing Household-Level Contamination

Preventing groundwater pollution from nitrates and fluorides can begin even at household levels. This includes being watchful of not throwing or abandoning chemical substance like disinfectants, toiletries, cleaning and laundry chemicals, cooking oils, etc. in kitchen sinks. All the water treatment plants and pumping stations should incorporate adequate and appropriate treatment strategies to remove all the contamination, especially nitrates and fluorides, which cause health impacts

Fig. 16.6 Ten years since the Deepwater Horizon spill. (Image captured on 19 April 2020, by The Maritime Executive 2020)

(Raj and Shaji 2017). Industry-owned autonomous pumping stations and water treatment plants should also be regulated using the right strategies as they may end up in the soil as waste disposal and the groundwater contamination cycle may continue again. Vulnerable aquifers and confined aquifers should be checked for constant movement, and, upon identification of any sluggish movement of groundwater, appropriate measures should be immediately applied to stop further contamination.

16.1.6 Nitrogen Mitigation Programmes

Enactment of land management processes, such as introduction of various mitigation strategies to reduce N contents in superficial waters, stresses the in-depth knowledge about the N range (Sutton et al., 2011). N concentration arrays are from (i) agricultural fields where N is battle-ready; (ii) the transport and retaining of N forms along various paths such as drainage, groundwater, and surface water streams; and (iii) the holding and destiny of N in squeezed water streams such as estuaries, rivers, lakes, and wetlands (Hinsby et al. 2012; Blicher-Mathiesen et al. 2014). Cheng et al. (2020) highlighted how individual wetlands are capable of improving groundwater quality despite these contaminations. A research was conducted on the nitrogen mass balance from 26 restored wells during which nitrogen was identfied as redued nitrates due to N loads levels from wetlands (Crumpton et al. 2020).

The relationship between policy, management, and the consequence of water streams with N is not forthright because of biophysical lagging and socioeconomic gaps (Wulff et al. 2014).

Mitigation methods comprise greater manure storing capability, rules for scheduling manure application, restrictions for the extreme quantity of N that may be used in various crops, and improved protection of catch crops in all seasons.

In Scotland, half of the nitrogen utilized has vanished to broader environs as it is unexplored. This is a waste of wealth and farmers' economy. Hence, Scotland has become the first nation to include a nitrogen budget in its 2019 Climate Change Bill. "The Bill's aim is to lessen Scotland's net greenhouse gas emissions to net zero by 2045." Approaches to nitrogen usage in Scotland are altering as stakeholders understand its extensive influences. In 2017, nitrous oxide made up 7.9% of Scotland's greenhouse gas emissions, from which 81% was from farming.

In India, a step towards this goal was taken by the Department of Biotechnology (DBT), India, in association with the Biotechnological and Biological Sciences Research Council, UK (BBSRC); and the Natural Environment Research Council (NERC), UK, decided to fund virtual joint centres (VJCs) to explore the techniques of handling agricultural Nr enlightening crop production and dropping energy involvements and reactive nitrogen Nr harms to the environment. To accomplish these goals, three key prospects were recognized:

1. The augmentation of nitrogen fixation in leguminous plants and initiation of its biological nitrogen fixation (BNF) prospective in non-leguminous crops.
2. The enhancement of nitrogen use efficiency (NUE) at a plant level.
3. The upgradation of NUE by means of agronomical customs.

Additionally, four VJCs were established in 2016, in association with the prominent N research societies of India and the UK. The four virtual centres are as follows:

1. The Cambridge-India Network for Translational Research in Nitrogen (CINTRIN), whose main objective is to diminish the usage of agricultural N by using native variants and plant genetics and updating on-farm choice sustenance.
2. The Indo-UK Center for the Improvement of Nitrogen Use Efficiency in Wheat (INEW), which explores the genetic controller and biochemical source of NUE in wheat as well as the exactitude use of N fertilizers.
3. The India-UK Nitrogen Fixation Center (IUNFC), which emphasizes on pigeon pea rhizobial N fixation and the impact of endophytes on N nutrition of rice.
4. The India-UK Newton-Bhabha Virtual Center on Nitrogen Efficiency of Whole-cropping Systems (NEWS) for enhanced performance and suppleness in agriculture, which reconnoitres the choices for improving NUE and improved planning of the whole crop rotation and, consequently, has set goals to relate the novel biological and agronomic progresses to the farm scale as well as to a nationwide scale.

In 2019, the United Nations Environment Programme (UNEP), along with the Sri Lankan government and the International Nitrogen Management System, promoted the United Nations Global Campaign on Sustainable Nitrogen Management under the guidance of President Sirisena. Similarly, many countries have come forward to manage N release and control greenhouse gas emissions.

16.1.7 Fluoride Mitigation Programmes

Innumerable fluorosis mitigation plans have been implemented globally by means of defluoridation divisions. Some of them are as follows:

1. The defluoridation sector in the Ngurdoto village, Tanzania, by means of a bucket defluoridator through bone char (Jecobson and Dahi 1997).
2. The fluorosis mitigation project in the village of Ban Sankayom, Thailand, by means of a bone char defluoridator (Bravo et al. 2000).
3. ICOH Mobile Bus Unit Project in Thailand (Nasakolnakorn 2004) gave an emphasis on generating awareness between societies about special effects of extreme intake of fluoride on healthiness, serving them with on-site exploration of fluoride substance in water samples and accessing them to resolve the difficult by means of probable options.

In India, the government has launched the National Drinking Water Mission to supply clean and safe portable water to its public and to battle fluorosis. Some of the accompanying mitigation plans are given below:

1. **SARITA Project in Dungarpur, Rajasthan** (Vaish and Vaish 2000): A pilot project of 800 defluoridation divisions by means of Nalgonda-based drum sets and activated alumina filters was spread in six villages of the Dungarpur district of Rajasthan under the support of the United Nations Children's Fund (UNICEF). The public regularly utilized the water from these divisions and experienced noteworthy relief from non-skeletal fluorosis signs.
2. **Sachetana Plus** Drinking Water Plus project (Mythri et al. 2012): The BAIF Institute for Rural Development, Karnataka (BIRD-K), a non-profit organization, implemented fluoride mitigation through rainwater harvesting in three districts of Karnataka, with a goal to resolve the problematic prevalence of fluorosis in the area, by offering an alternative source of water and by humanizing groundwater quality.
3. **Fluorosis mitigation project at Sonbhadra, UP** (Gautam and Tripathi 2009): In 2004, Peoples' Science Institute (PSI) in association with the Banwasi Seva Ashram (BSA) initiated an agenda of fluoride analysis and fluorosis mitigation in Sonbhadra. The chief modules comprised health valuation, water quality monitoring, awareness programmes among the public, nutritional involvements, etc.
4. Peoples' Science Institute (PSI) in association with Sahabhagi Vikash Abhiyan (SVA) initiated a fluorosis mitigation programme in four districts of western Orissa in 2005 (Fluoride Testing 2014) and surveyed all tube wells of portable

water to identify fluoride pollution and created an awareness programme among the public for prevention of fluorosis in drinking water.

5. **Fluorosis mitigation project in Dhar district, MP** (People's Science Institute 2011): In 2008, People's Science Institute, Dehradun, Vasudha Vikas Sansthan, Dhar, and WaterAid (UK) introduced a fluorosis mitigation programme in Dhar district, MP. Their fitted water supply systems, have a control to decline the urinary fluoride values. The effectiveness of these agencies has stimulated the executive authorities to carry out fluoride mitigation in six extra villages in the zone..
6. **Mitigation of fluorosis in Nalgonda district villages** (Narayana et al. 2004): The Sai Oral Health Foundation aided by the Government of Andhra Pradesh had implemented an approach of supplying low-fluoride water to affected villages of the Nalgonda district by means of utilizing bone char-based domestic defluoridators and rainwater harvesting schemes.
7. **Integrated Fluorosis Mitigation, Madhya Pradesh** (National Ground Water Association, NEERI 2007): The National Environmental Engineering Research Institute (NEERI) in association with UNICEF, the Regional Medical Research Centre for Tribals (ICMR), and the Public Health Engineering Department, Govt. of Madhya Pradesh, insisted the Integrated Fluorosis Mitigation project in two districts (Jhabua & Dhar) based on involvements, which involved water thinning and defluoridation and nutritional enhancements in the public and in schools.
8. **National Rural Drinking Water Program (NRDWP) (NRDWP 2010)**: In 2009, the Government of India, with an aim to supply each rural individual with suitable safe water for drinking, cooking, and other domestic basic necessities on a sustainable source, initiated this programme..

16.2 Future Perspectives

Battling nitrogen and fluorosis on a bulk quantity have persisted as a daydream till date due to scarcity of enormous communication agendas, shortage of awareness, lack of governmental motivation, etc. Frame media or social media can play a dynamic role in inhibiting and restricting the risks of fluorosis (Krishnan and Indu 2014). The intake of Nr in the practise of fertilizer has now been emerging with excessive speed and at the same time, the nitrogen use efficiency (NUE) of cropping schemes has been found declining. These trends have led to enhancing the ecological losses of Nr, endangering the feature of air, soils, and freshwater, thus threatening climate stability, bionetworks, and human health. Meanwhile, it has been predicted that fertilizer intake might increase by twofold by 2050. Hence, there is a vital need for scientific exploration to sustain improved nitrogen administration in the agricultural sector.

16.3 Conclusions

This chapter has focused on understanding groundwater vulnerability to both fluoride and nitrate; studying the various forms in which these contaminations occur is important to mitigate the impacts of these hazards on human health, while discussing how there is limited availability of vulnerable aquifers and how it takes a long time to improve the quality of those contaminated aquifers. Certainly, both fluoride and nitrate contaminations in groundwater have health impacts from infants to adults and animal to aquatic lives, thus costing the society. It is vital to address the nitrate and fluoride contamination in groundwater and is equally important to initiate a "call for action" towards improving the quality of groundwater, thereby eventually improving human health and well-being. A critical review of the existing mitigation measures provided insights into the vulnerability and susceptibility of countries in arid and semi-arid regions to fluoride and nitrate contamination in groundwater and their health impacts in the past and how they would be impacted in the future. Advanced wastewater treatment technologies such as advanced oxidation processes are categorized as changers of the chemical structure of pollutants, thus creating more environment-friendly products. One alteration to get ahead of nitrate and fluoride contamination in groundwater is to amend new technologies that provide proper mitigation measures and developing agronomics in a better way. Creating awareness could be beneficial to the communities, particularly to those developing countries with limited capacities in the arid and semi-arid regions.

References

P. Balamurugan, P.S. Kumar, K. Shankar, Dataset on the suitability of groundwater for drinking and irrigation purposes in the Sarabanga River region, Tamil Nadu, India. Data Brief **29**, 105255 (2020)

P. Balamurugan, P.S. Kumar, K. Shankar, R. Nagavinothini, K. Vijayasurya, Non-carcinogenic risk assessment of groundwater in southern part of Salem District in Tamilnadu, India. J. Chil. Chem. Soc. **65**(1), 4697–4707 (2020a)

P. Balamurugan, P.S. Kumar, K. Shankar, R. Nagavinothini, P.J. Sajil Kumar, Impact of climate and anthropogenic activities on groundwater quality for domestic and irrigation purposes in Attur region, Tamilnadu, India. Desalin. Water Treat. **208**, 172–195 (2020b)

G. Blicher-Mathiesen, H.E. Andersen, J. Carstensen, C.D. Børgesen, B. Hasler, J. Windolf, Mapping of nitrogen risk areas. Agric. Ecosyst. Environ. **195**, 149–160 (2014)

A. Bravo, H. Bregnhoj, J.P.N. Sakolnakorn, N. Rattanapibool, Demand responsive fluorosis prevention in a village in Thailand, in *Proceedings of the 3rd International Workshop on Fluorosis Prevention and Defluoridation of Water*, ed. by E. Dahi, S. Rajchagool, N. Osiriphan, (Chiang Mai, Thialand, 2000), pp. 111–119

K. Brindha, L. Elango, Fluoride in groundwater: Causes, implications and mitigation measures, in *Fluoride Properties, Applications and Environmental Management*, ed. by S. Monroy, (Nova Science Publishers, Inc., New York, 2011), pp. 111–136

D.M.R. Burkartaus, J.D. Stoner, Chapter 7. Nitrogen in groundwater associated with agricultural systems, in *Nitrogen in the Environment: Sources, Problems, and Management*, ed. by R. F. Hatfield, J. L. Follett, (Elsevier, Amsterdam, 2008), pp. 177–202

Centers for Disease Control and Prevention, (2015) Nitrate and Drinking Water from Private Wells. [Online] Available at: https://www.cdc.gov/healthywater/drinking/private/wells/disease/nitrate.html. Accessed 30 Apr 2021

F.Y. Cheng, K.J. Van Meter, D.K. Byrnes, N.B. Basu, Maximizing US nitrate removal through wetland protection and restoration. Nature **588**, 625–630 (2020)

A.L. Choi, G. Sun, Y. Zhang, P. Grandjean, Developmental fluoride neurotoxicity: A systematic review and meta-analysis. Environ. Health Perspect. **120**, 1362–1368 (2012)

M. Cohen, Water pollution from oil spills. Encyclo. Energy Nat. Res. Environ. Econ. **3**, 121–126 (2013)

P.D. Cohn, *A Brief Report on the Association of Drinking Water Fluoridation and the Incidence of Osteosarcoma among Young Males* (Department of Health Environment, Health Service, New Jersey, 1992)

W.G. Crumpton, G.A. Stenback, S.W. Fisher, J.Z. Stenback, D.I. Green, Water quality performance of wetlands receiving nonpoint-source nitrogen loads: Nitrate and total nitrogen removal efficiency and controlling factors. J. Environ. Qual. **49**(3), 735–744 (2020)

Cyclone Nargis, (2008). Hurricanes: Science and Society. http://www.hurricanescience.org/history/storms/2000s/cyclonenargis/

Ecosystems and biodiversity, (2019). https://www.unep.org/news-and-stories/story/why-nitrogen-management-key-climate-change-mitigation

European Nitrogen Assessment (ENA). http://www.nine-esf.org/node/204/ENA.html

J.F. Fathmawati, S. Evi Gravitiani, H. Adi Heru, Nitrate in drinking water and risk of colorectal cancer in Yogyakarta, Indonesia. J. Toxicol. Environ. Health **80**(2), 120–128 (2017)

Fluoride Testing and Fluorosis Mitigation nn Nuapada District Orissa: A Small Pilot-Scale Exercise [Internet]. [cited on 2014 12]. Available from: http://peoplesscienceinstitute.org/PDF's/EQMG/FLUORIDE%20TESTING%20AND%20FLUOROSIS%20MITIGATION%20IN%20NUAPADA.pdf

Gautam, A., Tripathi, R.C. (2009). Fluoride Testing and Fluorosis Mitigation In Sonebhadra District. Available from: https://ideas.repec.org/p/ess/wpaper/id1853.html

A.S. Giwa, A.G. Memon, J. Ahmad, T. Ismail, S.A. Abbasi, K. Kamran, B. Wang, B. Segun, H. Seydou, Assessment of high fluoride in water sources and endemic fluorosis in the North-Eastern communities of Gombe State, Nigeria. Environ. Pollut. Bioavail. **33**(1), 31–40 (2021)

Groundwater Foundation (2021). Groundwater Contamination. Available at: https://www.groundwater.org/get-informed/groundwater/contamination.html

R. Gupta, T. Khan, D. Agrawal, J. Kachhawa, The toxic effects of sodium fluoride on the reproductive system of male rats. Toxicol. Ind. Health **23**(9), 507–513 (2007)

B.K. Handa, Geochemistry and genesis of fluoride-containing ground waters in India. Groundwater **13**(3), 275–281 (1975)

K. Hinsby, S. Markager, B. Kronvang, J. Windolf, T. Sonnenborg, L. Thorling, Threshold values and management options for nutrients in a catchment of a temperate estuary with poor ecological status. Hydrol. Earth Syst. Sci. **16**(8), 2663–2683 (2012)

N.J.K. Howden, T.P. Burt, F. Worrall, S. Mathias, M.J. Whelan, Nitrate pollution in intensively farmed regions: What are the prospects for sustaining high-quality groundwater? Water Resour. Res. **47**(6), 1–13 (2011)

R. Indu, S. Krishnan, T. Shah, Impacts of groundwater contamination with fluoride and arsenic: Affliction severity, medical cost and wage loss in some villages of India. Int. J. Rural. Manag. **3**(1), 69–93 (2007)

International Association of Hydrogeologists, (2020). Groundwater – More about the hidden resource. Available at: https://iah.org/education/general-public/groundwater-hidden-resource

P. Jecobson, E. Dahi, Bone char based bucket defluoridator in tanzanian households, in *Proceedings of the 2nd International Workshop on Fluorosis Prevention and Defluoridation of Water. Nazreth, Ethiopia*, ed. by E. Dahi, J. M. Nielsen, (The International Society for Fluoride Research, New Zealand, 1997), pp. 156–159

P.K. Jha, P. Tripathi, Arsenic and fluoride contamination in groundwater: A review of global scenarios with special reference to India. Groundw. Sustain. Dev. **13**(1), 100576 (2021)
Krishnan, S., Indu, R. (2014) *How can we think ahead on Fluorosis Mitigation?* Available from: www.inrem.in/fluorosis/pdf/T1p3.pdf
S. Kumar, R. Singh, A.S. Venkatesh, G. Udayabhanu, P.R. Sahoo, Medical geological assessment of fluoride contaminated groundwater in parts of Indo-Gangetic Alluvial plains. Sci. Rep. **9**, 16243 (2019)
E. Kwon et al., Nitrate contamination of coastal groundwater: Sources and transport mechanisms along a volcanic aquifer. Sci. Total Environ. **768**, 145–204 (2021a)
E. Kwon, J. Park, W.-B. Park, B.-R. Kang, N.C. Woo, Nitrate contamination of coastal groundwater: Sources and transport mechanisms along a volcanic aquifer. Sci. Total Environ. **768**, 145204 (2021b)
R.L. Mahler, A. Colter, R. Hirnyck. (2007) *Nitrate and Groundwater* (University of Idaho Extension, Idaho)
D. Majumdar, The blue baby syndrome: Nitrate poisoning in human. Resonance **8**, 20–30 (2003)
N. Mohammed, (2009). *Learning from Cyclone Nargis: Investing in the Environment for Livelihoods and Disaster Risk Reduction* (United Nations Environment Programme, Kenya)
A. Móring, S. Hooda, N. Raghuram, T.K. Adhya, A. Ahmad, S.K. Bandyopadhyay, T. Barsby, G. Beig, et al., (2021). Nitrogen challenges and opportunities for agricultural and environmental science in India. Front. Sustain. Food Syst. (TSI) **5**, 1–16
H. Mythri, Dinesh, D. Bennadi, Current public health perspective of fluorosis mitigation project in Pavagada taluk, Karnataka. Indian J. Occup. Environ. Med. **16**(3), 124–126 (2012)
Narayana, A.S., Khandare, A.L., Krishnamurthi, M.V.R.S., Mitigation of fluorosis in Nalgonda District villages. In: Dahi, E., Rajchagool, S. (eds) Proceedings of the 4th International Workshop on Fluorosis Prevention and Defluoridation of Water. Colombo, SriLanka, pp. 98–106 (2004)
J.P. Nasakolnakorn, Means and mode of providing information to people in their fluorosis prevention. In: Dahi, E., Rajchagool, S., (eds) Proceedings of the 4th International Workshop on Fluorosis Prevention and Defluoridation of Water. Colombo, SriLanka, pp. 66–72 (2004)
National Ground Water Association, (2007). Information on Earth's Water. [Online] Available at: https://www.ngwa.org/what-is-groundwater/About-groundwater/information-on-earths-water NEERI, UNICEF. Report on Integrated Fluorosis Mitigation. pp. 86
National Ground Water Association, 2018. Groundwater facts. [online] Available at: https://www.ngwa.org/what-is-groundwater/About-groundwater/groundwater-facts [Accessed 25 April 2021]
Z. Nejatijahromi, H.R. Nassery, T. Hosono, M. Nakhaei, F. Alijani, A. Okumura, Groundwater nitrate contamination in an area using urban wastewaters for agricultural irrigation under arid climate condition, southeast of Tehran, Iran. Agric. Water Manag. **221**, 397–414 (2019)
NRDWP Department of Drinking Water Supply, Ministry of Rural Development, Government of India. Rajiv Gandhi National Drinking Water Mission, National Rural Drinking Water Programme. 2010. Pp. 46
B. Panneerselvam, K. Muniraj, C. Pande, N. Ravichandran, Prediction and evaluation of groundwater characteristics using the radial basic model in semi-arid region, India. Int. J. Environ. Anal. Chem., 1–17 (2021, 2021)
B. Panneerselvam, K. Muniraj, M. Thomas, N. Ravichandran, GIS-based legitimatic evaluation of groundwater's health risk and irrigation susceptibility using water quality index, pollution index, and irrigation indexes in semiarid region, in *Groundwater Resources Development and Planning in the Semi-Arid Region*, (Springer, Cham, 2021a), pp. 239–268
B. Panneerselvam, K. Muniraj, M. Thomas, N. Ravichandran, B. Bidorn, Identifying influencing groundwater parameter on human health associate with irrigation indices using the Automatic Linear Model (ALM) in a semi-arid region in India. Environ. Res. **202**, 111778 (2021b)
B. Panneerselvam, S.K. Paramasivam, S. Karuppannan, N. Ravichandran, P. Selvaraj, A GIS-based evaluation of hydrochemical characterisation of groundwater in hard rock region, South Tamil Nadu, India. Arab. J. Geosci. **13**(17), 1–22 (2020)

People's Science Institute, Dehradun. Vasudha Vikas Sansthan (Dhar). Water Aid, India. Report on Fluorosis mitigation program in Dhar District (M.P.) 2011; Pp. 40

D. Raj, E. Shaji, Fluoride contamination in groundwater resources of Alleppey, southern India. Geosci. Front. **8**(1), 117–124 (2017)

Richards M, Côrte-Real H, Forster P, et al. Paleolithic and neolithic lineages in the European mitochondrial gene pool [published correction appears in Am J Hum Genet 1996 Sep;5(3):747]. Am J Hum Genet. 1996;59(1):185–203.

B. Riemann, J. Carstensen, K. Dahl, H. Fossing, J.W. Hansen, H.H. Jakobsen, A.B. Josefson, D. Krause-Jensen, S. Markager, P.A. Stæhr, K. Timmermann, J. Windolf, J.H. Andersen, Recovery of Danish coastal ecosystems after reductions in nutrient loading: A holistic approach. Estuaries and Coast **39**, 82–97 (2016)

L.A. Schaider, L. Swetschinski, C. Campbell, R.A. Rudel, Environmental justice and drinking water quality: Are there socioeconomic disparities in nitrate levels in U.S. drinking water? Environ. Health **18**(3), 1 (2019)

C. Smith, K.A. Hill, L. Torrente-Murciano, Current and future role of Haber–Bosch ammonia in a carbon-free energy landscape. Energy Environ. Sci., 331–344 (2020)

R.F. Spalding, M.E. Exner, Occurrence of nitrate in groundwater – A review. J. Environ. Qual. **22**, 392–402 (1993)

S. Srivastava, S.J.S. Flora, Fluoride in drinking water and skeletal fluorosis: A review of the global impact. Curr. Environ. Health Rep. **7**(2), 140–146 (2020)

M.A. Sutton, C.M. Howard, J.W. Erisman, G. Billen, A. Bleeker, P. Grennfelt, H. Grinsven, B. Grizzetti, *European Nitrogen Assesment* (Cambridge University Press, Cambridge, 2011), p. 664

Sutton M.A., Howard C., Erisman J.W., Billen G., Bleeker A., Grennfelt P., van Grinsven H. and Grizzetti B. (2011) The European Nitrogen Assessment: Sources, Effects and Policy Perspectives (Eds.) Cambridge University Press. 612 pp.

P. Taneja, P. Labhasetwar, P. Nagarnaik, J.H.J. Ensink, The risk of cancer as a result of elevated levels of nitrate in drinking water and vegetables in Central India. J. Water Health **15**(4), 602–614 (2017)

The Maritime Executive, (2020) Ten Years Since the Deepwater Horizon Exploded. [Online] Available at: https://maritime-executive.com/article/ten-years-since-the-deepwater-horizon-exploded. Accessed 11 May 2021

B. Thole, Ground water contamination with fluoride and potential fluoride removal Technologies for East and Southern Africa. In: I. Ahmad & M. A. Dar, eds. *Perspectives in Water Pollution*. s.l.:IntechOpen (2013)

UNEP, *Frontiers 2018/19 Emerging Issues of Environmental Concern* (UN Environment Programme, Nairobi, 2019)

United Nations, (2021) Take action for the sustainable development goals. [Online] Available at: https://www.un.org/sustainabledevelopment/sustainable-development-goals/. Accessed 16 May 2021

US Environmental Protection Agency, *Getting Up to Speed: Ground Water Contamination* (US EPA Seminar Publication, Washington D.C., 2015)

US EPA, (2021a) Private Drinking Water Wells. [Online] Available at: https://www.epa.gov/privatewells. Accessed 13 May 2021

US EPA, (2021b) Protect Your Home's Water. [Online] Available at: https://www.epa.gov/privatewells/protect-your-homes-water#preventwellanchor. Accessed 18 May 2021

US Geological Survey, (2006). Pesticides in the Nation's Streams and Ground Water, 1992–2001—A Summary. [Online] Available at: https://pubs.usgs.gov/fs/2006/3028/pdf/fs2006-3028.pdf. Accessed 30 Apr 2021

A.K. Vaish, P. Vaish. *Proceedings of the 3rd International Workshop on Fluorosis Prevention and Defluoridation of Water.* Chiang Mai, Thialand: 2000 Nov 20–24. A Case Study of Fluorosis Mitigation in Dungarpur District, Rajasthan, India. In: Dahi E, Rajchagool S, Osiriphan N, editors; pp. 97–104

H.M. Ward, R.R. Jones, J.D. Brender, T.M. de Kok, P.J. Weyer, B.T. Nolan, C.M. Villanueva, S.G. van Breda, Drinking water nitrate and human health: An updated review. Int. J. Environ. Res. Public Health **15**(7), 1557 (2018)
Water Pollution, (2020). Eutrophication and water pollution: Causes of water pollution. [online] Available at: https://www.water-pollution.org.uk/eutrophication-and-water-pollution/. Accessed 09 Apr 2021
WHO, *Nitrate and Nitrite in Drinking-Water – Background Document for Development of WHO Guidelines for Drinking-Water* (World Health Organization (WHO), Geneva, 2011)
World Health Organization, (2019). Drinking-water. [Online] Available at: https://www.who.int/news-room/fact-sheets/detail/drinking-water#:~:text=Contaminated%20water%20and%20poor%20sanitation,A%2C%20typhoid%2C%20and%20polio. Accessed 03 May 2021
WorldAtlas, What Are The Major Sources Of Groundwater Contamination? [Online] (2018)
F. Wulff, C. Humborg, H.E. Andersen, G. Blicher-Mathiesen, M. Czajkowski, K. Elofsson, A. Fonnesbech Wulff, B. Hasler, B. Hong, et al., Reduction of Baltic Sea nutrient inputs and allocation of abatement costs within the Baltic Sea catchment. Ambio **42**, 11–25 (2014)
K.K. Yadav et al., Fluoride contamination, health problems and remediation methods in Asian groundwater: A comprehensive review. Ecotoxicol. Environ. Saf. **182**, 109362 (2019)

Chapter 17
Hydrological Modelling for Ungauged Basins: An Overview of the Past, Present, and Future Directions

Suresh Devaraj, C. Jenifa Latha, M. Geetha Priya, Colins Johnny Jesudhas, and Kiran Yarrakula

Abstract Hydrological modelling is an essential tool, in this century, for effective planning and management of water resources. Ground observations serve as the backbone of hydrological models in which inadequate field data obtained from a watershed have become a challenge to the research community for proper management of the watershed. Despite several progresses in the models, the heterogeneity of watersheds limits the measurement of hydrological parameters, resulting in uncertainty and affecting the applicability and confidence of the models. In this chapter, a detailed survey has been carried out to address the development in the hydrological models over the decades and the reliability of different models. Summarising the calibration and uncertainties of the hydrological models, adopting datasets obtained through space technology is highly recommended, and new approaches need to be developed with the integration of information technology, statistics, and space inputs to overcome the limitations.

Keywords Hydrological models · Calibration · Uncertainty · Ungauged stations · Hydrological advances

---

S. Devaraj (✉) · C. Jenifa Latha
Department of Civil Engineering, Saveetha Engineering College, Chennai, India

M. Geetha Priya
Centre for Incubation, Innovation, Research and Consultancy (CIIRC), Jyothy Institute of Technology, Bengaluru, India

C. J. Jesudhas
Department of Civil Engineering, University VOC College of Engineering, Thoothukudi, India

K. Yarrakula
Civil Engineering Department, GKCIET, West Bengal, Malda, India

B. Panneerselvam et al. (eds.), *Climate Change Impact on Groundwater Resources*, https://doi.org/10.1007/978-3-031-04707-7_17

17.1 Introduction

Over the millennia, hydrology has had a vast history of affecting human life and properties (Surampudi and Yarrakula 2020). The first hydrological model was developed by Mulvany in the year 1850 for computing peak discharge based on a rational method (Mulvany 1850). In 1856, an experiment was carried out by Darcy to study the flow of water through sands, which acts as the base for quantitative groundwater hydrology and is widely referred to as Darcy's law (Darcy 1856). Over the last century, several breakthroughs were achieved by researchers across the globe in modelling the hydrological cycle (Singh 2018), where few researches are based on physical models and others are based on statistical models (Vaishnavi et al. 2020). Hydrological modelling and analysis are indebted to the pre-1960 advancements that have served as the basis for their progress (Singh 2018). *Handbook of Applied Hydrology* by Chow, published in the year 1964, provides details of the hydrological models and their advances till the 1960s (Chow 1964). Development of hydrological models considering different components has witnessed a new revolution with the introduction of computers (Singh and Woolhiser 2002). With the development of computational power, several streams of hydrology such as numerical or digital hydrology and stochastic or statistical hydrology have emerged (Bredesen and Brown 2018).

The hydrological cycle was successfully simulated in 1966 by the construction of the "Stanford Watershed Model" (Kumar and Balamurugan 2018, 2019; Crawford and Linsley 1966), which has paved way for the development of various watershed models across the world over the decades (Singh 2017; Kouadri et al. 2021). The accuracy of the models played a key role in the development of optimisation techniques, which were widely adopted for calibration of the developed hydrological models (García-Romero et al. 2019). Advancements in computing methodologies led to the development of three-dimensional models, which assisted in modelling surface water, groundwater considering infiltration, and soil water flow (Bhola et al. 2013; Dibaj et al. 2021). Further development led to the simulation of various parameters including sediment and pollution and also modelling various stages of flow including liquids and gases (Azubuike et al. 2016; Manisalidis et al. 2020). Improvement in the scales of modelling has resulted in achieving outputs at a temporal scale of minutes or seconds (Ringkjøb et al. 2018). Focusing on the improvement of models, integration of the hydrological aspects with associated parameters such as climatic conditions (Sorooshian et al. 2008), geomorphology of the basin (Suresh et al. 2018), geology (Singh and Zhang 2018), soil texture (Singh 2006), and flow characteristics (Teng et al. 2017b) assisted in achieving models with higher accuracy. Combining the ecosystem with the hydrological models gave birth to the field of ecohydrology (Revelli and Porporato 2018). On witnessing the evolution in the development of hydrological models, climate change and global warming have also become a part of it (Krysanova et al. 2018).

With the exponential development in computation over the decades, statistically integrated hydrological tools have expanded both vertically and horizontally (Slater et al. 2019). Technology has enabled the development of a user-friendly work environment that has enabled researchers to acquire, store, analyse, and interpret results (Suresh and Yarrakula 2019). Introduction of space technologies has achieved different heights in hydrological analysis by providing near real-time datasets using satellites covering larger areas (Suresh and Yarrakula 2020). Similarly, development of geographic information systems (GISs) has also contributed to spatial modelling with the availability of huge raster and vector datasets (Gupta 2020). Recently, researchers have witnessed the influence of machine learning and artificial intelligence based on Fuzzy logic, artificial neural network, wavelet models, and genetic programming, leading to the development of precise predicted models (Mosavi et al. 2018). These models are said to play a dominant role in the near future.

17.2 History of the Hydrological Models

Initiated in 1850, significant development has been witnessed in the field of hydrology, focusing on planning and development of watersheds and management of floods, droughts, and other environmental aspects (Rosselló et al. 2020). Models based on rational methods were developed for every event based on unit hydrograph, infiltration, over land flow, etc. Computerised models have been in use since the 1960s, and one among the early models that is widely used is the SWM (Stanford Watershed Model) (Abbott et al. 1986). Over the decades, several developments have been made by researchers, focusing on the computation time, cost, and global climatic conditions. Models such as the Soil and Water Assessment Tool (SWAT) (Budamala and Baburao Mahindrakar 2021), Precipitation-Runoff Modelling System (PRMS) (Teng et al. 2017a), etc. are widely adopted and considered to be powerful modelling instruments for obtaining models with higher accuracy. For addressing the evolution of hydrological models, a snapshot of the major developments is listed below.

17.2.1 Modelling as a Mathematical Tool

The characteristics of river basins across the world are said to be identical, whereas parameters including climate, morphology, geology, and hydrology will vary based on the regional conditions and watershed considered for the study (Mahala 2019). Mathematical models with the combinations of equations expressing the natural conditions by considering the relationship between the parameters are widely adopted in modelling the hydrological process (Gamage et al. 2015). Considering the relationship between the hydrological parameters and adopting mathematical equations, a spatiotemporal structure is generated to produce a single variable, resulting in the development of a mathematical model, with the ability to reproduce the statistical characteristics of the hydrological cycle (Liu et al. 2021).

Modelling the natural environment using mathematical operations and equations has been widely accepted by the research community and is used for planning, development, and management of water resources (Archibald and Marshall 2018). Adopting mathematical models will assist in solving hydrological issues including runoff, watershed management, and water security considering both internal and external characteristics (Misra 2014).

17.2.2 Classification of Hydrological Modelling

Development of highly adoptable models can be achieved by gaining knowledge on the structure of the model through experiment, interaction, and observation (Aspers and Corte 2019). Models developed mathematically and physically represent the real-world scenario, which helps researchers understand the characteristics of the nature (Van Der Valk et al. 2007). Initially, analogue and physical models were widely adopted, and, with the introduction of mathematical models, a rapid development in modelling occurred due to their scientific applications (Siettos and Russo 2013). Hydrological models derived using mathematical methods are classified into (i) conceptual/empirical models, (ii) distributed or concentrated models, (iii) stochastic models, (iv) linear/non-linear models, (v) static/dynamic models, etc. (Budamala and Baburao Mahindrakar 2021; Balamurugan et al. 2020a, b). Based on simulation, hydrological models are generally classified into two categories as shown in Fig. 17.1.

Stochastic models are derived based on the probability and random components, which do not produce similar outputs upon execution (Raj and van Oudenaarden 2008). Stochastic models have attracted researchers' attention in recent decades due to their unique characteristics of modelling the varying atmospheric conditions and catchments considering time and space (Paschalis et al. 2013). Models derived based on the stochastic method are time- and chance-dependent, which requires time series analysis (Panneerselvam et al. 2020a, b; 2021b, c). The most widely used stochastic methodologies are NNs (neural networks), regression analysis, system identification, and transfer functions.

Deterministic models assist in achieving defined outputs based on the provided input datasets and boundary conditions (Loucks and van Beek 2017). Physical deterministic models develop responses based on the characteristics of the watershed considered (Farmer and Vogel 2016). Models developed using the deterministic method can be both continuously simulated and event-based (Irurzun-Arana et al. 2020). A continuous model executes simulation uninterruptedly and generates the required result, whereas an event-based model is executed only for a particular event (Camargo et al. 2020). Event-based models are usually carried out for a specific period of time (for example, days or hours) to analyse the influence of a disaster event (Fischer et al. 2019). Comparing the two deterministic models, continuous models are widely adopted for various hydrological analyses and assessments.

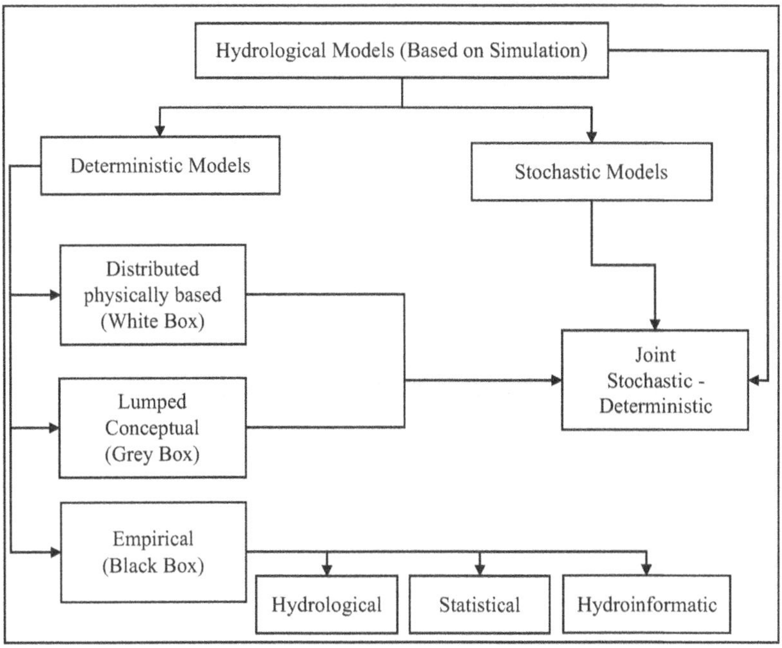

Fig. 17.1 Classification of hydrological models based on simulation

Lumped modelling approaches are executed by considering the river basin as a unit by omitting the spatial variability where the input datasets are directly associated with the outputs (Shakti et al. 2019). The limitations existing in lumped models include large volumes of assumptions and execution of simulations only at the outlet of the watershed, thus rendering the model unsuitable for flood forecasting, watershed management, land use, and climate change studies (Jeziorska and Niedzielski 2018). With the existence of certain limitations, several numerical complex models have been developed.

Distributed models such as SWE and SHE consider the spatial variability of input datasets in modelling the natural scenario (Müller et al. 2019). Several researchers have contributed to the development of distributed models for river flow modelling with the combination of physical and mathematical models (Irfan et al. 2019). Major disputes faced in implementation of the models for a watershed are due to the inadequacy of the models in the hydrological process and applying the models at varying scales over a catchment (Horan et al. 2021). In spite of the limitations, the models are highly adopted for hydrological modelling and analysing the pollutants and sediment yield along the river (Momm et al. 2021). The availability of high-resolution spatial datasets including rainfall, temperature, soil, land use, DEM (digital elevation model), and other parameters has led to the development of several sophisticated models that focus mainly on obtaining outputs based on parameterisation and calibration (Panneerselvam et al. 2021a, b, c; Priya et al. 2021;

Shunmugapriya et al. 2021). One of the major challenges faced by the models is error correction.

With the increase in the input parameters considered for hydrological modelling, the model is made complicated as many parameters have to be analysed and calibrated to generate an output (Wu et al. 2021). An uncertainty in the hydrological modelling is experienced during the estimation of several parameters with distributed models (Rafiei Emam et al. 2018). Based on the parameterisation of hydrological models, distributed models are classified into two categories, namely, homogeneous and heterogeneous (Rawat and Singh 2017).

17.3 Recent Developments in Hydrological Modelling

A drastic improvement has been noted in the field of hydrological modelling in understanding watershed systems and obtaining accurate output. Researchers across the globe are highly focusing on development of models for accurate prediction of hydrological parameters in ungauged catchment points.

17.3.1 Prediction in Ungauged Basins

Recently, several challenges have been experienced in obtaining the hydrological parameters over ungauged stations due to several uncertainties experienced in the watershed (Rawat and Singh 2017). Prioritising the challenges, the International Association of Hydrological Sciences (IAHS) initiated the Predictions in Ungauged Basins (PUB) programme to attract the attention of researchers across the globe towards modelling the uncertainties and assisting in the estimation of hydrological parameters over ungauged stations (Sivapalan et al. 2003). PUB aims at achieving hydrological models for an ungauged watershed by improving the characteristics of the models by considering the heterogenic effects and understanding the characteristics of the watershed (Narbondo et al. 2020; Zhu et al. 2021). With the focus on achieving this objective, PUB set two research targets: (i) examining and improving the existing models and (ii) developing a new innovative model for prediction of spatial variability.

Similar to PUB, the MOPEX (Model Parameter Estimation Experiment) has initiated and fetched better responses by encouraging researchers to contribute to the hydrological community (Duan et al. 2006). Models developed through the MOPEX have achieved higher levels by enhancing the insights of researchers into the prediction of hydrological parameters for ungauged watersheds (Darbandsari and Coulibaly 2020). The National Weather Service (NWS) has been the backbone of the MOPEX international project, which aims at the development of a highly precise model for the estimation of hydrological parameters for ungauged stations by considering various input parameters and improving the calibration techniques and evaluating them (Hrachowitz et al. 2013; Newman et al. 2015; Asurza-Véliz and Lavado-Casimiro 2020).

PUB and MOPEX have made several contributions towards the development of hydrological models for ungauged catchments by encouraging researchers towards their objective (van Emmerik et al. 2015; Panneerselvam et al. 2021a). Models developed for estimation of parameters for ungauged or poorly gauged stations are highly influenced by the inaccurate characteristics of the watershed and accuracy of the observed datasets, imperfect models, and input parameters, and it is highly impossible to define the nature of the errors in the input parameters (Raj et al. 2014; Jiang and Wang 2019). Researchers are also experiencing a challenge in adopting a model developed over a well-defined gauging station to an ungauged station considering the characteristics and responses of the watershed (Msaddek et al. 2020; Daniel et al. 2021).

17.3.2 Distributed Hydrological Modelling

The availability of satellite datasets has encouraged researchers across the world to make use of the distributed models to examine the characteristics of the watershed and model the river water flow (Wang et al. 2016; Groom et al. 2019). With a high spatial and temporal resolution, space datasets are considered as an alternate method for modelling the hydrological systems (Bhola et al. 2013; Khaki et al. 2020; Schaller et al. 2020). Space datasets distributed over the terrain are able to provide information over the varying topography but lack accuracy (Hurskainen et al. 2019; Kramm and Hoffmeister 2019). To overcome the accuracy issues and to make use of the distributed datasets, the space information obtained and calibrated with the collected ground information are widely used in modelling the hydrological conditions (Sun et al. 2018; Dembélé et al. 2020b; Suresh et al. 2021).

Adopting satellite-obtained observations for hydrological modelling has several advantages, whereas there exist few challenges in adopting the datasets (Dembélé et al. 2020a; Peng et al. 2021). The varying spatial resolutions of the input parameters for models directly influence the output generated, which renders the models unreliable (Fisher et al. 2018). Several researchers have indicated the uncertainties faced by adopting spatial datasets in hydrological modelling as they directly influence the output (Nkiaka et al. 2018; Her et al. 2019). Researchers across the world have carried out research for investigating the applicability of space-based inputs in comparison with their lumped counterparts (Razavi et al. 2021). The Distributed Model Intercomparison Project (DMIP) is one example, which was initiated by NOAA's National Weather Service (NWS) to compare 12 distributed models that use space-derived datasets as input with the lumped Sacramento Soil Moisture Accounting model, SAC-SMA (Madsen et al. 2020). Comparison results conclude that the distributed models are highly efficient compared to the lumped models (Paudel et al. 2011). Spatial distribution of the input parameters and channel routing contribution to distributed models have made them more accurate compared to lumped models (Al-Areeq et al. 2021). Although distributed models have been identified to be significant, the factors that influence the models are unclear as the results mainly depend on the catchment characteristics (Lintern et al. 2018). One of the major factors that contribute to the model is the continuous distribution of rainfall, and the model relies directly on the distributed information (Jaiswal et al. 2020). Errors and uncertainties should be considered in distributed models, which could help in improving their performance (Uusitalo et al. 2015). Lumped models are reliable over distributed models only when there is an inconsistency in the spatial datasets (Moustakas and Willems 2021).

17.4 Summary and Discussion

Hydrological modelling is one of the essential processes that has to be adopted for proper planning and management of water resources. With several hydrological models available in the market, selection of a model is based on the availability of datasets, characteristics of the watershed, spatial representation, and computational cost. An overview of the hydrological models is described in this study considering the characteristics of the models.

It is practically impossible to model all the hydrological aspects of a watershed, whereas a considerable aspect can be achieved with the integration of calibration in modelling. Model calibrations are adopted based on the ground information collected to minimise the errors in the models and to improve the accuracy where the results obtained are highly influenced by uncertainties.

Quantification and representation of uncertainties play a major role in the development of different hydrological models and in decision-making. One of the major setbacks due to the uncertainties in the models is the prediction of hydrological parameters as the output generated will be misleading. In order to overcome this issue, a model has to be developed considering all the uncertainties and biases for the final prediction. Several frameworks such as FUSE, IBUNE, and BATEA have been introduced by the scientific community to assess the uncertainties in the simulations, and it is recommended to put in more effort to the scientific understanding of the models. Researchers have suggested that the uncertainties in a model can be reduced by (i) obtaining higher spatial and temporal resolution datasets, (ii) understanding the characteristics of the hydrological watershed and developing a hydrological model well-suited for the observations, and (iii) developing models that could assist in the extraction and assimilation of parameters for the available datasets through models.

With several models used for hydrological modelling, recent advances have highlighted the significance of calibration and evaluation of models, which assist in the identification of the reasons for the failure of the models and the aspects that need to be focused on for improvement. Multi-objective optimisation algorithms assist in pointing out the reasons for the failure of the models and the area in which the model needs improvement, which improves the output generated and its reliability at various stages.

17.5 Conclusion

Several hydrological models have been developed by researchers focusing on the prediction of parameters for ungauged and poorly gauged stations and with the initiatives from PUB and MOPEX, challenges faced in the hydrological modelling are considered under environmental characteristics. The availability of space-based datasets has paved way for the development of several advanced models, yet the potential has not yet been fully achieved by distributed models. Adopting several calibration techniques and diagnosis of errors has contributed to the uncertainties in modelling, whereas development with the spatial information has to be maximised to improve the quality of the output. Prediction with respect to environmental changes plays a significant role in modelling, and studies carried out have concluded that the land use characteristics and climatic changes play a predominant role in the hydrological models. General circulation models (GCMs) are capable of quantifying the variation in the hydrological outputs caused due to climate change where further focus is required for downscaling the climatic scenario datasets. One of the major issues that affect the outcome of the models is the uncertainty, which limits the models based on spatial and temporal resolution.

With an increase in the integrated hydrological approaches in recent decades, modelling the hydrological parameters has improved as it has a significant impact on climate change and global warming, which directly influence the food and water security for human life. Recently, the boom in information technology and adopting statistical approaches have helped in the prediction of the near future, and it is recommended that the space datasets be widely explored to their maximum potential to obtain an effective hydrological model.

References

M.B. Abbott, J.C. Bathurst, J.A. Cunge, et al., An introduction to the European Hydrological System — Systeme Hydrologique Europeen, "SHE", 1: History and philosophy of a physically-based, distributed modelling system. J. Hydrol. **87**, 45–59 (1986). https://doi.org/10.1016/0022-1694(86)90114-9

A.M. Al-Areeq, M.A. Al-Zahrani, H.O. Sharif, The performance of physically based and conceptual hydrologic models: A case study for Makkah watershed, Saudi Arabia. Water (Switzerland) **13** (2021). https://doi.org/10.3390/w13081098

T.W. Archibald, S.E. Marshall, Review of mathematical programming applications in water resource management under uncertainty. Environ. Model. Assess. **23**, 753–777 (2018). https://doi.org/10.1007/s10666-018-9628-0

P. Aspers, U. Corte, What is qualitative in qualitative research. Qual. Sociol. **42**, 139–160 (2019). https://doi.org/10.1007/s11133-019-9413-7

F.A. Asurza-Véliz, W.S. Lavado-Casimiro, Regional parameter estimation of the SWAT model: Methodology and application to river basins in the Peruvian pacific drainage. Water (Switzerland) **12**, 1–25 (2020). https://doi.org/10.3390/w12113198

C.C. Azubuike, C.B. Chikere, G.C. Okpokwasili, Bioremediation techniques-classification based on site of application: Principles, advantages, limitations and prospects. World J. Microbiol. Biotechnol. **32**, 180 (2016). https://doi.org/10.1007/s11274-016-2137-x

P. Balamurugan, P.S. Kumar, K. Shankar, R. Nagavinothini, P.J. Sajil Kumar, Impact of climate and anthropogenic activities on groundwater quality for domestic and irrigation purposes in Attur region, Tamilnadu, India. Desalin. Water Treat. **208**, 172–195 (2020a)

P. Balamurugan, P.S. Kumar, K. Shankar, R. Nagavinothini, K. Vijayasurya, Non-carcinogenic risk assessment of groundwater in southern part of Salem District in Tamilnadu, India. J. Chil. Chem. Soc. **65**(1), 4697–4707 (2020b)

P. Bhola, A. Zabel, S.P. Rajaveni, et al., Integrated surface water and groundwater modeling for optimizing MAR structures in the Chennai region (2013)

A. Bredesen, C.J. Brown, Comparison of hydrologic model performance statistics using rain gauge and NEXRAD precipitation input at different watershed spatial scales and rainfall return frequencies for the upper St. Johns River, Florida USA. Proceedings **7**, 11 (2018). https://doi.org/10.3390/ecws-3-05806

V. Budamala, A. Baburao Mahindrakar, Enhance the prediction of complex hydrological models by pseudo-simulators. Geocarto Int. **36**, 1027–1043 (2021). https://doi.org/10.1080/10106049.2019.1629646

M. Camargo, M. Dumas, O. González-Rojas, Automated discovery of business process simulation models from event logs. Decis. Support. Syst. **134**, 113284 (2020). https://doi.org/10.1016/j.dss.2020.113284

V. Chow, *Handbook of Applied Hydrology* (McGraw-Hill Book Publishing Company, New York, 1964)

N.H. Crawford, R.K. Linsley, *Digital Simulation in Hydrology: Stanford Watershed Model IV* (Stanford University, 1966)

A. Daniel, R.R. Balieiro, N.R. Line, Gauging the ungauged: Regionalization of flow indices at grid level. J. Hydrol. Eng. **26**, 4021008 (2021). https://doi.org/10.1061/(ASCE)HE.1943-5584.0002067

P. Darbandsari, P. Coulibaly, Inter-comparison of lumped hydrological models in data-scarce watersheds using different precipitation forcing data sets: Case study of Northern Ontario, Canada. J. Hydrol. Reg. Stud. **31**, 100730 (2020). https://doi.org/10.1016/j.ejrh.2020.100730

H. Darcy, *Les Fontaines Publiques de la Ville de Dijon* (Dalmont, Paris, 1856)

M. Dembélé, N. Ceperley, S.J. Zwart, et al., Potential of satellite and reanalysis evaporation datasets for hydrological modelling under various model calibration strategies. Adv. Water Resour. **143**, 103667 (2020a). https://doi.org/10.1016/j.advwatres.2020.103667

M. Dembélé, M. Hrachowitz, H. Savenije, et al., Improving the predictive skill of a distributed hydrological model by calibration on spatial patterns with multiple satellite data sets. Water Resour. Res., 26 (2020b). https://doi.org/10.1029/2019WR026085

M. Dibaj, A.A. Javadi, M. Akrami, et al., Coupled three-dimensional modelling of groundwater-surface water interactions for management of seawater intrusion in Pingtung Plain, Taiwan. J. Hydrol. Reg. Stud. **36**, 100850 (2021). https://doi.org/10.1016/j.ejrh.2021.100850

Q. Duan, J. Schaake, V. Andréassian, et al., Model Parameter Estimation Experiment (MOPEX): An overview of science strategy and major results from the second and third workshops. J. Hydrol. **320**, 3–17 (2006). https://doi.org/10.1016/j.jhydrol.2005.07.031

W.H. Farmer, R.M. Vogel, On the deterministic and stochastic use of hydrologic models. Water Resour. Res. **52**, 5619–5633 (2016). https://doi.org/10.1002/2016WR019129

R.J. Fischer, E.P. Halibozek, D.C. Walters, Contingency planning emergency response and safety. Introd. Secur., 249–268 (2019). https://doi.org/10.1016/B978-0-12-805310-2.00011-1

J.R.B. Fisher, E.A. Acosta, P.J. Dennedy-Frank, et al., Impact of satellite imagery spatial resolution on land use classification accuracy and modeled water quality. Remote Sens. Ecol. Conserv. **4**, 137–149 (2018). https://doi.org/10.1002/rse2.61

S.H.P.W. Gamage, G.A. Hewa, S. Beecham, Modelling hydrological losses for varying rainfall and moisture conditions in South Australian catchments. J. Hydrol. Reg. Stud. **4**, 1–21 (2015). https://doi.org/10.1016/j.ejrh.2015.04.005

L. García-Romero, J. Paredes-Arquiola, A. Solera, et al., Optimization of the multi-start strategy of a direct- search algorithm for the calibration of rainfall- runoff models for water-resource assessment. Water (Switzerland) **11** (2019). https://doi.org/10.3390/w11091876

S. Groom, S. Sathyendranath, Y. Ban, et al., Satellite Ocean colour: Current status and future perspective. Front. Mar. Sci. **6**, 485 (2019). https://doi.org/10.3389/fmars.2019.00485

K. Gupta, Challenges in developing urban flood resilience in India. Philos. Trans. R. Soc. A Math. Phys. Eng. Sci. **378**, 20190211 (2020). https://doi.org/10.1098/rsta.2019.0211

Y. Her, S.-H. Yoo, J. Cho, et al., Uncertainty in hydrological analysis of climate change: Multi-parameter vs. multi-GCM ensemble predictions. Sci. Rep. **9**, 4974 (2019). https://doi.org/10.1038/s41598-019-41334-7

R. Horan, R. Gowri, P.S. Wable, et al., A comparative assessment of hydrological models in the upper cauvery catchment. Water (Switzerland) **13**, 1–25 (2021). https://doi.org/10.3390/w13020151

M. Hrachowitz, H.H.G. Savenije, G. Blöschl, et al., A decade of predictions in ungauged basins (PUB)—A review. Hydrol. Sci. J. **58**, 1198–1255 (2013). https://doi.org/10.1080/02626667.2013.803183

P. Hurskainen, H. Adhikari, M. Siljander, et al., Auxiliary datasets improve accuracy of object-based land use/land cover classification in heterogeneous savanna landscapes. Remote Sens. Environ. **233**, 111354 (2019). https://doi.org/10.1016/j.rse.2019.111354

S.A. Irfan, A. Shafie, N. Yahya, N. Zainuddin, Mathematical modeling and simulation of nanoparticle-assisted enhanced oil recovery – a review. Energies **12**, 1–19 (2019). https://doi.org/10.3390/en12081575

I. Irurzun-Arana, C. Rackauckas, McDonald TO, I.F. Trocóniz, Beyond deterministic models in drug discovery and development. Trends Pharmacol. Sci. **41**, 882–895 (2020). https://doi.org/10.1016/j.tips.2020.09.005

R.K. Jaiswal, S. Ali, B. Bharti, Comparative evaluation of conceptual and physical rainfall–runoff models. Appl Water Sci **10**, 48 (2020). https://doi.org/10.1007/s13201-019-1122-6

J. Jeziorska, T. Niedzielski, Applicability of TOPMODEL in the mountainous catchments in the upper Nysa Kłodzka river basin (SW Poland). Acta Geophys. **66**, 203–222 (2018). https://doi.org/10.1007/s11600-018-0121-6

D. Jiang, K. Wang, The role of satellite-based remote sensing in improving simulated streamflow: A review. Water (Switzerland) **11**, 10–14 (2019). https://doi.org/10.3390/w11081615

M. Khaki, H.-J. Hendricks Franssen, S.C. Han, Multi-mission satellite remote sensing data for improving land hydrological models via data assimilation. Sci. Rep. **10**, 18791 (2020). https://doi.org/10.1038/s41598-020-75710-5

S. Kouadri, C.B. Pande, B. Panneerselvam, K.N. Moharir, A. Elbeltagi, Prediction of irrigation groundwater quality parameters using ANN, LSTM, and MLR models. Environ. Sci. Pollut. Res., 1–25 (2021)

T. Kramm, D. Hoffmeister, A relief dependent evaluation of digital elevation models on different scales for northern Chile. ISPRS Int. J. Geo-Inf. **8** (2019). https://doi.org/10.3390/ijgi8100430

V. Krysanova, C. Donnelly, A. Gelfan, et al., How the performance of hydrological models relates to credibility of projections under climate change. Hydrol. Sci. J. **63**, 696–720 (2018). https://doi.org/10.1080/02626667.2018.1446214

P.S. Kumar, P. Balamurugan, Evaluation of groundwater quality for irrigation purpose in attur taluk, Salem, Tamilnadu, India. Water Energy Int. **61**(4), 59–64 (2018)

P.S. Kumar, P. Balamurugan, Suitability of ground water for irrigation purpose in Omalur Taluk, Salem, Tamil Nadu, India. Indian J. Ecol. **46**(1), 1–6 (2019)

A. Lintern, J.A. Webb, D. Ryu, et al., Key factors influencing differences in stream water quality across space. WIREs Water **5**, e1260 (2018). https://doi.org/10.1002/wat2.1260

Q. Liu, X. Zhao, H. Wang, Y. Sun, Establishment and analysis of spatiotemporal variation hydrological model of distributed rainfall and evaporation in Biliu River Basin. Complexity **2021**, 6639691 (2021). https://doi.org/10.1155/2021/6639691

D.P. Loucks, E. van Beek, System sensitivity and uncertainty analysis, in *Water Resource Systems Planning and Management*, ed. by D. P. Loucks, E. van Beek, (Springer International Publishing, Cham, 2017), pp. 331–374

T. Madsen, K. Franz, T. Hogue, Evaluation of a distributed streamflow forecast model at multiple watershed scales. Water, 12 (2020)

A. Mahala, The significance of morphometric analysis to understand the hydrological and morphological characteristics in two different morpho-climatic settings. Appl Water Sci **10**, 33 (2019). https://doi.org/10.1007/s13201-019-1118-2

I. Manisalidis, E. Stavropoulou, A. Stavropoulos, E. Bezirtzoglou, Environmental and health impacts of air pollution: A review. Front. Public Heal. **8**, 14 (2020). https://doi.org/10.3389/fpubh.2020.00014

A.K. Misra, Climate change and challenges of water and food security. Int. J. Sustain. Built Environ. **3**, 153–165 (2014). https://doi.org/10.1016/j.ijsbe.2014.04.006

H. Momm, R. Bingner, R. Wells, et al., Integrated technology for evaluation and assessment of multi-scale hydrological systems in managing nonpoint source pollution. Water (Switzerland), 13 (2021). https://doi.org/10.3390/w13060842

A. Mosavi, P. Ozturk, K. Chau, Flood prediction using machine learning models: Literature review. Water **10** (2018)

S. Moustakas, P. Willems, Testing the efficiency of parameter disaggregation for distributed rainfall-runoff modelling. Water (Switzerland) **13**, 1–24 (2021). https://doi.org/10.3390/w13070972

M. Msaddek, G. Kimbowa, A. El Garouani, Hydrological modeling of upper OumErRabia Basin (Morocco), comparative study of the event-based and continuous-process HEC-HMS model methods. Comput. Water Energy Environ. Eng. **09**, 159–184 (2020). https://doi.org/10.4236/cweee.2020.94011

C. Müller, J. Elliott, D. Kelly, et al., The global gridded crop model intercomparison phase 1 simulation dataset. Sci. Data **6**, 50 (2019). https://doi.org/10.1038/s41597-019-0023-8

T.J. Mulvany, On the use of self-registering rain and flood gauges in making observations of the relations of rain fall and flood discharges in a given catchment. Proc. Inst. Civ. Eng. **4**, 1–8 (1850)

S. Narbondo, A. Gorgoglione, M. Crisci, C. Chreties, Enhancing physical similarity approach to predict runoff in ungauged watersheds in sub-tropical regions. Water, 12 (2020). https://doi.org/10.3390/w12020528

A.J. Newman, M.P. Clark, K. Sampson, et al., Development of a large-sample watershed-scale hydrometeorological data set for the contiguous USA: Data set characteristics and assessment of regional variability in hydrologic model performance. Hydrol. Earth Syst. Sci. **19**, 209–223 (2015). https://doi.org/10.5194/hess-19-209-2015

E. Nkiaka, N.R. Nawaz, J.C. Lovett, Effect of single and multi-site calibration techniques on hydrological model performance, parameter estimation and predictive uncertainty: A case study in the Logone catchment, Lake Chad basin. Stoch Environ. Res. Risk Assess **32**, 1665–1682 (2018). https://doi.org/10.1007/s00477-017-1466-0

B. Panneerselvam, S. Karuppannan, K. Muniraj, Evaluation of drinking and irrigation suitability of groundwater with special emphasizing the health risk posed by nitrate contamination using nitrate pollution index (NPI) and human health risk assessment (HHRA). Hum. Ecol. Risk Assess. Int. J. **27**(5), 1324–1348 (2020a)

B. Panneerselvam, S.K. Paramasivam, S. Karuppannan, N. Ravichandran, P. Selvaraj, A GIS-based evaluation of hydrochemical characterisation of groundwater in hard rock region, South Tamil Nadu, India. Arab. J. Geosci. **13**(17), 1–22 (2020b)

B. Panneerselvam, K. Muniraj, M. Thomas, et al., Identifying influencing groundwater parameter on human health associate with irrigation indices using the Automatic Linear Model (ALM) in a semi- arid region in India. Environ. Res. **202** (2021a). https://doi.org/10.1016/j.envres.2021.111778

B. Panneerselvam, K. Muniraj, C. Pande, N. Ravichandran, Prediction and evaluation of groundwater characteristics using the radial basic model in Semi-arid region, India. Int. J. Environ. Anal. Chem., 1–17 (2021b)

B. Panneerselvam, K. Muniraj, M. Thomas, N. Ravichandran, GIS-based legitimatic evaluation of groundwater's health risk and irrigation susceptibility using water quality index, pollution index, and irrigation indexes in semiarid region, in *Groundwater Resources Development and Planning in the Semi-Arid Region*, 1st edn., (Springer International Publishing, 2021c), pp. 239–268

B. Panneerselvam, K. Muniraj, M. Thomas, N. Ravichandran, B. Bidorn, Identifying influencing groundwater parameter on human health associate with irrigation indices using the Automatic Linear Model (ALM) in a semi-arid region in India. Environ. Res. **202**, 111778 (2021d)

B. Panneerselvam, S. Pauline, K. Muniraj, M. Thomas, Assessment of inverse fluidized bed reactor on the treatment efficiency of distillery spent wash water. Int. J. Environ. Sci. Technol. (2021e). https://doi.org/10.1007/s13762-021-03650-2

B. Panneerselvam, K. Muniraj, C. Pande, N. Ravichandran, M. Thomas, S. Karuppannan, Geochemical evaluation and human health risk assessment of nitrate-contaminated groundwater in an industrial area of South India. Environ. Sci. Pollut. Res., 1–18 (2021f)

A. Paschalis, P. Molnar, S. Fatichi, P. Burlando, A stochastic model for high-resolution space-time precipitation simulation. Water Resour. Res. **49**, 8400–8417 (2013). https://doi.org/10.1002/2013WR014437

M. Paudel, E. Nelson, C. Downer, R. Hotchkiss, Comparing the capability of distributed and lumped hydrologic models for analyzing the effects of land use change. J. Hydroinf. **13**, 461 (2011). https://doi.org/10.2166/hydro.2010.100

J. Peng, T. Liu, Y. Huang, et al., Satellite-based precipitation datasets evaluation using gauge observation and hydrological modeling in a typical arid land watershed of central asia. Remote Sens. **13**, 1–26 (2021). https://doi.org/10.3390/rs13020221

K. Priya, B. Panneerselvam, K. Muniraj, et al., Integration of multi criteria decision analysis and GIS for evaluating the site suitability for aquaculture in southern coastal region, India. Mar. Pollut. Bull. **172**, 112907 (2021). https://doi.org/10.1016/j.marpolbul.2021.112907

A. Rafiei Emam, M. Kappas, S. Fassnacht, N.H.K. Linh, Uncertainty analysis of hydrological modeling in a tropical area using different algorithms. Front. Earth Sci. **12**, 661–671 (2018). https://doi.org/10.1007/s11707-018-0695-y

A. Raj, A. van Oudenaarden, Nature, nurture, or chance: Stochastic gene expression and its consequences. Cell **135**, 216–226 (2008). https://doi.org/10.1016/j.cell.2008.09.050

C. Raj, A. Pavizham, K. Sudheer, I. Chaubey, Application of distributed hydrological models for predictions in ungauged basins: A method to quantify predictive uncertainty. Hydrol Proc., 28 (2014). https://doi.org/10.1002/hyp.9721

K.S. Rawat, S.K. Singh, Estimation of surface runoff from semi-arid ungauged agricultural watershed using SCS-CN method and earth observation data sets. Water Conserv. Sci. Eng. **1**, 233–247 (2017). https://doi.org/10.1007/s41101-017-0016-4

S. Razavi, A. Jakeman, A. Saltelli, et al., The future of sensitivity analysis: An essential discipline for systems modeling and policy support. Environ. Model. Softw. **137**, 104954 (2021). https://doi.org/10.1016/j.envsoft.2020.104954

R. Revelli, A. Porporato, Ecohydrological model for the quantification of ecosystem services provided by urban street trees. Urban Ecosyst. **21**, 489–504 (2018). https://doi.org/10.1007/s11252-018-0741-2

H.-K. Ringkjøb, P.M. Haugan, I.M. Solbrekke, A review of modelling tools for energy and electricity systems with large shares of variable renewables. Renew. Sust. Energ. Rev. **96**, 440–459 (2018). https://doi.org/10.1016/j.rser.2018.08.002

J. Rosselló, S. Becken, M. Santana-Gallego, The effects of natural disasters on international tourism: A global analysis. Tour. Manag. **79**, 104080 (2020). https://doi.org/10.1016/j.tourman.2020.104080

N. Schaller, J. Sillmann, M. Müller, et al., The role of spatial and temporal model resolution in a flood event storyline approach in western Norway. Weather Clim. Extrem. **29**, 100259 (2020). https://doi.org/10.1016/j.wace.2020.100259

P.C. Shakti, T. Nakatani, R. Misumi, The role of the spatial distribution of radar rainfall on hydrological modeling for an urbanized river Basin in Japan. Water, 11 (2019). https://doi.org/10.3390/w11081703

K. Shunmugapriya, B. Panneerselvam, K. Muniraj, N. Ravichandran, P. Prasath, M. Thomas, K. Duraisamy, Integration of multi criteria decision analysis and GIS for evaluating the site suitability for aquaculture in southern coastal region, India. Mar. Pollut. Bull. **172**, 112907 (2021)

C.I. Siettos, L. Russo, Mathematical modeling of infectious disease dynamics. Virulence **4**, 295–306 (2013). https://doi.org/10.4161/viru.24041

V.P. Singh, Review of water flow in soils, 2nd Ed., by Tsuyoshi Miyazaki. J Hydrol Eng **11**, 291 (2006). https://doi.org/10.1061/(ASCE)1084-0699(2006)11:3(291)

V.P. Singh, Kinematic wave theory of overland flow. Water Resour. Manag. **31**, 3147–3160 (2017). https://doi.org/10.1007/s11269-017-1654-1

V.P. Singh, Hydrologic modeling: Progress and future directions. Geosci. Lett. **5**, 15 (2018). https://doi.org/10.1186/s40562-018-0113-z

V.P. Singh, D.A. Woolhiser, Mathematical modeling of watershed hydrology. J. Hydrol. Eng. **7**, 270–292 (2002). https://doi.org/10.1061/(ASCE)1084-0699(2002)7:4(270)

V.P. Singh, L. Zhang, Copula–entropy theory for multivariate stochastic modeling in water engineering. Geosci. Lett. **5**, 6 (2018). https://doi.org/10.1186/s40562-018-0105-z

M. Sivapalan, K. Takeuchi, S.W. Franks, et al., IAHS decade on predictions in ungauged basins (PUB), 2003–2012: Shaping an exciting future for the hydrological sciences. Hydrol. Sci. J. **48**, 857–880 (2003). https://doi.org/10.1623/hysj.48.6.857.51421

L.J. Slater, G. Thirel, S. Harrigan, et al., Using R in hydrology: A review of recent developments and future directions. Hydrol. Earth Syst. Sci. **23**, 2939–2963 (2019). https://doi.org/10.5194/hess-23-2939-2019

S. Sorooshian, K. Hsu, E. Coppola, et al., *Hydrological Modelling and the Water Cycle*, 1st edn. (Springer, Berlin Heidelberg, 2008)

Q. Sun, C. Miao, Q. Duan, et al., A review of global precipitation data sets: Data sources, estimation, and intercomparisons. Rev. Geophys. **56**, 79–107 (2018). https://doi.org/10.1002/2017RG000574

S. Surampudi, K. Yarrakula, Mapping and assessing spatial extent of floods from multitemporal synthetic aperture radar images: A case study on Brahmaputra River in Assam state, India. Environ. Sci. Pollut. Res. **27**, 1521–1532 (2020). https://doi.org/10.1007/s11356-019-06849-6

D. Suresh, K. Yarrakula, InSAR based deformation mapping of earthquake using Sentinel 1A imagery. Geocarto Int, 1–10 (2019). https://doi.org/10.1080/10106049.2018.1544289

D. Suresh, K. Yarrakula, Evaluation of Sentinel 1–derived and open-access digital elevation model products in mountainous areas of Western Ghats. India. Arab J. Geosci. (2020). https://doi.org/10.1007/s12517-020-06108-w

D. Suresh, J. Collins Johnny, B.K. Jayaprasad, et al., Morphometric analysis for identification of groundwater recharge zones: A case study of Neyyar river basin. Indian J. Geo-Marine Sci. **47**, 1969–1979 (2018)

D. Suresh, M.G. Priya, B. Vaishnavi, Comparison of streams delineated using various SRTM, ASTER and Cartosat DEM. Indian J. Ecol. **48**, 757–759 (2021)

F. Teng, W. Huang, Y. Cai, et al., Application of hydrological model PRMS to simulate daily rainfall runoff in Zamask-Yingluoxia Subbasin of the Heihe River Basin. Water **9** (2017a). https://doi.org/10.3390/w9100769

J. Teng, A.J. Jakeman, J. Vaze, et al., Flood inundation modelling: A review of methods, recent advances and uncertainty analysis. Environ. Model. Softw. **90**, 201–216 (2017b). https://doi.org/10.1016/j.envsoft.2017.01.006

L. Uusitalo, A. Lehikoinen, I. Helle, K. Myrberg, An overview of methods to evaluate uncertainty of deterministic models in decision support. Environ. Model. Softw. **63**, 24–31 (2015). https://doi.org/10.1016/j.envsoft.2014.09.017

B. Vaishnavi, K. Yarrakula, J. Karthikeyan, Flood inundation mapping of lower Godavari River basin using remote sensing and GIS. Indian J. Ecol. **47**, 30–35 (2020)

T. van Emmerik, G. Mulder, D. Eilander, et al., Predicting the ungauged basin: Model validation and realism assessment. Front. Earth Sci. **3**, 62 (2015). https://doi.org/10.3389/feart.2015.00062

T. Van Der Valk, J.H. Van Driel, W. De Vos, Common characteristics of models in present-day scientific practice. Res. Sci. Educ. **37**, 469–488 (2007). https://doi.org/10.1007/s11165-006-9036-3

G. Wang, S. Mang, H. Cai, et al., Integrated watershed management: Evolution, development and emerging trends. J. For. Res. **27**, 967–994 (2016). https://doi.org/10.1007/s11676-016-0293-3

H. Wu, B. Chen, X. Ye, et al., An improved calibration and uncertainty analysis approach using a multicriteria sequential algorithm for hydrological modeling. Sci. Rep. **11**, 16954 (2021). https://doi.org/10.1038/s41598-021-96250-6

Y. Zhu, L. Liu, F. Qin, et al., Application of the regression-augmented regionalization approach for BTOP model in ungauged basins. Water **13** (2021). https://doi.org/10.3390/w13162294

Chapter 18
Appraising the Groundwater Potential of Liddar Sub-Basin (Western Himalayas) Using Geospatial Techniques

Suhail A. Lone and Ghulam Jeelani

Abstract Groundwater is one of the essential freshwater sources after the cryosphere in the Himalayas, which plays an important role in terrestrial and aquatic life systems. Evaluating the potential groundwater recharge zones is vital to check water quality and for sustainable development. In this chapter, geospatial methods were used to recognize and map the prospective zones to assess the groundwater availability in the Liddar sub-basin. The various thematic layers, namely, geology, geomorphology, soil, land use/land cover (LU/LC), lineaments, drainage network, and slope, were generated using Landsat Thematic Mapper (TM) and Google Earth™ imageries and transferred into raster data to identify the potential groundwater zones and generate a groundwater potential map. The satellite data, lithology, LU/LC database, and field validations were used to derive the weight of elements contributing to the groundwater recharge. The score and weight calculated from multi-influencing element techniques were allocated to the raster maps of these elements. It is observed that 105 km^2, 577 km^2, and 601 km^2 are categorized as low, medium, and good groundwater prospective zones, accounting for 8.1%, 45%, and 46% of the total area of the sub-basin, respectively. The results imply that the primary recharge areas lie in the downstream region. In contrast, the least potential zones are found in the upstream area of the basin owing to the steep gradient and less infiltration rates due to massive hard rock massifs. Due to continuous increase in temperatures and the retreating trend of the Himalayan glaciers, such studies are extremely crucial for sustainably managing groundwater in the Himalayas.

Keywords Liddar sub-basin · Spatial analysis · Landsat · Himalayas · Groundwater

S. A. Lone (✉) · G. Jeelani
Department of Earth Sciences, University of Kashmir, Srinagar, India

B. Panneerselvam et al. (eds.), *Climate Change Impact on Groundwater Resources*, https://doi.org/10.1007/978-3-031-04707-7_18

18.1 Introduction

Due to the inadequacy of potable waters for domestic and other allied purposes, groundwater is attracting increasing attention (Lone et al. 2016; Hussein et al. 2017; Shah and Lone 2019; Lone et al., 2020, 2021, 2021a). It has become vital not only in monitoring and preserving this significant resource but also in delineating the groundwater recharge potential zones (Lone et al. 2016, 2019; Shah and Lone 2019; Jeelani et al. 2020; Lone et al. 2020). Groundwater exploration, particularly in massive hard rocks, involves ample knowledge of the physical geology, surface geomorphology, and structural geology of the region, influenced by topographical features like the rate of mass wasting, degree of fractures, porosity, gradient, drainage, landscape, land use/land cover (LU/LC), and climate (Yeh et al. 2009; Lone et al. 2016; Shah and Lone 2019; Jeelani et al. 2020, 2021a, b). The availability of groundwater in hard rock terrains is extremely less. In such terrains, groundwater is confined to secondary fractures/cracks and weathered horizons (Jeelani et al. 2020, 2021a). A prompt and economical allocation of groundwater potential zones is performed using a methodical integration of numerous datasets followed by hydrogeomorphological analysis. Hence, an integrated methodology for groundwater delineation is necessary. Geospatial technologies (remote sensing and geographic information system (GIS)) are widely used for managing different natural resources (Dar et al. 2010; Magesh et al. 2011). Demarcating the prospective groundwater zones using geospatial means is highly effective (Lone et al. 2016; Shah and Lone 2019). In last few decades, the widespread use of geospatial technologies along with ancillary data has made it easy to collect basic information about groundwater resources (Tweed et al. 2007; Chowdhury et al. 2010; Jaan et al. 2015; Lone et al. 2016). Remote sensing is also extensively used to distinguish the physiography of Earth's surface (e.g. tectonic structures, drainage networks) and also to demarcate the groundwater prospective areas for better water resource management (Sener et al. 2005).

Groundwater is one of the substantial sources of freshwater for domestic and other recreational purposes in the upper Jhelum basin (UJB). Due to a steeper gradient and hard rock terrains surrounding the valley, precipitation at high altitudes drains directly and quickly into tributaries and streams. Spatial as well as temporal water resources and water demands are unevenly disseminated. Water scarcity has become a significant issue due to global climate change variations (Lone et al., 2019, 2021b). An escalation in water consumption for the development of the economy has resulted in the scarcity of surface water resources. Therefore, groundwater exploration has increased, leading to its overexploitation, instigating serious environmental issues such as a decline in the water table, and groundwater contamination, leading to stagnation of water quality. Groundwater in the Liddar sub-basin is extracted from shallow springs, dug wells, and bore wells for domestic and agricultural purposes. The groundwater recharge has reduced due to variability in climatic and anthropogenic activities (Jeelani et al. 2021a, b). The average minimum depth to the water table is 5 m in areas along the river valleys; however, the water level in high-altitude areas is found to vary from 50 to 60 m, thus causing serious health problems and threatening the population and, consequently, the economy of the region. Hence, it is essential to systematically comprehend the water resources of the upper Jhelum basin, particularly the groundwater, for better planning and management at the local and regional scales. Hence, the main purpose of this study is to classify and map the groundwater potential zones of the Liddar sub-basin of the upper Jhelum basin using geospatial methods (remote sensing and GIS). This study will provide a better understanding for identifying the potential groundwater zones that will ultimately help decision-makers and hydrogeologists to effectively plan for better groundwater management and ecosystem restoration.

18.2 Study Area

Liddar is an upland snow- and glacier-dominated mountainous sub-basin of the upper Jhelum basin (UJB) and lies in the Kashmir Valley (in the district of Anantnag) between geographical coordinates of latitudes 33°45′ and 34°15′ N and longitudes 75°15′ and 75°30′ E, covering an area of 1283 km^2 (Fig. 18.1).

The climate of the Liddar sub-basin is of the temperate-type. The sub-basin receives an average annual precipitation of 1240 mm (Lone et al. 2017) with the maximum in winter and early spring mostly in solid form ((snow; Lone et al. 2017)). The mean maximum and minimum precipitation values are recorded in March (183 mm) and November (36 mm), respectively, whereas the mean maximum and minimum temperatures are recorded in July (24 °C) and January (1.2 °C), respectively (Lone et al. 2017). The vegetation of the Liddar sub-basin is variable, ranging from rich marsh and aquatic vegetation at low-lying areas to lush evergreen conifers at higher altitudes. Conquering the major portion of its land area, its terrestrial vegetation ranges from domesticated plant systems at lower elevations to natural green forests and pastures at the mountain level. The Liddar sub-basin is covered by many glaciers, with two of the largest glaciers, viz., Kolahoi and Sheshram, covering an area of 16.38 km^2 (Lone et al. 2017).

18.3 Data Collection

The Landsat Thematic Mapper (TM)s, Enhanced Thematic Mapper Plus (ETM+) data (24 October 2019, Path-148-49/Row-36-37) images (Path-106/Row-058) and Advanced Spaceborne Thermal Emission and Reflection Radiometer (ASTER) Digital Elevation Model (DEM) of 30 m resolution downloaded from the Earth Explorer website (https://earthexplorer.usgs.gov/) were used. In this study, Landsat satellite images of the month of October were used due to the presence of the least snow cover during this period. The data acquired from the satellite images were validated in the field using a hand-held Global Positioning System (GPS). Survey of

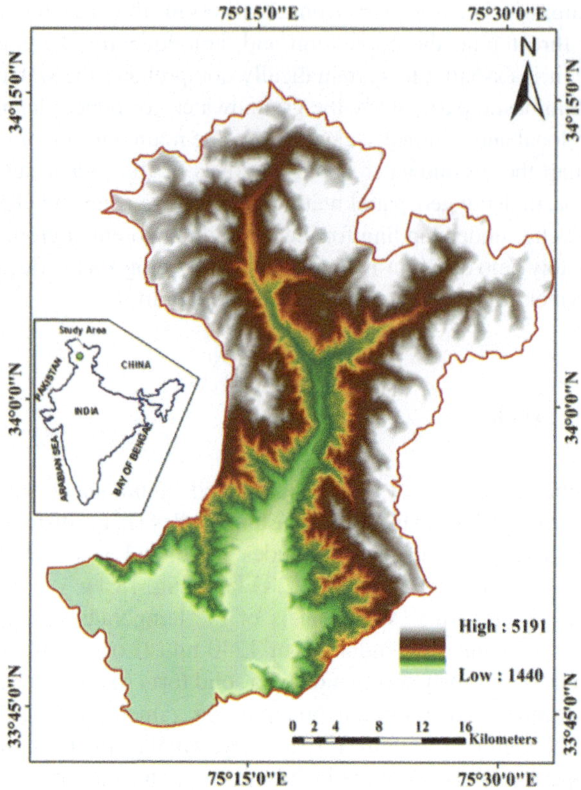

Fig. 18.1 Study area (Liddar sub-basin)

India (SOI) toposheets (No. 43 N/4, 43 N/8 at 1: 50,000 scales) were also used. A lithological map was digitized from Middlemiss 1910, whereas groundwater table data were obtained from the report of the Central Ground Water Board (CGWB 2018). Geometic_2015 software was used to delineate the lineaments from satellite imagery. ERDAS 9.1 software was used for image processing, whereas the geographic information system (GIS)-based software ArcGIS 10.4.1 was used to generate different thematic layers. After collecting the rainfall data from the Indian Meteorological Department (IMD), Srinagar, a rainfall map was generated using the Thiessen polygon method.

18.4 Materials and Methods

Geospatial techniques (remote sensing and GIS) along with field validations and multiple types of satellite imageries were employed to demarcate and map the potential groundwater areas of the Liddar sub-basin. The satellite data (Landsat TM,

ETM+, and OLI) from the Earth Explorer (30-metre resolution), supported by Google Earth™ imagery (5 m resolution), were employed. The base map of the sub-basin was created using ASTER DEM (30 m resolution), overlain by a false-colour composite (FCC) of satellite images. The false-colour composite (FCC) properties, standard image features, and on-screen digitization were used to extract different elements (lithology, geomorphological, drainage, slope, and soil types) in the ArcGIS 10.4.1 environment. The structural features (fractures, faults) were extracted using Geometic_2015. A land use/land cover (LU/LC) map was generated using the image classification properties of ArcGIS 10.4.1. The LU/LC information of the study area was obtained using interactive supervised classification that involves selection of training sites as reference for the categorization using the maximum likelihood classification method on Landsat satellite images. On an average, 15 reference sites were selected for each class. The accuracy assessment of the classes obtained was performed with the groundwater truth data collected using the Global Positioning System (GPS). The Arc Hydro tool and raster surface tool properties of ArcGIS were used to prepare the slope and drainage maps, respectively. ASTER DEM was the first to be analysed for a stream grid in the Arc Hydro tool, and, then, the drainage map was generated using the hydrology module in the Spatial Analyst tool of the Arc Toolbox. Similarly, the slope map was generated using the raster surface property of the 3D Analyst tool in the Arc Toolbox. The soil map of the study area was generated from the soil map of the Kashmir Valley provided by the Department of Agriculture using the on-screen digitization method. The factors affecting the recharge processes/properties and the interaction between the elements were studied. Weightage to elements was assigned depending on the importance of each element. All the thematic layers were overlapped in a weighted overlay property of the ArcMap to generate a groundwater potential map of the Liddar sub-basin.

18.4.1 Relationship Among the Elements

A weighted overlay method based upon integrating the influential factor layers in the GIS environment has proved beneficial for delineating groundwater potential areas. Authors across the globe have used the weighted overlay method to incorporate different thematic layers in order to delineate potential groundwater zones in massive hard rock and alluvial basins (Shaban et al. 2006; Yeh et al. 2009; Subagunasekar and Sashikkumar, 2012; Krishnamurthy et al. 2000). A similar methodology has been adopted in different parts of the world to identify potential zones in hard and soft rock landscapes (Saraf and Choudhury, 1998, Sashikkumar and Metilda, 2012). In this study, the lithology, geomorphology, structural data, land use/land cover data, drainage network, soil, and gradient affecting the groundwater potential of the high-altitude snow- and glacier-dominant sub-basin are displayed in Table 18.1. In this study, we used seven different factors that affect the groundwater potential of the Liddar sub-basin, with a view that there exists a relationship between influencing factors, which affects the recharge potential of the

Table 18.1 Effects of influencing factors, relative rates, and score for each potential factor

Factors	Major effect	Minor effect	Proposed relative effect (A + B)	Proposed score of each influencing factor
Lithology	1 + 1	0	2	20
Land use	1 + 1	0.5 + 0.5	3	22
Drainage	1	0.5	1.5	9
Slope	1 + 1	0.5	2.5	16
Rainfall	1	0.5	1.5	9
Soil	1	0	1	6
Geomorphology	1 + 1	0.5	2.5	5
Lineaments	1 + 1	0	2	13
			Σ16	**Σ100**

sub-basin, revealing the major and minor relationships among the different elements influencing the groundwater potential of the sub-basin. Each association is weighted based upon its asset. The characteristic weight of an element is the total sum of all weights from each factor. An element with a greater weighted value exhibits a larger effect on groundwater recharge. Spatial assimilation and examination were achieved in a GIS environment to validate the groundwater potential zones. This study examined the hydrological and geographical characteristics of the Liddar sub-basin and recognized seven major elements affecting the recharge potential. Each element was examined and allocated a suitable weight.

18.5 Results and Discussion

18.5.1 Geomorphological Features

The geomorphology of a particular area significantly determines its groundwater potential and occurrence. The main geomorphological features identified and mapped in the sub-basin include river deposits, Piedmont deposits, moderately fractured and glaciated landscapes/highly fractured hills, river valleys, glacial lakes, and denudational hills (Figs. 18.2 and 18.3; field evidences). River deposits, moderately fractured hills, and glaciated landscapes are the main geomorphological features of the sub-basin.

Recently, alluvium and Piedmont deposits have been the potential groundwater sites because their unconsolidated nature leads to the infiltration of surface water into groundwater; however, due to higher slope and less petrophysical properties, massive glaciated terrains and mountain ridges lead to more surface runoff than infiltration (Lone et al. 2016). However, these hard rock terrains can act as good reservoirs of groundwater if secondary porosity is well-developed and interconnected.

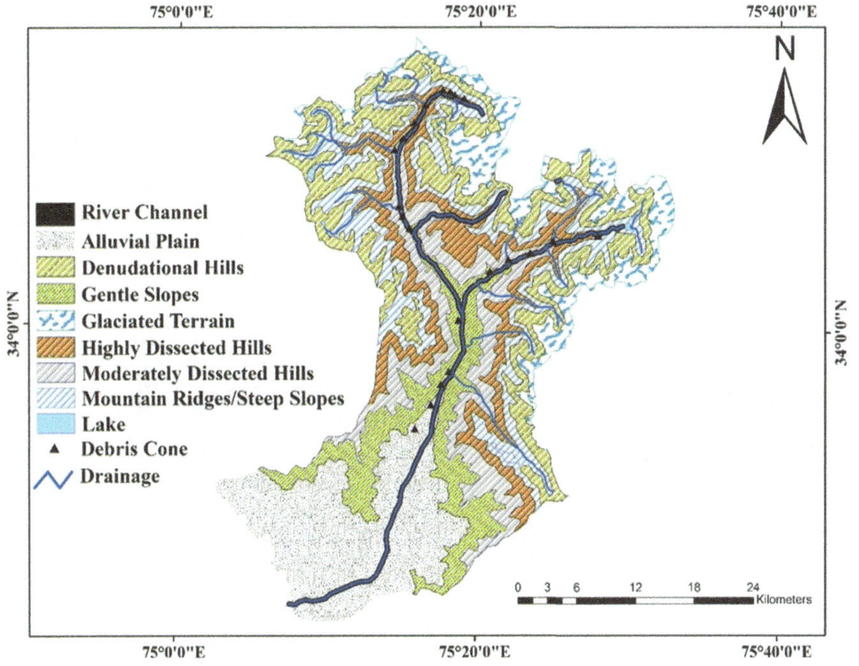

Fig. 18.2 Geomorphology of the study area

Fig. 18.3 Field photographs of various geomorphic units in the Liddar sub-basin. (**a**) River channel; (**b**) highly dissected hills; (**c**) debris cones; (**d**) barren land; (**e**) forests; and (**f**) denudational hills

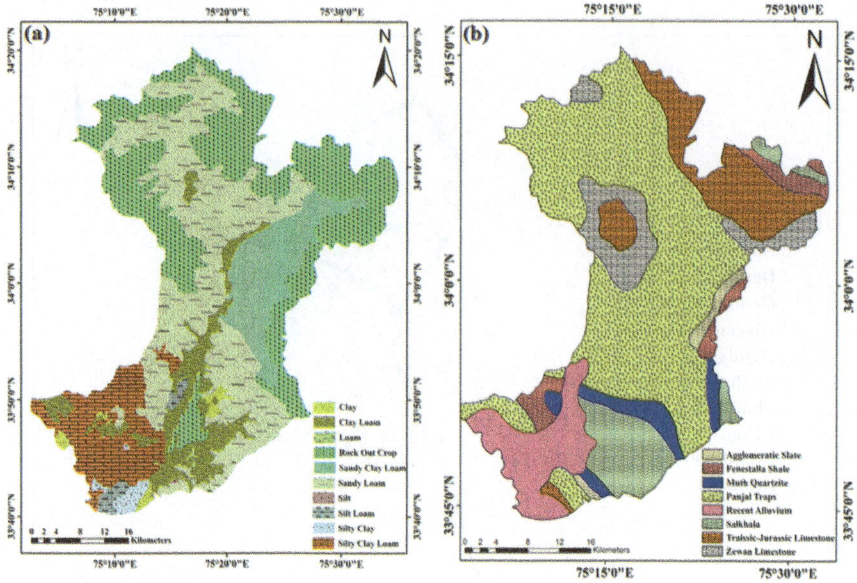

Fig. 18.4 (**a**) Soil map of the Liddar sub-basin. (**b**) Lithological map of the Liddar sub-basin

18.5.2 Soil

Soil acts as a dominant factor in developing groundwater potential zones of a particular area due to its good porosity and permeability characteristics. The downward movement of water and then percolation through different layers of soil is highly dependent upon the porosity of different soil types. The different soil types, which were identified and mapped in the Liddar sub-basin of the upper Jhelum basin, includes clay, clay with loam, loam, rock outcrop, clay loam with sand, sandy loam, silt, silt loam, silty clay, and silty clay loam (Fig. 18.4a). It can be well established from the soil map of the sub-basin that the lower area has more infiltration rates due to the dominance of alluvial soil (sand, sandy loam, silty clay) brought down by different streams. However, the upper area can have less infiltration rates due to the dominance of hard rock outcrops, which leads to surface runoff. The different soil types dominant in the Liddar sub-basin of the upper Jhelum basin are shown in Fig. 18.4a.

18.5.3 Lithological Information

Geology is one of the main factors controlling the occurrence, movement, and storage of groundwater. The types of water-bearing discontinuities, depth, weathering rate, and petrophysical properties are dominantly controlled by the particular rock type. In addition, the geological features impact the production and nature of water-bearing zones along which water can move easily. The lithology of a particular area is an important factor affecting the quantitative and qualitative occurrence of groundwater (Bhuvaneswaran et al. 2015; Shah and Lone 2019). In the Liddar sub-basin, Panjal Traps ((basalt/rhyolite (Carboniferous)) are the dominant lithology (Fig. 18.4b), followed by Triassic–Jurassic Limestone, Archean Salkhala, Devonian Muth Quartzite, Carboniferous Agglomeratic Slate and Fenestella Shale, Permian Zewan Limestone, and recent alluvium. The lithology affects the porosity and permeability of the confined and unconfined aquifers. The groundwater zonation map showed that the most appropriate groundwater prospective area encompassed the lithological units of limestone and alluvium because of their highly developed secondary porosity and permeability. However, each lithological unit behaves differently in determining and controlling groundwater recharge and discharge in the basin.

18.5.4 Drainage Density

Drainage density defines the closeness of the spacing of stream channels. It measures the total length of the stream segments per unit area. Drainage density is the total drainage length/unit area and is calculated as:

$$D_d = \frac{\sum_{i=0}^{i=n} Si}{A} \tag{18.3}$$

where $\sum_{i=n}^{i=0} Si$, is the total drainage length (L) and A is the unit area (L^2).

Drainage density is an opposite function of intrinsic permeability. Permeability dominantly controls the infiltration of rainfall. Due to highly developed, secondary interconnected porosity, the high-drainage-density area shows a zone with a high infiltration rate. The Liddar watershed is dominated by sedimentary rocks like limestone. The karstified geomorphology, prominent in the area in the form of numerous springs, blind valleys, etc., has resulted in well-developed, interconnected secondary pores. The drainage density is sometimes greater than 2.5 km/km^2, characteristic of the best groundwater recharge zones. The upper and middle streams are typically 0.50–1.5 km/km^2. In this study, two drainage density assemblies were recognized (Fig. 18.5a).

(1) An extremely fine drainage texture (drainage density) is observed and delineated in the southern part of the Liddar sub-basin, whereas coarse, textured zones are towards the eastern, western, and northern flanks of the Liddar sub-basin. (2) The moderate- and low-drainage-density classes occupy the south-eastern part of the Liddar sub-basin. In the southern part of the Liddar sub-basin (Fig. 18.5a), lithological settings and structural features dominantly control the drainage texture and pattern of the sub-basin. In the Liddar sub-basin, generally, a dendritic to sub-dendritic drainage pattern is observed. The promising groundwater potential regions are

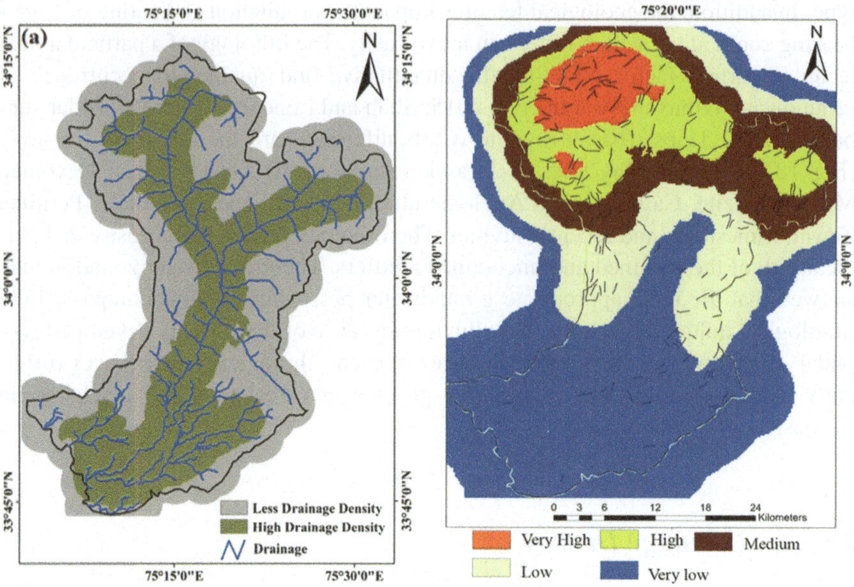

Fig. 18.5 (**a**) Drainage density cum drainage map of the Liddar sub-basin. (**b**) Lineament density map of the Liddar sub-basin

the least in those parts of the sub-basin where extremely high drainage density exists, which allows less water to infiltrate into the subsurface and vice versa.

18.5.5 Lineament Density

The structurally controlled linear features identified on the satellite imagery by their relatively linear alignments are known as lineaments. Lineaments depict the surface morphology of the subsurface features. These features reveal porosity and permeability instigated due to the faulting and fracturing of rocks. Lineaments are extremely vital because they provide the subsurface flowpaths for groundwater movement, and, therefore, their existence defines a permeable zone. The lineament density is the total lineament length/unit area and is calculated as:

$$L^d = \frac{\sum_{i=0}^{i=n} Li}{A} \tag{18.4}$$

where $\sum_{i=n}^{i=0} Li$ is the total lineament length (L) and A is the unit area (L^2).

The course lineament density depicts high porosity, demonstrating promising areas of groundwater (Fig. 18.5b). In high-lineament-density areas, the surface water easily infiltrates into the subsurface due to the interconnectivity of pores, and, hence, these have significant potential for groundwater development. Such areas act as good reservoirs of groundwater, and it has been reported that most of the springs are found along lineaments.

18.5.6 Slope

Slope is an another factor used to delineate groundwater potential zones. Higher slopes result in quick runoff and increased erosion rate with less infiltration. The slope map of the sub-basin was created using ASTER DEM in the ArcGIS environment. The slope grid is defined as "the maximum rate of change in value from each cell to its neighbors". It is obvious from the map that the basin can be classified into four classes. The downstream area of the Liddar sub-basin is covered by man-made structures and agricultural/horticultural lands, with slope values ranging from 0 to 10 (Fig. 18.6a). The slope angle value varying between 10 and 30 is categorized as gentle-to-moderate, covered by natural plantations such as forests. The slope values of 30–60 lying towards the north-eastern and northern parts of the Liddar sub-basin are dominated by barren and glaciated settings. The slope of the region is one of the governing factors determining the groundwater potential and occurrence. The plainer areas with lesser slope hold the surface runoff (rainfall) for longer durations and therefore enable a greater proportion of recharge than do steep gradients with rapid runoff (Lone et al. 2016; Shah and Lone 2019).

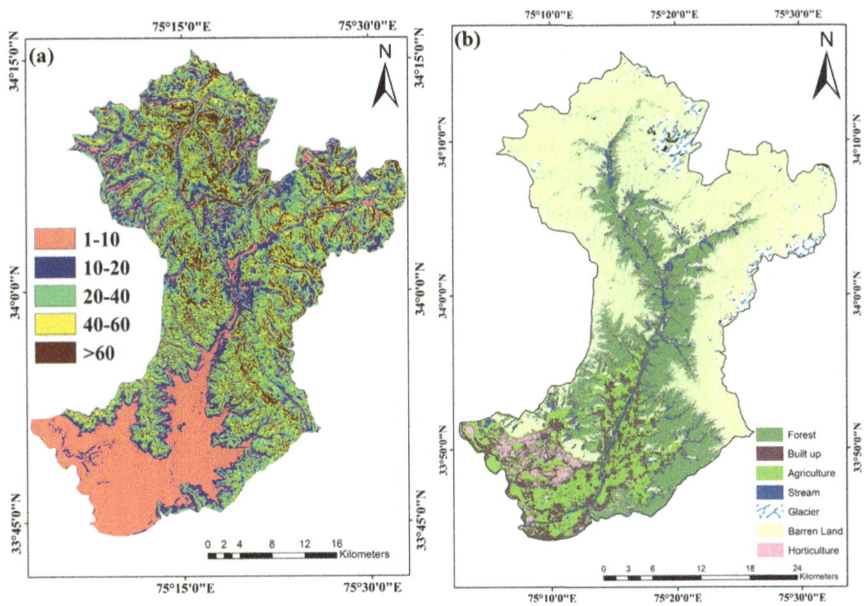

Fig. 18.6 (**a**) Slope map of the Liddar sub-basin. (**b**) Land use/land cover of the Liddar sub-basin

18.5.7 *Land Use/Land Cover (LU/LC)*

The climate, geomorphological features, and agroecological and man-made built-up dominantly impact the LU/LC (land use/land cover) of the region, which is an important element influencing the potential and occurrence of groundwater. In this study, seven classes were delineated including man-made structures, jungles, cultivation area, barren land, water bodies, snow, and glaciers. These classes were delineated using a supervised image classification method that involves selection of training sites as reference for the categorization in ArcMap 10.4.1 (Fig. 18.6b). The accuracy assessment of the classes obtained was performed with the groundwater truth data collected using the Global Positioning System (GPS). The overall accuracy of the maximum likelihood classification was 88.8%, and the Kappa coefficient was 0.82. The variety and environment of LU/LC controlling groundwater in the Liddar sub-basin of the upper Jhelum basin are forest, agriculture/horticulture, barren land, and man-made structures. Due to the high infiltration rate, the forests and agricultural/horticultural areas are most feasible areas for groundwater occurrence. Crop cultivation is dominantly controlled by the availability of water. Hence, the presence of a good amount of groundwater in an area intensifies the degree of cultivation.

18.5.8 *Rainfall*

Rainfall plays an important role in recharging groundwater directly. It governs the quantity of water available that would infiltrate the subsurface (Fig. 18.7). The annual average rainfall in the Liddar watershed of the UJB is around 1137 mm (Lone et al. 2017). The mid-latitude circulations (western disturbances) account for 71% of the total precipitation mostly in solid form during winter. However, the rest of precipitation is brought about by the Indian summer monsoon (ISM). The perennial flow in the Liddar stream is dominantly influenced by meltwaters generated from the winter precipitation and glacier melt. The Liddar watershed depends mainly on precipitation, which is brought about by the mid-latitude circulations. The distribution of precipitation along with the gradient of the basin affects the infiltration rate of runoff and hence escalates the probability of groundwater prospective zones.

18.6 Groundwater Prospecting and Validation

In this study, the groundwater map showing potential zones was generated based on the combination of different thematic layers in ArcGIS 10.4.1 (Fig. 18.8). The spatially distributed potential zones (Fig. 18.8) were assessed and classified as good,

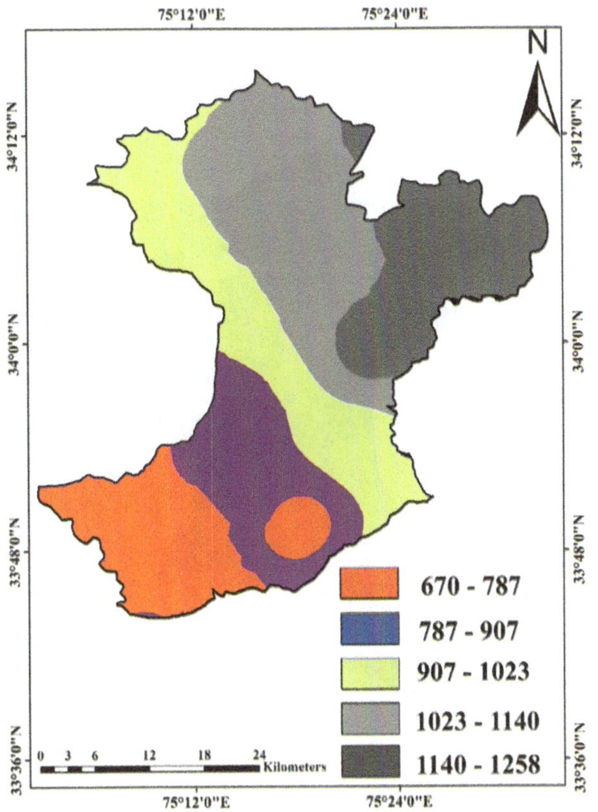

Fig. 18.7 Rainfall distribution map of the Liddar sub-basin, showing that higher rainfall is received at higher altitudes than at lower altitudes

moderate, and low. In the Liddar sub-basin, 105 km^2, 577 km^2, and 601 km^2 were categorized as low, moderate, and good potential zones, which account for 8.1%, 45%, and 46% of the total area of the sub-basin, respectively. The study further suggested that the promising potential groundwater zones lie towards the downstream part of the basin due to the presence of huge amounts of alluvium, gentle slopes, and cultivated land areas.

Besides, the drainage density in a region suggests that groundwater recharge largely depends upon streamflow. The upper high-altitude region of the sub-basin is the least significant, dominated by massive hard rocks with sharp gradients. The potential groundwater map generated was substantiated with ground observations ((the depth to the water table data was measured in different seasons by the Central Ground Water Board (CGWB, 2018)), and it was seen that the areas with poor-to-moderate groundwater potential have an extremely deep water table, whereas those with high-to-very-high potential have shallow depth to the water table and show less variability in different seasons. The average yearly water table fluctuation varies from 1 to 9 mbgl. The water level in wells is flat and shallow (<3mbgl) in high

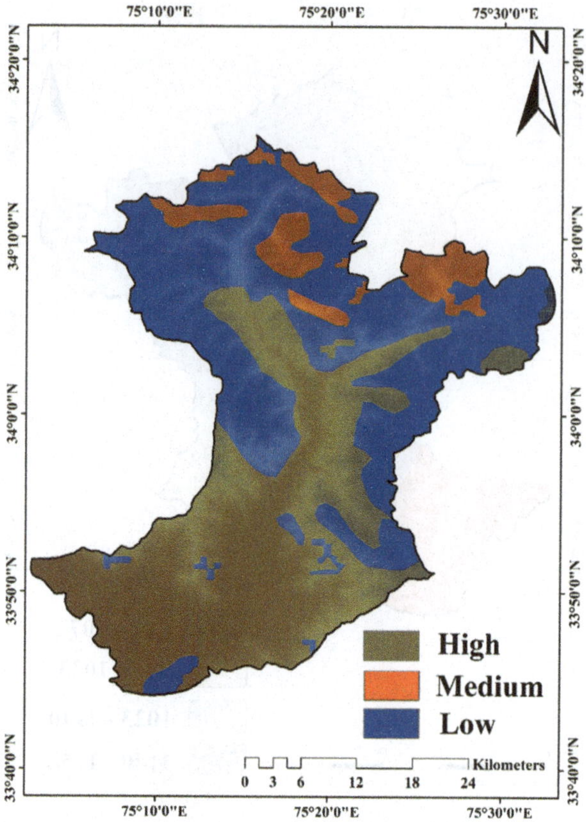

Fig. 18.8 Groundwater potential map of the Liddar watershed. The different colours of the map depict different potential zones. The map is overlapped with ASTER DEM of 30 m resolution

groundwater prospective zones, deep (>3–6 mbgl) in moderate prospective zones, and deeper (>6mbgl) in low prospective zones.

18.7 Conclusion

The geospatial methods were used to prepare a potential groundwater map of the Liddar sub-basin. The physiographic maps, satellite images, and ancillary data were used to produce thematic layers of different elements. The different thematic layers were allocated with proper weighted values and then overlayed in the ArcGIS to generate a groundwater potential map. The results suggested that 105 km^2, 577 km^2, and 601 km^2 are categorized as low, moderate, and good potential zones, which account for 8.1%, 45%, and 46% of the total area of the sub-basin. The results indicated that the high-altitude areas were least effective for groundwater recharge due to the dominance of complex hard rock topography. The results suggested that the Liddar sub-basin could be classified into different zones depending upon the ability to infiltrate the recharging water. The results can assist in policies for sustainably managing the groundwater resources in the Himalayas. This method can be widely applied in high-altitude catchments for groundwater exploration at suitable sites.

Acknowledgement The authors are highly thankful to the USGS (https://earthexplorer.usgs.gov/) for free access to satellite data.

References

C. Bhuvaneswaran, A. Ganesh, S. Nevedita, Spatial analysis of groundwater potential zones using remote sensing, GIS and MIF techniques in uppar Odai sub-sub-basin, Nandiyar, Cauvery basin, Tamilnadu. Int. J. Curr. Res. **7**(09), 20765–20774 (2015)

A. Chowdhury, M.K. Jha, V.M. Chowdary, Delineation of groundwater recharge zones and identifcation of artifcial recharge sites in West Medinipur district, West Bengal, using RS, GIS and MCDM techniques. Environ. Earth Sci. **59**, 1209 (2010). https://doi.org/10.1007/s12665-009-0110-9

Central Ground Water Board, March (2018) Ground water yearbook 2016–17. http://cgwb.gov.in/Regions/GW-year-Books/GWYB-%202016-17/J&K.pdf

I.A. Dar, K. Sankar, M.A. Dar, Deciphering groundwater potential zones in hard rock terrain using geospatial technology. Environ. Monit. Assess. **173**, 597–610 (2010)

A.A. Hussein, V. Govindu, A.G.M. Nigusse, Evaluation of groundwater potential using geospatial techniques. Appl Water Sci **7**(5), 2447–2461 (2017)

O. Jaan, S.A. Lone, R. Malik, A. Lone, W. Mir, A. Nissa, Morphotectonic and Morphometric analysis of Vishav Basin left bank Tributary of Jhelum River SW Kashmir Valley India. Int. J. Econ. Environ. Geol. **6**(2), 17–26 (2015)

G. Jeelani, S.A. Lone, A.U. Nisa, A. Mukherjee, R.D. Deshpande, Sources and processes of groundwater arsenic mobilization in upper Jhelum basin, western Himalayas. J. Hydrol. **591**, 125292 (2020)

G. Jeelani, S.A. Lone, A.U. Nisa, R.D. Deshpande, V. Padhya, Use of stable water isotopes to identify and estimate the sources of groundwater recharge in an alluvial aquifer of Upper Jhelum Basin (UJB), western Himalayas. Hydrol. Sci. J. (2021b). https://doi.org/10.1080/02626667.2021.1985126

G. Jeelani, S.A. Lone, A. Lone, R.D. Deshpande, Groundwater resource protection and spring restoration in upper Jhelum Basin (UJB), western Himalayas. Groundw. Sustain. Dev. **15**, 100685 (2021a)

J. Krishnamurthy, A. Mani, V. Jayaraman, M. Manivel, Groundwater resources development in hard rock terrain-an approach using remote sensing and GIS techniques. Int. J. Appl. Earth Obs. Geoinf. **2**(3–4), 204–215 (2000)

S.A. Lone, A.A. Lone, G. Jeelani, Characterization of groundwater potential of Sindh watershed Western Himalayas. J Res Dev **16**, 29–41 (2016)

S.A. Lone, G. Jeelani, R.D. Deshpande, R.A. Shah, Evaluating the sensitivity of glacier to climate by using stable water isotopes and remote sensing. Environ. Earth Sci. **76**, 598 (2017). https://doi.org/10.1007/s12665-017-6937-6

S.A. Lone, G. Jeelani, R.D. Deshpande, A. Mukherjee, Stable isotope ($\delta^{18}O$ and δD) dynamics of precipitation in a high altitude Himalayan cold desert and its surroundings in Indus river basin, Ladakh. Atmos. Res. **221**, 46–57 (2019)

S.A. Lone, G. Jeelani, A. Mukherjee, P. Coomar, Geogenic groundwater arsenic in high altitude bedrock aquifers of upper Indus river basin (UIRB) higher Himalayas. Appl. Geochem. **113**, 104497 (2020)
S. Lone, G. Jeelani, R.D. Deshpande, A. Mukherjee, S. Jasechko, A. Lone, Meltwaters dominate groundwater recharge in the cold arid desert of upper Indus River Basin (UIRB), western Himalayas. Sci. Total Environ. **786**, 147514 (2021)
S. Lone, G. Jeelani, A. Mukherjee, P. Coomar, Arsenic fate in upper Indus river basin (UIRB) aquifers: Controls of hydrochemical processes, provenances and water-aquifer matrix interaction. Sci. Total Environ. **795**, 148734 (2021a)
S.A. Lone, G. Jeelani, A. Alam, M.S. Bhat, H. Farooq, Effects of changing climate on the water resources of upper Jhelum basin (UJB), India, in *Riverine Systems: Understanding the Hydrological, Hydrosocial and Hydroheritage Connections*, ed. by A. Mukherjee, (Springer Capital Publishing, New Delhi, 2021b), pp. 139–154
C.S. Middlemiss, Revision of Silurian-Trias sequence of Kashmir. Rec. Geol. Surv. India **40**(3), 206–260 (1910)
N.S. Magesh, N. Chandrasekar, J.P. Soundranayagam, Morphometric evaluation of Papanasam and Manimuthar watersheds, parts of Western Ghats, Tirunelveli district, Tamil Nadu India: A GIS approach. Environ. Earth Sci. **64**, 373–381 (2011)
A.K. Saraf, P.R. Choudhury, Integrated remote sensing and GIS for groundwater exploration and identification of artificial recharge sites. Int. J. Remote Sens. **19**(10), 1825–1841 (1998)
E. Sener, A. Davraz, M. Ozcelik, An integration of GIS and remote sensing in groundwater investigations: A case study in Burdur, Turkey. Hydrogeol. J. **13**(5–6), 826–834 (2005)
A. Shaban, M. Khawlie, C. Abdallah, Use of remote sensing and GIS to determine recharge potential zones: The case of occidental Lebanon. Hydrogeol. J. **14**(4), 433–443 (2006)
M.C. Sashikkumar, L.D. Metilda, GIS approach for the demarcation of groundwater recharge potential zones in hard rock terrain. Int. J. Earth Sci. Eng. **5**(1), 117–120 (2012)
M. Subagunasekar, M.C. Sashikkumar, GIS for the assessment of the groundwater recharge potential zone in Karunkulam block, Thoothukudi district, Tamil Nadu, India. Int. J. Curr. Sci., 159–162 (2012, 2012)
R.A. Shah, S.A. Lone, Hydrogeomorphological mapping using geospatial techniques for assessing the groundwater potential of Rambiara river basin, western Himalayas. Appl Water Sci **9**(3), 64 (2019)
S.O. Tweed, M. Leblanc, J.A. Webb, M.W. Lubczynski, Remote sensing and GIS for mapping groundwater recharge and discharge areas in salinity prone catchments, southeastern Australia. Hydrogeol. J. **15**, 75–96 (2007)
H.F. Yeh, C.H. Lee, K.C. Hsu, P.H. Chang, GIS for the assessment of the groundwater recharge potential zone. Environ. Geol. **58**(1), 185–195 (2009)

Chapter 19
Groundwater Potential Zones (GPZs) Delineation in Dhamani River Basin in Kolhapur District, Maharashtra, India, Using Remote Sensing (RS), Geographical Information System (GIS) and Multi-Criteria Decision Analysis (MCDA) Techniques

Gurav Chandrakant, Md Babar, J. B. Pishte, and Gurav Netra

Abstract Despite being a part of the high rainfall zone of Maharashtra, India (average annual rainfall 5000 mm), the Dhamani river basin and adjoining area in Kolhapur district face water scarcity soon after February. An attempt in this work is made to identify and delineate Groundwater potential zones (GPZs) in this area by integrated use of remote sensing (RS), geographical information system (GIS) and multi-criteria decision analysis (MCDA)-analytical hierarchy process (AHP). Dhamani river basin occupies an area of about 196.4 sq. km. The entire basin is covered by three types of rock formations, namely, basalts of Deccan Volcanic Province (DVP), laterite capping over basalts and Quaternary deposits along the stream courses. This investigation is carried out using morphometric analysis of Dhamani river basin and its 33 tributary watersheds. Various thematic layers are prepared, namely, geology, slope, soil, normalised difference vegetation index (NDVI), drainage system, drainage density, lineament density, hydrogeomorphology and well inventory to facilitate these studies. It was inferred that merely 0.35% of the total area has good, while 53.5% has moderate and 46.1% has poor groundwater potential.

M. Babar

G. Chandrakant (✉)
Department of Geology, Dnyanopasak College, Parbhani, Maharashtra, India

J. B. Pishte
Department of Geology, Gopal Krishan Gokhale College, Kolhapur, Maharashtra, India

G. Netra
Department of Environmental Science, YCMU, Centre Vivekanand College, Kolhapur, Maharashtra, India

B. Panneerselvam et al. (eds.), *Climate Change Impact on Groundwater Resources*, https://doi.org/10.1007/978-3-031-04707-7_19

Keywords AHP · GIS · Groundwater · Hydrogeomorphology · MCDA · Morphometry · Watershed

19.1 Introduction

Groundwater is one of the largest available natural resources of freshwater on the earth (Rokade et al. 2004; Sharma and Kajur 2012; Neelakantan and Yuvaraj 2012; Kumar 2013). More than 90% of the rural and nearly 30% of the urban Indian population depends on groundwater for agricultural, domestic and industrial needs (Todd and Mays 2005). In the last two decades, increasing demand for water has adversely affected the supply of both groundwater and surface water. If this continues in future, this can lead to acute water crisis (Thakur et al. 2011).

An integrated approach of geographical information system (GIS) and remote sensing (RS) technology in hard rock terrain lends itself as an economically effective result-oriented method for planning, development and management of water resources (Rao and Jugran 2003). Remote sensing-based satellite data of the earth's surface presents a first-hand tool for demonstrating conditions as they exist at a particular time in a given area (Dury 1986; Gupta 1991). It also helps to study the hydrogeomorphology, lineaments (fracture/ joints), land use and land cover, etc., which are indicatives of groundwater movement (Babar and Jadhav 2017). Knowledge of interrelationship between hydrogeomorphic units and topographic features can be of great help in demarcating the groundwater potential zones (GPZs) (Sakthivel et al. 2003). GIS technique is comparatively new, powerful and excellent tool for prospecting groundwater aquifers in basaltic hard rock terrain (Baker et al. 2003; Cherkauer 2004). Interpretation of satellite imagery with ground truth confirmation helps in identification of geomorphic features, geological structures and their hydraulic characters (Das et al. 1997). This technique also plays an important role in groundwater prospecting and opens new paths for proper management of the water resources (Pani et al. 2016). One of the greatest advantages of this technique for hydrological investigations and monitoring is its ability to generate information in spatial and temporal domain, which is very crucial for successful analysis, prediction and validation (Sarma and Saraf 2002). Many researchers have applied integrated RS and GIS technique to delineate GPZs and groundwater recharge sites in Maharashtra, namely, Varade et al. (2011), Dutta (2015), Pande et al. (2017), Gurav and Babar (2018), Gurav and Babar (2019) and Gurav et al. (2019). Nag and Ghosh (2013) used hydrogeomorphology, slope and lineament study along with remote sensing and GIS techniques for delineation of groundwater potential zones in Chhatna Block, Bankura district in West Bengal. RS, GIS and multi-criteria evaluation (MCE) techniques were used in hilly terrain of Pavagarh, Gujarat by Gupte and Srivastava (2010). Four thematic layers, namely, lineament density, drainage density, slope and land use land cover were used for assessment of GPZs.

Identification of GPZs in Maharashtra is very difficult because of its hard-rock terrain (Singh and Singh 2009). In such areas, groundwater is confined to fractures and weathered horizons (Babar and Shah 2011; Babar and Gurav 2014). Geological structures (extent of secondary porosity), topography (relief, slope and drainage), landforms, land use land cover, climatic condition and interrelationship between these factors control groundwater movement (Rastogi and Sharma 1976; Mukharjee 1996).

Analytic hierarchy process (AHP) is a method of measurement through pairwise comparisons and relies on the judgement of experts to derive priority scale (Saaty 2008). Delineation of GPZs using integrated RS, GIS and AHP in hard rock aquifer system was carried by Muralitharan and Palanivel (2015) and Nithya et al. (2019). Several other researchers have applied AHP for assessing GPZs, namely, Chandio et al. (2011), Bunruamkaew and Murayama (2011), Jhariya et al. (2016). Multi-Influencing Factors (MIF), remote sensing and GIS analysis have been performed to find the suitable sites for GPZs in Theni district, Tamil Nadu (Magesh et al. 2012). GIS-based AHP has been effectively performed to find GPZs in the hard rock terrain and groundwater management in the Chittar basin, Tirunelveli district, Tamil Nadu (Nithya et al. 2019). According to Adiat et al. (2012), 'GIS based elementary multi-criteria decision analysis (MCDA) technique is capable of producing accurate and reliable predictions particularly if the set of criteria used for the prediction are coherent'. Groundwater resources analysis and their suitability for urban residential planning have been carried using RS, GIS and MCDA in the Bhopal, Madhya Pradesh by Tiwari et al. (2019).

Dhamani river is a tributary of river Kumbhi in Kolhapur district, Maharashtra, India. This river basin receives nearly 5000 mm rainfall every year. As the area is with steep slope, the large amount water goes off in the form of runoff and at few places where water harvesting structures are made the recharge takes place. Therefore, the main objective of this research is to delineate GPZs and attempt to develop a novel approach in groundwater prospecting using integrated study of RS, GIS and AHP. The lineament, geology and geomorphology of the area are the useful parameters for this research.

19.2 Study Area

Dhamani river basin is bounded by latitude 16°24′47"N to 16°41'18"N and longitude 73°52′54"E to 74°03'06"E and covered in the Survey of India (SOI) toposheet no. 47H/14, 47 H/15 and 47 L/2 on the scale 1:50000. It constitutes a total area of 196.4 sq. km in Kolhapur district, Maharashtra, India (Fig. 19.1). The climate in the area is tropical, and receives rainfall from SW monsoon from June to September (nearly 5000 mm/year). Nearly 80% precipitation in India occurs in the short duration of these months (National Disaster Management Authority [NDMA]-2010). The average maximum temperature in the area is 40 °C in May and the average minimum is 13 °C in January. Geologically, the study area belongs to the Deccan

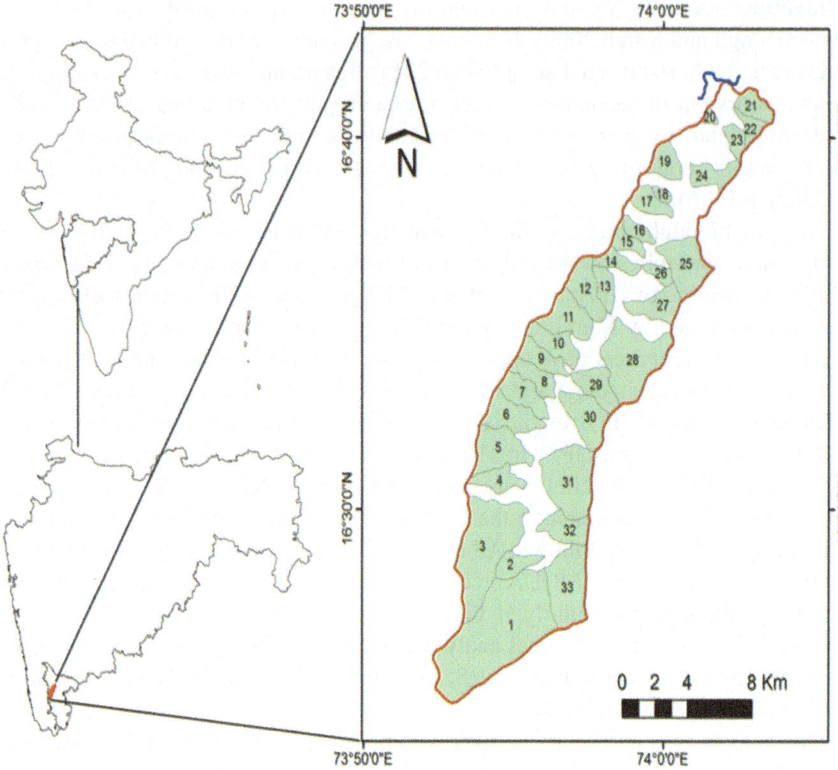

Fig. 19.1 Location map of Dhamani river and its tributary watershed

basalts of late Cretaceous to early Eocene age, along with the Lateritic uplands and narrow fringe of Quaternary deposits along the river channel. Besides the Quaternary deposits, two types of basalt flows, namely, compact basalt ('aa' type flow) and vesicular- amygdaloidal basalt ('pahoehoe' type flow) are the main aquifers in the study area.

19.3 Methodology

GIS-based terrain analysis is carried in this work for delineating GPZs using AHP as in the MCDA. The step-by-step methodology adopted for this study is represented in flowchart (Fig. 19.2). Morphometric parameters of Dhamani river basin and its 33 watersheds were calculated using Survey of India (SoI) topographic maps on 1:50000 scale. The stream-ordering method as suggested by the Strahler (1957) has been employed. For this purpose, onscreen digitisation technique is carried out in ArcGIS 10.3 software. Geology of the area is studied with reference to the field visits and the well inventory study at different locations of geomorphic surfaces.

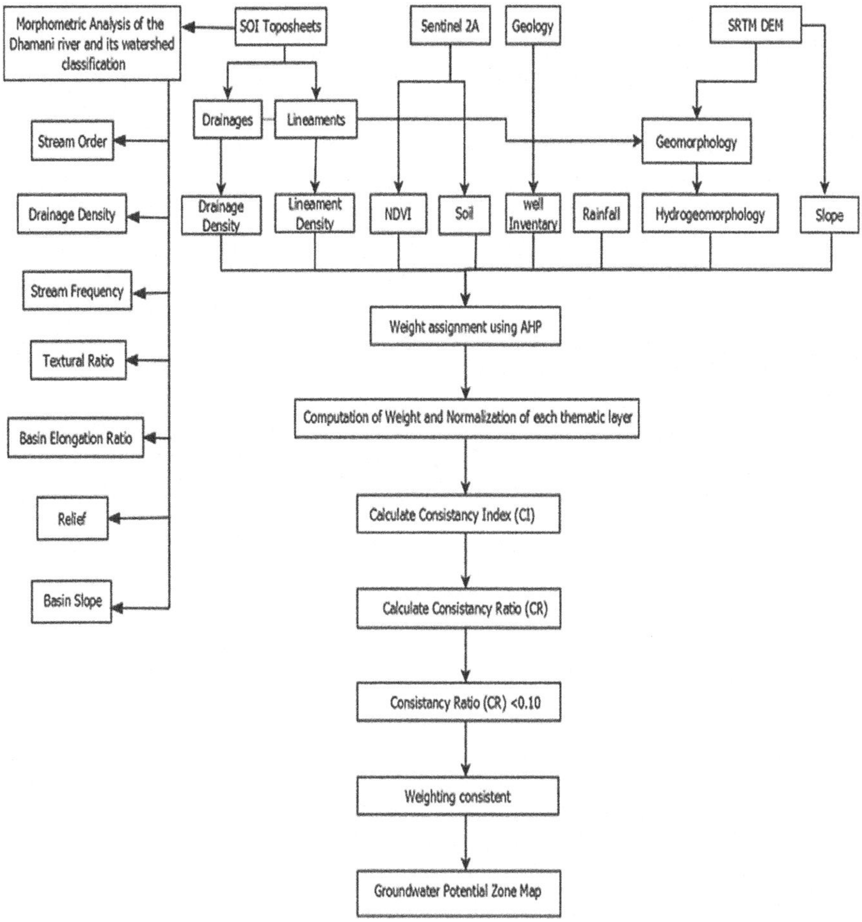

Fig. 19.2 Flowchart of the methodology

Lineaments are marked based on straight course of the streams and fracture zones (Kaplay et al. 2016) using topographic maps in conjunction with the satellite imagery and limited ground-based (field) observations. On this basis, a lineaments density map was also prepared. Shuttle Radar Topography Mission Digital Elevation Model (SRTM DEM) having 30 m resolution is used for estimation of slope, relief and terrain characteristics. The Sentinel 2A satellite imagery (Feb. 2019) with 10 m resolution downloaded from United States Geological Survey (USGS) official site (https://earthexplorer.usgs.gov/) is used to determine the soil cover and normalised difference vegetation index (NDVI). The results of these studies are presented in the form of tables and maps at appropriate places in the main body of the text.

Results of morphometric analysis of watersheds are explained in the 'Result' section. Based on the values of each parameter, the watersheds were classified and corresponding maps were prepared, namely, drainage network and drainage density, geology, slope, rainfall, water depth, Lineament and Lineament density, NDVI, soil, hydrogeomorphology and GPZs.

The AHP on every thematic layer was applied and correlated with the groundwater recharge conditions in the area. Rating factor method was adopted in this study. The first step in AHP is to assign a scale for pairwise comparison of the thematic layer matrix (Saaty 2005 and 2008). Nine × nine matrix thematic layers are used in this study. For this purpose, thematic layer of rainfall (TL1) is taken in first rank and layers with relatively lesser influence are added later gradually with the layer on water depth (TL9) at last. Two thematic layers that inherit equal contribution for groundwater are clubbed in one rank. All the thematic layers are arranged according to their influence on groundwater. The most influential layer (Rainfall) is ranked TL-1 and the layer with least importance (water depth) is ranked as TL-9 (Gebriu et al. 2020).

For hydrogeomorphology and well inventory study, this study area is classified into seven layers, namely, younger alluvial plain, pediplain, pediment, moderately dissected plateau, highly dissected plateau, denudational hill, and lateritic plateau. The calculations in AHP depend upon final value of the consistency ratio. If it is less than 0.10, it indicates suitability for GPZ analysis.

19.4 Results

For GPZs delineation, morphometric analysis of Dhamani river basin and its 33 watersheds and studies on geology, lineaments, slope, NDVI, soil, hydrogeomorphology and well inventory were carried out. All the formulae adopted for AHP analysis are listed in Table 19.1. It is further used to calculate the normalised weight

Table 19.1 Formulae adapted for AHP

Class	Formula
Normalised weight (NW)	Rating on each category/total rating of the layer (Table 19.5)
Pairwise NW	Elements values in each column/sum of the column
Criteria weight (CW)	Sum value of all the elements in row/number of thematic layers
Accuracy	Each value in the column (Table 19.7) × by criteria weight (Table 19.8)
Ratio	Weight sum value (WSV)/CW (Table 19.8)
λ max	Sum of ratio value in the column/number of thematic layers (Table 19.7)
Consistency index (CI	λ max-1/n-1
Consistency ratio (CR)	Consistency index/ random index (RI) Random index (RI) is the consistency index of randomly generated pairwise matrix shown in Table 19.3 up to 10 criteria

(NW) in Table 19.1. According to Magesh et al. (2012), higher NW indicates high rate of infiltration of the surface water and vice versa. The Ordinal scale preference of judgements is given in Table 19.2, and the Random Consistency Index (Saaty 1980) is used for calculating the final consistency ratio for 10 thematic layers as tabulated in Table 19.3.

Dhamani river is a tributary of river Kumbhi and Kumbhi is the main tributary of Bhogavati. Bhogavati meets Kasari near Chikhali village and thereafter this river is called Panchaganga until it drains into Krishna River. Thus, this area is constituting a part of Krishna Basin of Peninsular India. Nearly 126.6 sq. km area of the river receives more than 5000 mm rainfall every year (Table 19.5).

The river Dhamani is a fifth-order stream, flowing over DVB. In DVB terrains, availability of groundwater is limited and confined to fractured and weathered horizons (Babar 2003; Babar and Shah 2011; Babar and Gurav 2014). Morphometric parameters of Dhamani river basin, namely, relief 464 m, Dd 2.75 km/km^2, Fs 4.89 streams/km^2, Rt 11.43 km^{-1} and Re 0.26 point towards the basin shape being highly elongated (Gurav et al. 2020), situated in hilly area with impermeable lithology, high runoff and fine drainage texture. The Dd and Fs values of third-order streams are higher than those of the fourth-order streams and of Dhamani river basin as a whole. They depend on the stream order, relief, and shape of the watersheds and local influence of lithology. It can be seen that the Dd and Fs values in the area decrease as the stream order increases and vice-versa. 21.0% area (Table 19.5) of the Dhamani river basin has lower drainage density and it is good sign of groundwater infiltration in the area. Similar observation can be made regarding the textural ratio (Rt) of the basins. The Rt value for Dhamani river basin as a whole is high, that is, 11.43 km^{-1}, that for fourth- and third-order watersheds is 4.25 km^{-1} and 1.86 km^{-1}, respectively. This shows that Rt value decreases with increase in stream order and vice versa (Singh et al. 2019). Most of the area in Dhamani river basin has dendritic to sub-dendritic drainage pattern. The deep soil area (15.4%) is good for infiltration; while area without soil (31.5%) is poor (Table 19.5).

Table 19.2 Ordinal scale preference of judgements (Saaty 2005)

Ordinal scale	Degree of preference	Remarks
1	Equally	Factors inherit equal contribution
3	Moderately	One factor moderately favoured over other
5	Strongly	Judgement strongly favours over other
7	Very strongly	One factor very strongly favoured over other
9	Extremely	One factor favoured over other in highest degree
2,4,6,8	Intermediate	Compensation between weights 1,3,5,7 and 9
Reciprocals	Opposite	Refers inverse comparison

Table 19.3 Random Consistency Index for 10 layers (Saaty 1980)

N	1	2	3	4	5	6	7	8	9	10
RI	0.00	0.00	0.58	0.90	1.12	1.24	1.32	1.41	1.45	1.49

N number of layers and *RI* Random Consistency Index

19.4.1 Morphometric Analysis of Dhamani River Basin and Its Watersheds

Morphometric analysis gives quantitative results related to the area, slope, shape, length, etc. of the basin. These parameters affect stream flow pattern through their influence on concentration time of the water (Jones 1999). The drainage characteristics of Dhamani river basin and its watersheds have been examined with reference to stream order (u), drainage density (Dd), stream frequency (Fs), texture ratio (Rt), basin elongation ratio (Re), basin relief (H) and basin slope (BS).

19.4.2 Stream Order (U)

Stream ordering is the 'successive assimilation of the streams within a drainage basin' (Strahler, 1957). Based on Strahler (1957) method of stream ordering, it is found that Dhamani river is a fifth-order stream. In this river basin, total of 33 watersheds are demarcated (Table 19.4). Out of these, 28 are third-order basins and five are fourth-order basins (Fig. 19.3a).

19.4.3 Drainage Density (Dd)

The Drainage density (Dd) is defined as 'ratio of total length of all streams to the total area of the basin' (Horton 1932). It depends on lithology, runoff, infiltration capacity, surface roughness, climate conditions and vegetation (Verstappen 1983; Reddy et al. 2004). The *Dd* value is calculated using the formula,

$$Dd = Lu / A \tag{19.1}$$

where '*Lu*' is the total stream length of all orders in km and '*A*' is area of the basin in sq. km. The *Dd* values in this work are grouped into five categories namely, low (<1), medium (1–2), medium to high (2–3), high (3–4) and very high (>4). Lower *Dd* values indicate coarse permeable sub-soil and low relief of the basin, whereas higher values indicate impermeable rock, high relief and fine drainage texture (Strahler 1964; Manjunath et al. 2017). The *Dd* values of watersheds are classified into three categories (Fig. 19.3b).

Table 19.4 Morphometric parameters of Dhamani river watersheds

W	A	P	BL	u	Re	Dd	Fs	Rt	H	(BS %)
01	25.3	22.3	8.62	04	0.65	3.63	5.69	6.46	400	4.86
02	2.28	7.17	3.49	03	0.34	3.66	4.39	1.40	392	11.8
03	14.3	18.4	8.72	04	0.64	3.25	6.08	4.73	432	5.19
04	2.59	8.61	3.90	03	0.52	4.02	6.95	2.09	408	11.0
05	5.54	11.2	4.78	03	0.77	3.56	5.05	2.50	403	8.85
06	4.41	10.4	3.94	03	0.67	3.45	4.99	2.12	319	8.50
07	3.57	8.89	3.69	03	0.63	3.87	5.60	2.42	332	9.45
08	2.67	8.25	3.53	03	0.60	3.82	5.93	1.94	327	9.73
09	2.70	8.18	3.55	03	0.58	3.89	4.82	1.59	336	9.95
10	3.39	9.33	4.02	03	0.60	3.78	4.72	1.72	341	8.91
11	4.66	9.92	3.38	03	0.91	3.03	4.94	2.32	324	10.1
12	4.22	9.16	4.44	03	0.77	3.48	5.21	2.40	307	7.26
13	2.04	6.29	3.06	03	0.61	3.54	5.39	1.75	295	10.1
14	1.94	7.37	2.47	03	0.52	3.56	4.64	1.22	307	13.1
15	1.66	6.59	2.80	03	0.67	4.19	7.28	1.82	285	10.7
16	1.31	5.43	2.45	03	0.56	4.08	6.87	1.66	290	12.5
17	2.19	6.61	2.60	03	0.54	4.09	5.02	1.66	282	11.4
18	0.63	3.25	1.45	03	0.65	4.30	11.1	2.15	196	14.3
19	2.55	6.55	2.31	03	0.84	3.49	5.49	2.14	255	11.6
20	0.47	2.87	1.01	03	0.86	4.89	14.9	1.08	114	11.9
21	2.48	6.51	3.02	03	0.71	3.05	4.44	1.69	320	11.2
22	1.47	6.98	2.64	03	0.67	3.58	4.76	1.00	290	11.6
23	1.52	5.39	2.01	03	0.80	2.75	4.61	1.30	292	15.3
24	2.69	8.59	2.57	03	0.73	3.03	4.46	1.40	287	11.8
25	5.71	10.3	3.95	04	0.75	3.33	4.55	2.52	395	10.5
26	1.10	4.66	1.99	03	0.60	3.84	6.36	1.50	287	15.2
27	3.60	9.28	3.65	03	0.65	3.47	3.89	1.51	375	10.8
28	10.0	14.4	5.09	03	0.83	3.15	4.10	2.85	301	6.20
29	3.62	8.51	3.67	03	0.67	3.79	5.53	2.35	315	9.01
30	4.87	9.47	3.87	03	0.78	3.82	5.34	2.75	391	10.6
31	8.91	11.8	4.36	04	0.93	3.67	4.71	3.56	415	10.0
32	3.47	8.61	3.15	03	0.68	3.17	4.32	1.74	397	13.3
33	7.93	11.8	3.55	04	0.91	3.56	5.93	3.98	389	11.5
Dhamani river			34.9	05	0.26	2.75	4.89	11.4	464	1.40
Avg. values of third-order watersheds					0.67	3.66	5.75	1.86	313	10.7
Avg. values of fourth-order watersheds					0.77	3.49	5.39	4.25	406	10.6

W watershed number, *A* area in sq. km, *P* perimeter of the watersheds in km, *BL* Basin length in km, *u* stream order, *Re* basin elongation ratio, *Dd* drainage density in km/km^2, *Fs* stream frequency in stream/ km^2, *Rt* textural ratio in km^{-1}, *H* relief of the watershed in m and *BS* Basin slope in %

Dd map of Dhamani river basin is also prepared to show variation in drainage density (Fig. 19.4b) and the corresponding variation in potential for groundwater storage (PGS) is shown in Table 19.5.

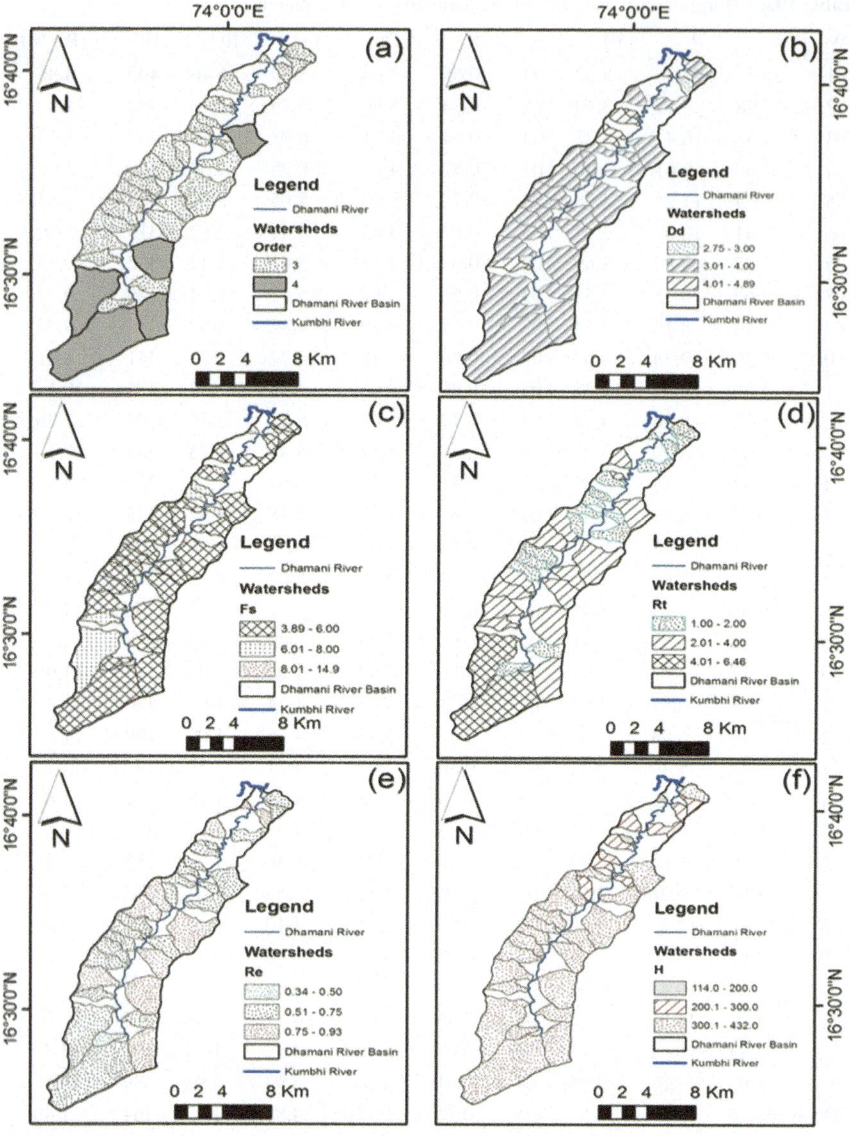

Fig. 19.3 Watersheds classification maps of Dhamani river. (**a**) Stream order (u), (**b**) drainage density (Dd), (**c**) stream frequency (Fs), (**d**) textural ratio (Rt), (**e**) basin elongation ratio (Re) and (**f**) relief (H)

19.4.4 Stream Frequency (Fs)

The Stream frequency (Fs) is defined as 'the ratio of total number of stream segments of all orders to the total basin area' (Horton 1945). The Fs value is calculated using the formula,

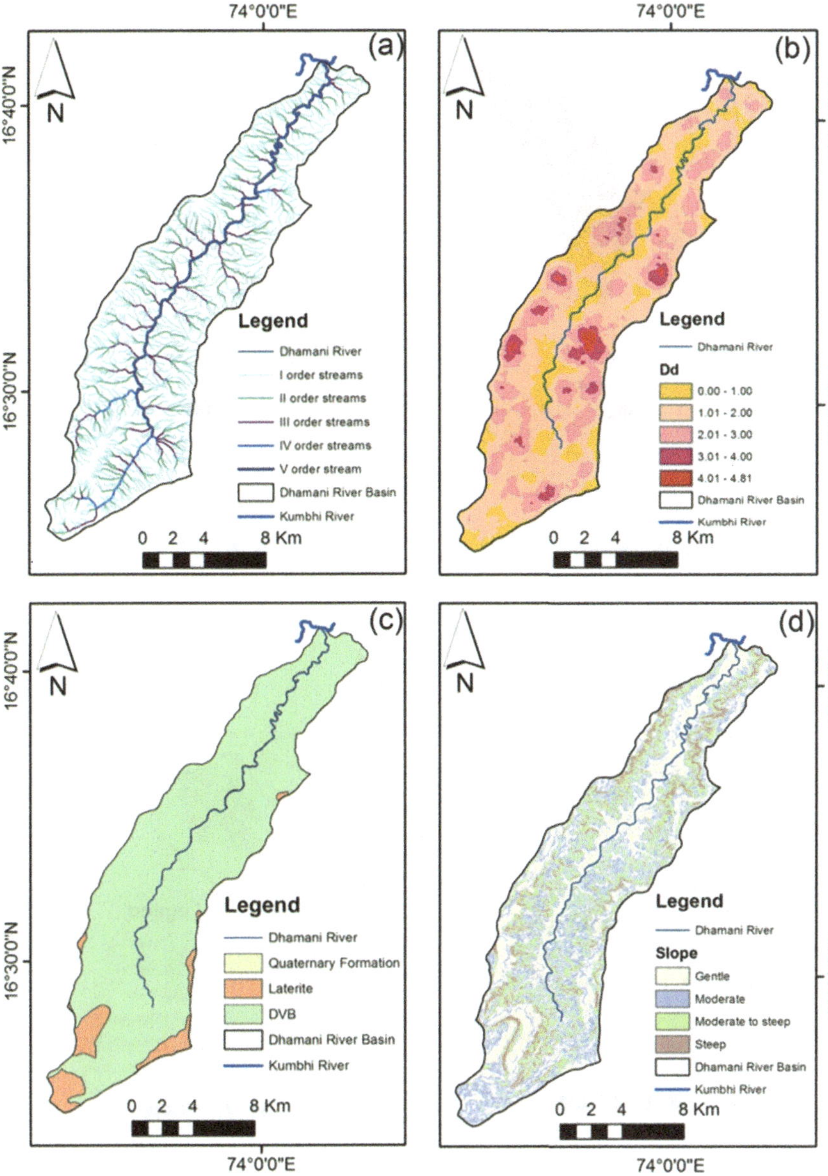

Fig. 19.4 Thematic layer maps of Dhamani river. (**a**) Drainage, (**b**) drainage density (Dd), (**c**) geology, (**d**) slope

$$Fs = Nu / A \tag{19.2}$$

where ‘Nu’ is total number of streams of all orders and ‘A’ is total area of the basin in sq. km. The Fs value of Dhamani basin as a whole is 4.89 streams/ km^2. The

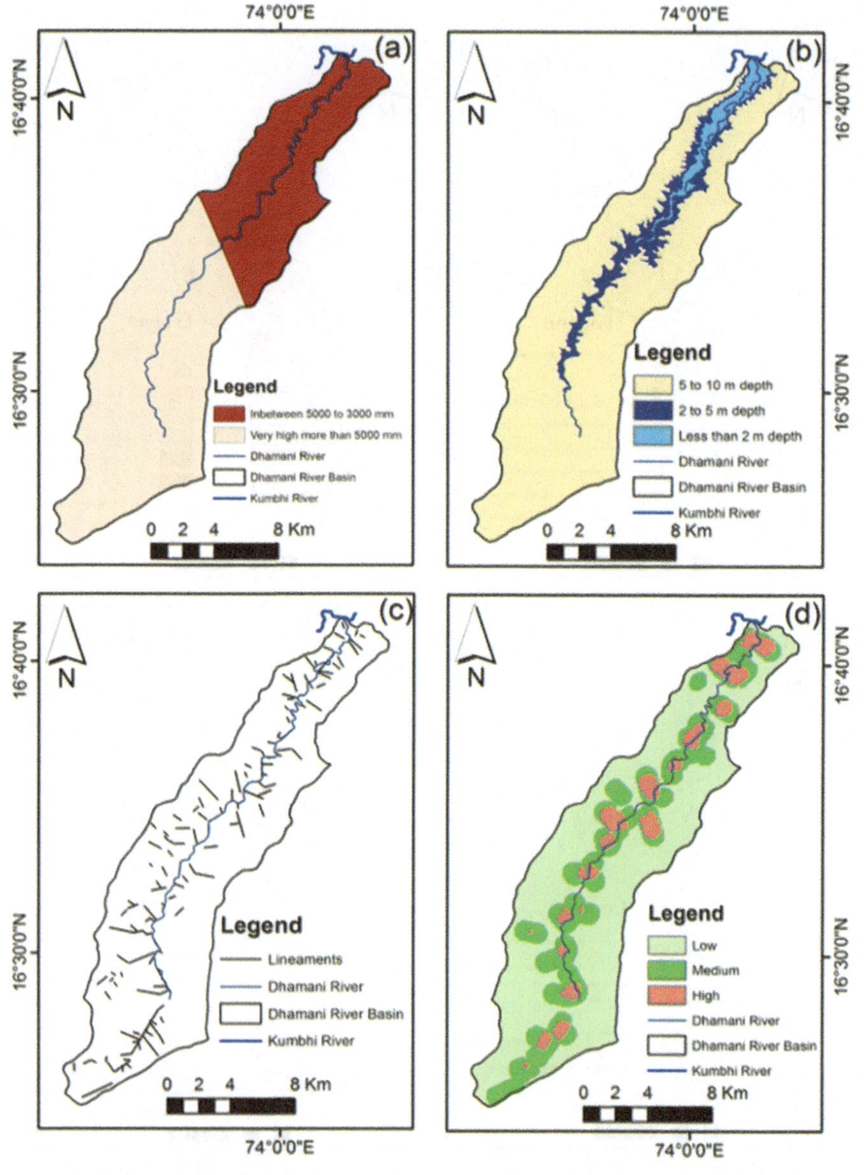

Fig. 19.5 Thematic layer maps of Dhamani river. (**a**) Rainfall, (**b**) water depth, (**c**) lineament and (**d**) lineament density (Ld)

average Fs value of all third-order basins is 5.75 streams/ km^2 and that for all fourth-order basins is 5.39 streams/ km^2 (Table 19.4). The Fs values are classified into three categories (Fig. 19.3c).

Table 19.5 Assigned and normalised weight of thematic layers for GPZs

TL	Category	PGS	Rating	Area	A in %	NW
TL1	More than 5000 mm	Good	5	126.6	64.5	33
	5000–3000 mm	Good	5	69.4	35.3	
TL2	Alluvial plain	Good	5	1.00	0.51	16
	Basalt	Poor	2	182.9	93.1	
	Laterite	Poor	2	12.5	6.37	
TL3	0.00–1.00	Very good	5	41.2	21.0	10
	1.01–2.00	Good	4	99.5	50.7	
	2.01–3.00	Moderate	3	45.9	23.4	
	3.01–4.00	Mod. To poor	2	8.40	4.28	
	4.01–4.81	Poor	1	1.26	0.64	
TL4	Low	Poor	1	129.6	65.9	08
	Moderate	Moderate	3	46.4	23.6	
	High	Good	5	20.4	10.4	
TL5	Deep soil	Very good	5	30.3	15.4	11
	Moderate soil	Good	4	49.3	25.1	
	Shallow reddish	Moderate	3	54.9	28.0	
	Without soil	Poor	1	61.9	31.5	
TL6	0.00–4.00	Good	5	113.1	57.7	08
	4.01–6.00	Moderate	4	65.0	33.2	
	6.01–8.00	Moderate	4	16.2	8.27	
	8.01–10.2	Poor	2	1.71	0.87	
TL7	Water	Very good	5	0.02	0.01	06
	Barren land	Poor	1	0.02	0.01	
	Sparse vegetation	Poor	1	78.8	40.2	
	Moderate vegetation	Moderate	3	63.9	32.5	
	Dense vegetation	Good	4	53.5	27.2	
TL8	YAP	Very good	5	1.00	0.51	04
	PP	Good	5	9.00	4.58	
	PD	Moderate	3	32.4	16.5	
	MDP	Poor	1	47.3	24.5	
	HDP	Moderate	2	55.2	28.1	
	DH	Poor	1	39.0	19.9	
	LP	Moderate	2	12.5	6.36	
TL9	Less than 2 m	Very good	5	8.99	4.58	03
	2–5 m	Good	4	18.47	9.40	
	5–10 m	Poor	2	168.9	85.9	

TL Thematic layer, *TL1* rainfall, *TL2* geology, *TL3* drainage density, *TL4* lineament density, *TL5* soil, *TL6* slope, *TL7* NDVI, *TL8* hydrogeomorphology, *TL9* water depth, *PGS* potential for groundwater storage, *A* area, *NW* normalised weight, *YAP* younger alluvial plain, PP pediplain, *PD* pediment, *MDP* moderately dissected plateau, *HDP* highly dissected plateau, *DH* denudational hill and *LP* lateritic plateau.

19.4.5 Texture Ratio (Rt)

The Texture Ratio is the ratio of 'total number of streams of all orders to the perimeter of the basin' (Horton 1945). The Rt value is calculated using the formula,

$$Rt = Nu / P \tag{19.3}$$

where 'Nu' is total number of streams of all orders and 'P' is total perimeter of the basin in km. Drainage texture is classified into 5 different classes, namely, very coarse (<2), coarse (2–4), moderate (4–6), fine (6–8) and very fine (>8) (Smith 1950). The Rt value of Dhamani river basin as a whole is 11.43 km^{-1} (very fine). The average Rt value of third-order basins is 1.86 km^{-1} (very coarse) and that of fourth order is 4.25 km^{-1} (moderate) (Table 19.4). Watersheds are also classified using the Smith (1950) classification and are shown in (Fig. 19.3d).

19.4.6 Basin Elongation Ratio (re)

According to Schumm (1956), Basin Elongation Ratio (Re) is a 'ratio between diameter of a circle of the same area as the drainage basin and the maximum length of the basin'. It is calculated using the formula,

$$Re = \left(2\sqrt{A / \pi}\right) / LB \tag{19.4}$$

where 'A' is area of the drainage basin and 'LB' is the maximum length of the basin and π is 3.14 (Bull and McFadden 1977). For Dhamani river basin, the overall elongation ratio is 0.26, which indicates that the basin is highly elongated. The watersheds are classified into two types, namely, 'elongated basins' and 'circular basins' on the basis of the value of Re. Re value is less than 0.75 indicates elongated basins and that more than 0.75 indicates circular basins shown in Fig. 19.3e and tabulated in Table 19.4.

19.4.7 Basin Relief (H)

The elevation difference between the highest and lowest elevation is the relief of basin. The H value is calculated using the formula,

$$H = \text{Highest elevation} - \text{lowest elevation in m} \tag{19.5}$$

The highest elevation in Dhamani river basin is 1011 m and the lowest is 547 m; thus, the relief amounts to 464 m. The average H value of third-order watersheds is

313 m and that of fourth-order watersheds is 406 m (Table 19.4). The H values are classified into three categories (Fig. 19.3f, Table 19.4).

19.4.8 Basin Slope (BS)

The basin slope is calculated using the formula;

$$BS^{\circ} = \left(H / BL\right) \times 60 \tag{19.6}$$

where H is the relief of the area and BL is the basin length (Farhan and Anaba 2016; Mahmood and Rahman 2019). Based on this formula basin slope of the watersheds and Dhamani river is calculated as shown in Table 19.4. The degree slope is also converted in percentage format using the formula %slope = degree slope/360 × 100. The basin slope of Dhamani river is 0.74° (1.36%), third-order watersheds slope is 6.10° (10.69%) and fourth order is 6.02° (10.55%). Higher angle of basin slope causes rapid surface runoff, and heightened basin susceptibility to flooding (Green and Nelson 2002).

19.4.9 Geology

Geologically, there are three types of rock formations, namely, basalts of Deccan Volcanic Province (DVP), plateau laterite and quaternary formation in middle and upper reach of Dhamani river (Gurav et al. 2020, 2021). Geologically, three types of rock formations occur in Dhamani river basin, namely, basalts of Deccan Volcanic Province (DVP) and laterite capping over basalts in the upper reaches of the basin; and Quaternary deposits along the stream courses (Gurav et al. 2020). Basaltic rocks are of two types: compact basalt ('aa' type flow) and vesicular-amygdaloidal basalt ('pahoehoe'-type flow). The top surface of compact basalt flow is vesicular and fragmentary; the lower surface is vesicular with volcanic breccia while the middle portion is thick and massive (Babar 2003). The vesicular-amygdaloidal basalt flows are formed by the outpouring of viscous lava in small quantities through a large number of outlets (Gupte 1971; Kulkarni et al. 1980). Based on fieldwork seven lava flows were identified in the area, which at some places, are separated by red bole beds or Tachylitic beds (Athavale and Natu 2005). Lithostratigraphically, present area comes under Mahabaleshwar Formation of Upper Deccan Traps (Godbole et al. 1996; Gurav et al. 2020, 2021). The hill tops above 920 m height in the area are covered by plateau laterite rock (Pleistocene age) formation (Athavale and Natu 2005). The Quaternary sediments along stream courses consist of cobbles, pebbles, sand, silt and clay deposited over the basaltic substratum. It is apparent from the depositional features and lineaments observed along the river channel, that the basin is uplifted in recent geological past (Gurav et al. 2020). The cobble pebble formation provides evidences of high energy flow, the high flood level while silty clay formation are the indicators of arid climate in the past (Gurav et al. 2021). The Quaternary sediment deposits along the Dhamani river are reddish in colour indicates oxidising condition in the area.

According to hydrogeological characters, the aa-type basalt flows can be differentiated into two parts. The top part is vesicular, un-jointed and watertight in fresh condition, but produces sheet jointing due to weathering. The middle and lower parts of the flow are jointed. These joints (fractures) in the aa type basalt flows are closely spaced linear along preferred orientations and the continuity of these joints is traceable over considerable distances. These fracture zones act as recharge conduits (lateral as well as down ward transmission of water) for deeper Deccan basaltic aquifers (Kulkarni et al. 1980; Babar 2001). However, quantity of percolation of water depends upon joint spacing and interconnection of joints. Water can percolate through closely spaced joints faster as compared to broadly spaced and non-interconnected joints.

In amygdaloidal basalt (Pahoehoe type) flow, the top portion is vesicular and at places the gas cavities are filled up with secondary minerals by eliminating the original vesicular nature. Owing to the presence of amygdales, the fresh amygdaloidal basalt is free from joints and occurs as homogeneous, watertight mass (Babar, 2001, 2011; Babar and Jadhav, 2017). On the contrary, the vesicular amygdaloidal basalt unit is more susceptible to weathering and exhibits deep weathering profile, which is characterised by the formation of sheet joints. Sheet joints are also developed at the contact between the vesicular amygdaloidal and the compact basalt that is within the contact zone. Such weathered amygdaloidal basalt holds groundwater (Babar 2001 and Babar 2011). However, capacity of groundwater occurrence depends upon the thickness of weathered zone.

19.4.10 Lineament

Joints and fractures develop during the cooling and crystallisation of lava or magma (Babar 2003; Babar and Jadhav 2017); and lineament like surface features due to tectonic stress and strain, provide important clue on infiltration of surface run off and also for sub-surface movement and storage of groundwater (Subbarao et al. 2001). The lineaments are very important for determining GPZs because they reflect the secondary porosity and permeability of the sub-surface rocks (Sreedevi et al. 2005). They also play an important role in identifying suitable sites for artificial recharge of groundwater as they reflect rock structures through which water can percolate and travel down to several meters (Babar 2001). Lineaments are marked for present study based on straight course of the streams. Using these lineaments (Fig. 19.5c), a lineament density (LD) map is prepared (Fig. 19.5d). LD map helps to identify high, medium and low lineament density areas in the basin. Area having high LD is good for GPZ (Haridas et al. 1998). Based on this, areas having high, medium and low LD are recognised as follows – high LD – 20.38 sq. km (10.38%), medium LD – 46.43 sq. km (23.64%) and low LD – 129.61sq. km (65.99%) (Table 19.5).

Lineament density analysis shows 65.9% of the total area has low lineament density which points towards impermeable subsurface lithology and 10.4% (Table 19.5) area having high lineament density shows permeable subsurface lithology.

19.4.11 Slope Analysis

According to Tiwar et al. (2019), slope is 'inclination of a surface with the horizontal in a landscape'. It is a very important factor for GPZs delineation and watershed management because it controls velocity of surface water flow and rain water infiltration. For this study, slope analysis was carried by using the SRTM DEM-Arc~30 m resolution with the help of ArcGIS 10.3 software. Total area is classified into four classes of slope; namely <4%, 4–6%, 6–8%, and > 8%. The areas falling in each class are as follows:< 4% – 113.1 sq. km (57.7%), 4–6% – 65.0 sq. km (33.1%), 6–8% – 16.21 sq. km (8.2%) and > 8% – 1.71 sq. km (0.9%) as shown in Table 19.5 (Fig. 19.4d). IMSD (1995) classification is followed for the slope to be considered in percentage.

Slope is a very important aid in determining infiltration; particularly runoff, in relation to groundwater potential. Infiltration rate is inversely related to slope, that is, gentle slope facilitates higher infiltration and lesser runoff and vice versa (Pande et al. 2017; Morbidelli et al. 2018). In the Dhamani river basin, 57.7% area is gentle sloping and 43.3% area is under moderate to steep sloping surface (Table 19.5). It can be pointed out that the area in the third-order basins is comparatively steeper than that in fourth-order basins and Dhamani river basin as a whole.

19.4.12 Normalised Difference Vegetation Index (NDVI)

Any change in vegetation cover over the earth surface has direct impact on water and energy budgets through plant transpiration, surface albedo, emissivity and roughness (Aman et al. 1992). During last two decades NDVI has been one of the most widely used techniques to study, assess and monitor global vegetation cover (Huete and Liu 1994; Leprieur et al. 2000). It is defined as

$$\mathrm{NDVI} = \left(\mathrm{NIR} - \mathrm{Red}\right) / \left(\mathrm{NIR} + \mathrm{Red}\right) \tag{19.7}$$

where R is the surface reflectance averaged over visible (red band) (λ ~0.6μm) and N represents surface reflectance averaged over Near-infrared (NIR) (λ ~0.8μm). In this study, NDVI is derived from Sentinel 2A satellite imagery using the bands 4 and 8. Based on this, the equation is rewritten as,

$$\mathrm{NDVI} = (\rho 8 - \rho 4)/(\rho 8 + \rho 4) \tag{19.8}$$

where ρ8 and ρ4 are the surface reflectance of Sentinel 2A bands in the Near-infrared and red spectrum ranges, respectively. The ranges of NDVI, for present study are set as <0 for water bodies, 0–0.03 for barren land, 0.03–0.3 for sparse vegetation, 0.3–0.5 for moderate vegetation and > 0.5 for dense vegetation (Table 19.5) and shown in Fig. 19.6a.

NDVI of Dhamani river is classified into five classes and ranks based on vegetation index. The dense vegetation area in general shows good potential for groundwater and sparse and barren land area of poor potential for groundwater. Based on this classification in this study it is found that 27.2% area has good and 72.8% has under moderate to poor groundwater potential.

19.4.13 Soil

A large part of the study area is covered by reddish lateritic, reddish soil which is important for raising sugarcane production. For this study, four categories of soil cover are recognised as area without soil cover – 61.86 sq. km (31.50%), shallow soil – 54.96 sq. km (27.98%), moderate soil – 49.30 sq. km area (25.10%) and, deep soil – 30.30 sq. km area (15.43%) (Table 19.5 and Fig. 19.6b). Textural characteristics of these soils are the amount of soil particles less than 2 mm in diameter, that is, sand, silt and clay. It is observed in this study, that majority of the soils are in the clay and clay loam category. The clay content varies from 10.25% to 55.20%. Though, the clay and clay loam type of soils were reported predominantly from the downstream part and in the catchment of Dhamani river. This is probably due to insufficient drainage and unfavourable topography and siltation along the river channel and in the reservoirs. Besides this, the presence of alluvial deposits showing low permeability may be due to increased clay content. The silt content of the soils ranged from 8.5% to 35.0%. On the other hand, higher values of slit are observed for the soils from irrigated areas and in the areas of lesser elevation. The sand particles ranged from 7.20% to 40.50%. The higher content of these sands was recorded in the vicinity and downstream part of Dhamani river. Conversely, the lower values are found in the areas with flat topography and alluvial soils.

19.4.14 Hydrogemorphology

Hydrogeomorphology is interdisciplinary research focused on the relationship between hydrologic process and earth materials; and interaction of geomorphic process with surface or groundwater flow regime. For this study, various geomorphic units are interpreted from the satellite imagery and represented in Table 19.5 and

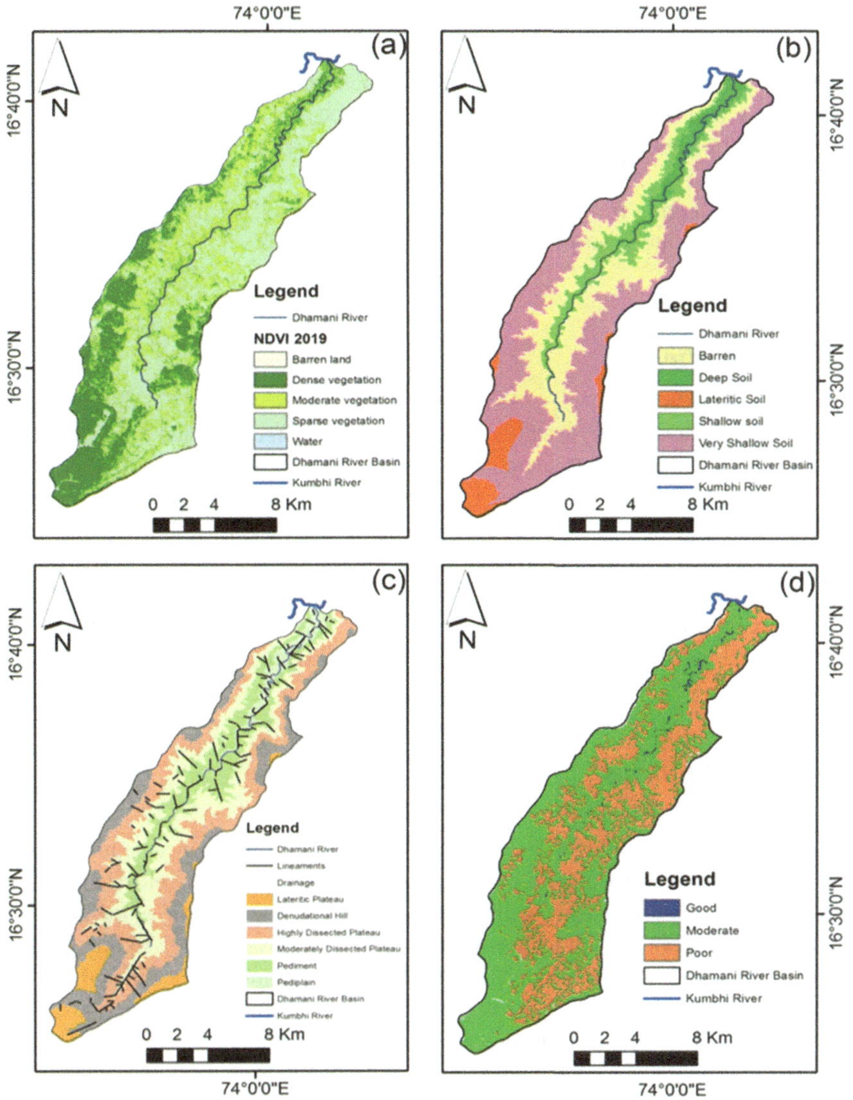

Fig. 19.6 Thematic layer maps of Dhamani river. (**a**) NDVI, (**b**) soil, (**c**) hydrogeomorphology and (**d**) groundwater potential zone map

shown in Fig.19.6c. The hydrogeomorphic units that are recognised include younger alluvial plain, pediplain, pediment, moderately dissected plateau, highly dissected plateau, denudational hill and lateritic plateau. The hydrogeomorphological study shows that younger alluvial plain, pediplain and pediment area having moderate to good groundwater potential while moderately dissected, highly dissected, denudational hill and lateritic plateau is poor.

19.4.15 Younger Alluvial Plain

The younger alluvial plain occurs as narrow belts on either side of the Dhamani river course and its tributaries. It has a gentle slope (0.15%), very low Dd, Fs and coarse Rt ratios and low relative relief (53 m). The groundwater potential is very good in this region, mostly used for agriculture and domestic use. The water table fluctuation is ranging from 0.40 to 1.10 m in dug wells. The range of yield of dug wells is 240–320 liters per minute (lpm).

19.4.16 Pediplain

Pediplain surface is highly weathered basaltic formation controlled by gentle sloping surface. It lies parallel to main stream course occupying 9.00 sq. km (4.58%) of the total area. Its drainage density is 4.11 km/ km^2, stream frequency is 8.78 streams/ km^2, textural ratio is 1.73 km^{-1}; and the elevation range is 547–560 m above msl. It is mostly covered by agricultural land. The pediplain surface has very good groundwater potential because of the gently sloping, thick soil cover and weathered rock formation; hence good infiltration. Water table fluctuation ranges from 1.00 to 1.60 m above msl, range of yield in wells is 180–280 lpm.

19.4.17 Pediments

A major part of Dhamani river basin is hilly and steep sloping, hence pediment zone is very difficult to demarcate. It is marked on the basis of slightly elevated, weathered surface of basalt, shallow soil and moderate to gentle (1.30%) slope observed on satellite imagery as well as during field checks. It covers 32.41 sq. km, that is, 16.50% area of the entire basin. The morphometric parameters calculated for this area are Dd 3.78 km/ km^2, Fs 9.10 streams/ km^2, Rt 1.69 km^{-1} and the relief is 80 m. The groundwater availability in this zone is good and mostly used for agricultural purpose. The water table fluctuation in dug well ranges from 4.00 to 5.40 m above msl, range of yield in wells is 65–110 lpm (Table 19.6).

19.4.18 Moderately Dissected Plateau

The moderately dissected plateau occurs between highly dissected plateau and pediment units. It is mostly a weathered basaltic terrain having steep sloping surface without soil cover. It covers 47.32 sq. km, that is, 24.09% area of the basin and has Dd value 3.82 km/ km^2, Fs 10.95 streams/km^2, Rt 1.90 km^{-1} and relief of 80 m. The

Table 19.6 Hydrogeomorphic unit and well inventory data

Geomorphic Units	ODW	Range of depth (m)	Range of depth to water (mbgl)		Water table fluctuation (m)	Range of yield (lpm)
			Pre monsoon	Post monsoon		
Laterite	1	12.5	7.5	3.2	4.3	55
DH	1	10.5	8.6	3.8	4.8	40
HDP	2	15.2–17.5	7.2–10.5	2.8–5.2	4.4–5.3	52–80
MDP	8	16.5–18.5	5.6–8.0	2.5–3.2	3.1–4.8	140–230
PD	3	13.4–17.6	6.6–9.5	2.6–4.1	4.0–5.4	65–110
PP	18	12.0–19.5	1.5–5.0	0.5–3.4	1.0–1.6	180–280
YAP	5	10.5–13.2	0.5–1.5	0.1–0.4	0.4–1.1	240–320

A area sq. km, *YAP* younger alluvial plain, *PP* pediplain, *PD* pediment, *MDP* moderately dissected plateau, *HDP* highly dissected plateau, *DH* denudational hills and *ODW* observed dug wells

water table fluctuation in dug well ranges from 3.10 to 4.80 m above msl, range of yield in wells is 140–230 lpm.

19.4.19 Highly Dissected Plateau

This zone of moderate to steeply sloping surface with or without soil cover occurs at the foot of denudational hills in the area. In this zone, most of the dug wells are seasonal, that is, they can yield groundwater hardly till the end of February. This zone covers an area of about 55.22 sq. km (i.e. 28.12% of the total area). The morphometric parameters of this area are Dd 3.53 km/ km^2, Fs 9.74 streams/ km^2, Rt 2.28 km^{-1} and the relief is 120 m. The water table depth is relatively deeper in dug well hence lesser quantity of water available in this zone. The water table fluctuation in dug well ranges from 4.4 to 5.3 m above msl, range of yield in wells is 52–80 lpm.

19.4.20 Denudational Hills

Denudational hills are massive hillocks with resistant rocky terrain. Dug wells are not observed in this zone. In the study area, this zone is covered by columnar basalt. It is demarcated based on sudden rise in elevation and steep slope. It covers 40.0 sq. km (20.37% of the total area). The relief is 120 m and Dd value is 3.10 km/ km^2, Fs is 8.38 streams/ km^2 and Rt is 2.16 km^{-1}. The water table depth in dug well 4.8 m above msl, and the yield in well is 40 lpm.

Table 19.7 Pairwise comparative matrix of AHP

TL	TL1	TL2	TL3	TL4	TL5	TL6	TL7	TL8	TL9
TL1	1	7	5	6	5	5	2	5	7
TL2	0.14	1	3	6	2	3	2	4	5
TL3	0.2	0.2	1	2	2	2	2	4	3
TL4	0.17	0.17	0.5	1	2	2	2	3	4
TL5	0.2	0.5	0.5	0.5	1	4	6	5	2
TL6	0.2	0.33	0.5	0.5	0.25	1	6	4	3
TL7	0.5	0.5	0.5	0.5	0.17	0.17	1	4	2
TL8	0.2	0.25	0.25	0.33	0.2	0.25	0.25	1	6
TL9	0.14	0.2	0.33	0.25	0.5	0.33	0.5	0.17	1
Sum	2.75	10.15	11.58	17.08	13.12	17.75	21.75	30.17	33

TL thematic layer, *TL1* rainfall, *TL2* geology, *TL3* drainage density, *TL4* lineament density, *TL5* soil, *TL6* slope, *TL7* NDVI, *TL8* hydrogeomorphology and *TL9* water depth.

Table 19.8 Normalised pairwise comparative matrix for analysis for calculating criteria weight in AHP

TL	TL1	TL2	TL3	TL4	TL5	TL6	TL7	TL8	TL9	CW
TL1	0.36	0.69	0.43	0.35	0.38	0.28	0.09	0.17	0.21	0.33
TL2	0.05	0.10	0.26	0.35	0.15	0.17	0.09	0.13	0.15	0.16
TL3	0.07	0.02	0.09	0.12	0.15	0.11	0.09	0.13	0.09	0.10
TL4	0.06	0.02	0.04	0.06	0.15	0.11	0.09	0.10	0.12	0.08
TL5	0.07	0.05	0.04	0.03	0.08	0.23	0.28	0.17	0.06	0.11
TL6	0.07	0.03	0.04	0.03	0.02	0.06	0.28	0.13	0.09	0.08
TL7	0.18	0.05	0.04	0.03	0.01	0.01	0.05	0.13	0.06	0.06
TL8	0.07	0.02	0.02	0.02	0.02	0.01	0.01	0.03	0.18	0.04
TL9	0.05	0.02	0.03	0.01	0.04	0.02	0.02	0.01	0.03	0.03

CW criteria weight

19.4.21 Plateau Laterite

Plateau laterite occurs on top of the columnar basalt of the denudational hills. It is formed due to in situ chemical alteration of pre-existing basaltic rock formation. It occurs above 920 m and covers an area of about 12.45 sq. km, that is, 6.34% of the total. The soil cover is reddish to dark brown in colour, mostly dry without vegetation. The water table depth in dug well 4.3 m above msl, and the yield in well is 55 lpm.

RS and GIS have been proved as efficient tools in watershed studies, as well as in updating and augmenting the hydro-geomorphological analyses of drainage basins (Babar 2011). This study also provides very accurate results in very short time span for groundwater prospecting (Gurav and Babar 2018). The GPZ delineation in Dhamani river basin is carried out based on remote sensing, GIS and AHP. The first stage in AHP is rating the thematic layers. It is done based on their influence on groundwater recharge. For this work, 1–5 hierarchy ratings were

Table 19.9 Pair-wise comparative matrix for analysis for calculating the Ratio in AHP

TL	TL1	TL2	TL3	TL4	TL5	TL6	TL7	TL8	TL9	WSV	Ratio
TL1	0.29	1.02	0.45	0.47	0.62	0.45	0.12	0.18	0.15	0.42	1.26
TL2	0.04	0.15	0.27	0.47	0.25	0.27	0.12	0.14	0.11	0.20	1.25
TL3	0.06	0.03	0.09	0.16	0.25	0.18	0.12	0.14	0.07	0.12	1.24
TL4	0.05	0.02	0.05	0.08	0.25	0.18	0.12	0.11	0.09	0.10	1.24
TL5	0.06	0.07	0.05	0.04	0.12	0.36	0.35	0.18	0.04	0.14	1.28
TL6	0.06	0.05	0.05	0.04	0.03	0.09	0.35	0.14	0.07	0.10	1.16
TL7	0.14	0.07	0.05	0.04	0.02	0.02	0.06	0.14	0.04	0.06	1.03
TL8	0.06	0.04	0.02	0.03	0.02	0.02	0.01	0.04	0.13	0.04	0.95
TL9	0.04	0.03	0.03	0.02	0.06	0.03	0.03	0.01	0.02	0.03	1.17
λ max											**1.17**

WSV Weight Sum Value

adopted. Using this rating normalised weight of each thematic layer is calculated (Table 19.5). Based on this, the study area is classified into three categories, namely, area having good, moderate and poor recharge potential. It was found that only 0.35% area (0.69 sq. km) has good recharge potential, 53.5% area (105.1 sq. km) has moderate and 46.1% area (90.6 sq. km) has poor recharge potential (Fig. 19.6d).

A pair-wise comparative matrix is also derived using Saaty's (1980) nine-point importance scale. Step-by-step calculations and results are tabulated in Tables 19.7–19.9 based on eqations shown in Table 19.1. Saaty (2005) has given a measure of consistency called Consistency Index (CI), which is calculated using the ratio equation shown in Table 19.1.

$$\lambda_{\max} = 1.26 + 1.25 + 1.24 + 1.24 + 1.28 + 1.16 + 1.03 + 0.95 + 1.17 / 9 = 1.17$$

$$\text{CI} = (1.17 - 9) / (9 - 1) = -0.98$$

Consistency Ratio (CR) is a measure of consistency of pair-wise comparison matrix (Table 19.8) shown in Table 19.1. For $n = 9$, CR is −0.98 which is <0.10, it implies that there is a reasonable level of consistency in the pair-wise comparison and the inconsistency is acceptable. If it is >10%, there is a need to revise the subject judgement (Agarwal et al. 2013; Jhariya et al. 2016).

19.5 Conclusion

Dhamani river basin is highly elongated, having impermeable lithology, with high velocity of run-off resulting in less infiltration of surface water. Due to this, only 0.35% area is found to be of good groundwater potential. The values of Dd, Fs, Rt, etc., of Dhamani river and its tributary watersheds show impermeable to permeable nature of the lithology and topography less suitable for infiltration of surface water. The integrated use of remote sensing, GIS and AHP along with field evidence has helped in obtaining reliable results in this study. As far as surface water is concerned, the authors recommend early completion of Dhamani river reservoir to overcome water scarcity in this area, especially in summer months. Enhanced groundwater recharge and reduced flood hazards by Kumbhi river will be added benefits of this reservoir.

References

K.A.N. Adiat, M.N.M. Nawawi, K. Abdullah, Assessing the accuracy of GIS-based elementary multi criteria decision analysis as a spatial prediction tool – A case of predicting potential zones of sustainable groundwater resources. J. Hydrol. **440-441**, 75–89 (2012). https://doi.org/10.1016/j.jhydrol.2012.03.028

E. Agarwal, R. Agarwal, R.D. Garg, P.K. Garg, Delineation of groundwater potential zone: An AHP/ANP approach. J. Earth Syst. Sci. **122**(3), 887–898 (2013). https://doi.org/10.1007/s12040-013-0309-8

A. Aman, H.P. Randriamanantena, A. Podaire, R. Froutin, Upscale integration of normalized difference vegetation index: The problem of spatial heterogeneity. IEEE Trans. Geosci. Remote Sens. **30**, 326–338 (1992)

T. Anuj, A. Ahuja, B.D. Vishwakarma, K. Jain, Groundwater potential zone (GWPZ) for urban development site suitability analysis in Bhopal, India. J Indian Soc. Remote Sens. **47**(11), 1793–1815 (2019). https://doi.org/10.1007/s12524-019-01027-0

H. Athavale, S.R. Natu, Report on transvers mapping in toposheet 47L/1, 2 and 3 in Sangali and Kolhapur District, Maharashtra 050/MAP/CR/MH/2004/2009 (2005)

Md. Babar, Hydrogeomorphological studies by remote sensing applications in Akoli watershed (Jintur), Parbhani district, Maharashtra, India. In: Spatial Information Technology: Remote Sensing and GIS-ICOGG, edited by Murali Krishna IV, 2, pp. 137–143 (2001)

M. Babar, Geological and geomorphological mapping of Akoli watershed in Jintur Tahsil of Parbhani district, Maharashtra, Indian. J. Geomorphol. **8**(1 and 2), 87–94 (2003)

M. Babar, Hydrogeomorphological analysis for watershed development in Jintur Tahsil, Parbhani district, Maharashtra. Indian Streams Res. J. **1**(5), 168–173 (2011)

Md. Babar, G. Chandrakant, Influence of Geological and Geomorphological Characteristics on Groundwater Potential in Lendi River Sub-Basin of Manjra River, Maharashtra, India. 4th International conference on Hydrology and Watershed Management (ICHWAM-2014) pp. 103–109 (2014)

M. Babar, S.I. Jadhav, Groundwater Potential Studies of Lower Part of Sindphana River Basin Beed District, Maharashtra, India Using remote Sensing and GIS, in *First Indian National Groundwater conference on Sustainable Development and Management of Groundwater Resources in Arid and Semiarid Regions*, (Jawaharlal Nehru Technological University, Hyderabad, 2017), pp. 101–110

M. Babar, I.I. Shah, Remote sensing and GIS applications for groundwater potential zones in Tawarja River Sub-Basin, Latur District, Maharashtra, India. Int. J. Earth Sci. Eng. (IJEE) **4**(3), 71–79 (2011)

M.E. Baker, M.J. Wiley, M.L. Carlson, P.W. Seelbach, A GIS model of subsurface water potential for aquatic resource inventory, assessment, and environmental management. Environ. Manag. **32**(6), 706–719 (2003)

W.B. Bull, L.D. McFadden, Tectonic geomorphology north and south of Garlock fault, California, in *Geomorphology in Arid Regions Eighth Annual Geomorphology Symposium*, ed. by D. O. Doehring, (State University New York, Binghamton, 1977), pp. 115–138

I.A. Chandio, A.-N. Matori, D.U. Lawal, S. Sabri, GIS- based land suitability analysis using AHP for public parks planning in Larkana city. Mod. Appl. Sci. **5**(4), 177 (2011). https://doi.org/10.5539/mas.v5n4p177

G. Chandrakant, Md. Babar, Identification of groundwater potential zones and artificial recharge sites in Vedganga River Sub-Basin-using remote sensing and GIS techniques. American Society of Civil Engineering (ASCE) conference proceeding-urbanization challenges in emerging economic, pp. 189–199 (2018) https://doi.org/10.1061/9780784482025

G. Chandrakant, M. Babar, J. Anilraj, Morphotectonics study of Dhamani River basin in Kolhapur District, Maharashtra, India. Bull. Pure Appl. Sci. **39F**, 131–149 (2020). https://doi.org/10.5958/2320-3234.2020.00001.31

G. Chandrakant, M. Babar, J. Anilraj, Morphostratigraphic and Lithostratigraphic studies of quaternary sediments to decipher climate change in Dhamani river basin, Kolhapur district, Maharashtra, India. J. Geosci. Res. Gondwana Geolog. Soc. Nagpur **6**(2), 155–170 (2021)

D.S. Cherkauer, Quantifying ground water recharge at multiple scales using PRMS and GIS. Ground Water **42**(1), 97–110 (2004)

S. Das, S.C. Behera, Kar, Hydrogeomorphological mapping of groundwater exploration using remote sensing data- A case study in Keaunjhar district, Orissa, j. Indian Soc. Remote Sens. **25**(4), 247–260 (1997)

S.A. Drury, Remote sensing of geological structure in temperate agricultural terrains. Geol. Magaz. **123**(02), 113–121 (1986)

P.Y. Dutta, Assessment of Groundwater Potential in a part of Vedganga River Basin of Kolhapur District, Maharashtra Using GIS 36th IIG Congress and International Seminar, (2015). https://www.researchgate.net/publication/317901976

Y. Farhan, Anaba o Flash flood risk estimation of Wadi Yutum (southern Jorden) watershed using GIS based morphometric analysis and remote sensing techniques. Open J Mod Hydrol 6(2), 9 (2016). https://doi.org/10.4236/ojmh.2016.62008

S.M. Godbole, R.S. Rana, S.R. Natu, Lava stratigraphy of Deccan basalt of Western Maharashtra. Gondwana Geol. Mag. Spl. Vol. **2**, 125–134 (1996)

J.I. Green, E.J. Nelson, Calculation of time of concentration for hydrologic design and analysis using geographic information system vector object. J. Hydroinf. **4**(2), 75–81 (2002)

R.P. Gupta, Remote Sensing Geology, Geological applications, Springer, Chapter-16, pp. 223–309 (1991)

R.B. Gupte, Engineering aspects of Deccan trap basalt. Maharashtra Publication Work J. **7**(3–4), 7–20 (1971)

M. Gupte, P.K. Srivastava, Integrating GIS and remote sensing for identification of groundwater potential zones in the hilly terrain of Pavagarh, Gujarat India. Water Int. **35**(2), 233–245 (2010). https://doi.org/10.1080/02508061003664419

C. Gurav, M. Babar, Hydro-geomorphological studies for groundwater potential of MangyalNala watershed of Lendi River, Nanded district, Maharashtra. Challenges in Groundwater and Surface water in India. J. Geosci. Res. (JGSR) Gondwana Geol. Soc. Nagpur, Spec. **2**, 61–66 (2019)

C. Gurav, M. Babar, I. Khan, Identification of groundwater potential zones in Manar river Sub-Basin, Maharashtra using remote sensing and GIS techniques. J. Geosci. Res. (JGSR) Gondwana Geolog. Soc. **4**(1), 31–38 (2019)

V.R. Haridas, S. Aravindan, G. Girish, Remote sensing and its applications for groundwater favourable area identification. Quart. J. GARC **6**, 18–22 (1998)

G. Hindeye, G. Tesfamichael, H. Ermias, Identification of groundwater potential zones using analytical hierarchy process (AHP) and GIS- remote sensing interaction, the case of Golina river basin, northern Ethiopia. Cloud Publ. Int. J. Adv. Remote Sens. GIS **9**(1), 3289–3311 (2020). https://doi.org/10.23953/cloud.ijarsg.460

R.E. Horton, Drainage basin characteristics. Trans. Amr. Geophy. Union **13**, 350–361 (1932)

R.E. Horton, Erosional development of streams and their drainage basins: Hydrophysical approach to quantitative morphology. Geol. Soc. Am. Bull. **56**, 275–370 (1945)

A.R. Huete, H. Liu, An error and sensitivity analysis of the atmospheric- and soil-correcting variants of the NDVI for the MODIS-EOS. IEEE Trans. Geosci. Remote Sens. **32**, 897–905 (1994)

IMSD, *Integrated Mission for Sustainable Development: Technical Guidelines* (NRSA, Hyderabad, India, 1995), pp. 1–127

D.C. Jhariya, K. Tarun, M. Gobinath, P. Diwan, N. Kishore, Assessment of groundwater potential zone using remote sensing, GIS and multi criteria decision analysis techniques. J. Geol. Soc. India **88**(4), 481–492 (2016). https://doi.org/10.1007/s12594-016-0511-9

J.A.A. Jones, *Global Hydrology: Processes* (Longman, Resources and Environmental Management, 1999), p. 399

R.D. Kaplay, M. Babar, S. Mukherjee, T. Vijaykumar, Morphotectonic expressions of geological Strucures in the eastern part of south East Deccan Volcanic Province (around Nanded, Maharashtra, India). Tectonics of the Deccan Large Igneous Province. Geolog. Soc. London Special Publication **445**, 317–335 (2016)

B. Khwanruthai, Y. Murayam, Site suitability evaluation for ecotourism using GIS & AHP: A case study of Surat Thani Province, Thailand. Procedia. Soc. Behav. Sci. **21**, 269–278 (2011). https://doi.org/10.1016/j.sbspro.2011.07.024

S.R. Kulkarni, B.N. Karmarkar, R.B. Gupte, Variation in the nature of Deccan trap volcanicity of western Maharashtra in time and space. Memoir Geolog. Soc. India **3**, 143–152 (1980)

C.P. Kumar, Assessment and strategies for development potential of deeper confined aquifers in India. Asian Acad Res. J. Multidiscip. **1**(8), 247–258 (2013)

C. Leprieur, Y.H. Kerr, S. Mastorchio, J.C. Meunier, Monitoring vegetation cover across semi-arid regions: Comparison of remote observa- tions from various scales. Int. J. Remote Sens. **21**, 281–300 (2000)

N.S. Magesh, N. Chandrasekar, J.P. Soundranayagam, Delineation of groundwater potential zones in Theni District, Tamil Nadu, using remote sensing, GIS and MIF techniques. Geosci. Front. **3**(2), 189–196 (2012). https://doi.org/10.1016/j.gsf.2011.10.007

S. Mahmood, A.U. Rahman, Flash flood susceptibility modelling using geomorphometric approach in the Ushairy Basin, eastern Hindu Kush. J. Earth Syst. Sci. **128**(4) (2019). https://doi.org/10.1007/s12040-019-1111-z

S. Manjunath, M. Dalwai, S. Raju, J. Davithuraj, Morphometric Analysis of Karanja River Basin, Bidar District, Karnataka, India, using Remote Sensing and GIS techniques. J. Geosci. Res. **2**(1), 45–53 (2017)

R. Morbidelli, C. Saltalippi, A. Flammini, R.S. Govindaraju, Role of slope on infiltration: A review. J. Hydrol. (2018). https://doi.org/10.1016/j.jhydrol.2018.01.019

S. Mukharjee, Targeting saline aquifer by remote sensing and geophysical methods in an part of Hamirpur-Kanpur, India. Hydrol. J. **19**(1), 53–64 (1996)

J. Muralitharan, K. Palanivel, Groundwater targeting using remote sensing, geo- graphical information system and analytical hierarchy process method in hard rock aquifer system, Karur district, Tamil Nadu. India. Earth Sci. India **8**(4), 827–842 (2015). https://doi.org/10.1007/s12145-015-0213-7

S.K. Nag, P. Ghosh, Delineation of groundwater potential zone in Chhatna block, Bankura District, West Bengal, India using remote sensing and GIS techniques. Environ. Earth Sci. **70**(5), 2115–2127 (2013)

NDMA (2010). National Disaster Management Guidelines; Management of Urban Flooding ISBN: 978–93–80440-09-5. National Disaster management Authority New Delhi

R. Neelakantan, S. Yovaraj, Evaluation of groundwater using geospatial data-a case study from Salem taluka, Tamil Nadu India. Int J Remote Sens Geosci **1**(2), 7 (2012)

C.N. Nithya, Y. Srinivas, N.S. Magesh, S. Kaliraj, Assessment of groundwater potential zones in Chittar basin, southern India using GIS based AHP technique. Remote Sens. Appl. Soc. Environ. (2019). https://doi.org/10.1016/j.rsase.2019.100248

B. Pande Chaitanya, S.F.R. Khadri, N. Moharir Kanak, R.S. Patode, Assessment of groundwater potential zonation of Mahesh River basin Akola and Buldhana districts, Maharashtra, India using remote sensing and GIS techniques. Sustain. Water Res. Manag., 1–18 (2017). https://doi.org/10.1007/s40899-017-0193-5

S. Pani, A. Chakrabarty, S. Bhadury, Groundwater potential zone identification by analytical hierarchy process (AHP) weighted overlay in GIS environment: A case study of Jhargram block, Paschim Medinipur. Int J Remote Sens. Geosci. **5**(3), 1–10 (2016)

Y.S. Rao, D.K. Jugran, Delineation of groundwater potential zones and zones of groundwater quality suitable for domestic purposes using remote sensing and GIS. Hydrol. Sci. **48**, 821–833 (2003)

R.A. Rastogi, T.C. Sharma, Quantitative analysis of drainage basin characteristics. J. Soil Water Conserv. India **26**(1–4), 18–25 (1976)

G.P. Reddi, M.A.K. Obi, K.S. Gajbhiye, Drainage morphometry and its influence on landform characteristics in a basaltic terrain, Central India-a remote sensing and GIS approach. Int. J. Appl. Earth Obs. Geoinf. **6**(1), 1–16 (2004)

V.M. Rokade, P. Kundal, A.K. Joshi, Water resources development action plan for Sasti watershed, Chadrapur district, Maharashtra using remote sensing and geographic information system. J. Indian Soc. Remote Sens. **32**(4), 359–368 (2004)

T.L. Saaty, (1980) The Analytic Hierarchy Process, Decision Analysis. https://doi.org/10.3414/ME10-01-0028

T.L. Saaty, *Analytic Hierarchy Process* (Wiley Online Libr. Encycl. Biostat., Hoboken, 2005). https://doi.org/10.1002/0470011815.b2a4a002

T.L. Saaty, Decision making with the analytic hierarchy process. Int. J. Serv. Sci. **1**, 83–98 (2008)

R. Sakthivel, M. Manivel, A. Alagappa Moses, A. Chandramohan, D.A. Vijay, Hydrogeomorphological Mapping in Kalrayan hills, Tamil Nadu. Indian J. Geomorphol **8**(1 and 2), 1–6 (2003)

B. Sarma, A.K. Saraf, Study of Landuse-groundwater relationship using an in- tegrated remote sensing and GIS approach. In: Proceedings of map Asia 2002. Asian conference on GIS, GPS, aerial photography and remote sensing, organised by Asian Institute of Technology, Bangkok and CSDMS, New Delhi, held in Bangkok between 7-9 August (2002)

S.A. Schumm, Evolution of drainage system and slope in Badlands at perthAmoboy, New Jersey. Geologic. Soc. Am. Bull. **67**, 597–646 (1956)

M.P. Sharma, A. Kujur, Application of remote sensing and GIS for groundwater recharge zone in and around Gola block, Ramgargh district, Jharkhand, India. Int. J. Sci. Res. Publ. **2**(2), 1 (2012)

P.K. Singh, U.C. Singh, Water resource evaluation and management for Morar River basin, Gwalior District, Madhya Pradesh, using GIS. E-J. Earth Sci. India **3**, 174–186 (2009)

S. Singh, S. Kanhaiya, S. Ankita, K. Chaubey, Drainage network characteristics of the Ghaghghar River basin (GRB), Son Valley, India. Geol. Ecol. Landsc. **3**(3), 159–167 (2019). https://doi.org/10.1080/24749508.2018.1525670

K.G. Smith, Standard for grading texture of erosional topography. Am. J. Soc. **5**(298), 655–668 (1950)

P.D. Sreedevi, K. Subrahmanyam, S. Ahmed, The significance of morphometric analysis for obtaining groundwater potential zones in a structurally controlled terrain. Environ. Geol **47**(3), 412–420 (2005)

A.N. Strahler, Quantitative analysis of watershed geomorphology. Trans. Am. Geophys. Union **38**, 913–920 (1957)

A.N. Strahler, Quantitative Geomorphology of Drainage Basins and Channel Networks. In: Chow, V., Ed., Handbook of Applied Hydrology, McGraw Hill, New York, 439–476 (1964)

N. Subbarao, G.K.J. Chakradhar, V. Srinivas, Identification of groundwater potential zones using remote sensing techniques in and around Guntur town, Andhra Pradesh, India. J Indian Soc. Remote Sens. **29**(1), 69–78 (2001)

J.K. Thakur, R.K. Thakur, A. Ramanathan, M. Kumar, S.K. Singh, Arsenic con-tamination of groundwater in Nepal – An overview. Water **3**(1), 1–20 (2011)

D.K. Todd, L.W. Mays, *Groundwater Hydrology*, 3rd edn. (Willey, New York, 2005)

M. Varade Abhay, W. Priyanka, D. Kartik, Y.D. Khare, K. Hemant, Integrated approach using remote sensing and GIS for assessment of groundwater situation in part of Chandrapur and Godchiroli Districts of Maharashtra. J. Ind. Geophys. Union **15**(4), 195–206 (2011)

H.I. Verstappen, *Applied Geomorphology; Geomorphological Surveys for Environmental Development* (Elsevier Science Ltd, Amsterdam, 1983), p. 450

[illegible]

Chapter 20
Assessment of Groundwater Potential Zone Mapping for Development of Semi-Arid Region Through AHP and GIS Techniques

Kanak N. Mohairi, Omprakash Madguni, and Chaitanya B. Pande

Abstract Estimation of the groundwater regime and the soil and water conservation structure is a very challenging task, especially in an arid region. Currently, the remote sensing approach is a very useful technique to provide accurate data on the surface and sub-surface. We are to find out the groundwater potential zones in the PTK4 watershed. The multi-criteria approach depending on the remote sensing and GIS using six factors was used based on the AHP method and assigned to the weightage while different parameters ranked as per the significance in groundwater potential zone. The morphometry analysis was also used to provide a quantitative description of the drainage basin. The results of AHP have been given better information on suitable zones and prioritisation. The groundwater suitable zones are more realistically identified and, on the basis of this, conservation structures are recommended.

Keywords Groundwater potential zone · Analytic hierarchical process (AHP) · Artificial recharge structure · RS and GIS techniques

20.1 Introduction

Water and land resources are very important factors for maintaining the ecological and groundwater system (Kandekar et al. 2021). In most countries, they have been used and are helpful for agriculture, ecosystem, and human purposes; day by day they play a critical role and population pressure continues to rise (Balamurugan et al. 2020a, 2020b). Water is both a life source and a life necessity. Groundwater is

K. N. Mohairi · O. Madguni
Indian Institute of Forest Management, Bhopal, Madhya Pradesh, India

C. B. Pande (✉)
Primus Partners Private Limited, Delhi, India

B. Panneerselvam et al. (eds.), *Climate Change Impact on Groundwater Resources*, https://doi.org/10.1007/978-3-031-04707-7_20

the basic supply of water for home and agricultural purposes, and its demand is growing by the day, particularly in desert regions where rainfall is scarce (Panneerselvam et al. 2020a, 2021a). So many natural resources expansion schemes have been implemented on a watershed level; therefore, priority is critical to an effective natural resource planning and management for long-term development (Food and Agricultural Organization 1985, 1987; Doke et al. 2018). Agriculture uses 42% of the world's total groundwater. In the rural and urban climate, groundwater is a crucial natural resource for dependable, cost-effective development water plans. It is essential for water and soil resources, but its use is inadequate due to monsoon disasters, unreliable surface water, and rapid urbanisation and industrialisation, all of which have posed a serious danger to this valuable resource (Ganapuram et al. 2016; Parameswari and Padmini 2018; Das and Pal 2019). Geological and geomorphic setup affect a region's hydrological setting and groundwater occurrence (Das 2019). The physiographic elements, on the other hand, control the rate of runoff, infiltration, percolation, and so on (Das et al. 2017; Maity and Mandal 2019; Pande et al. 2019). GIS and RS techniques are used to supply vast amounts of geospatial data (Das and Mukhopadhyay 2018). Satellite images accurately and efficiently offer baseline information on groundwater (Shekhar and Pandey 2014). When compared to RS data, conventional data (hydrogeological surveys) are more expensive, particularly in dry land area (Jha et al. 2010). GIS is a powerful and dependable technology for managing large amounts of spatial data for long-term natural resource planning (Murthy 2000; Balamurugan et al. 2020a, 2020b; Kumar and Balamurugan 2018).

One of the most prominent approaches for obtaining spatial plans, resource allocation, and other information is the analytical hierarchy process (AHP) (Saaty 1980a, 82, 91, 2000). However, only a few examines have utilised the method for precisely groundwater potential zones mapping by combined and multi-criteria analysis with GIS methods for producing a groundwater suitable sites map (Pande and Moharir 2017; Pande et al. 2017, 2021; Rajesh et al. 2021). The goal was to use GIS and multi-criteria analysis to find out the suitable zones for the development of the groundwater and analyse the mapping of the groundwater planning (Panneerselvam et al. 2021c; Kouadri et al. 2021; Shunmugapriya et al. 2021). In this chapter we have been utilising remote sensing and GIS techniques applied for the preparation of various thematic maps for the identification of groundwater suitable zones for development in the poor aquifer zones of basaltic rock formation.

20.2 Study Area

The Melghat hills and forest region surround the Akola district. Unconsolidated sediments, black cotton soil, and Deccan Traps basaltic rocks cover the majority of the watershed area. The Gavilgarh range, which is limited on the north by the southern foothills of the Gavilgarh range, is bounded on the south by high hills that stretch through the heart of the district. The Purna, Katepurna, Morna, and Mankarna rivers drain the entire region, making it a more or less rich alluvial tract. The amount of work done so far to understand the hydrogeological parameters that govern water quality is relatively low. An effort was made to understand the groundwater regime using geomorphology, LU, soil, slope, lineament and lineament density in the PKT4 watershed.

20.3 Materials and Methods

The criteria for this study were chosen based on six layers that were developed in a GIS environment. The following steps were performed to prioritise the watershed using the AHP technique. Using the Sentinel 2 satellite picture and the SRTM DEM, create remote sensing data such as LULC, lineament, slope, and lineament density. Soil, topo-sheet, and geomorphology are examples of secondary data. Drainage density is digitised and calculated using topo-sheet drainage (Fig.20.1). Groundwater recharge is closely linked to physiographic elements (Zolekar 2018). The water

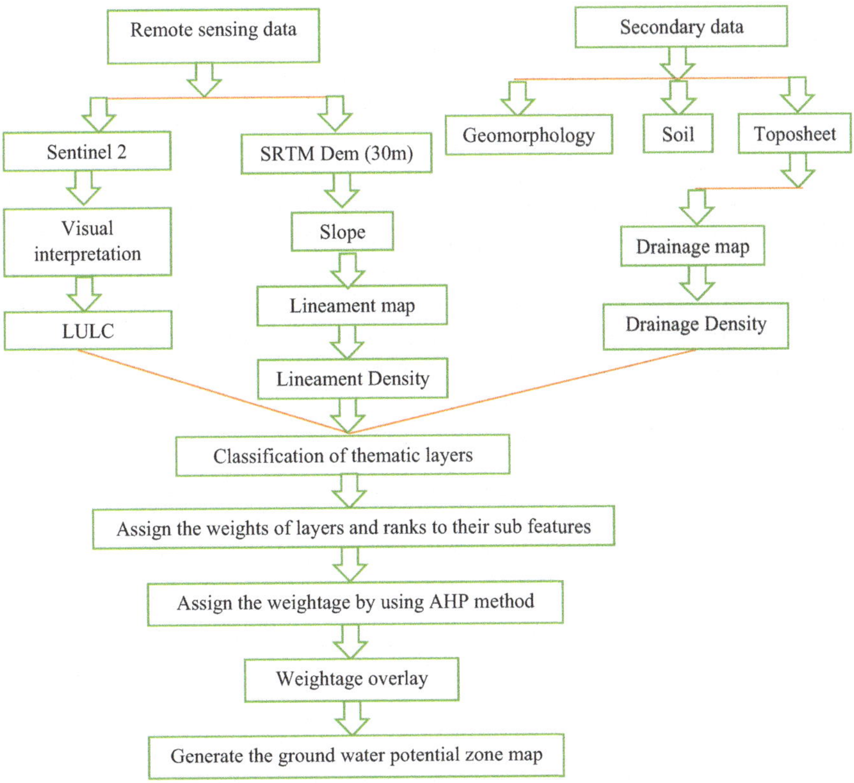

Fig. 20.1 Flow chart of methodology

level variation depends on the various factors such as geology, rainfall, and land use factors (Panneerselvam et al. 2020a, 2020b, 2021b; Choudhari et al. 2018). For the demarcation of the groundwater zone, most scientific research has employed various thematic layers criteria (Prasad et al. 2008). The combined influence of these factors has identified the groundwater potential zone (Yeh et al. 2016; Das and Pal 2019).

The entire technique in this study relies on multi-layered or natural resources data which was calculated for the groundwater potential zone (Maity and Mandal 2019). These overlays were created from satellite images, and they were used to produce the existing maps for the current investigation. The spatial resolution of ASTER Dem was 30 m. The various types of satellite data were downloaded by the United States Geological Survey Earth Explorer (earthexplorer.usgs.gov). It is used to calculate the drainage density and slope. Unsupervised categorisation of Sentinel 2 satellite images in a GIS environment was used to create the LULC map. The GWPZ was developed by the weighted overlay analysis method.

20.4 Analytical Hierarchy Process (AHP)

Saaty proposed the AHP, which is an MCDM approach (1980, 1997). It assists decision-makers in resolving the spatially complicated challenge associated with environmental concerns (Zolekar and Bhagat 2015). To detect a probable groundwater zone, multi-parameters have been explored (Adimalla and Taloor 2020; Chenini et al. 2010; Dunning et al. 2000; Flug et al. 2000; Kumar and Balamurugan 2019; Nagarajan and Singh 2009). The weights of each parameter were computed based on their respective importance in the decision-making process for locating groundwater zones (Tables 20.1 and 20.2). The essential qualities of AHPs are hierarchical formulation, cost-effective and time-saving results, and precision (Das 2019). The AHP approach is successfully used for water resource management in such situations all over the world. AHP's quantitative and qualitative approach is one of its strongest assets (Forman 1993).

Table 20.1 Pairwise comparison matrix for AHP

Factors	Geomorphology	Slope	Soil	Lineament density	LULC	Drainage density	Weightage
Geomorphology	6	5	4	3	2	1	0.40
Slop	6/2	5/2	4/2	3/2	2/2	1/2	0.20
Soil	6/3	5/3	4/3	3/3	2/3	1/3	0.13
Lineament density	6/4	5/4	4/4	3/4	2/4	1/4	0.10
LULC	6/5	5/5	4/5	3/5	2/5	1/5	0.081
Drainage density	6/6	5/6	4/6	3/6	2/6	1/6	0.067
Total	1						

Table 20.2 Weightage on PTK4 watershed

Factor	Weightage	Rank	Over all
Geomorphology			
Alluvial plain	40	5	200
Plateau		3	120
Denudational hill		2	80
Habitation mask		3	120
Water body		4	160
Slope			
1–3	20	5	100
3–5		4	80
Habitation mask		3	60
Water body		5	100
Soil			
Clay	13	4	52
Clay loam		3	39
Gravelly clay		3	39
Gravelly sandy clay loam		2	26
Gravelly sandy loam		1	13
Gravelly clay loam		3	39
Sandy clay loam		2	26
Habitation mask		2	26
Silty loam		2	26
Water body		5	65
Lineament density			
Low	10	1	10
Medium		2	20
High		3	30
Very high		4	40
Excellent		5	50
LULC			
Water body	8.1	5	40.5
Forest land		3	24.3
Agriculture land		2	16.2
Settlement		1	8.1
West land		3	24.3
Drainage density			
Low	6.7	5	33.5
Medium		4	26.8
High		3	20.1
Very high		2	13.4
Excellent		1	6.7

20.5 Weighted Overlay Analysis (WOA)

This technique is designed to tackle a difficult problem that occurs when determining site suitability using common measures of several inputs. The WOA technique was crucial in the discovery of GPZ. As previously explained, the evaluation scale of 1–10 is selected in WOA for awarding a score to sub-parameters (Saaty 1982; Saaty and Vargas 1991; Saaty and Vargas 2000; Saaty 1980b). The WOA tool was used to overlay all reclassified raster maps of selected parameters. The evaluation scale assigned a score to the cell values for each input raster in the evaluation. The weights calculated were assigned to the raster layer. Finally, the cell score of each input raster is multiplied by the weighted values of each raster layer (Eq. (20.2)).

$$S = \sum ni = 1\left(wi.xi\right) \tag{20.2}$$

Where S is total GPZ score, wi denotes weight of GPZ criteria, xi Indicates sub-criteria score of i GPZ criteria, and n represents total number of GPZ criteria.

20.6 Result and Discussion

Six criteria are used to identify the groundwater availability by using AHP method. These six criteria are as follows: (1) Geomorphology. (2) Slope. (3) Soil. (4) Lineament density. (5) Land use land cover. (6) Drainage density. For each parameter, we prepared the map. All the weightage which was assigned to each parameter help to prepared the groundwater potential zone.

20.6.1 Geomorphology

Geomorphology is an influence of the groundwater circulation in the aquifer zones and sub-surface area, while this factor is important for the preparation of the groundwater potential zones and conservation of the rain water into fractured zones (Andualem and Demeke 2019). The principal geomorphic features in the research area include alluvial plains, denudation hills, habitation masks, plateaus, and water bodies. More than 60% of the land is covered by a plateau with a medium level of groundwater availability. Nearly a quarter of the land is covered by the alluvial plain, which has a lot of groundwater. In the accompanying diagram, 10% of the area is covered by water, while the remaining 5% is occupied with habitation masks (Khadri and Pande 2014).

20.6.2 Slope

Slope is an important criterion, since it has a direct impact on the research area's surface runoff, percolation rate, and infiltration rate (Das 2019, 2021). A 1–3% slope in the research area indicates that groundwater is abundant. On the basis of this criteria, a weightage of 3–5% is assigned, indicating a medium amount of groundwater and habitation mask with low availability (Fig.20.2).

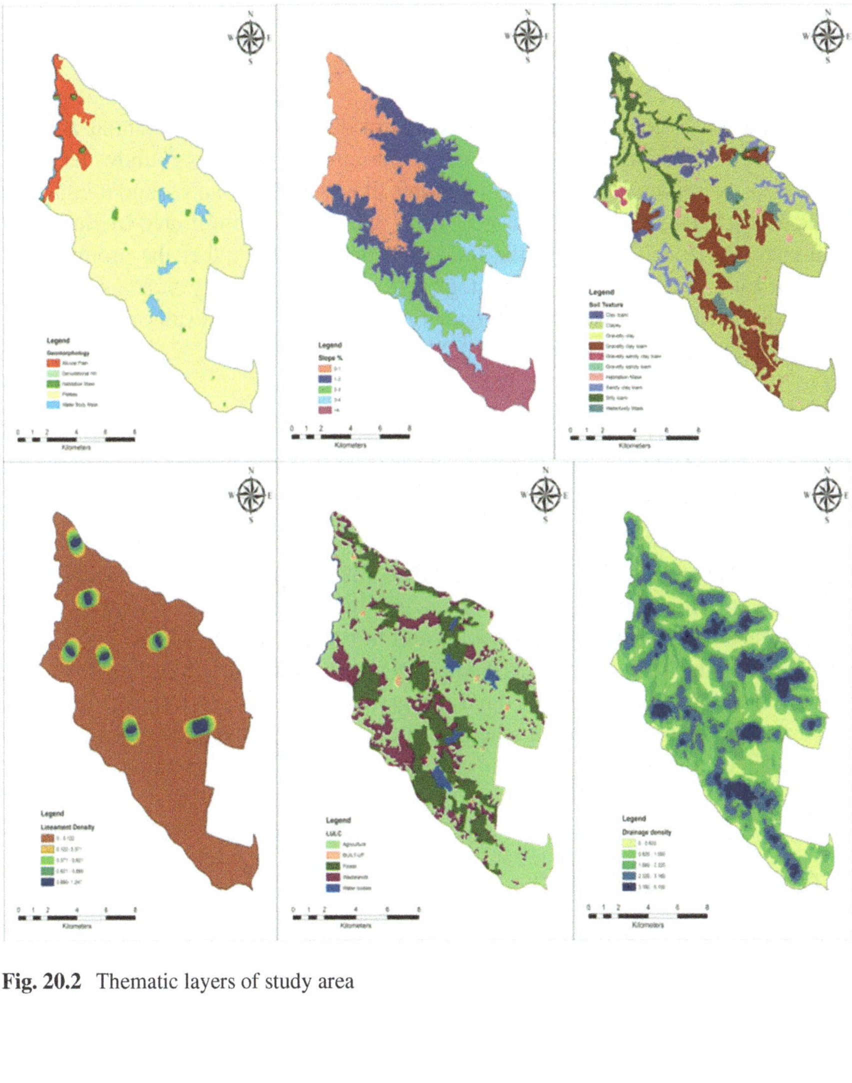

Fig. 20.2 Thematic layers of study area

20.6.3 Soil Texture

Soil features, such as soil type, soil texture, and so on, appear to play a significant effect in groundwater recharge spatial variation (Mehra et al. 2016; Zomlota et al. 2015). Clay soils have a favourable relationship with groundwater, while habitation mask has a negative relationship with groundwater. Clay soil has a higher porosity and permeability rate, as well as a high water retention capacity due to its fine texture. The main soil textures in the PTK4 watershed are loam and clay.

20.6.4 Lineament Density

Another key hydrological indicator that aids in determining the potential zone of groundwater is lineament (Kumanan and Ramasamy 2003). The groundwater prospect zone is determined by lineament features like as faults, fissures, and joints (Das and Pardeshi 2018). Groundwater movement is completely infiltrated by the lineament characteristic. The higher the lineament density, the higher the groundwater potential zone, and vice versa. Weightage is assigned in the study area based on this theory. In the lineament density map, a large volume of groundwater is seen in the centre, which is coloured blue (Fig. 20.2).

20.6.5 Land Use

Land use provides consistent and extremely valuable information for water resource management (Pande, C. 2014). It obstructs the flow of water and the permeability of the soil. Agriculture land covers the most area in the PTK4 watershed (more than 50%), followed by built-up land (5%). Land in the forest (20%), land in the west (15%), and water bodies (15%) (10%). As a result, agriculture has a positive link with groundwater, but built-up land has a negative correlation. As a result, farm land is given a higher weighting than other parameters, with the exception of water bodies, which have a positive relationship and are also given a high weighting (Fig. 20.2).

20.6.6 Drainage Density

The infiltration of river water and underlying rock formation has been obtained by drainage density, which is a crucial indication of hydrological landscapes (Murmu et al. 2019). The rate of groundwater recharge varies depending on the drainage density of the research area. Due to a low possibility of groundwater recharge, a high value of drainage density is associated with a low amount of groundwater availability, and vice versa (Fig. 20.2).

20.6.7 Valuation of Groundwater Potential Zone

Six parameters were determined using GIS technology, and weightage was assigned to each. Using the AHP approach, a groundwater potential zone map was created based on these spatial mapping results. The final output map was created and was separated into five different groups. Excellent GWPZ, high GWPZ, medium GWPZ, low GWPZ, and very Low GWPZ are all examples of GWPZ.

Water availability is very high in outstanding GWPZ. Because the area does not have a steep slope and the surface is flat, rainwater runoff is moderate, and percolation is high, it is regarded as an outstanding groundwater potential zone. Areas with poor, medium, and high slopes are found in the high, medium, and low suitable zones. The aim of this research we are prepared groundwater potential map for identification of locations within any geographic location and to provide greater suitable sites for the development of groundwater regime (Diaz-Alcaide and Martinez-Santos 2019). The final groundwater map is created in such a way that it can be easily understood by anyone without a scientific background (Rajaveni et al. 2017; Murmu et al. 2019; Andualem and Demeke 2019; Shekhar and Pandey 2014). However, generating such exact potential maps requires extensive understanding of the hydrological, geological, and other advanced technologies and methods (Bagyaraj et al. 2012; Das et al. 2017; Agarwal et al. 2013). This approach has already been widely used in landscapes and environments that are comparable to those in our study area (Razandi et al. 2015; Shahid et al. 2000; Kumar et al. 2020; Saranya and Saravanan 2020).

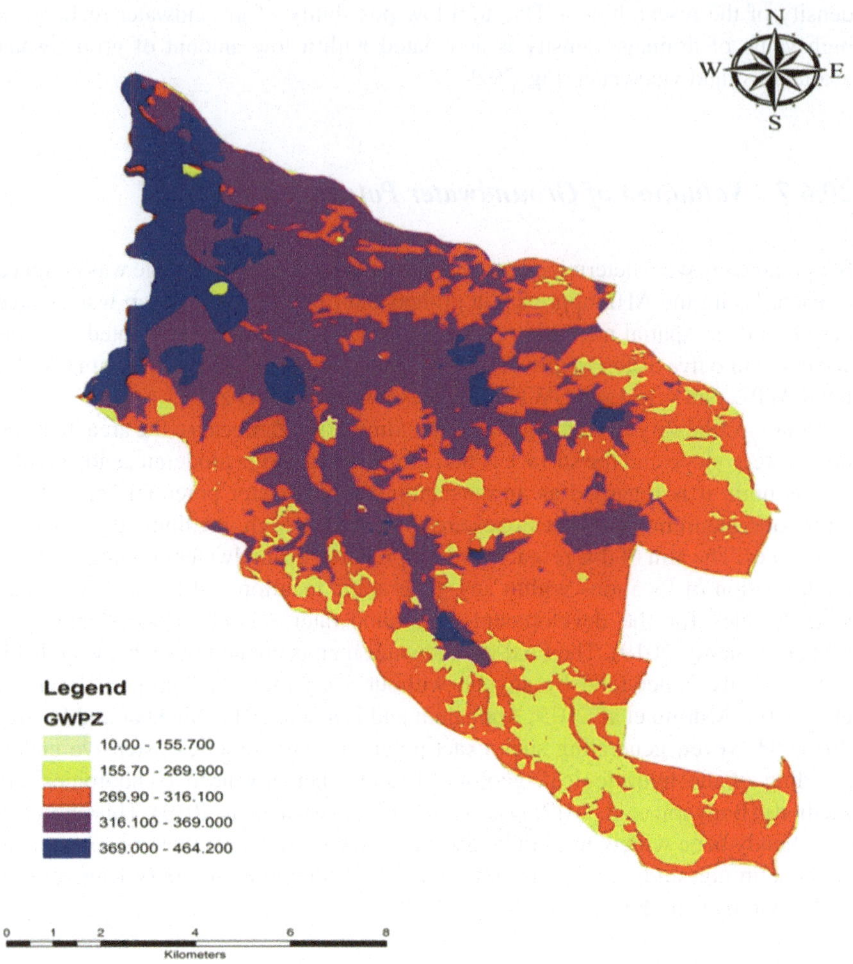

Groundwater potential zone

20.7 Conclusion

The determination of the topic investigations was to calculate the groundwater potential zone of PTK4 watershed in Maharashtra's Akola district. The six criteria of groundwater potential zones are classified as geomorphology, LU, stream density, soil, lineament density, and slope. The study area was split into five classes: excellent, high, medium, low, and extremely low. Regions identified as having a very low GWPZ require effective management and planning in perspectives of conservation, soil erosion, agriculture, plantation, and appropriate economic activities.

Because waste land is prevalent in the current study area, these areas will require particular attention and planning in terms of water resource extraction and utilisation. This approach offers a large database of groundwater management and planning for agricultural irrigation, as well as a suggested artificial recharge system for soil and water conservation issues. This is a very effective study, and the results are quite accurate; hence, it may be used in other places with similar geologic, climatic, and environmental characteristics.

References

N. Adimalla, A.K. Taloor, Hydrogeochemical investigation of groundwater quality in the hard rock terrain of South India using Geographic Information System (GIS) and groundwater quality index (GWQI) techniques. Groundwater Sustain. Dev. **10**, 100288 (2020). https://doi.org/10.1016/j.gsd.2019.100288

E. Agarwal, R. Agarwal, R.D. Garg, P.K. Garg, Delineation of groundwater potential zone: An AHP/ANP approach. J. Earth Syst. Sci. **122**(3), 887–898 (2013). https://doi.org/10.1007/s12040-013-0309-8

T.G. Andualem, G.G. Demeke, Groundwater potential assessment using GIS and remote sensing: A case study of Gunatana landscape, upper blue Nile Basin. Ethiopia J. Hydrol. Reg. Stud. **24**, 100610 (2019). https://doi.org/10.1016/j.ejrh.2019.100610

M. Bagyaraj, T. Ramkumar, S. Venkatramanan, B. Gurugnanam, Application of remote sensing and GIS analysis for identifying groundwater potential zone in parts of Kodaikanal taluk, South India. Front. Earth Sci. **7**(1), 65–75 (2012)

P. Balamurugan, P.S. Kumar, K. Shankar, R. Nagavinothini, P.J. Sajil Kumar, Impact of climate and anthropogenic activities on groundwater quality for domestic and irrigation purposes in Attur region, Tamilnadu, India. Desalin. Water Treat. **208**, 172–195 (2020a)

P. Balamurugan, P.S. Kumar, K. Shankar, R. Nagavinothini, K. Vijayasurya, Non-carcinogenic risk assessment of groundwater in southern part of Salem District in Tamilnadu, India. J. Chil. Chem. Soc. **65**(1), 4697–4707 (2020b)

I. Chenini, A.B. Mammou, M.E. May, Groundwater recharge zone mapping using GIS-based multi-criteria analysis: A case study in Central Tunisia (Maknassy basin). Water Resour. Manag. **24**, 921–939 (2010)

P.P. Choudhari, G.K. Nigam, S.K. Singh, S. Thakur, Morphometric based prioritization of watershed for groundwater potential of Mula river basin, Maharashtra, India. Geol. Ecol. Landscape **2**(4), 256–267 (2018). https://doi.org/10.1080/24749508.2018.1452482

B. Das, S.C. Pal, Assessment of groundwater recharge and its potential zone identification in groundwater- stressed Goghat-I block of Hugli District, West Bengal, India. (2019) https://doi.org/10.1007/s10668-019-00457-7

N. Das, S. Mukhopadhyay, Application of multi-criteria decision making technique for the assessment of groundwater potential zones: A study on Birbhum district, West Bengal, India. (2018) https://doi.org/10.1007/s10668-018-0227-7

S. Das, Comparison among influencing factor, frequency ratio, and analytical hierarchy approach process techniques for groundwater potential zonation in Vaitarna basin, Maharashtra, India. Groundwater Sustain. Dev. **8**, 617–629 (2019)

S. Das, Hydro-geomorphic characteristics of the Indian (Peninsular) catchments: based on morphometric correlation with hydro-sedimentary data. (2021) https://doi.org/10.1016/j.asr.2021.01.043

S. Das, A. Gupta, S. Ghosh, Exploring groundwater potential zones using MIF technique in semi-arid region: A case study of Hingoli district, Maharashtra. (2017) https://doi.org/10.1007/s41324-017-0144-0

S. Das, S.D. Pardeshi, Comparative analysis of lineaments extracted from Cartosat, SRTM and ASTER DEM: a study based on four watersheds in Konkan region, India. (2018) https://doi.org/10.1007/s41324-017-0155-x

S. Diaz-Alcaide, P. Martinez-Santos, Review: Advances in groundwater potential mapping. Hydrogeol. J. **27**, 2307–2324 (2019)

A. Doke, S.D. Pardeshi, S.S. Pardeshi, S. Das, Identification of morphogenetic regions and respective geomorphic processes: A GIS approach. Arab. J. Geosci. **11**, 1–13 (2018). https://doi.org/10.1007/s12517-017-3358-5

D.J. Dunning, Q.E. Ross, M.W. Merkhofer, Multi attribute utility analysis for addressing section 316(b) of the Clean Water Act. Environ. Sci. Pol. **3**, 7–14 (2000)

M. Flug, H.L.H. Seitz, J.F. Scott, Multi criteria decision analysis applied to Glen Canyon Dam. J. Water Res. Pl-ASCE **126**(5), 270–276 (2000)

Food and Agricultural Organization, Soil and water conservation in semi-arid areas. Soil Bull., No. 57, FAO, Rome. Food and Agricultural Organization (1985).

Food and Agricultural Organization, Watershed development with special reference to soil and water conservation, Soil Bull., No. 44, FAO, Rome (1987)

E.H. Forman, Facts and fictions about the analytic hierarchy process. Math. Comput. Model. **17**(4–5), 19–26 (1993). https://doi.org/10.1016/0895-7177(93)90172-U

S. Ganapuram, R. Nagarajan, G. Chandrasekhar, Identification of groundwater drought prone zones in Pedda vagu and Ookachetti vagu watersheds, tributaries of the Krishna River, India. Geocarto Int. **31**(4), 385–407 (2016)

M.K. Jha, V.M. Chowdary, A. Chowdhury, Groundwater assessment in salboni block, west Bengal (India) using remote sensing, geographical information system and multi-criteria decision analysis techniques. Hydrogeol. J. **18**, 1713–1728 (2010). https://doi.org/10.1007/s10040-010-0631-z

V.U. Kandekar, C.B. Pande, J. Rajesh, A.A. Atre, S.D. Gorantiwar, S.A. Kadam Bhau Gavit, Surface water dynamics analysis based on sentinel imagery and Google Earth Engine Platform: a case study of Jayakwadi dam, Sustain. Water Resour. Manag. **7**, 44 (2021). https://doi.org/10.1007/s40899-021-00527-7

S.F.R. Khadri, C.B. Pande, Remote sensing and GIS Applications of geomorphological mapping of Mahesh River Basin, Akola & Buldhana Districts, Maharashtra, India using Multispectral Satellite Data. Indian Streams Res J **4**(5) (2014)

S. Kouadri, C.B. Pande, B. Panneerselvam, K.N. Moharir, A. Elbeltagi, Prediction of irrigation groundwater quality parameters using ANN, LSTM, and MLR models. Environ. Sci. Pollut. Res. **29**, 1–25 (2021)

C.J. Kumanan, S.M. Ramasamy, Fractures and the transmissivity behaviour of the hard rock aquifer systems in parts of Western Ghats, Tamil Nadu, India. Escap Water Resour. J., 53–59 (2003)

P.S. Kumar, P. Balamurugan, Evaluation of groundwater quality for irrigation purpose in attur taluk, Salem, Tamilnadu, India. Water Energy Int. **61**(4), 59–64 (2018)

P.S. Kumar, P. Balamurugan, Suitability of ground water for irrigation purpose in Omalur taluk, Salem, Tamil Nadu, India. Indian J. Ecol. **46**(1), 1–6 (2019)

V.A. Kumar, N.C. Mondal, S. Ahmed, Identification of groundwater potential zones using RS, GIS and AHP techniques: A case study in a part of Deccan Volcanic Province (DVP), Maharashtra. India. J. Indian Soc. Remote Sens. **48**, 497–511 (2020)

D.K. Maity, S. Mandal, Identification of groundwater potential zones of the Kumari river basin, India: An RS & GIS based semi-quantitative approach. Environ. Dev. Sustain. **21**, 1013–1034 (2019). https://doi.org/10.1007/s10668-017-0072-0

M. Mehra, B. Oinam, C. Singh, Integrated assessment of groundwater for agricultural use in Mewat District of Haryana. India Using Geographical Information System (GIS). (2016). https://doi.org/10.1007/s12524-015-0541-6

P. Murmu, M. Kumar, D. Lal, I. Sonker, S.K. Singh, Delineation of groundwater potential zones using geospatial techniques and analytical hierarchy process in Dumka District, Jharkhand, India. Groundwater Sustain. Dev. **9**, 100239 (2019). https://doi.org/10.1016/j.gsd.2019.100239

K. Murthy, Groundwater potential in a semi-arid region of Andhra Pradesh – A geographical information system approach. Int. J. Remote Sens. **21**(9), 1867–1884 (2000)

M. Nagarajan, S. Singh, Assessment of groundwater potential zones using GIS technique. J. Indian Soc. Remote Sens. **37**, 69–77 (2009). https://doi.org/10.1007/s12524-009-0012-z

C. Pande, Change detection in land use/land cover in Akola Taluka using remote sensing and GIS technique. Int. J. Res. **1**(8), 1–11 (2014)

C.B. Pande, K.N. Moharir, GIS based quantitative morphometric analysis and its consequences: a case study from Shanur River Basin, Maharashtra India. Appl Water Sci **7**, 861–871 (2017) https://doi.org/10.1007/s13201-015-0298-7

C.B. Pande, K.N. Moharir, S.K. Singh, A.M. Varade, An integrated approach to delineate the groundwater potential zones in Devdari watershed area of Akola district, Maharashtra, Central India. Environ. Dev. Sustain. **22**, 4867–4887 (2019). https://doi.org/10.1007/s10668-019-00409-1

C.B. Pande, K.N. Moharir, B. Panneerselvam, et al., Delineation of groundwater potential zones for sustainable development and planning using analytical hierarchy process (AHP), and MIF techniques. Appl. Water. Sci. **11**, 186 (2021). https://doi.org/10.1007/s13201-021-01522-1

C.B. Pande, S.F.R. Khadri, K.N. Moharir, et al., Assessment of groundwater potential zonation of Mahesh River basin Akola and Buldhana districts, Maharashtra, India using remote sensing and GIS techniques. Sustain. Water Resour. Manag. **4**, 965–979 (2017). https://doi.org/10.1007/s40899-017-0193-5

B. Panneerselvam, S. Karuppannan, K. Muniraj, Evaluation of drinking and irrigation suitability of groundwater with special emphasizing the health risk posed by nitrate contamination using nitrate pollution index (NPI) and human health risk assessment (HHRA). Hum. Ecol. Risk Assess. Int. J. **27**(5), 1324–1348 (2020a)

B. Panneerselvam, K. Muniraj, C. Pande, N. Ravichandran, Prediction and evaluation of groundwater characteristics using the radial basic model in semi-arid region, India. Int. J. Environ. Anal. Chem. **2021**, 1–17 (2021a)

B. Panneerselvam, K. Muniraj, C. Pande, N. Ravichandran, M. Thomas, S. Karuppannan, Geochemical evaluation and human health risk assessment of nitrate-contaminated groundwater in an industrial area of South India. Environ. Sci. Pollut. Res., 1–18 (2021c, 2021)

B. Panneerselvam, K. Muniraj, M. Thomas, N. Ravichandran, B. Bidorn, Identifying influencing groundwater parameter on human health associate with irrigation indices using the Automatic Linear Model (ALM) in a semi-arid region in India. Environ. Res. **202**, 111778 (2021b)

B. Panneerselvam, S.K. Paramasivam, S. Karuppannan, N. Ravichandran, P. Selvaraj, A GIS-based evaluation of hydrochemical characterisation of groundwater in hard rock region, South Tamil Nadu. India. Arab. J. Geosci. **13**(17), 1–22 (2020b)

K. Parameswari, T.K. Padmini, Assessment of groundwater potential in Tirukalukundram block of southern Chennai metropolitan area. Environ. Dev. Sustain. **20**(9) (2018). https://doi.org/10.1007/s10668-017-9952-6

R.K. Prasad, N.C. Mondal, P. Banerjee, M.V. Nandakumar, V.S. Singh, Deciphering potential groundwaterzone in hardrock through the application of GIS. Environ. Geol. **55**, 467–475 (2008). https://doi.org/10.1007/s00254-007-0992-3

S.P. Rajaveni, K. Brindha, L. Elango, Geological and geomorphological controls on groundwater occurrence in a hard rock region. Appl Water Sci **7**, 1377–1389 (2017). https://doi.org/10.1007/s13201-015-0327-6

J. Rajesh, C.B. Pande, S.A. Kadam, et al., Exploration of groundwater potential zones using analytical hierarchical process (AHP) approach in the Godavari river basin of Maharashtra in India. Appl Water Sci **11**, 182 (2021). https://doi.org/10.1007/s13201-021-01518-x

Y. Razandi, H.R. Pourghasemi, N.S. Neisani, O. Rahmati, Application of analytical hierarchy process, frequency ratio, and certainty factor models for groundwater potential mapping using GIS. Earth Sci. Inf. **8**, 867–883 (2015)

T.L. Saaty (1980a) The Analytic Hierarchy Process, New York: McGraw Hill. International, Translated to Russian, Portuguese, and Chinese, Revised editions, Paperback (1996, 2000), Pittsburgh: RWS Publications

T.L. Saaty, The analytic hierarchy process: Planning, priority setting, resource allocation. McGraw Hill International, New York, NY, USA. Saaty, T.L., 1997. Ascaling method for priorities in hierarchical structures. J. Math. Psychol. **15**, 234–281 (1980b). https://doi.org/10.1016/0022-2496(77)90033-5

T.L. Saaty (1982) Decision Making for Leaders: The Analytical Hierarchy Process for Decisions in a Complex World, Belmont, CA: Wadsworth. Translated to French, Indonesian, Spanish, Korean, Arabic, Persian, and Thai, latest edition, revised, (2000), Pittsburgh: RWS Publications

T.L. Saaty, L.G. Vargas, *Prediction, Projection and Forecasting* (Kluwer Academic, Boston, 1991)

T.L. Saaty, L.G. Vargas, *Models, Methods, Concepts and Applications of the Analytic Hierarchy Process* (Kluwer Academic Publishers, Boston, 2000)

T. Saranya, S. Saravanan, Groundwater potential zone mapping using analytical hierarchy process (AHP) and GIS for Kancheepuram district, Tamilnadu. India. Modell. Earth Syst. Environ. **6**, 1105–1122 (2020)

S. Shahid, S.K. Nath, J. Roy, Groundwater potential modelling in a soft rock area using a GIS. Int. J. Remote Sens. **21**, 1919–1924 (2000)

S. Shekhar, A.C. Pandey, Delineation of groundwater potential zone in hard rock terrain of India using remote sensing, geographical information system (GIS) and analytic hierarchy process (AHP) techniques. Geocarto Int. **30**(4), 402–421 (2014). https://doi.org/10.1080/10106049.2014.894584

K. Shunmugapriya, B. Panneerselvam, K. Muniraj, N. Ravichandran, P. Prasath, M. Thomas, K. Duraisamy, Integration of multi criteria decision analysis and GIS for evaluating the site suitability for aquaculture in southern coastal region. India. Marine Pollut. Bull. **172**, 112907 (2021)

H.F. Yeh, Y.S. Cheng, H.I. Lin, C.H. Lee, Mapping groundwater recharge potential zone using a GIS approach in Hualian River. Taiwan. Sustain. Environ. Res. **26**(1), 33–43 (2016). https://doi.org/10.1016/j.serj.2015.09.005

R.B. Zolekar, Integrative approach of RS and GIS in characterization of land suitability for agriculture: A case study of Darna catchment. Arab. J. Geosci. **11**, 780 (2018). https://doi.org/10.1007/s12517-018-4148-4

R.B. Zolekar, V.S. Bhagat, Multi-criteria land suitability analysis for agriculture in hilly zone: Remote sensing and GIS approach. Comput. Electron. Agric. **118-C**, 300–321 (2015). https://doi.org/10.1016/j.compag.2015.09.016

Z. Zomlota, B. Verbeirena, M. Huysmansa, O. Batelaana, Spatial distribution of groundwater recharge and base flow: Assessment of controlling factors. J. Hydrol. Reg. Stud. **4**, 349–368 (2015). https://doi.org/10.1016/j.ejrh.2015.07.005

Chapter 21
Assessment of Groundwater Prospective Zone in Adigrat Town and Its Surrounding Area Using Geospatial Technology

Muthukumarasamy Ranganathan, Shankar Karuppannan, Bagyaraj Murugasen, Grmay Kassa Brhane, and Balamurugan Panneerselvam

Abstract This study has been conducted in Adigrat Town and its surrounding area, Ethiopia, to identify suitable groundwater potential zones and methods of proper groundwater management and find suitable groundwater potential zones using recent scientific approaches to remote sensing and GIS techniques. To identify the groundwater potential of the study area, the parameters of geology, slope, drainage, geomorphic units, lineament density, and land use/land cover were generated and integrated with an inverse distance weighted (IDW) model based on GIS data. For each of these parameters, appropriate weightage factors were assigned. Weightage factors were assigned to the various geomorphic units based on their ability to store groundwater. This procedure was repeated for each of the remaining layers, and the resulting layers were reclassified. After that, the reclassified layers were integrated to demarcate zones as very good, good, moderate, low, and poor. The analysis identified groundwater potential zones with very good, good, moderate, low, and poor prospects covering 7.19 km^2, 55.13 km^2, 28.03 km^2, 29 km^2, and 28.45 km^2, respectively. This groundwater potential data could be used to effectively identify suitable locations for potable water extraction for rural populations.

M. Ranganathan · G. K. Brhane
Department of Geology, College of Natural and Computational Sciences, Adigrat University, Adigrat, Ethiopia

S. Karuppannan (✉)
Department of Applied Geology, School of Applied Natural Science, Adama Science and Technology University, Adama, Ethiopia

B. Murugasen
Department of Geology, College of Natural and Computational Sciences, Debre Berhan University, Debre Berhan, Ethiopia

B. Panneerselvam
Department of Civil Engineering, M.Kumarasamy College of Engineering, Karur, India

B. Panneerselvam et al. (eds.), *Climate Change Impact on Groundwater Resources*, https://doi.org/10.1007/978-3-031-04707-7_21

Keywords Groundwater potential zones · GIS · Remote sensing · IDW · Thematic map

21.1 Introduction

Groundwater is critical to global food security and economic growth, and it is even more critical to the sustainability of human life. Groundwater is vitally important for drinking, irrigation, and domestic use (Hosseinifard and Mirzaei Aminiyan 2015; Panneerselvam et al. 2020a, b). Globally, increasing population and human activities, such as effluent discharge into surface and groundwater resources, have made appropriate water resource management a difficult task. Rapid urbanization and population growth have resulted in over-exploitation of this resource, resulting in a significant decrease in groundwater levels and a definite deterioration of groundwater resources (Singh et al. 2013; Kavitha et al. 2019a, b; Balamurugan et al. 2020).

Groundwater is found in subterranean rocks known as aquifers, which store and transport water. Groundwater is a type of water that fills all the gaps in geological strata. The water-bearing rocks of the lithosphere serve as transmission channels as well as reservoirs for water storage. The presence of groundwater in a rock formation and the potential for its extraction are primarily determined by the porosity of the formation. The quantity and dispersion of groundwater are determined by the amount of open space available and the unique nature of these rocks. In turn, the behavior of these rocks is determined by their creation and the geological events that formed their condition (Bashe 2017; Das and Pardeshi 2018). Elevated topography and steep slopes enhance runoff, while geographical depressions improve infiltration. When compared to a low drainage density region, a high drainage density region has more water flow. Surface water features, such as rivers and ponds, may serve as recharge zones, increasing the groundwater potential in the surrounding region (Kumar and Shankar 2014; Shankar et al. 2011a, b).

In developing countries such as Ethiopia, uncontrolled population growth and inadequate resource management have directly and indirectly impacted groundwater quality and quantity. As a result, unscientific exploitation and inappropriate use of water policy are also potential factors. As a consequence, determining sustainable groundwater availability is important for long-term management (Shankar and Kawo 2019; Abebe 2020). As a result, identifying and quantifying these characteristics is critical for developing a groundwater potential model of a study region.

Groundwater is a vital natural resource that supports human health and economic growth. It has become a significant source of water supply in both urban and rural regions of any country because of its constant availability and outstanding natural quality (Hussein et al. 2017; Haji et al. 2021a, b; Kawo and Shankar 2018). Groundwater development is advantageous to low discharge springs, surface, and rainfall growth since they are discovered to be insufficiently yielding and no longer drought resistant, while most of them may be totally dried up and never readily restored after such a lengthy dry season (Serele et al. 2020). Traditional field surveys utilizing existing water sources as a guideline guided the selection of a site for groundwater exploration in Ethiopia. Most of the time, a systematic approach to groundwater development was inadequate. As a result, a comprehensive knowledge of groundwater occurrence and availability is needed for groundwater resources planning and also the supply of sufficient water to satisfy the increasing population demand (Solomon and Quiel 2003).

Numerous studies have utilized Geographic Information System (GIS) technology and remotely sensed data to manage water resources, evaluate groundwater, and simulate them. The purpose of this research is to create methods for identifying potential groundwater zones employing remote sensing and GIS technologies. The approach to consolidated remote sensing (RS) and geographic data framework has ended up being a valuable tool in groundwater research (Elewa and Qaddah 2011; Jasrotia et al. 2013; Kumar and Shankar 2014; Rahmati et al. 2015; Jenifer and Jha 2017). RS and GIS advances have arisen as successful devices for water asset contemplates. Quite possibly, the main advantage of utilizing RS and GIS techniques for hydrological exploration and observing is their ability to deliver information in both the geological and transient spaces, which is needed for powerful investigation, forecast, and approval (Sarma and Saraf 2002; Andualem and Demeke 2019; Das et al. 2019; Kumar et al. 2016; Magesh et al. 2012). Geology, geomorphology, lithology, structures, lineaments, porosity, incline, drainage, precipitation, LULC, water quality, level of water, net groundwater recharge, and environment are generally factors that impact groundwater events and streams (Andualem and Demeke 2019; Elewa and Qaddah 2011; Jaiswal et al. 2003; Jha et al. 2007; Murthy and Mamo 2009). As a consequence, the factors utilized vary between research, and the results change (Magesh et al. 2012). The use of remote sensing and GIS technologies to extract comprehensive drainage, slope, and geomorphic characteristics, as well as land use and land cover information in sections of Adigrat, proposes approaches for groundwater potential zone investigations. The aim of this research is to determine appropriate groundwater potential zones and methods for appropriate groundwater resources, as well as to locate suitable groundwater potential zones utilizing current scientific approaches to remote sensing and GIS techniques.

21.2 Materials and Methods

21.2.1 Description of the Study Area

The area, which includes the town of Adigrat and its surrounding areas, is located in Tigray Regional State, northern Ethiopia (Fig. 21.1). It is 900 km north of Addis Ababa and about 120 km north of Mekelle City, in zone 37N UTM grid of latitude 14°11′ 40″N to 14 °21 ′00″N and longitude 39° 24′ 00″E to 39°33′ 20″E. The extent of

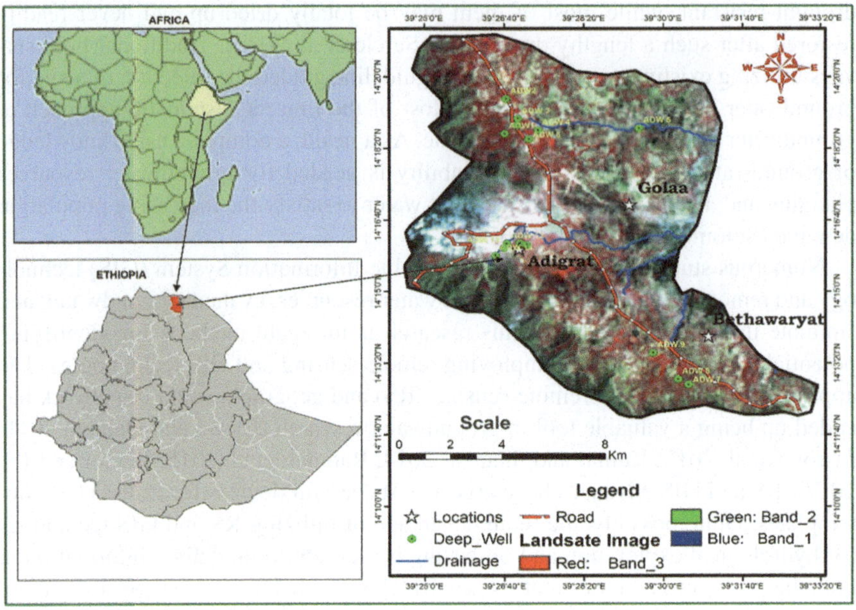

Fig. 21.1 Study area

the study area is 148.52 km^2. It is accessible through all the asphalt roads running from Addis Ababa to Asmara and the newly constructed asphalt road from Wukro to Adwa. The study area is connected with Mekelle city through an asphalt road. The study area is accessible by seasonal gravel roads and trails.

21.2.2 Topography

The area is characterized by continuous rugged mountain topography that is dissected by many minor tributaries of the surrounding river, flatlands in the center, and some small plateaus toward the eastern boundary. Its altitude ranges from 2000 m in the valleys of the eastern portion to above 3000 m above sea level in the steep western boundary and mountains in the southeast and south, like the Andel ridge. The steep slope characterizes the western and eastern boundaries of the study area. The Adigrat ridge, which is the area's western boundary, divides the plateau into the northern (Eritrea) part and the southern (Tigrean part). The ridge is a triple water shed between the Tekeze (south), Mereb-Gasha (north) river basin, and the escarpment in the east (Danakil river basin). The area is drained by many small and medium-scale intermittent rivers originating from the Western highlands toward the Danakil depression. The main source of recharge for the rivers is precipitation during the rainy season.

21.2.3 Land Use, Land Cover, and Soil

The area constitutes four land cover classes, namely the grassland, thorn bushes and shrubs, sparse vegetation of eucalyptus trees and juniper, cultivated land, and settlement. Cultivated land constitutes the larger part of the study area and is mainly located in the central part and eastern plateau. It generally rests on the Enticho sandstone. The grasslands are small, lens-type fields confined to the major rivers. The sparse vegetation of juniper and eucalyptus trees is located almost in all parts of the study area. Scattered settlements within the cultivated land and dense settlements in the town of Adigrat on Kandaro Mountain are also part of the land cover and land use. The major crops in the area are barley, teff, wheat, and sorghum. Three soil classes are present in the study area. This is clay which occupies the relatively steep part of the study area and is thin in thickness. The second soil class is loam clay. It is generally confined to the major rivers and the gentle slopes of the basaltic formation. The third one is the loam Sand which lies on the central and eastern plateau of the study area.

21.2.4 Climate

According to the Ethiopian temperature zoning, the study area is located in the Woina Dega climate zone, where the mean annual temperature is between 15 and 20 degrees centigrade. In the study area, the hottest month is June, with a mean maximum temperature of 15.7 degrees centigrade and a mean minimum of 10.4 degrees centigrade. The coldest month, December, is about 10.4 degrees centigrade cooler, with the corresponding maximum and minimum values being 23degree centigrade and 6.6degree centigrade respectively. The variation in the temperature cycle from year to year is slight. As far as precipitation is concerned, from the long-term rainfall record, the highest average value of rainfall is recorded in the month of August with a value of 195 mm and the lowest in the month of January with an average rainfall value of 8 mm.

21.3 Groundwater Field Investigations Study

21.3.1 Springs

They have been important sources of water for people (Soujanya et al. 2020). Springs have been important in the social, cultural, and economic history of mankind. Most springs in the study area are seasonal and intermittent. They are gravity springs caused by different permeabilities, that is, at geological contacts (mainly manifested at the contact between the Adigrat sandstone and the Enticho sandstone).

The basaltic dykes in the Enticho sandstone act as an aquiclude and cause springs (Dibla Yohannes spring). Many springs are fracture controlled, as observed in the area around Beati Michael spring. Concerning temperature, the springs in the study area are cold.

21.3.2 *Wells*

The geographic distribution of water points in the study area is uneven, and is mainly determined by the hydrogeological characteristics of the geological units. More than 90% of the water points lie in the Enticho sandstone and alluvial sediments with respect to this. Few water points are inventoried at the Aiba basalt with favorable hydrogeological parameters. The wells in the study area are hand-dug wells, shallow wells, and deep wells. Most hand-dug wells are constructed at a shallow depth by excavations of 2–3.2 m in diameter. The depth of the hand-dug wells ranges from 3 m to 13 m. They are mainly located in the shallow overburden or alluvial sediments intervening in rock formations near the mountains. The bore wells constructed by The down Hole Hammer (DTH) rigs are dominantly concentrated at the Enticho sandstone. The depth of the bore wells ranges from 126 m to 202 m. According to data from Tigray waterworks construction enterprise, the depth of water strikes also ranges from 10 m to 54 m and the yield of the wells is between 0.8 l/sec and to >10 l/sec.

21.4 Methods

The base map was prepared by using a geological survey of Ethiopia toposheet no. 139A and Google Earth on a 1:50000 scale and Landsat satellite data. GIS and remote sensing technologies were used to create groundwater thematic maps such as drainage density, lineament density, geomorphology, slope, land use, and land cover. It categorized the inverse distance weighted (IDW) model, and its ranking was employed to determine the best groundwater potential zone. These thematic maps are then ranked and weighted for analysis in the GIS domain. Using GIS, all of the thematic layers were superimposed to get the final integrated result of the prospective groundwater zones. Geomorphology, land use and land cover, slope, drainage density, and lineament density were all taken into account in this research to identify groundwater potential zones.

21.4.1 GIS Analysis for Identification of Groundwater Potential Zones

There are plenty of methods for locating and mapping the presence and distribution of groundwater. This research primarily focuses on the use of satellite images to better precisely detect and define surface features. This technique is perhaps more efficient than traditional methods for mapping groundwater potential zones and produces superior results (Krishanmurthy et al. 1996). Remote sensing is used as a preliminary inventory technique to assess groundwater potential and aids in defining regions where additional investigation using hydrogeological and geophysical methods is required. Furthermore, the use of remote sensing methods in conjunction with GPS in a unified platform, as well as the integration of GIS tools, allowed improved data processing and interpretation. The IDW model gave various rankings and weights to key geosystem characteristics. The topics were given specific weights based on the relative significance of groundwater exploration, as stated. For ranking purposes, the land use, land cover, and drainage density were given higher rankings, while the lineament density and drainage density were assigned lower ranks. Following the assignment of rankings (maximum weightages) to the various geosystem, system parameters, individual weights were assigned to sub-variables of the aforementioned major six geosystem parameters.

Similarly, different weights were assigned to each geosystem parameter's unique sub-variable. The GIS layers on lineament density, geomorphology, slope, and drainage density were thoroughly examined throughout this procedure, and weightages were given to each sub-variable (Butler et al. 2002; Skubon 2005; Asadi et al. 2007; Yammani 2007). The GIS layers with final weightages for six geosystem characteristics were also added, and the research region was finally split into various groundwater potential zones relying on the additional weights. Particular theme and feature evaluations were weighted and added to the layers dependent on their ability to contain groundwater (Table 21.1). This is frequently alluded to as multi-measures assessment. There are a few strategies for deciding interclass/between map reliance; a likelihood weighted methodology has been embraced, which considers a straight blend of the likelihood loads of each topical guide (W_t), and various classifications of determined topical guides have been allocated scores (W_i), contingent upon their aptness to embrace groundwater. The greatest value is assigned to the feature with the highest groundwater potential, while the lowest value is assigned to the feature with the lowest potential. In raster-based GIS software systems, the weighted linear combination method is dominant. The ArcGIS spatial analyst extension has been used to convert the features to raster and for final analysis. The total weights of the final integrated map were calculated using this technique as the sum or product of the weights given to the various layers based on their appropriateness. Finally, a map of the groundwater potential zone was produced.

Table 21.1 Ranks and weights of controlling parameters

Parameters	Classes	Area (km²)	Weights (%)	Rank
Geology	Basalt	33	5	3
	Adigrat sandstone	12		2
	Edaga Arbi tillite	14.28		4
	Enticho sandstone	77.17		2
	Metamorphic rocks	11.62		5
Geomorphology	Dissected plateau	20.18	5	1
	Pedimont plateaus	59.45		4
	Structural hill	35.67		1
	Structural valley	9.53		2
	Valley fill	23.67		3
Slope	Gentle slope	92.87	15	3
	Moderate slope	21.69		2
	Moderate steep slope	14.7		2
	Steep slope	8.77		1
	Very steep slope	10.41		1
Lineament density	Low	89.74	20	1
	Moderate	51.97		2
	High	6.79		3
Drainage density	Low	64.41	25	3
	Moderate	51.14		2
	High	32.96		1
Land use/ land cover	Agricultural lands	30.97	30	4
	Dry lands	68.81		2
	Settlements	20.72		1
	Wetland/Hill shade	27.78		3

21.4.2 Methodology for Ranking Weights

For the present study, the different ranks and weights were assigned to relevant different geosystem parameters described as geology, lineaments, geomorphology, slope, drainage density, land use/land cover, and their corresponding sub-variables, and the same are shown in Table 21.1 for the identification of groundwater potential zones.

Drainage density, land use/land cover, lineament density, and slope were given greater weightages, whereas geology and geomorphology were awarded lower weightages, resulting in a maximum of 10. Following the assignment of rankings (maximum weightages) to the various geosystem parameters, individual weights were assigned to sub-variables of the six major geosystem parameters listed above. As a result, lineament densities were given weights ranging from 3 to 1, with the lowest values corresponding to extremely low lineament density and medium and higher weights corresponding to rises in density. Similarly, different weights were given to each geosystem parameter's unique sub-variable. The GIS layers on geology, lineament density, geomorphology, slope, drainage density, and land use/land cover were thoroughly evaluated throughout this procedure, and weights were given to each of their sub-variables. The GIS layers with the final weights of six geosystem characteristics were also added, and the research region was ultimately split into various groundwater potential zones based on the additional weights.

21.5 Result and Discussion

21.5.1 Deterministic Factors for Groundwater Occurrence and Movements

21.5.1.1 Geology

The geology of the study area (Fig. 21.2) mainly composed of Enticho sandstone present in Bethawaryat area, Golaa area and Adigrat area. While preparing the geology data, the same may be converted into a GIS environment. The geology GIS layer was examined with particular attention to the groundwater retaining and conducting ability of different rock types and classified into five classes: basalt, Adigrat sandstone, tillite, Enticho sandstone, and metamorphic rock, as indicated in the figure (Table 21.1).

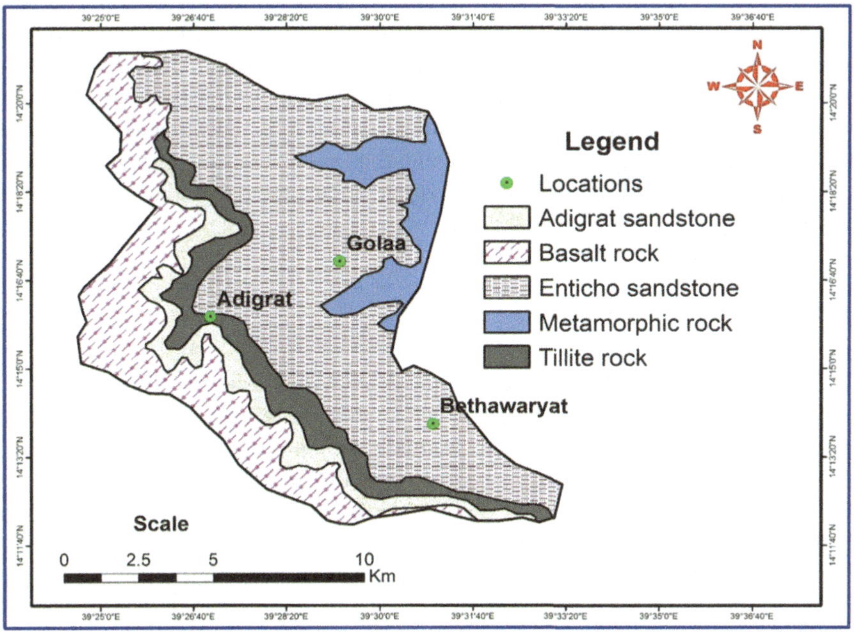

Fig. 21.2 Geology map

Depending on the groundwater holding and conducting behavior of these different classes of rocks, the least weightage 2 was given to Sandstone/ Shale, and the highest weightages of 3 was assigned to Alkali olivine basalt and tuffs rare rhyolites. Such weightages-based GIS layer having two rock types/sub-variables generally water-bearing present in sandstone.

21.5.1.2 Lineament Density

In hard rock terrain, secondary porosity is primarily responsible for groundwater storage and circulation, that is, the presence of lineaments and fractures. A lineament analysis of the region based on remotely sensed data offers significant information on subsurface fractures that may influence groundwater flow and storage. The fracture density of rocks affects subterranean permeability (Münch and Conrad 2007; Vijith 2007). As a result, identifying lineaments in the landscape using satellite imagery is increasingly important. The majority of the lineaments are identified by anomalies related to characteristics such as straight drainage channels, plant patterns, terrain, and so on. The region is crisscrossed by major and tiny lineaments. There are two lineaments in the area: NNW–NS and NE–SW. The study of lineament density may aid in the identification of weathered zones in a region, which is important in research related to groundwater exploration, soil erosion, and so on.

Weights were also given to lineament density (Fig. 21.3). It is classified into three categories based on density: high, moderate, and low. The lineament density polygons were divided into three groups based on the weightage provided as low (1), moderate (2), and high (3), as indicated in (Table 21.1), and a weighted GIS layer was created as a result. Low lineament density zones will have poor groundwater prospects, whereas higher and intermediate lineament density zones will have improved groundwater prospects. According to that, the Bethawaryat region has the least quantity of high groundwater potential zone. There are high groundwater potential zones accessible in the Bethawaryat, Golan, and Adigrat and Adigrat east regions.

21.5.1.3 Geomorphology

The geomorphic characteristics include the dissected plateau, structural hill, structural valley, valley fill, and pediment plateau, which were interpreted from Landsat data. The pediments exhibit a wide, gradually sloping erosion surface and a low relief plain. They are supported by bedrock, and this pediment takes up the bulk of the entire research area (Fig. 21.4). Structural hills are found in almost every section of the research area and are identified using image features. Satellite photography is used to define valley fills. In the valley regions, there was more plant cover. The intricacy of geomorphic evolution results in several different geomorphic characteristics. The prevalence and size of these geomorphic zones vary by location. Leblanc et al. (2007) reported similar findings.

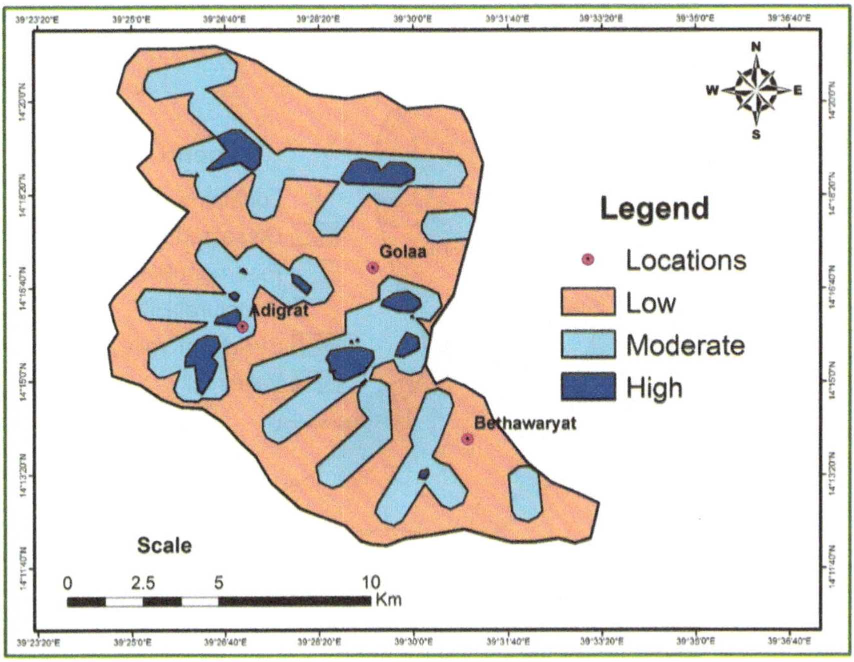

Fig. 21.3 Lineament density

As illustrated in (Table 21.1), the region's geomorphic characteristics were weighted based on their groundwater prospects, and the different geomorphic characteristics were dissolved into five classes employing GIS tools. The greatest significant contribution to groundwater possibilities comes from pediment plateaus and valley fill. Accordingly, these two geomorphic units were dissolved in the geomorphology GIS layer and assigned 1 & 2, and the least weightage was assigned for the rock exposure feature. Adigrat central place mostly covered pediment, valley-fill deposits present Adigrat and Adigrat north. So from this, we can identify groundwater potential zone.

21.5.1.4 Drainage Density

The drainage pattern reflects both surface and subterranean development. The drainage density (in kilometers per square kilometer) indicates the closeness of channel spacing as well as the kind of surface material (Prasad et al. 2008). The higher the flow of water in a drainage density, the more the runoff. As a result, drainage density specifies runoff in an area, or the amount of relative rainfall that may have penetrated. As a consequence, the higher the probability of recharging or prospective groundwater zone, the lower the drainage density. As previously mentioned, the bulk of the drainage originates in the northwest and southwest of the region.

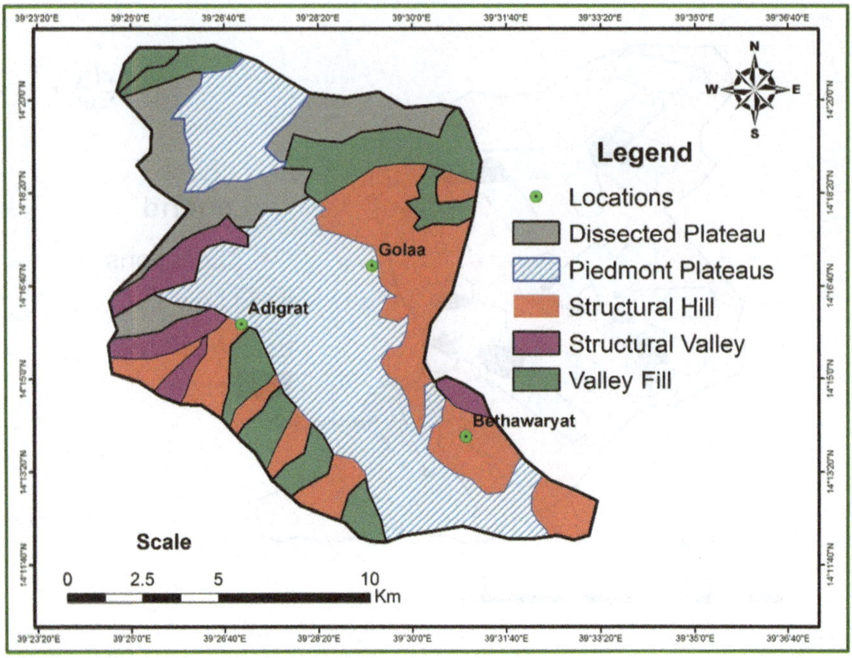

Fig. 21.4 Geomorphology map

High-density areas have a low infiltration rate, while low-density areas have a high infiltration rate (Saha et al. 2010, Vasanthavigar et al. 2011). The drainage density polygons were weighted and split into three groups, as shown in (Table 21.1 and Fig. 21.5), and a weighted GIS layer was produced.

21.5.1.5 Slope

The slope is also an important factor in determining the occurrence and recharging conditions of groundwater in a certain region. The slope is expressed in degrees (Dawoud et al. 2005; Vittala et al. 2005; Solomon and Quiel 2006). The slope of the study area was classified into five classes (Table 21.1 and Fig. 21.6), such as <5 plain areas, slope zone 5–10°, 10–20°, 20–25° & >25°, and weightages of 3, 2, and 1 were, respectively, assigned to them based on their groundwater prospects for a gentle slope, moderate slope, and moderate steep slope, whereas steep slope and very steep slope are not assigned because there is no groundwater potential and corresponding GIS layer was prepared. In this case, higher weightage was given to gentle slopes because of plain area, and gradually lesser and lesser weights were assigned steeper and steeper slopes because runoff is directly proportional to the slope. Gentle slope more percentage covered at Adigrat & Golaa area, compared to another place, and the gentle slope will have high potential.

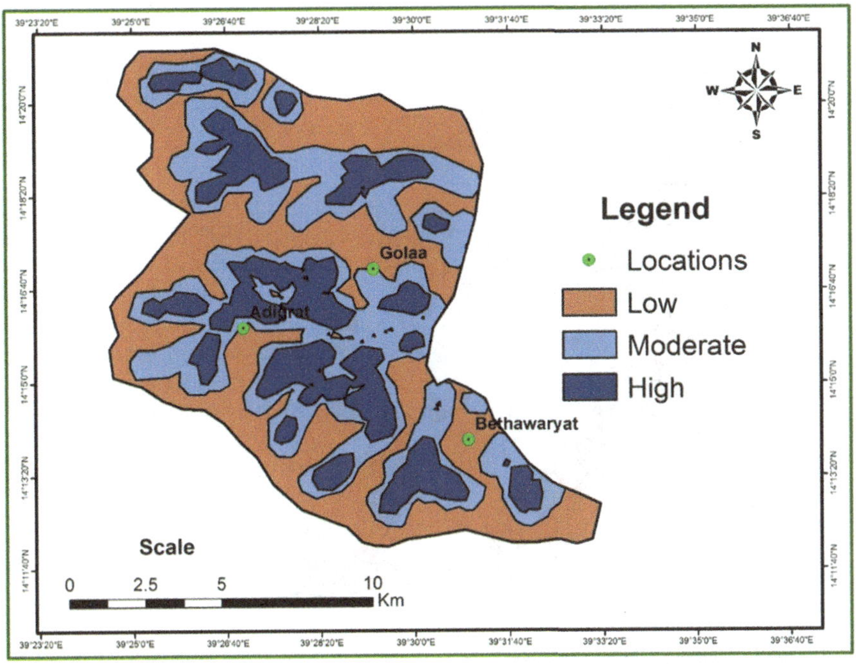

Fig. 21.5 Density map the area

21.5.1.6 Land Use and Land Cover

The land use/land cover cram vicinity has been attempted in order to discover and map the assorted types of land use/land cover curriculum in the vicinity by image explanation, photo-interpretation elements developed the classification system (Hussain and Karuppannan 2021). The term "land usage" refers to "man's activities on land that are intrinsically related to land" (Anderson et al. 1976; Athick et al. 2019; Athick and Shankar 2019). The study of land cover is dedicated to the flora and man-made structures that cover the ground surface. The word land cover refers to the kind of feature found on the earth's surface. Buildings, lakes, forests, trees, and glacial ice are all examples of land cover types. The following land use/land cover practices were observed, such as agricultural lands, drylands, settlements, wetland/water shade, and the same has been digitized in the GIS environment (Fig. 21.7).

Agricultural Lands

The tonal contrast of cropland varies from deep red to pink depends on the health of the crop. In the area, the cropland has mostly observed Golaa N-E and Adigrat S-W.

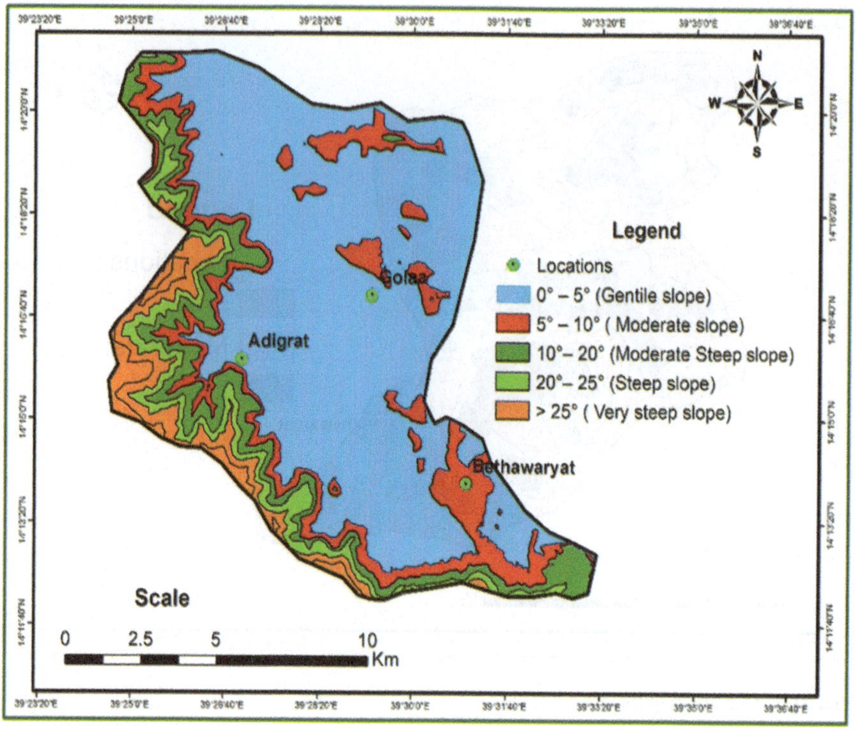

Fig. 21.6 Slope map of the area

Drylands

In the area equally, drylands are demarcated based on the tonal variation. Golaa N-E area under agricultural tree crop and mountain, some valley also, planted adopting certain agricultural management techniques.

Settlements

It is described as a human settlement established as a result of non-agricultural usage, with a cover of buildings, transportation, and communication, utilities associated with water, vegetation, and empty land. The built-up terrain is mostly visible in the Adigrat region and its surrounding areas, with pinkish blue and square patterns.

The land use/land cover characteristics were classified into four classes, and the associated polygons were dissolved. A new GIS layer was constructed with just four classes, as shown in (Table 21.1), with settlements and dry-lands receiving the least weightage.

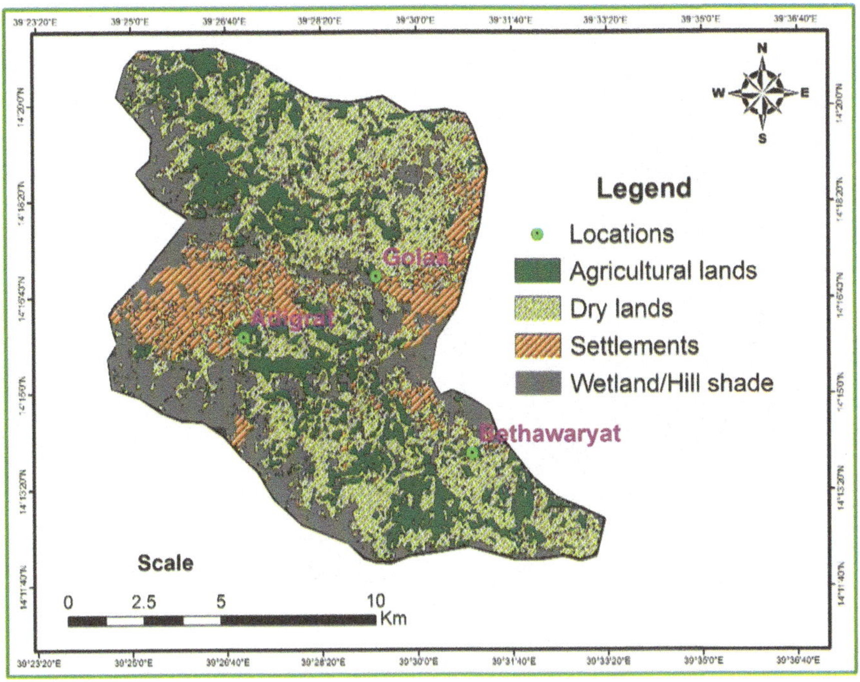

Fig. 21.7 Land use/land cover map of the area

21.5.1.7 GIS Integration and Delineation of Groundwater Potential Zones

After generating the GIS layers on land use/land cover, lineament density, drainage density, geomorphology, Geology, and slope were integrated using the GIS Add function. Such GIS integration has resulted in 1391 polygon classes and having weightage from 1 to 15. Such dynamic range of weightage of these 1391 polygon classes was grouped into three via less than 0–3 poor groundwater prospects, 03–06 low potential zone, 06–10 moderate potential, 10–13 good potential zones, and 13–15 very good potential zones. The analysis revealed groundwater potential zones with very good, good, moderate, low, and poor prospects (Fig. 21.8) covering an area of 7.19 km^2, 55.13 km^2, 28.03 km^2, 29 $km^{2.}$ and 28.45 km^2, respectively.

21.6 Conclusion

The remote sensing and GIS method are quite beneficial since it assorted diverse geospatial information, particularly groundwater potential zone mapping. The research focused on the effectiveness of remote sensing and GIS in identifying and

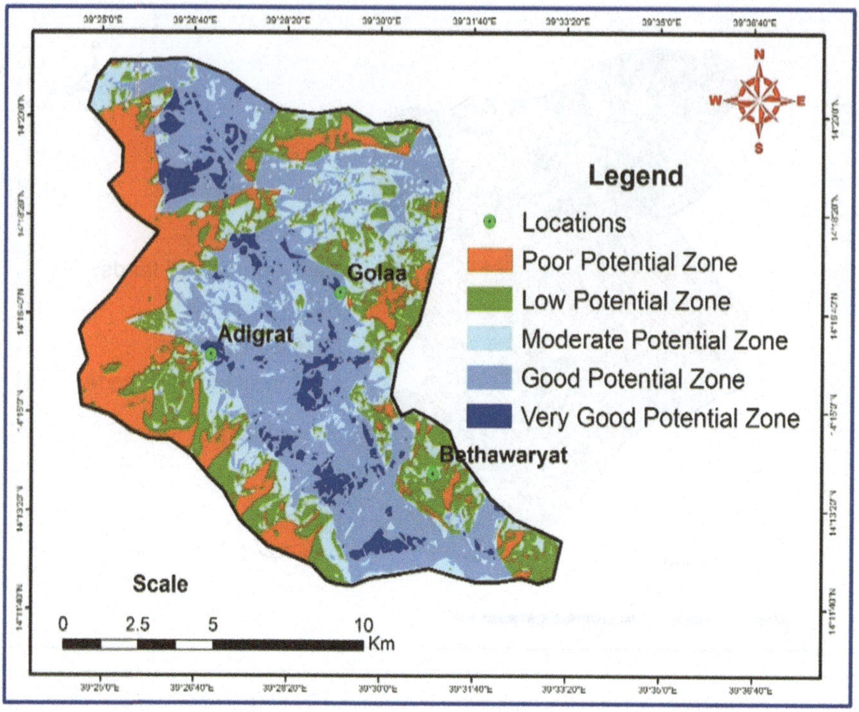

Fig. 21.8 Groundwater potential zone map of the area

delineating groundwater prospective zones in the study region. This study comprehends the groundwater potential zone in Adigrat and pictorially represents it utilizing GIS. The analysis revealed groundwater potential zones with very good, good, moderate, low, and poor prospects covering an area of 7.19 km^2, 55.13 km^2, 28.03 km^2, 29 km^2, and 28.45 km^2, respectively. As per the findings of the current research, a slope area of 0°–10°, lineament density, land use/land cover, structural valleys, valley fills, pediment agriculture, wetland, and the slope are suitable zones for groundwater investigation. These areas should be prioritized for future watershed development efforts. This critical information may be successfully utilized to identify appropriate sites for the extraction of drinkable water for urban and rural populations. The present multi-parametric method based on GIS and remote sensing is holistic in nature and will save time and money, particularly when it comes to finding groundwater potential zones.

References

Abebe Debele Tolche. Groundwater potential mapping using geospatial techniques: A case study of Dhungeta-Ramis sub-basin, Ethiopia Geol. Ecol. Landscapes, https://doi.org/10.1080/24749508.2020.1728882 (2020)

J.R. Anderson, et al. A Land Use and Land Cover Classification System for Use with Remote Sensor Data. Geological Survey Professional Paper No. 964, U.S. Government Printing Office, Washington DC, 28 (1976)

T.G. Andualem, G.G. Demeke, Groundwater potential assessment using GIS and remote sensing: A case study of Guna tana landscape, upper Blue Nile Basin, Ethiopia. J. Hydrol. Regional Stud. **24**, 100610 (2019)

S.S. Asadi, P. Vuppala, M.A. Reddy, Remote sensing and GIS techniques for evaluation of groundwater quality in municipal corporation of Hyderabad (zone-V), India. Int. J. Environ. Res. Public Health **4**(1), 45–52 (2007). https://doi.org/10.3390/ijerph2007010008

A.A.S.M. Athick, & K. Shankar. Data on land use and land cover changes in AdamaWereda, Ethiopia, on ETM+, TM and OLI- TIRS landsat sensor using PCC and CDM techniques, Data in Brief, 24. https://doi.org/10.1016/j.dib.2019.103880 (2019)

A.M.A. Athick, K. Shankar, H.R. Naqvi, Data on time series analysis of land surface temperature variation in response to vegetation indices in twelve Wereda of Ethiopia using mono window, split window algorithm and spectral radiance model. Data Brief **27**, 104773 (2019). https://doi.org/10.1016/j.dib.2019.104773

P. Balamurugan, P.S. Kumar, K. Shankar, R. Nagavinothini, K. Vijayasurya, Non-carcinogenic risk assessment of groundwater in southern part of Salem District in Tamilnadu, India. J. Chil. Chem. Soc. **65**(1), 4697–4707 (2020). https://doi.org/10.4067/S0717-97072020000104697

B.B. Bashe, Groundwater potential mapping using remote sensing and GIS in Rift Valley Lakes Basin, Weito Sub Basin, Ethiopia. Int. J. Sci. Eng. Res. **8**(2), 43–50 (2017) Available at: http://www.ijser.org

M. Butler, J. Wallace, M. Lowe. Ground-water quality classification using GIS contouring methods for Cedar Valley, Iron County, Utah. In: Proceedings of Digital Mapping Techniques, 2002 Workshop, US Geological Survey Open-File Report02–370 (2002)

S. Das, S.D. Pardeshi. Integration of different influencing factors in GIS to delineate groundwater potential areas using IF and FR techniques: A study of Pravara basin, Maharashtra, India. App. Water Sci. https://doi.org/10.1007/s13201-018-0848-x (2018)

B. Das, S.C. Pal, S. Malik, R. Rabin Chakrabortty, Modeling groundwater potential zones of Puruliya district, West Bengal, India using remote sensing and GIS techniques. Geol. Ecol. Landscapes **3**(3), 223–237 (2019)

M.A. Dawoud, M.M. Darwish, M.M. El-Kady, GIS-based groundwater management model for western Nile delta. Water Resour. Manag. **19**(5), 585–604 (2005). https://doi.org/10.1007/s11269-005-5603-z

H.H. Elewa, A.A. Qaddah, Groundwater potentiality mapping in the Sinai Peninsula, Egypt, using remote sensing and GIS-watershed-based modeling. Hydrgeol. J. **19**(3), 613–628 (2011)

M. Haji, S. Karuppannan, D. Qin, H. Shube, N.S. Kawo, Potential human health risks due to groundwater fluoride contamination: A case study using multi-techniques approaches (GWQI, FPI, GIS, HHRA) in Bilate River Basin of Southern Main Ethiopian Rift, Ethiopia. Arch. Environ. Contam. Toxicol. **80**, 277–293 (2021a). https://doi.org/10.1007/s00244-020-00802-2

M. Haji, D. Qin, Y. Guo, L. Li, D. Wang, S. Karuppannan, H. Shube. Origin and geochemical evolution of groundwater in the Abaya Chamo basin of the Main Ethiopian Rift: Application of multitracer approaches. Hydrgeol. J. 1–20. https://doi.org/10.1007/s10040-020-02291-y (2021b)

S.J. Hosseinifard, M. Mirzaei Aminiyan, Hydrochemical characterization of groundwater quality for drinking and agricultural purposes: A case study in Rafsanjan Plain. Iran. Water Quality Exposure Health **7**(4), 531–544 (2015)

S. Hussain, S. Karuppannan. Land use/land cover changes and their impact on land surface temperature using remote sensing technique in district Khanewal, Punjab Pakistan. Geol. Ecol. Landscapes, 1–13. https://doi.org/10.1080/24749508.2021.1923272 (2021)

A.-A. Hussein, V. Govindu, A.G.M. Nigusse, Evaluation of groundwater potential using geospatial techniques. Appl. Water Sci. **7**(5), 2447–2461 (2017). https://doi.org/10.1007/s13201-016-0433-0

R.K. Jaiswal, S. Mukherjee, J. Krishnamurthy, R. Saxena, Role of remote sensing and GIS techniques for generation of groundwater prospect zones towards rural development – An approach. Int. J. Remote Sens. **24**(5), 993–1008 (2003)

A.S. Jasrotia, B.D. Bhagat, A. Kumar, R. Kumar, Remote sensing and GIS approach for delineation of groundwater potential and groundwater quality zones of Western Doon valley, Uttarakhand. India. J. Indian Soc. Remote Sens. **41**(2), 365–377 (2013)

M.A. Jenifer, M.K. Jha, Comparison of analytical hierarchy process, catastrophe and entropy techniques for evaluating groundwater prospect of hard-rock aquifer systems. J. Hydrol. **548**, 605–624 (2017)

M.K. Jha, A. Chowdhury, V.M. Chowdary, S. Peiffer, Groundwater management and development by integrated remote sensing and geographic information systems: Prospects and constraints. Water Resour. Manag. **21**(2), 427–467 (2007)

M.T. Kavitha, K. Shankar, R. Divahar, T. Meenambal, R. Saravanan, Impact of industrial wastewater disposal on surface water bodies in Kalingarayan canal, Erode district, Tamil Nadu, India. Arch. Agric. Environ. Sci. **4**(4), 379–387 (2019a). https://doi.org/10.26832/24566632.2019.040403

M.T. Kavitha, R. Divahar, T. Meenambal, K. Shankar, R. VijaySingh, T.D. Haile, C. Gadafa, Dataset on the assessment of water quality of surface water in Kalingarayan Canal for heavy metal pollution, Tamil Nadu. Data Brief **22**, 878–884 (2019b). https://doi.org/10.1016/j.dib.2019.01.010

N.S. Kawo, K. Shankar, Groundwater quality assessment using water quality index and GIS technique in Modjo River Basin, Central Ethiopia. J. Afr. Earth Sci. **147**, 300–311 (2018). https://doi.org/10.1016/j.jafrearsci.2018.06.034

J. Krishnamurthy, N. Venkatesa Kumar, V. Jayaraman, M. Manivel, An approach to demarcate groundwater potential zones through remote sensing and a GIS. Int. J. Remote Sens. **17**(10), 1867–1884 (1996). https://doi.org/10.1080/01431169608948744

G.R.S. Kumar, K. Shankar, Assessment of groundwater potential zones using GIS. Front. Geosci. **2**(1), 1–10 (2014)

P. Kumar, S. Herath, R. Avtar, K. Takeuchi, Mapping of groundwater potential zones in Killinochi area, Sri Lanka, using GIS and remote sensing techniques. Sustain. Water Res. Manage. **2**(4), 419–430 (2016)

M. Leblanc, G. Favreau, S. Tweed, C. Leduc, M. Razack, L. Mofor, Remote sensing for groundwater modeling in large semiarid areas: Lake Chad basin. Africa. Hydrogeol J **15**(1), 97–100 (2007). https://doi.org/10.1007/s10040-006-0126-0

N.S. Magesh, N. Chandrasekar, J.P. Soundranayagam, Delineation of groundwater potential zones in Theni district, Tamil Nadu, using remote sensing, GIS and MIF techniques. Geosci. Front. **3**(2), 189–196 (2012). Malczewski, J. (1999). GIS and multicriteria decision analysis. New York: Wiley

Z. Münch, J. Conrad, Remote sensing and GIS based determination of groundwater dependent ecosystems in the Western Cape. South Africa. Hydrogeol J **15**(1), 19–28 (2007). https://doi.org/10.1007/s10040-006-0125-1

K.S.R. Murthy, A.G. Mamo (2009). Multi-criteria decision evaluation in groundwater zones identification in Moyale–Teltele subbasin, South Ethiopia. Int. J. Remote Sens., 30(11), 2729–2740

B. Panneerselvam, S.K. Paramasivam, S. Karuppannan, et al., A GIS-based evaluation of hydrochemical characterisation of groundwater in hard rock region, South Tamil Nadu. India. Arab J. Geosci. **13**(17), 1–22 (2020a). https://doi.org/10.1007/s12517-020-05813-w

B. Panneerselvam, S. Karuppannan, K. Muniraj. Evaluation of drinking and irrigation suitability of groundwater with special emphasizing the health risk posed by nitrate contamination using nitrate pollution index (NPI) and human health risk assessment (HHRA). Hum. Ecol. Risk Assess. Int. J. 1–25. https://doi.org/10.1080/10807039.2020.1833300 (2020b)

R.K. Prasad, N.C. Mondal, P. Banerjee, M.V. Nandakumar, V.S. Singh, Deciphering potential groundwater zone in hard rock through the application of GIS. Environ. Geol. **55**(3), 467–475 (2008). https://doi.org/10.1007/s00254-007-0992-3

O. Rahmati, A.N. Samani, M. Mahdavi, H.R. Pourghasemi, H. Zeinivand, Groundwater potential mapping at Kurdistan region of Iran using analytic hierarchy process and GIS. Arab. J. Geosci. **8**(9), 7059–7071 (2015)

D. Saha, Y.R. Dhar, S.S. Vittal, Delineation of groundwater development potential zones in parts of marginal Ganga alluvial plain in south Bihar, eastern India. Environ Monitor **165**(1–4), 179–191 (2010). https://doi.org/10.1007/s10661-009-0937-2

B. Sarma, A.K. Saraf. Study of land use-groundwater relationship using an integrated remote sensing and GIS approach, http://www.gisdevelopment.net (2002)

C. Serele, A. Pérez-Hoyos, F. Kayitakire, Mapping of groundwater potential zones in the drought-prone areas of south Madagascar using geospatial techniques. Geosci. Front. **11**(4), 1403–1413 (2020). https://doi.org/10.1016/j.gsf.2019.11.012

K. Shankar, N.S. Kawo, Groundwater quality assessment using geospatial techniques and WQI in North East of Adama Town, Oromia Region. Ethiopia. Hydrospatial Anal. **3**(1), 22–36 (2019). https://doi.org/10.21523/gcj3.19030103

K. Shankar, S. Aravindan, S. Rajendran, Hydrogeochemistry of the Paravanar river sub-basin, Cuddalore District, Tamilnadu. India. E-J Chem. **8**(2), 835–845 (2011a). https://doi.org/10.1155/2011/107261

K. Shankar, S. Aravindan, S. Rajendran, Hydrochemical profile for assessing the groundwater quality of Paravanar River Sub-Basin, Cuddalore district, Tamil Nadu. India. Curr. World Environ. **6**(1), 45–52 (2011b). https://doi.org/10.12944/CWE.6.1.05

A.K. Singh, B. Raj, A.K. Tiwari, M.K. Mahato, Evaluation of hydrogeochemical processes and groundwater quality in the Jhansi district of Bundelkhand region, India. Environ. Earth Sci. **70**(3), 1225–1247 (2013). https://doi.org/10.1007/s12665-012-2209-7

B.A. Skubon, Groundwater quality and GIS investigation of a shallow sand aquifer, Oak opening region, Northwest Ohio. Geol. Soc. Am. Abstr. Programs **7**, 37–94 (2005)

S. Solomon, F. Quiel Integration of remote sensing and GIS for groundwater assessment in Eritrea, geoinformation for European-wide integration (2003)

S. Solomon, F. Quiel, Groundwater study using remote sensing and geographic information systems (GIS) in the central highland s of Eritrea. Hydrgeol. J. **14**(6), 1029–1041 (2006). https://doi.org/10.1007/s10040-006-0096-2

B. Soujanya Kamble, P.R. Saxena, R.M. Kurakalva, K. Shankar, Evaluation of seasonal and temporal variations of groundwater quality around Jawaharnagar municipal solid waste dumpsite of Hyderabad city. India. SN Appl. Sci. **2**(3), 498 (2020). https://doi.org/10.1007/s42452-020-2199-0

M. Vasanthavigar, K. Srinivasamoorthy, K. Vijayaragavan, S. Gopinath, S. Sarma, Groundwater potential zoning in Thirumani- muttar sub-basin Tamilnadu, India—A GIS and remote sensing approach, geo-spatial. Inform. Sci. **14**(1), 17–26 (2011)

H. Vijith, Groundwater potential in the hard rock terrain of western Ghats: A case study from Kottayam district, Kerala using resourcesat (Irs-P6) data and GIS techniques. J. Indian Soc. Remote Sens. **35**(2), 163–171 (2007). https://doi.org/10.1007/BF02990780

S. Vittala, S. Govindaiah, H.H. Gowda, Evaluation of groundwater potential zones in the sub-watersheds of North Pennar River basin around Pavagada, Karnataka, India, using remote sensing and GIS techniques. J. Indian Soc. Remote Sens. **33**(4), 483–493 (2005). https://doi.org/10.1007/BF02990733

S. Yammani, Groundwater quality suitable zones identification: Application of GIS, Chittoor area, Andhra Pradesh. India. Env. Geol. **53**(1), 201–210 (2007). https://doi.org/10.1007/s00254-006-0634-1

Chapter 22
Computation of Rainfall Infiltrates into Coastal Soil of Andhra Pradesh, India

Dinagarapandi Pandi, Anand Balasubramanian, and Praveenraj Durai

Abstract In the water cycle, infiltration plays a main role in the freshwater intrusion as a downward direction of the ground. Thus, the basic source of water resources is the rainfall that occurs in a tropical climate. The tropical zone of East Coastline stretches from Kanyakumari to Ganga delta in India. A choice of Andhra Pradesh Coast happens with the dynamic coastal morphology that tempts to estimate the soil infiltration. Therefore, PERSIANN gridded rainfall is collected for the pre- and post-monsoon over Andhra Pradesh. Lithology coefficient is applied in the rainfall to contribute to soil infiltration. The rainfall recharge into soil and groundwater distribution on a spatial scale is analyzed using GIS techniques. For data evaluation, hydrological soil groups are segregated from soil classes, and groundwater monitored data are compared to estimate soil infiltration. As per this study, the most likely rainfall infiltration capacity occurs in the topsoil of the Andhra Pradesh coast, resulting in coastal hazards such as seawater intrusion, climate change, and storm surge. Infiltration capacity has been compared with the groundwater distribution along with the thematic layers such as soil, geomorphology, and slope. Coastal regulation zone III and IV are adopted for the sustainable coastal environment.

Keywords PERSIANN · Infiltration · Andhra Pradesh · Coastal regulation zone

---

D. Pandi (✉)
School of Civil Engineering, Vellore Institute of Technology, Chennai, Tamil Nadu, India

A. Balasubramanian
Department of Agriculture Engineering, KIT-Kalaignarkarunanidhi Institute of Technology, Kannampalayam, Tamil Nadu, India

P. Durai
Department of Civil Engineering, SRM Institute of Science and Technology, Chengalpattu, Tamil Nadu, India

B. Panneerselvam et al. (eds.), *Climate Change Impact on Groundwater Resources*, https://doi.org/10.1007/978-3-031-04707-7_22

22.1 Introduction

In the infiltration process, soil always retains the moisture content at the top layer for about 5 meter (m) width due to its gravity. It is one of the major parameters of hydrological components; it facilitates the groundwater resources (Mahapatra et al. 2020). Thus, from the infiltration process, one can find through the rainfall rate archives at the particular area/region. Soil infiltrate is processed with the porosity and standard value of hydraulic conductivity from the nature of the soil (Saxton and Rawls 2006). Uncompacted and compacted soils are described within their physical properties of the soil (Vand et al. 2018). Some soils retain their field capacity/bulk density, allowing for the calculation of infiltration in the residue soil cover (Patle et al. 2019). Tillage, irrespective of the physical characteristic, causes an abnormality in the infiltration rate as a result of anthropogenic activity. Hence, estimating the infiltration process is extremely beneficial for agriculture engineering (Patle et al. 2019), water resource engineering, and for studying the soil desalination purposes.

The geomorphology of the coastal region is more variable due to the presence of seawater (Rizzo and Anfuso 2020). Mostly, shorelines analysis, seawater intrusion, environment impact assessment, etc., play vital roles in the coastal studies. Thus, environmental impact assessment is the part to exaggerate the water influence in the hydrological components. So it deals with the thermohaline circulation from the ocean to land. In India, monsoon season gives more water resources, i.e., rainfall. Rainfall is part of the orographic effect from the sea to land. Thus, it produces the infiltration into soil and then percolates toward the coastal aquifer. As a result, cyclonic rainfall occurs due to abnormal nature. Rainfall causes the floods over the coastal region comparatively in the hinterland. The main reason for cyclonic rainfall is the environmental impact, i.e., climatic change. East Coastline of India is applied in such studies as coastal vulnerability index and shoreline analysis that occur in Andhra Pradesh (Basheer Ahammed and Pandey 2019; Rao et al. 2009). These coastal landforms are dynamic due to sea-level rise and coastal erosion (Rani et al. 2018). Tides, wind, waves, currents, and storms are the major forces on the coast.

To calculate the infiltrate rate in real time, mostly double ring infiltrometers were used; hence, the Horton's, Philip's, Kostiakov model, etc., were used to estimate the infiltration rate over the NIT Kurukshetra campus (Singh et al. 2018). Double ring infiltrometer was used to calculate the infiltrate rate at College of Agricultural Engineering and Post Harvest Technology, Ranipool, Gangtok, Sikkim, India, and multiple regression analysis are carried out for five soil depth classes to find out the correlation between the physical soil parameters (Patle et al. 2019). Gaussian process regression, Support Vector Machine, and Multi Linear Regression were predicted the infiltration rate of the soil using software Weka 3.9 (Vand et al. 2018). Precipitation Estimation from Remotely Sensed Information using Artificial Neural Networks (PERSIANN) discovered by the Center for Hydrometeorology and Remote Sensing with the help of GridSat-B1 infrared satellite data (Nguyen et al. 2019). PERSIANN algorithm trained with the Artificial Neural Network. The spatial resolution of 0.25 o x0.25o gridded data is collected as a raster format for the pre- and post-monsoon. Rainfall intimates the process of infiltration. Thus, spatiotemporal rainfall in the coast region multiplied with the maximum coefficient of the litho unit for the infiltration process. Groundwater data are distributed with cloud points. The cloud points with the nonspatial data give the details of groundwater on a temporal basis. Using GIS tools, archive the spatial distribution of Groundwater adopting the interpolation method such as Inverse Distance Weighted (Arslan 2014; Sun et al. 2009), Spline (Boer et al. 2001), Kriging (Ahmadi and Sedghamiz 2007). However, rainfall is a preliminary source of water resources, and infiltration, percolation, runoff, etc., contribute to the hydrological cycle. Therefore, infiltration process also performs due to rainfall contribution in the respective area. Rainfall hits the surface; it causes soil water outflow or inflow components. By the soil properties, it gives the main characteristics such as bulk density, porosity, permeability, transmissivity, and storage conductivity. In this chapter, the main objectives are probable litho properties involving the infiltration process and validating with the hydrological soil groups and groundwater distribution. This study focused to analyze the spatial and temporal variability of soil infiltration that occurs in the Andhra Pradesh coast. Finally, coastal vulnerable zone is to predict for coastal sustainability.

22.2 Materials and Methodology

22.2.1 Study Area

The study area is located on the East coast of India. The second long stretch of the east coast is Andhra Pradesh coast (Prusty and Farooq 2020). It is parallel to the Bay of Bengal and lies in the mid portion of the East coastline. The two states of India are Orissa of North and Tamilnadu of South. Andhra coastline is buffer to 25 km from the left side. This coastline has been subjected to accretion and erosion for a long time. The Godavari and the Krishna delta are the major influence on the Southeast coast (Pramanik et al. 2016). It covers the seven districts along the coast. Geographical coordinates of the study area extend from 79° E to 85° E longitude and 13°30' N to 19°30' N latitude. It starts from the Pazhaverkadu (Pulicat lake) of Tiruvallur district, Tamilnadu to Jhadupudi of Srikakulam district, Andhra Pradesh. It covers a distance of approximately 976 km in length and covers about the 24,127 sq.km (Fig. 22.1).

The ecosystem areas of coast occupied with the Pulicat Lake in the south, the Mangrove ecosystem of Krishna Delta, and Coringa at Godavari estuary. Alluvial plains, tidal mudflows, mangrove swamps, and lagoons are examples of prominent geomorphology in the coastal landscapes (Rani et al. 2018). Fishing and agriculture are the two major economic activities along with a huge population. The maximum and minimum temperatures in the area are 41.5 °C and 11.1 °C, respectively. The monsoons from the north-east and south-west bring a lot of rain. The annual mean rainfall varies from about 74 cm in the South to about 200 cm in the North with considerable fluctuations. 68.5% is experienced during south-west monsoon (June

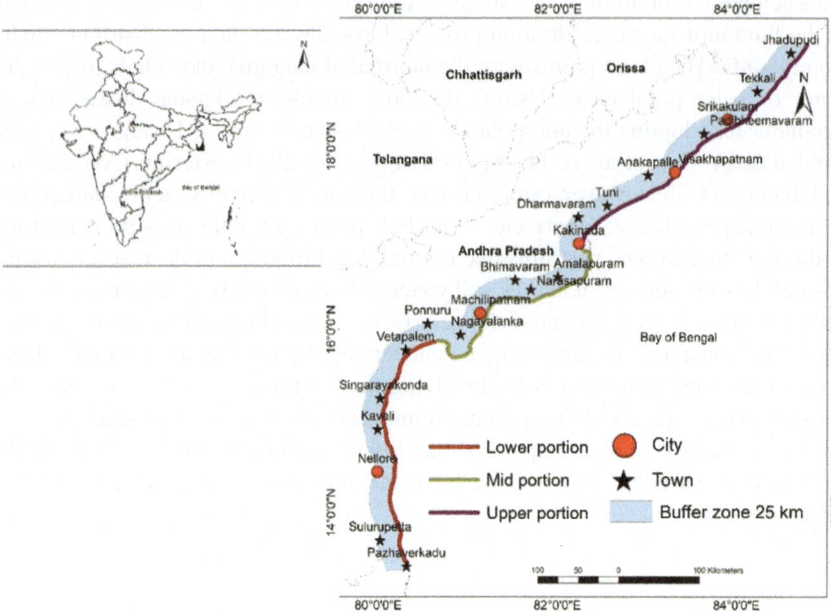

Fig. 22.1 Location map of the Andhra Pradesh coast with the 25 km buffer zone

to September), about 22.3% during north-east monsoon (October–December), and the remaining 9.2% during winter and summer (Tyagi and Rai 2020).

22.2.2 Infiltration Method

The infiltration method, which is followed by the rock properties, gives the suitability infiltrate zone. Litho units are overlaid with the use of rainfall distribution by giving the infiltration coefficient (f) as referred from the study of Subramanya 2013. Rainfall distribution is obtained by the gridded format with the spatial resolution of 0.25 × 0.25. Infiltration contribution stimulates the rainfall distribution as spatial and infiltration coefficient, namely, nonspatial add in the litho units data (Table 22.1).

The coefficient simply explains the maximum efficiency of an entity. So GIS tools are being more optimized for finding the spatial and temporal contribution of an entity. Likewise, the rational method is developed by Mulvany (Peel and McMahon 2020). Then, infiltration coefficient is carried by the artificial knowledge-based learning such as analytical hierarchical process and multi-influencing factor. The same concept applies to lithomorphology, which includes alluvial areas, hard rock terrain, and soft rock terrain. From that recommended value of infiltration, coefficient is prepared from the spatial litho units. The litho units have been interpreted with the district resources map of geology, Andhra Pradesh. By the way, the infiltration computes by the use of rainfall as a probable. But it is compared with the

Table 22.1 Litho units contribute to the Andhra Pradesh coast and its recommended valves of infiltration coefficient

Litho units	f
Acid to Intermediate Charnockite	0.05
Black clay (active tidal flat)	0.03
Black clay underlain by coarse sand (Palaeo)	0.04
Black silty clay (Active flood plain)	0.05
Brown fine sand (coastal dune)	0.05
Brown sandy clay (active lagoon)	0.05
Brown silt (active levee)	0.06
Brown silty clay (active flood plain)	0.05
Calc granulite	0.05
Clayey sand (tidal channel)	0.05
Coarse sand with rock fragments (active channel)	0.06
Garnet gneiss	0.06
Garnet-Biotite gneiss	0.06
Garnet-Sillimanite gneiss +Graphite+Cordierite	0.03
Granite	0.07
Granite gneiss	0.08
Granitoid	0.07
Grey fine sand (active beach ridge and spit)	0.05
Grey hornblende biotite gneiss	0.06
Grey sand (Tidal Channel)	0.05
Laterite	0.13
Manganiferous gneiss	0.08
Migmatite gneiss	0.07
Mud (mud flat)	0.03
Pyroxene granulite	0.05
Pyroxene granulite/ Basic Charnockite	0.04
Quartz-Garnet-Sillimanite-Graphite Schist/ gneiss	0.06
Quartzite	0.07
Sand (channel bar/ point bar)	0.09
Sand and silt (Palaeo channel bar)	0.09
Sand, silt, and gravel	0.1
Sand, silt, and clay	0.07
Sandstone	0.08
Shale	0.12

soil properties of the study area. Then, it mapped out the hydrological soil groups, soil depth, soil texture, etc. The hydrological soil groups are varied with the percentage of clay sand and silt with respect of soil texture also (Patle et al. 2019). Infiltration capacity is estimated by the following Eq. 22.1.

$$R_{rf} = f.A.P_{nm} \quad (22.1)$$

where R_{rf} is the infiltration in the rainfall period; f is the infiltration coefficient; A is study area (24,127 sq.km); $P_{nm} = R_i$; R_i is the rainfall contribution in spatial (A. P_{nm}) and temporal scale (i). Additionally, cloud points of groundwater data are prepared by the spatial analyst tool, i.e., Inverse Distance Weighting Method. Monitoring groundwater data influences the study in the way of percolation rate that happens in the Andhra Pradesh coast. Slope and geomorphology are used to determine the interpretation of groundwater distribution.

22.3 Results

22.3.1 Geology

Andhra Pradesh coast belongs to the Archaean to Quaternary age. The Supergroup of this coast covers the Dharwar, Cuddapah, Deccan trap, Gondwana, Peninsular Gneissic Complex, and Eastern Ghat. It comprises 34 litho units as shown in Fig. 22.2. It can be interpreted with three portions as upper (Kakinada to Jhadupudi), mid (Ponnuru to Kakinada), and lower (Pazhaverkadu to Ponnuru). The 15 litho units cover 59% of the study area. The mid and lower portions of coast occurred

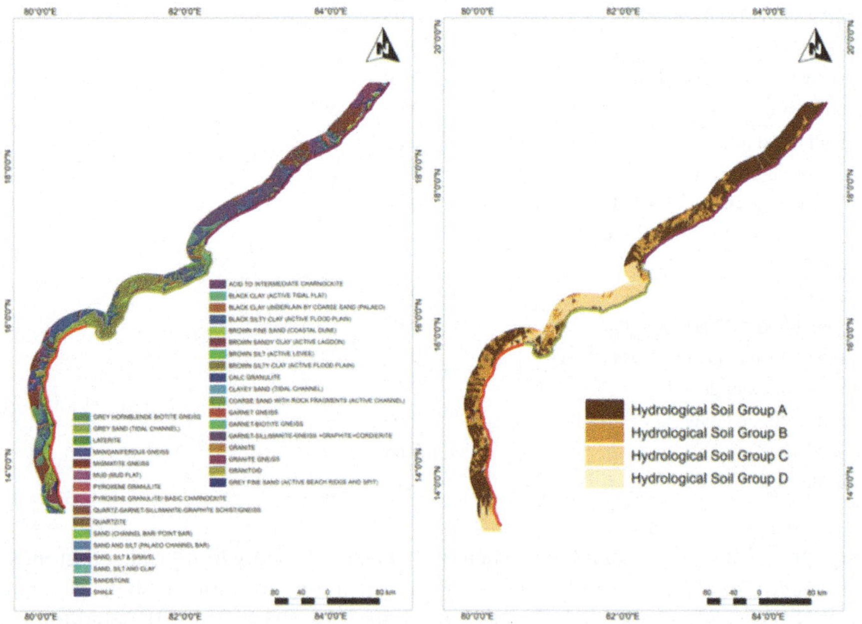

Fig. 22.2 Lithology and soil map of a study area

with the beach ridges and beach, i.e., sand, clay, and silt with various soil textures. Thus it has major intrusion-related rock; more porosity, and transmissivity of rocks explore the huge amount of groundwater recharge. When the rainfall contribution falls on any kind of rock type, it gets slow to infiltrate in the regolith. Then rain water gets the percolate for the groundwater recharge. The Migmatite gneiss, Granite gneiss, Sandstone, Garnet gneiss, and Acid to Intermediate Charnockite cover the 20% of the study area. Granite gneiss and garnet gneiss occurred in the upper portion of the coast. Migmatite gneiss, Sandstone, and Acid to Intermediate Charnockite occurred in the upper and lower portion of the coast. The remaining litho units cover about 21% of the study area.

22.3.2 Soil

Soil map is provided by the Andhra Pradesh Space application Centre, Government of Andhra Pradesh, India. Soil classes that occurred within the study area are about 13 along with the settlement and water bodies as shown in Fig. 22.2. Soil properties varied with the sand silt clay percentage, soil depth, soil texture, hydraulic conductivity, pore materials, etc. Soil data are given as soil texture. It is classified into four classes as hydrological soil groups (HSG). So it defines the maximum level of soil properties to estimate the rate of infiltration (Ross et al. 2018; Woltemade 2010). The maximum contribution of HSG is A because upper and lower coast area is covered by gravel particles more than clay particles. Thus, it indicates a high infiltrate possibility. Here, accretion in mid portion replicates the minor issues at the upper and lower coast. This accretion creates the high saline at Krishna and Godavari deltaic region (Pramanik et al. 2016). Hencce, it came to be low infiltration, i.e., C group of HSG: 24% of the study area covers the C group of HSG. The B group covers away from the shoreline. So, soil infiltration can percolate the rain water in this B group. Soil features play a vital role to understand the landscape morphology of the Andhra Pradesh coast.

22.3.3 Infiltration Coefficient

Based on the litho unit, the coefficient is a frame with the basic soil properties in the coastal region. It is described with the eroded weather particle allies on the shallow fine pediment. In their coastal region, 11 litho units are varied with their soil properties where its weathered particles make the different O–D horizon layers, where field capacity and welding point are used to find out the all soil horizon layer (Ross et al. 2018). Mostly, O horizon layers help to understand the variability of infiltration capacity, i.e., field capacity. So, it is used to estimate the infiltration coefficient, i.e., probability of infiltration values.

In the case study, eroded and deposit sand of river delta have a precise contribution with the 59% as 14,000 sq.km. With the maximum coefficient of 0.13 to 0.1, the high probability rate of infiltration process is obtained between these coefficients itself. Laterite and Shale have a high coefficient because the porosity and transmissivity are much higher than the other. So it lies far away from the shore. Lower portion of hinter coast, Laterite, and Shale are present with 0.13 coefficient; this covers the 550 sq.km. Andhra Pradesh coast comprises of Pennar, Krishna, Godavari, Vamsadhara, and Nagavali deltaic plain; hence, it is fully covered with the sand, silt, clay, etc., parallel toward the shoreline. Krishna and Godavari delta give the major impact on the entire coast (Rao et al. 2009). Most of the coast area covers with 0.05 coefficient of up to 39%.

Soil infiltration coefficient has been modified into the gridded format as similar to PERSIANN gridded data. The 80 tiles cover the study area. The mean value of the infiltration coefficient is considered to single 0.25 × 0.25 gridded format named tiles. Then, the range has changed to 0.08 to 0.03. More than 0.05 coefficients are occupied in 48 tiles, which is part of the lower and upper portion. Machilipatnam and its surroundings are covered with the 32 tiles below 0.05.

22.3.4 Soil Infiltration

Soil infiltration process is the downward movement of water on the top soil layer due to friction or gravity action. This process mainly contributes to the surface seepage of rainfall influence. Rainfall is the main source of the Indian coast region; it is particularly caused by monsoon winds. It leads to calamities such as cyclone and flood. These issues are eradicated by understanding the soil infiltration probability of the Indian coastal region.

In the case study, pre- and post-monsoon are taken for a period of 2004, 2009, 2014, and 2019 (Fig. 22.3). Infiltration capability is much higher than the post-monsoon. Soil infiltration mainly depends upon the rainfall. Due to a highly-populated area, estimation of soil infiltration gives the prediction uncertainty. To remove the uncertainty, elevation data are required to analyze the slope and aspect. The tone of dark blue color indicates the high infiltrate; it is shown in the upper portion of the coast. The upper portion has covered by the metamorphic rocks, i.e., Migmatite gneiss, Granite gneiss, and Garnet gneiss. The C HSG in the upper portion resembles the fine clay particles that transmit the rainwater slowly (Fig. 22.4). The 2004 and 2014 of pre monsoon look similar. This coast is dynamic, which causes the deposition of river sediment and tidal effect. Soil infiltration in Krishna and Godavari delta is quite low compared to others. The 2009 and 2019 of pre- and post-monsoon have low soil infiltration, i.e., light blue tone. The range of soil infiltration has slightly increased the trend movement.

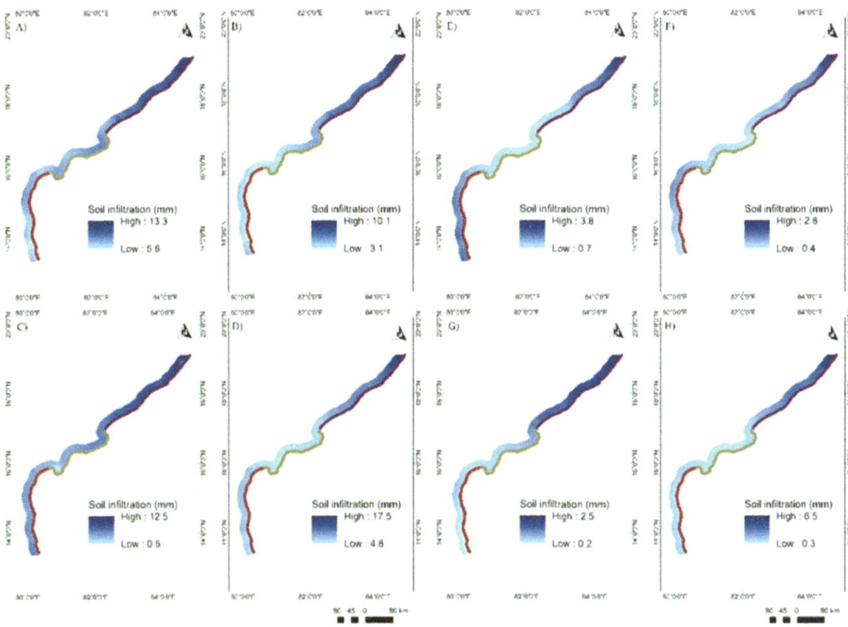

Fig. 22.3 Soil infiltration of pre-monsoon (**a**) 2004, (**b**) 2009, (**c**) 2014 and (**d**) 2019; followed by post-monsoon (**e**) 2004, (**f**) 2009, (**g**) 2014 and (**h**) 2019

22.3.5 Groundwater

Soil infiltrates the rain into top soil and then percolates toward the aquifer. Coastal aquifers are controlled by the geological factor. The fluvial and marine origins are mainly influenced by this coastal geomorphology (Fig. 22.4). The aquifer in the coastal region is mostly affected by seawater intrusion due to sea-level rise (Rao et al. 2009). Pre- and post-monsoon of groundwater (Fig. 22.5) are planned to eradicate the similarity of soil infiltration in this coast. The upper and lower portions of coast are higher than the mid portion throughout pre- and post-monsoon. The upper is slightly higher from the lower coast. 2014 and 2019 of pre-monsoon get a rise in the post-monsoon. Nearby Kakinada, groundwater is highly contributed. The steep and very steep slope, dissected hills and valleys lies in the upper portion of Andhra Pradesh coast; which performs the structural form like fault and fractures. These structures and pediment pediplain complex induce the aquifer recharge in the upper portion of the coast. Deltaic plain and coastal plain with nearly level appear in the mid portion; hence, groundwater is low. The lower portion has covered the pediment pediplain complex with a gentle slope that helps to percolate into aquifers at a period of heavy rainfall. Hence, groundwater in the lower portion is slightly higher than the mid portion of the coast.

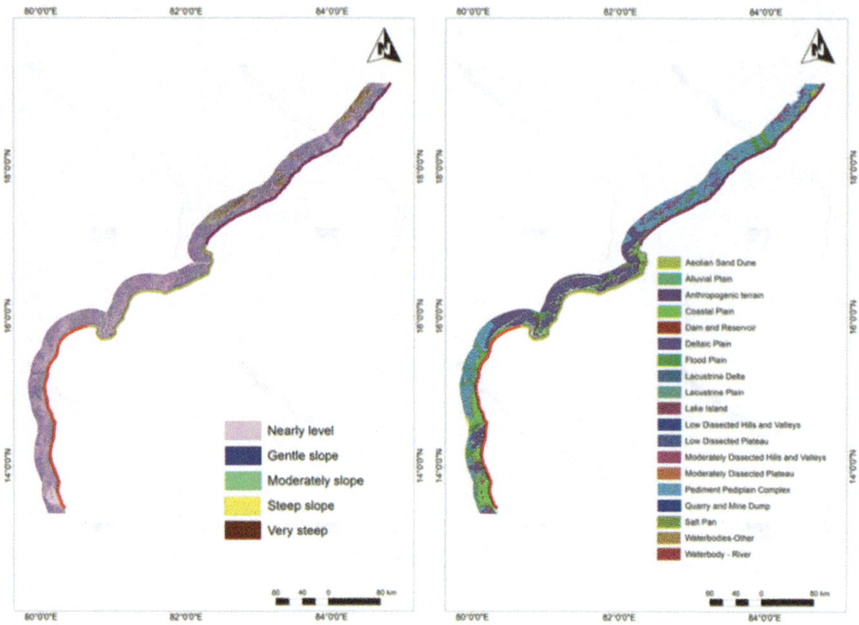

Fig. 22.4 Slope and geomorphology of a study area

22.3.6 Discussion

Rehabilitation of coastal regions takes place as large-scale industrial, small farming, salts panning, and fishing. As per environmental concern, coastal regions are varied with the natural calamities such as storm surge, seawater intrusion, eutrophication, tidal effects, and coastal erosion (Arslan 2014; Pramanik et al. 2016; Prusty and Farooq 2020; Rani et al. 2018; Rao et al. 2009; Rizzo and Anfuso 2020; Tyagi and Rai 2020). In this study, Andhra Pradesh coast experiences the soil water content, i.e., soil infiltration rate for the period of 2004 to 2019 by means of field capacity and welting point (Pramanik et al. 2016). The geology condition deals with the main role of finding the soil infiltration rate with the proper coefficient of each rock. The litho units are defined as the coefficient that substantiates with the soil data also. Then, PERSIANN rainfall distribution is used to archive the amount of infiltration capacity in the Andhra Pradesh coast.

The coast is divided into three portions, namely, upper, mid, and lower for interpreting the soil infiltration. Coastal influences rainfall to infiltration on the monsoonal period. Thus, pre- and post-monsoon are segregated to understand the spatial distribution of infiltration. At pre-monsoon, the maximum rate of infiltration is about 17.5 mm and the minimum rate is about less than 0.5 mm. Post-monsoon receives a maximum of 3.8 mm and minimum of 0.2 mm. Rainfall contribution plays the main role in the rise and fall of infiltration. Accretion occurs in Krishna and Godavari deltaic plain which causes the coastal vulnerability on this coast. It lies in the mid portion. The mid portion results from the infiltration rate as lower

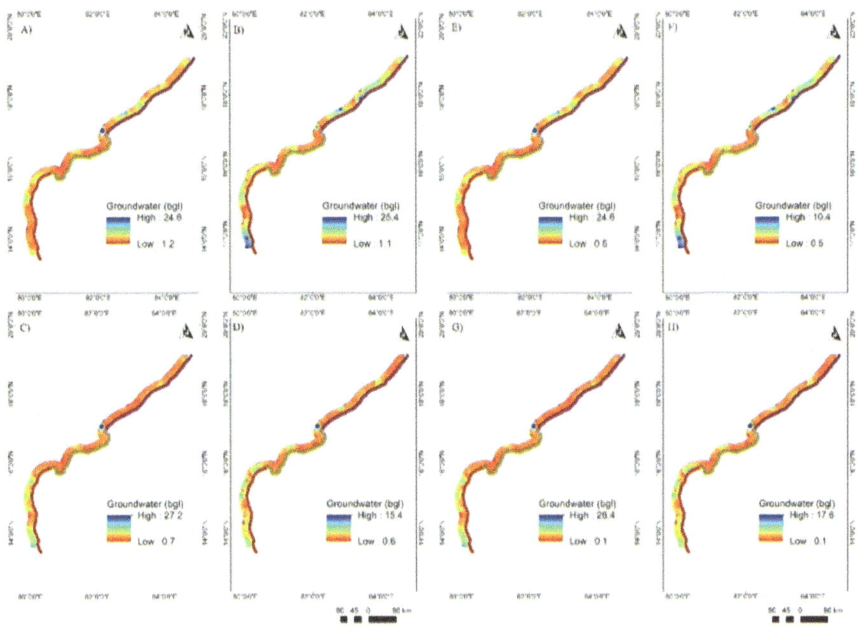

Fig. 22.5 Groundwater distribution of Pre monsoon (**a**) 2004, (**b**) 2009, (**c**) 2014 and (**d**) 2019; followed by Post monsoon (**e**) 2004, (**f**) 2009, (**g**) 2014 and (**h**) 2019

than the upper and lower portions. It resembles the groundwater distribution as well. Groundwater distribution is compared with the slope and geomorphology to understand the morphology behavior in this Andhra Pradesh coast.

22.4 Conclusion

Andhra Pradesh coast resembles the three-portion coastline. Pre- and post-monsoon have influenced this coast. The results of hydrological soil group, soil infiltration, and groundwater distribution describe the high infiltration rate at upper portion. It is not under coastal vulnerability zone compared to others. Accretion of mid portion leads to seawater intrusion. Erosion of lower portion causes the low groundwater level. The rainfall of this tropical coast changes the coastal ecology. Therefore, it needs coastal sustainability at mid and lower portions. Here, coastal regulation zones III and IV lead to the improvement of surface and groundwater.

Acknowledgments The author is thankful to CHRS-portal for the PERSIANN precipitation data. Also, the author extends the thanks for Geological Survey of India and Andhra Pradesh State Remote Sensing Application Centre. Finally, the author is also thankful to our Research Institute.

Conflict of Interest Statement Authors should declare no competing or conflict of interest.

References

S.H. Ahmadi, A. Sedghamiz, Geostatistical analysis of spatial and temporal variations of groundwater level. Environ. Monit. Assess. **129**(1–3), 277–294 (2007). https://doi.org/10.1007/s10661-006-9361-z

H. Arslan, Estimation of spatial distrubition of groundwater level and risky areas of seawater intrusion on the coastal region in Çarşamba Plain, Turkey, using different interpolation methods. Environ. Monit. Assess. **186**(8), 5123–5134 (2014). https://doi.org/10.1007/s10661-014-3764-z

K.K. Basheer Ahammed, A.C. Pandey, Shoreline morphology changes along the Eastern Coast of India, Andhra Pradesh by using geospatial technology. J. Coast. Conserv. **23**(2), 331–353 (2019). https://doi.org/10.1007/s11852-018-0662-5

E.P.J. Boer, K.M. de Beurs, A.D. Hartkamp, Kriging and thin plate splines for mapping climate variables. Int. J. Appl. Earth Obs. Geoinf. **2001**(2), 146–154 (2001)

S. Mahapatra, M.K. Jha, S. Biswal, D. Senapati, Assessing variability of infiltration characteristics and reliability of infiltration models in a tropical sub-humid region of India. Sci. Rep. **10**(1), 1–18 (2020). https://doi.org/10.1038/s41598-020-58333-8

P. Nguyen, E.J. Shearer, H. Tran, M. Ombadi, N. Hayatbini, T. Palacios, P. Huynh, D. Braithwaite, G. Updegraff, K. Hsu, B. Kuligowski, W.S. Logan, S. Sorooshian, The CHRS data portal, an easily accessible public repository for PERSIANN global satellite precipitation data. Scientific Data **6**, 1–10 (2019). https://doi.org/10.1038/sdata.2018.296

G.T. Patle, T.T. Sikar, K.S. Rawat, S.K. Singh, Estimation of infiltration rate from soil properties using regression model for cultivated land. Geol. Ecol. Landscapes **3**(1), 1–13 (2019). https://doi.org/10.1080/24749508.2018.1481633

M.C. Peel, T.A. McMahon, Historical development of rainfall-runoff modeling. Wiley Interdiscip. Rev. Water **7**(5), 1–15 (2020). https://doi.org/10.1002/wat2.1471

M.K. Pramanik, S.S. Biswas, B. Mondal, R. Pal, Coastal vulnerability assessment of the predicted sea level rise in the coastal zone of Krishna–Godavari delta region, Andhra Pradesh, east coast of India. Environ. Dev. Sustain. **18**(6), 1635–1655 (2016). https://doi.org/10.1007/s10668-015-9708-0

P. Prusty, S.H. Farooq, Seawater intrusion in the coastal aquifers of India - a review. HydroResearch **3**, 61–74 (2020). https://doi.org/10.1016/j.hydres.2020.06.001

M. Rani, S. Rehman, H. Sajjad, B.S. Chaudhary, J. Sharma, S. Bhardwaj, P. Kumar, Assessing coastal landscape vulnerability using geospatial techniques along Vizianagaram–Srikakulam coast of Andhra Pradesh, India. Nat. Hazards **94**(2), 711–725 (2018). https://doi.org/10.1007/s11069-018-3414-9

K.N. Rao, P. Subraelu, T.V. Rao, B.H. Malini, R. Ratheesh, S. Bhattacharya, A.S. Rajawat, Ajai, Sea-level rise and coastal vulnerability: An assessment of Andhra Pradesh coast, India through remote sensing and GIS. J. Coast. Conserv. **12**(4), 195–207 (2009). https://doi.org/10.1007/s11852-009-0042-2

A. Rizzo, G. Anfuso, Coastal dynamic and evolution: Case studies from different sites around the world. Water (Switzerland) **12**(10) (2020). https://doi.org/10.3390/w12102829

C.W. Ross, L. Prihodko, J. Anchang, S. Kumar, W. Ji, N.P. Hanan, HYSOGs250m, global gridded hydrologic soil groups for curve-number-based runoff modeling. Scientific Data **5**, 180091 (2018). https://doi.org/10.1038/sdata.2018.91

K.E. Saxton, W.J. Rawls, Soil water characteristic estimates by texture and organic matter for hydrologic solutions. Soil Sci. Soc. Am. J. **70**(5), 1569–1578 (2006). https://doi.org/10.2136/sssaj2005.0117

B. Singh, P. Sihag, K. Singh, Comparison of infiltration models in NIT Kurukshetra campus. Appl. Water Sci **8**(2), 1–8 (2018). https://doi.org/10.1007/s13201-018-0708-8

K. Subramanya, *Engineering Hydrology* (The McGraw-Hill Companies, 3rd, 2013)

Y. Sun, S. Kang, F. Li, L. Zhang, Comparison of interpolation methods for depth to groundwater and its temporal and spatial variations in the Minqin oasis of northwest China. Environ. Model Softw. **24**(10), 1163–1170 (2009). https://doi.org/10.1016/j.envsoft.2009.03.009

S. Tyagi, S.C. Rai, Monitoring shoreline changes along Andhra coast of India using remote sensing and geographic information system. Indian J. Geo-Marine Sci. **49**(2), 218–224 (2020)
A.S. Vand, P. Sihag, B. Singh, M. Zand, Comparative evaluation of infiltration models. KSCE J. Civ. Eng. **22**(10), 4173–4184 (2018). https://doi.org/10.1007/s12205-018-1347-1
C.J. Woltemade, Impact of residential soil disturbance on infiltration rate and stormwater runoff. J. Am. Water Resour. Assoc. **46**(4), 700–711 (2010). https://doi.org/10.1111/j.1752-1688.2010.00442.x

Chapter 23
Feasibility Assessment of Low-Cost Filter to Adopt in Rooftop Rainwater Harvesting (RWH)

C. R. Suribabu, J. S. Sudarsan, and S. Nithayanntham

Abstract Our environment is being subjected to drastic changes over the past couple of years. The increasing population in urban areas and their demands in basic resources like quality of food, air, water, and hygiene surroundings are challenging. In urban cities like Pune, these requirements of human are not yet fully organized and managed. The city is in a zone of earnest mismanagement. Lifestyle of users and increasing per-capita demand are projecting future years of water scarcity. Rigorous expansion in the urban cities are destroying existing local sources and polluting the water table. In addition, industrialization is adding additional damage to water eco system.

At the present rate of uncontrolled growth rate, expansions, and behavior of society, the water resources will be in danger of depletion. To overcome such situations, the awareness level among people about rainwater harvesting (RWH) system should be improved, and the same must be implemented in an effective manner. As this one of the effective water management strategies, it is necessary that cost-effective techniques need to evolve so that it can be adopted implemented in the rural and urban clusters as well. The maintenance of the RWH system is also not proper because the quality of water collected through the RWH system is poor. To spread the importance of RWH and ascertain this laboratory setup for improving the quality of collected surface water, the following study was initiated. A trail unit of low-cost filter unit with different filter media combinations was fabricated, and the efficiency of the function was also assessed. Based on different filter media combinations, the

C. R. Suribabu
School of Civil Engineering, SASTRA Deemed to be University, Thanjavur, Tamilnadu, India

J. S. Sudarsan
National Institute of Construction Management and Research (NICMAR), Pune, India

S. Nithayanntham (✉)
Post Graduate and Research Department of Physics, (Ultrasonic / NDT and Bio-Physics Divisions), Thiru.Vi.KalyanasundaramGovt Arts and Science College
(Affiliation - Bharadhidasan University, Thiruchirapalli), Thiruvarur, Tamilnadu, India

B. Panneerselvam et al. (eds.), *Climate Change Impact on Groundwater Resources*, https://doi.org/10.1007/978-3-031-04707-7_23

low-cost natural filter media combination of Light Expanded Clay Aggregate (LECA) seeds as a filter media works effectively in improving the quality of water. Approximately, for a normal residential apartment or house with 750–1000 sqft this system works effectively, and the cost approximately for installing the system will be around 5–10 k Indian rupees. If it was properly monitored, the cost spent toward installation can be gained back in a couple of years.

Keywords Rainwater harvesting (RWH) · Water management · Filter media · LECA seeds · Low cost

23.1 Introduction

Freshwater could be a terribly precious resource, and it's wide according that the world reserve of the water is quick depleting (Bevan and Germann 1982; McConnell et al. 1984; Sukop et al. 2000; Maria et al. 2021; Suzanne et al. 2021). Life on earth is incomprehensible without water. The temperature change contemplations across the world has prognosticated extreme water deficiency due to corrupting eco-framework administrations and World Business Council for property improvement (WBCSD) in its Vision 2050 explanation has assessed an additional heap of 500,000,000 people underneath water-stress conditions inside the BRIC (Brazil-Russia-India-China) nations by 2030. In India, with the National Action setup for temperature change (NAPCC) and its constituent National Water Mission, an objective of two hundredth expansion in water-use productivity has been set to counter the test (Corey et al. 2015; Zhang et al. 2015; Ranran and Julie 2015; Eric et al. 2019). The National Water Policy of Asian countries recognizes the need of planning the exercises such as horticulture, businesses, and substantial advancement in congruity with availability of this imperative asset and has guided water-drafting. It is noticed that urban clusters and building fraternity are believed to be a critical customer of water asset, in a few cases, due to overexploitation through bore wells, the precious resource is getting depleted rapidly (Pacey and Cullis 1986; NDWC; Peter et al. 2016). Moreover, there had been reports of significant subsidence dangers as a result of over-abuse of this precious resource (Norman and Amilcare 2016; Haque et al. 2016; Waso et al. 2016). In the complete life cycle of a building, water and energy region units are considered to be significant factors that help human activities in it (Pacey and Cullis 1989; Eregno et al. 2017; Li et al. 2017). So it is high time to find an effective cost-effective management strategy to mitigate this crisis. One of the simple and sustainable concepts is rainwater harvesting (RWH). In addition, the authors incorporated systems used for RWH in various countries over the globe.

23.2 Rainwater Harvesting

RWH is a technology that collects, conveys, and retains rain from reasonably clean surfaces such as a roof, land surface, or rock catchment for later use. The water is usually collected in a storm water tank or is led to groundwater recharge. Infiltration of rainwater is another aspect of RWH which plays an important role in storm water management and groundwater level replenishment as represented in Figs. 23.1, 23.2, and 23.3. RWH has been practiced around the world for over 4000 years, historically in arid and semi-arid regions, supplying potable water, household water, and water for livestock and minor irrigation (Jay et al. 2017; Ren et al. 2017; Prasanna et al. 2017; Pradhan et al. 2019).

The method of rainwater collection from rainfall events can be divided into two different types based on land and roof such as (i) land-based and (ii) roof-based. In the first kind, rainwater gathered through as the furrow dikes, dams, tanks, and reservoirs. Otherwise in the second type RWH refers to the gathering of rainwater runoff from roof surfaces which typically provide much pure water source that can utilize for domestic purpose (Prasanan et al. 2017; Andrea et al. 2018).

In RWH, water can be assembled from rooftops, dams, and lakes can be built to hold gigantic amounts of water so that even on days where there is almost no precipitation; this is sufficient to accessible to inundate crops (Andini et al. 2017; Nasrin et al. 2018; Xueer et al. 2018). The pollutants can be reduced altogether by redirecting the underlying progression of run-off water to first drain off. Improved water quality can likewise be gotten by utilizing a skimming draw-off component (instead of from the foundation of the tank) and by utilizing a progression of tanks, pull out toward from the end in arrangement (Sartory et al. 2017; Ting and David 2019; Zhang et al. 2020; Juneseok et al. 2019).

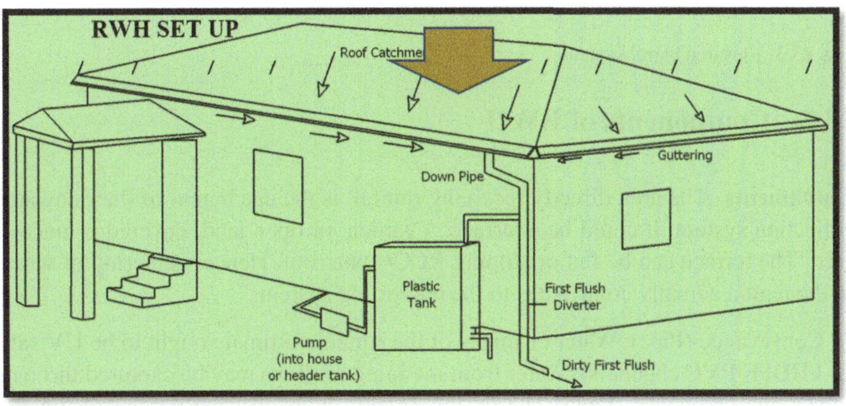

Fig. 23.1 Rooftop RWH schematic sketch

Fig. 23.2 Rainwater harvesting collection system

Fig. 23.3 Elevated barrel system

23.3 Components of RWH

Catchments The area directly receiving rainfall is the catchment of the rainwater collection system. It could be a terrace, a garden, or open land, covered or uncovered. The terrace can be flat or sloping RCC/stone roof. Hence, gathering of water in the region actually contributes to the harvesting system

Conveyance (Pipe): Water channels of the suitable estimate ought to be UV safe (ISI HDPE/PVC channels). Water from inclining rooftops may be captured through canals and channeled down. The drain's mouth ought to have wire work on the porches to restrain drifting fabric.

Filter: Filters are used in water treatment to efficiently reduce turbidity, color, and microorganisms. Water can pass through filters or be deposited directly in tank and filters prior to use after first flushing of rainfall (Donchyts et al. 2016; Zaman and Sizemore 2017). Depending on the use, gravel, sand, and "screen" mesh is built and put on top of the storage tank or near the tank. Slow Sand Filter (SSF) has been incorporated in our setup. The rate of filtration for Slow Sand Filter is 100–200 l/hr/m^2. In addition, not one or the other development nor operation and support require more than essential aptitudes. Consequently, moderate sand filtration could be a promising filtration strategy for little to medium-sized, provincial communities with a reasonably great quality of the starting surface water source (Ghanashyam et al. 2020; Fausto et al. 2020). By filtering the rainwater, the impurities accompanied with it can be reduced which in turn helps in improving the quality of groundwater. The filter materials are different from conventional materials and are lower in cost which help in easy installation.

This research study aims to utilize the rainwater effectively, thereby increasing the groundwater table by allowing it to pass through different layers of soil, to analyze the percolation time through the different layers of soil media. This study also aims to determine the efficient infiltration rate for the particular combination of soil layers and to compare quality parameters of inflow and outflow water for different soil setup.

23.4 Methodology

The methodology for carrying out the design of effective one-dimensional water threshold for filtration of groundwater was represented in Fig. 23.4, and the basic procedure adopted for design are based on the CPCB guidelines and water quality parameters are assessed based on the normal procedure in vogue (Ghisi and Ferreira 2007; Fausto et al. 2020; Joanna et al. 2020).

The setup consists of a drum of 120 l capacity fitted with a valve at bottom as represented in Fig. 23.5. The drum is filled with soil, coarse materials, and fine materials (filter layers) from bottom to top. These materials change from one setup to the other.

The soil and other materials are filled for around 15 cm constituting about 45 cm which is critical for root growth of any type of vegetation. To avoid intermingling of layers a steel mesh of varied size is provided between each layer. The layers are left for self-compaction.

The measured quantity of water that is filled in the drum with around 15 cm of water is allowed to stagnant above the fine materials. During the filling process, the valve fitted with a pipe is kept closed. This setup is left undisturbed for half an hour for the pores to get filled with water as represented in Fig. 23.6. Then the water is allowed to drain from the valve at certain velocity which is made constant and with the help of stop watch, and the time for outflow of water is noted.

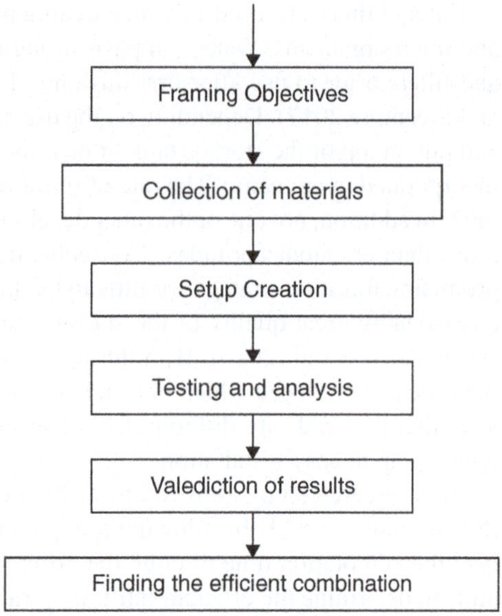

Fig. 23.4 Flow chart representing methodology

Fig. 23.5 The setup drum with valve fixation

This process is repeated for 5 times a day for 2 weeks, and the time is noted for each trial. From this tabulation, the infiltration rate is calculated. The inflow and outflow water is collected and given for water test. The setup is changed for various types of filler materials (filter layers) and the process is repeated (Igboekwe and Ruth 2011; Sudarsan et al. 2021).

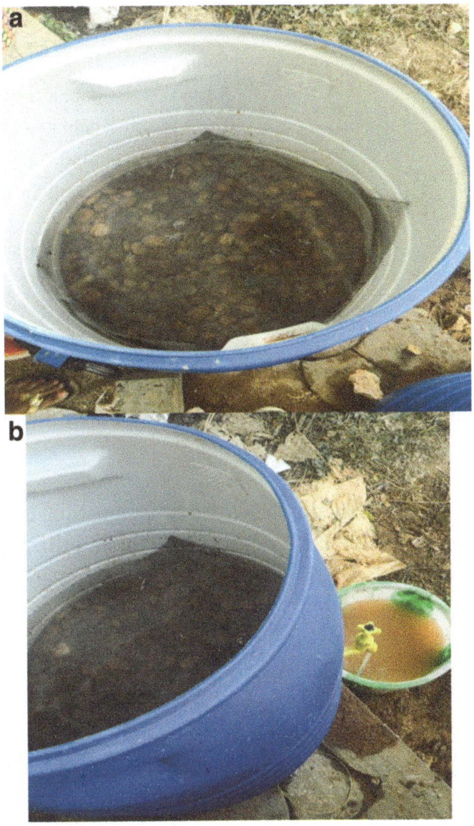

Fig. 23.6 (a) Filling of materials and (b) draining of water

23.5 Results and Discussion

The setup has various combinations of filter layers, and their efficiency is tested through water test results. The type of soil is made constant, and the filter layers are varied. Different combinations are represented in Table 23.1.

Filter materials used in the setup are pebbles, activated carbon, and recycled aggregate, silica, laterite soil, and LECA.

Pebble A stone may a clast of shake with a molecule to estimate 2–64 millimeters. Stones are by and large considered bigger than granules (2–4 millimeters breadth) and littler than cobbles (64–256 millimeters distance across (Jay et al. 2017).

Activated Carbon Activated or inacted carbon could be a shape of carbon prepared to have little, low-volume pores that increment the surface region accessible for adsorption or chemical responses. Actuated carbon is as a rule inferred from charcoal and is some of the time utilized as biochar (Pradhan et al. 2019).

Table 23.1 Combinations implemented in the setup

Combination	Setup
1	Surface soil + pebble + carbon + sand
2	Laterite soil + pebble + carbon + sand
3	Laterite soil + sand
4	Laterite soil sand (mixture 1:3) + recycled aggregate
5	Laterite soil + LECA + sand

Silica Sand Filter sand is naturally occurring, river washed, glacial deposit products. Filter sand is graded specifically for water filtration plants. It can be used in municipal, industrial or residential applications for sediment filtration (Eregno et al. 2017).

Recycled Aggregate Reused total is fabricated by smashing concrete, and in some cases black-top, to recover the total. Reused total can be utilized for numerous purposes (Andrea et al. 2018).

Laterite Soil Laterite soils have an extraordinary clay matter, which cruel they have way better Cation Trade Capacity and water-holding capacity than sandy soils. This consolidates by implies of innate water-harvesting strategies. The critical portrayals of the lateritic soils are lateritic soils are their special color, destitute richness, and tall clay substance and lower cation trade capacity (Xueer et al. 2018).

LECA LECA is a wholly natural product made from clay, which boasts thermal insulation properties and has a plethora of horticultural/agricultural uses. An acronym for Light Expanded Clay Aggregate, LECA is an environmentally-friendly product derived from base clay. To manufacture LECA, clay is extruded into a kiln in pellets, heated, expanded and burned at extremely high temperatures. During the firing process, any organic materials are burned off, leaving a lightweight yet strong product which is impervious to dry rot, wet rot, and insects – hence its popularity in plant applications. It can be purchased in the standard brown/terracotta pellet form or as a LECA Lightweight Soil, which is premixed as shown in Fig. 23.7, 23.8, and 23.9. While LECA is lightweight, it is incredibly strong. In addition, where the expansion process creates cavities in the individual pellets, LECA is also incompressible. It will not decompose and one need not have to worry about watering the soil as often as normal, provided the product has been given a good soaking beforehand. As it allows increased levels of oxygen into the soil, LECA can lead to faster growth and can be used on structures that cannot take heavy loads – providing a green solution where soil is not an option. What is more, LECA's characteristics mean that it will not change overtime and so can be stored for a long while – meaning one need not panic about it molding in the shed. The soil and LECA combination, which is not packed as tightly as pure soil, leaves lots of room for roots to grow

Fig. 23.7 Balls of clay

and expand, without tainting the plants with any unnatural substrates (Sartory et al. 2017; Andini et al. 2017; Andrea et al. 2018; Nasrin et al. 2018; Xueer et al. 2018; Ting and David 2019; Zhang et al. 2020).

Properties of Filter Material Initially, the filter materials are checked for the suitability of application of the same in trail unit. Mainly the soil plays an important role as filter media in the RWH unit the following properties permeability coefficient of surface soil and laterite soil was studied and the same was reported in (Tables 23.2 and 23.3). Based on the study, it is perfect that the selected soil is best suited to act as a filter media.

Permeability Coefficient A unit of measurement for the rate at which a molecule can penetrate a membrane, such as a lipid bilayer; equal to the diffusion coefficient divided by the membrane width in cm/s.

Experimental Trail Study of RWH Filter Unit Once after studying the characteristics of the filter media. The unit was allowed for trail assessment. Based on the several combination trail assessments, the combination of filter media with clay ball and LECA found to be more efficient (Shahbaz et al. 2020).

Component Combination Five kinds of combinations are made from the components that are listed in Table 23.1. Component combinations 4 and 5 are very effective and low-cost filter unit. Several series of trails were carried out in these units, and the following parameters such as pH, Turbidity, Conductivity, Hardness, TDS are assessed and reported in Table 23.4. Based on the consecutive trail results, the combinations 4 and 5 work effective and improve the quality of water, and the results represented in Table 23.4 show that efficiency of the combination of filter media and it evident from results combinations 4 and 5 are very effective. Parameters

Fig. 23.8 Clay balls being soaked

Fig. 23.9 LECA in water

represented in Table 23.4 are the most common parameter which is considered as the representative parameters to know about the quality of water. Normally, the surface water and rainwater is pure, and it gets polluted only if it gets in contact with some external factors like concrete surface, floating impurities, etc., In this research study, the main objective is to find an effective filtration media to treat this collected

Table 23.2 Permeability coefficient for surface soil

S.No	Initial head (cm)	Final head (cm)	Time (s)	Coefficient of permeability (10^(−4)) (m/s)
1	90	85	4.63	5.72
2	85	80	4.09	6.82
3	80	75	4.74	6.26
4	75	70	5.11	6.21
5	70	65	5.08	6.71
6	65	60	6.09	6.05

Table 23.3 Permeability coefficient for laterite soil

S.No	Q (m^3/s)	Time (minutes)	Coefficient of permeability (10^-2) (m/s)
1	50	36.67	0.10
2	100	59.07	0.12
3	150	53.70	0.20
4	200	52.88	0.27
5	250	59.63	0.30

rainwater and surface water and same was proved with different trails and results are interpreted in Tables 23.4 and 23.5 and in the following figures.

Turbidity Turbidity may be a degree of the degree to which the water loses its straight forwardness due to the nearness of suspended particulates. The turbidity values were found to be follows for both influx and surge water (Patricia et al. 2020; Yuk et al. 2021).

Conductivity Conductivity may be a degree of water's capability to pass electrical flow. The electrical conductivity of water gauges the overall sum of solids broken down in water-TDS, which stands for Add up to broken down solids. TDS is measured in ppm (parts per million) or in mg/l. This capacity is specifically related to the concentration of particles within the water (Patricia et al. 2020; Yuk et al. 2021).

Hardness Hardness is caused by compounds of calcium and magnesium, and by an assortment of other metals. Water is an amazing dissolvable and promptly breaks down minerals it comes in contact with. The hardness of water is alluded to by three sorts of estimations; gains per gallon, milligrams per liter (mg/L), or parts per million (ppm). Common rules for classification of waters are; up to 60 mg/L as calcium carbonate is deliate; 61–120 mg/L as tolerably difficult, 121 to 180 mg/L as difficult, and more than 180 mg/L as exceptionally difficult. Similar trend of value obtained by some researchers (Patricia et al. 2020; Yuk et al. 2021).

TDS Add up to broken up solids (TDS) are a degree of the combined substance of all inorganic and natural substances contained in a fluid in atomic, ionized or micro-

Table 23.4 Efficiency of different filter media in treatment of Water

Combination	1	2	3	4	5
P^H	8.2	8.3	8.4	7.8	7.5
Turbidity (NTU)	1.00	0.75	0.55	0.35	0
Conductivity(mS)	0.528	0.453	0.526	0.302	0.255
Hardness(mg/l)	948.34	998	953	856	742
TDS (mg/l)	342	406	426	216	201

Table 23.5 Infiltration rate of various setup combinations

Component combination	Inflow water (L)	Outflow water (L)	Time taken (min)	Infiltration rate (m/s)
Soil+ pebble+ carbon+ sand	30	29.5	7.29	4
Soil+ pebble+ carbon+ sand	30	29.5	9.01	3.33
Soil + sand	30	30	9.15	3.27
Soil sand (mixture 1:3) + recycled aggregate	30	30	9.49	3.16
Soil + LECA + sand	30	28.5	15.26	2.0

granular (colloidal sol) suspended frame. In any case, Water containing TDS concentrations underneath 1000 mg/liter is ordinarily worthy to shoppers, in spite of the fact that adequacy may change a greening to circumstances (Patricia et al. 2020; Yuk et al. 2021; Sara et al. 2021). TDS of the inflow water = 785 ppm.

Infiltration Calculation

Invasion is the method by which water on the ground surface enters the soil. It permits the soil to briefly store water making, it is accessible for take-up by plants and soil living beings. In most cases, the sum of penetration that occurs amid may give precipitation or snowmelt occasion speaks to the major component misfortune of precipitation to surface runoff or, the potential sum of water which may go to groundwater revive. Penetration is caused by two powers, gravity and capillary action. Both powers act within the vertical course to cause permeation descending (Mehran et al. 2017; Patricia et al. 2020; Yuk et al. 2021; Sara et al. 2021).

Effective Design From the obtained results, graphs were plotted and the same is represented in Figs. 23.10, 23.11, 23.12, and 23.13 with the combinations on x-axis and the test results on y-axis and are provided.

From pH results, the combination 5 is found to be effective in reducing the p^H from 7.82 (pH of inflow water) to 7.11 and thus stabilized.

From hardness results, the combination 5 is found to be effective in reducing the hardness from 910.34 ppm (hardness of inflow water) to 756 ppm. Though the result does not fit in through the permissible limits, it has been efficient in reducing. Hence combination 5 is considered as effective.

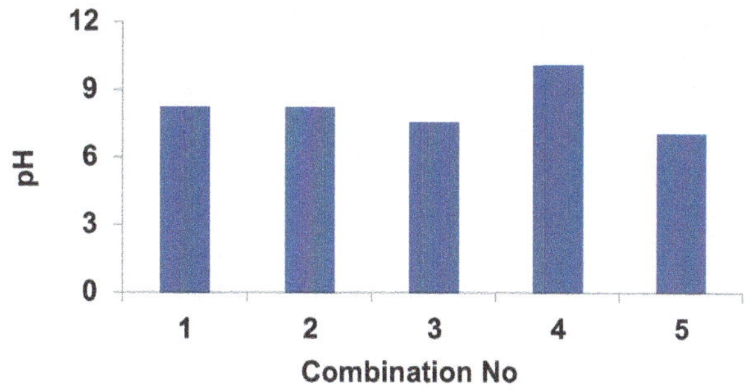

Fig. 23.10 Variation in pH values for outflow water with respect to setups

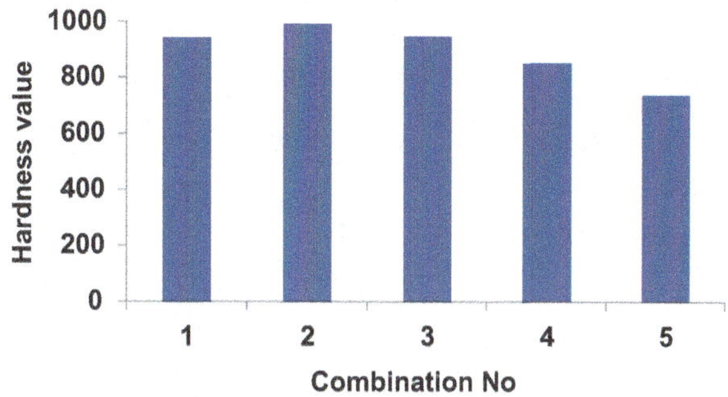

Fig. 23.11 Variation in hardness values for outflow water with respect to setups

From the conductivity results, combination 5 is found to be effective and is within permissible limits.

From TDS results, combination 5 is found to be effective and is within permissible limits.

Based on Figs. 23.8, 23.9, 23.10, and 23.11, it is proved that filter media with combination 4 is more efficient and the same was proved based on the improvement in the efficiency of water quality for the main basic parameters.

To check the efficiency of the filter media the infiltration rate with respect to different combination of filter media also assessed and same was reported in Fig. 23.14.

From infiltration results as represented Fig. 23.13, it is found that combination 1 is considered to have high infiltration rate which means it drains out water with easily (Donchyts et al. 2016; Zaman and Sizemore 2017; Ghanashyam et al. 2020; Fausto et al. 2020). But water-holding capacity of the combinations 4 and 5 seems to be good.

Cost Estimation Estimation is an important process in calculating the cost efficacy. The cost of each material used is considered with certain miscellaneous costs that are explained below. Using their weights and cost, the total cost incurred in each setup is calculated and are tabulated below. Based on the local cost of the material, the rate is taken from the local vendors and the same is represented in Table 23.6. Some of the major components needed for fabricating the filter unit were considered for cost estimation.

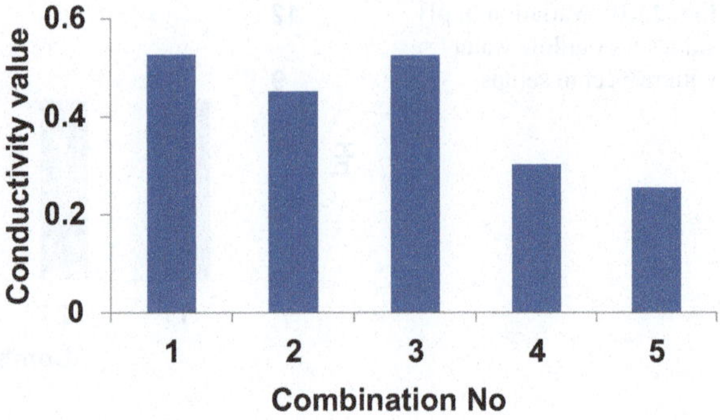

Fig. 23.12 Variation in conductivity values for outflow water with respect to setups

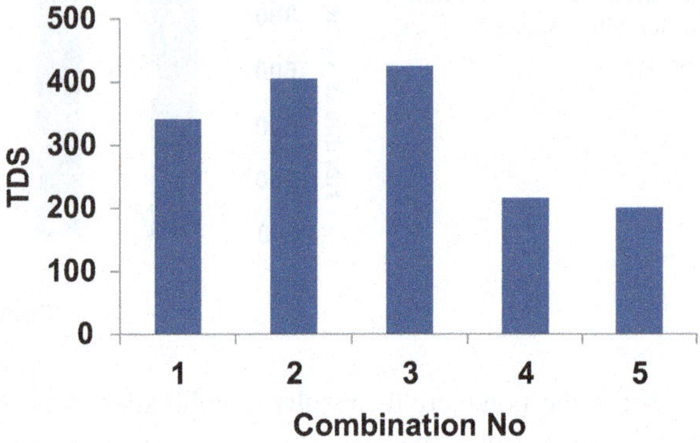

Fig. 23.13 Variation in TDS values for outflow water with respect to setups

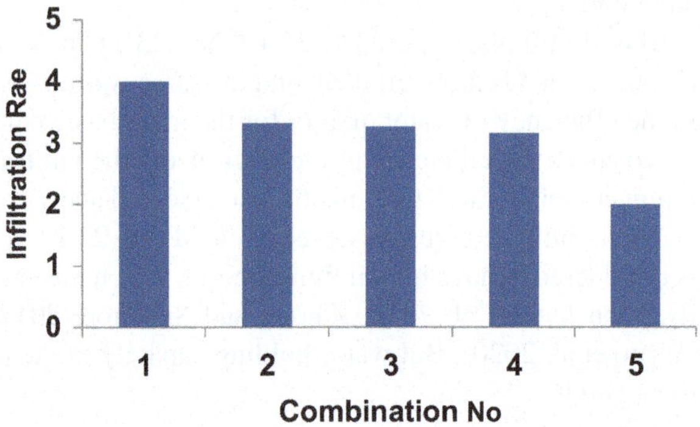

Fig. 23.14 Variation in infiltration rate for outflow water with respect to setups

Table 23.6 Miscellaneous costs for components

Component	Cost (Rs)
Drum	550
Labor and transportation	500
MESH	109.5
Total miscellaneous cost	1159.5

Table 23.7 Cost estimation of various setup combinations

S. No.	Component combination	Weight per item	Cost per item	Total cost = (weight × cost) + miscellaneous cost
1	Soil+ pebble+ carbon+ sand	Pebble: 5.5 Carbon: 2.5 Sand: 4.0	Pebble: 80/kg Carbon: 25/kg Sand: 40/kg	1822.0
2	Soil+ pebble+ carbon+ sand	Pebble: 5.5 Carbon: 2.5 Sand: 4.0	Pebble: 80/kg Carbon: 25/kg Sand: 40/kg	1822.0
3	Soil + sand	Sand: 12.0	Sand: 40/kg	1640.0
4	Soil sand (mixture 1:3) + recycled aggregate	Sand: 15.0	Sand: 40/kg	1760.0
5	Soil + LECA + sand	LECA: 6.0 Sand: 4.0	LECA: 55/kg Sand: 40/kg	1650.0

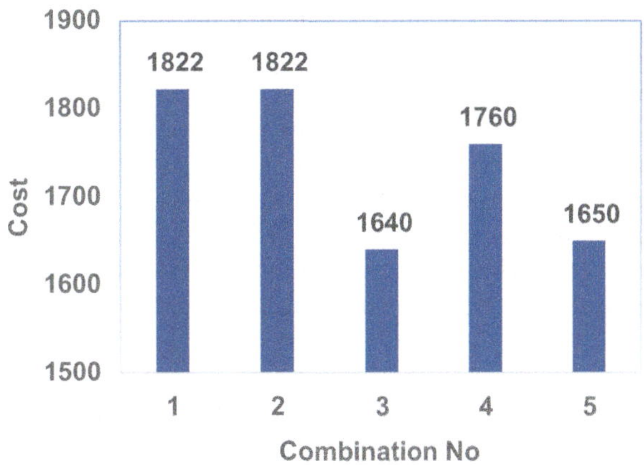

Fig. 23.15 Cost estimation of setups

The cost estimation of the combination of the different filter media unit was also carried out and the same was represented in Table 23.7.

Thus, as per the study, after the detailed analysis of the water quality test results, combination 4 (laterite soil+LECA+sand) is found to be effective. In materials point of view, combination 3 (soil + sand) is found to be cost-effective. Considering infiltration rate combination 1 (soil+pebble+carbon+sand) has found to have better result but considering its cost, and this combination cannot be used for large-scale purpose. Thus, from the above discussion we infer that combination 4 is found to be convincing in all the aspects both in terms of quality and cost (Sudarsan et al. 2018; Shahbaz et al. 2020; Patricia et al. 2020).

Based on the representation in Fig. 23.15, it is evident that combinations 3, 4, and 5 are found to be of less cost, and difference among these three also very minimum. Combinations 4 and 5 are the filter unit whose efficiency of functioning is also good. It is clear from the trail study that local filter media combination unit with natural seeds and clay balls are quite effective in treatment and also, they are cost-effective in adoptability of the same in the filter unit. If the filter unit is properly maintained and properly managed, it will help in treating the collected rainwater effectively, and cost spent toward this process can be retrieved within couple of years and in a maximum period of 5 years the amount spent can be gained back. This kind of locally developed technique is the need hour to adopt the local concepts in achieving the maximum efficiency of the water management process (Mehran et al. 2017; Patricia et al. 2020; Yuk et al. 2021; Sara et al. 2021).

23.6 Conclusion

After the detailed analysis of the water quality test results, combination 4 (laterite soil+LECA+sand) is found to be effective. Rainwater gathering is more often than not taken to cruel the quick collection of water running from surfaces, rooftop, etc. The viability of rainwater collecting lies in its capacity to meet the location prerequisites. The rainwater collecting presents the finest alternative for time to come, rainwater collecting presents the finest alternative for time to come. Application of RWH in urban water framework gives a considerable advantage for both water supply and wastewater subsystems. This combination can be implemented in filters for rainwater harvesting because of its high-water-holding capacity. The use of this material in filters can improve the quality of water, and this water can be made suitable for drinking purpose after some disinfection process. Water which is allowed to pass through this setup has better pH, hardness, conductivity, turbidity, and TDS.

23.6.1 Recommendations

LECA has high water-holding capacity, and from the experimental analysis, it is clear that it absorbs more water. To avoid this water loss, LECA can be allowed to submerge in water initially for 4 to 5 days before using it. From this study, it is suggested to use this setup for residential use to support vegetation. In addition to this, it is recommended for pavements to improve the groundwater table where the top layer could be made with porous concrete.

References

A.P. Andini, P. Purwanto, S. Sudarno, Assessing water quality of Ciujung River in Lebak regency by using pollution index. Adv. Sci. Lett. **23**(3), 2462–2464 (2017)

T. Andréa, T. Celimar, G. Enedir, C. Ernani, Impact of rainwater harvesting on the drainage system: Case study of a condominium of houses in Curitiba, Southern Brazil. Water **10**(8), 1100 (2018)

K. Beven, P. Germann. Macropores and water flow in soils, – water resources research, Wiley online library. First published: October 1982 Full publication history. https://doi.org/10.1029/WR018i005p01311 (1982)

D.W. Corey, T.B. Ryan, A. Mazdak, Rainwater catchment system design using simulated future climate data. J. Hydrol. **529**, 1798–1809 (2015).

G. Donchyts, F. Baart, H. Winsemius, N. Gorelick, J. Kwadijk, D.G.N. Van, Earth's surface water change over the past 30 years. Nat. Clim. Chang. **6**(9), 810–813 (2016)

F.E. Eregno, M.E. Moges, A. Heistad, Treated greywater reuse for hydroponic lettuce production in a green wall system: Quantitative health risk assessment. Water **9**(7), 454–473 (2017)

R. Eric, R. Weaver, H.N. Mahmood, Parameters sensitivities for sustainable urban infrastructure, proceedings of the Institution of Civil Engineers. Munic. Eng. **172**(2), 73–82 (2019)

A.C. Fausto, G.M. Joanna, J. Piotr, S.S. Joanna, W. Katarzyna, W. Marcin, K. Bartosz, Long-term trends in 20-day cumulative precipitation for residential rainwater harvesting in Poland. Water **12**(7), 1932–1948 (2020)

K. Ghanashyam, T. Anusha, D. Niranjan, R.P. Udaya, A review on harvesting and harnessing rainwater: An alternative strategy to cope with drinking water scarcity. Water Supp. **20**(8), 2951–2963 (2020)

E. Ghisi, D. F. Ferreira, Potential for potable water savings by using rainwater and greywater in a multi-storey residential building in Southern Brazil. Build. Environ. **42**(7):2512–2522 (2007)

M.M. Haque, R. Ataur, S. Bijan, Evaluation of climate change impacts on rainwater harvesting. J. Clean. Prod. **137**, 60–69 (2016)

M. Igboekwe, A. Ruth, Groundwater recharge through infiltration process: A case study of Umudike, Southeastern Nigeria. J. Water Resour. Protect **3**(5), 295–299 (2011)

P.D. Jay, J.B. Steven, T.D. Hassan, S. Defne, Introducing demand to supply ratio as a new metric for understanding life cycle greenhouse gas (GHG) emissions from rainwater harvesting systems. J. Clean. Prod. **163**, 274–284 (2017)

S.S. Joanna, G.M. Joanna, J. Piotr, B. Andrzej, O. Piotr, W. Marcin, K. Bartosz, The quality of stored rainwater for washing purposes. Water. **12**(1), 252–268 (2020)

L. Juneseok, G.-D. Gamze, K. Jonathan, Drinking water infrastructure, encyclopedia of water, (1-19), (2019). https://doi.org/10.1002/9781119300762

Y.H. Li, C.P. Tung, P.Y. Chen, Stormwater management toward water supply at the community scale—A case study in northern Taiwan. Sustainability. **9**(7), 1206–1221 (2017)

L.K. McConnell, R.C. Sims, B.B. Barnett, Reovirus removal and inactivation by slowrate sand filtration, Appl Environ Microbio. **48**(4), 818–825 (1984)

D. Maria, A. Céu, C. Nelson, S. João, E. Saeid. Risk analysis of water harvesting systems, handbook of water harvesting and conservation, 177–189 (2021)

N. Mehran, N. Chris, M. Mahdi, J. Nicole, R.B. Brittany, A.M. Michael, Storm water management model: Performance review and gap analysis. J. Sustain. Water Built Environ. **3**(2), 04017002 (2017)

A. Nasrin, J.S. David, L. Jia, R. Andrew, Assessing climate change impacts on the reliability of rainwater harvesting systems. Resour. Conser. Recyc. **132**, 178–189 (2018)

P. Norman, P. Amilcare, Sizing a rainwater harvesting cistern by minimizing costs. J. Hydrol. **541**, 1340–1347 (2016)

A. Pacey, A. Cullis, *Rainwater Harvesting: The Collection of Rainfall and Runoff in Rural Areas* (Intermediate Technology Publications, London, 1989)

A. Pacey, A. Cullis, *Rainwater Harvesting, 1986; The Collection of Rainfall and Runoff in Rural Areas* (Intermediate Technology Publications, London, 1986)

M.S. Peter, W. Sarah, B. David, Rainwater Harvesting Typologies for UK Houses: A Multi Criteria Analysis of System Configurations, Water, 8, 4, (129), (2016). https://doi.org/10.1016/j.scitotenv.2018.10.226

A.M. Patricia, M.S. Peter, S.W. Jy, Fee credits as an economic incentive for green infrastructure retrofits in stormwater-impaired urban watersheds. J. Sustain. Water Built Environ. **6**(4), 04020015 (2020)

S. Pradhan, S.G. Al-Ghamdi, H.R. Mackey, Greywater recycling in buildings using living walls and green roofs: A review of the applicability and challenges. Sci. Total Environ. **652**, 330–344 (2019) https://doi.org/10.1016/j.scitotenv.2018.10.226

K. Prasanna, J.S. Sudarsan, S. Nithiyanantham, Wastewater treatment using combined biological and constructed wetlands technique in paper mills. Sustain. Water Resour. Manag. **3**(4), 431–439 (2017)

W. Ranran, B. Julie, Zimmerman, economic and environmental assessment of office building rainwater harvesting systems in various U.S cities. Environ. Sci. Tech. **49**(3), 1768–1778 (2015)

N. Ren, Q. Wang, Q. Wang, H. Huang, X. Wang, Upgrading to urban water system 3.0 through sponge city construction. Fron. Environ. Sci. Eng. **11**(4), 9 (2017). https://doi.org/10.1007/s11783-017-0960-4

M. Sara, N. Mukunth, K. David, Green infrastructure performance in arid and semi-arid urban environments. Urban Water J. **18**(4), 275–285 (2021)

D.P. Sartory, K. Spies, B. Lange, S. Schneider, B. Langer, Evaluation of a most probable number method for the enumeration of legionella pneumophila from potable and related water samples. Lett. Appl. Microbiol. **64**(4), 271–275 (2017)

A. Shahbaz, Z. Shouhong, Y. Tongjia, Environmental and economic assessment of rainwater harvesting systems under five climatic conditions of Pakistan. J. Clean. Prod. **259**, 120829 (2020)

J.S. Sudarsan, J. Kakuru, R. Priyanka, A.H. Haseeb, P.K. Biyabani, S. Swati, Impact of fragile water management strategies and mitigation-a case study of Pune City, India. IOP Conf. Ser.: Mater. Sci. Eng. **1026 01200**, 3–12 (2021)

J.S. Sudarsan, R. Annadurai, M. Mukhopadhyay, P. Chakraborty, S. Nithiyanantham, Domestic wastewater treatment using constructed wetland: an efficient and alternative way. Sustain. Water Resour. Manag. **4** 781–787 (2018)

M. Sukop, M. VanDijk, G.J. Perfect, E. Van Loon. Percolation thresholds in dimensional prefractal models of porous media. Agronomy Abstracts (2000)

D. Suzanne, M.C. Anita, K. Misgana, J.L. Muleta, Is rainwater harvesting worthwhile? A benefit–cost analysis. J. Water Resour. Plan. Manag. **147**(4), 04021011 (2021)

M. Ting, H. David, Stated preferences for smart green infrastructure in stormwater management. Landsc. Urban Plan. **187**, 1–10 (2019)

M.T. Waso, P.H. Ndlovu, S. Dobrowsky, W.K. Khan, Presence of microbial and chemical source tracking markers in roof-harvested rainwater and catchment systems for the detection of fecal contamination. Environ. Sci. Poll. Res. **23**(17), 16987–17001 (2016)

J. Xueer, Z. Shouhong, Z. Jianjun, W. Yujie, W. Yunqi, Y. Tongjia, Analysis and modelling of stormwater volume control performance of rainwater harvesting systems in four climatic zones of China. Water Resour. Manag. **32**(8), 2649–2664 (2018)

S.L. Yuk, M.D. Safari, N.M. Md, L.T. Mou, A.Z. Nor, K.C. Chun, Assessing the effectiveness of mitigation strategies for flood risk reduction in the Segamat River basin, Malaysia. Sustainability **13**(6), 3286–3309 (2021)

M.S. Zaman, R.C. Sizemore, Freshwater resources could become the Most critical factor in the future of the earth. J. Mississippi Acad. Sci. **62**(4), 348–352 (2017)

K. Zhang, A. Randelovic, L.M. Aguiar, D. Page, D.T. McCarthy, Methodologies for pre-validation of biofilters and wetlands for stormwater treatment. PLoS One **10**(5), e0125979 (2015)

K. Zhang, P.M. Bach, J. Mathios, C.B.S. Dotto, A. Deletic, Quantifying the benefits of stormwater harvesting for pollution mitigation. Water Res. **171**, 115395 (2020)

Chapter 24
Prioritization of Subwatershed Through Morphometric Analysis of Amaravathi Watershed Using Geoinformatic Techniques

S. Arivazhagan, A. Karthi, M. Lavanya, and M. Kirubakaran

Abstract Prioritization of subwatersheds in a river basin using morphometric parameters is one of the elegant approaches for the development of watersheds and its management. The concept of prioritization plays a key role in identifying areas which require immediate intervention for natural resources conservation plans in the subwatershed level. In this study, an attempt has been made to prioritize subwatershed based on morphometric parameters using remote-sensing and GIS techniques to understand the role of channel network on the geo-hydrological behavior of Amaravathi watershed. This study also describes geology, geomorphology, and soil characteristics of the watershed. The ASTER DEM data with a spatial resolution of 30 m were utilized to prepare thematic maps such as drainage, stream order, watershed, and drainage density map. The drainage morphometric analysis was performed through linear, aerial, and relief aspects. The predominantly dendritic drainage pattern is noticed and Amaravathi watershed, which has been divided into 31 subwatersheds. Prioritization of subwatershed was carried out based on compound parameter analysis by adding the values of linear and shape parameters of each subwatershed. In this study, it is observed that out of 31 subwatersheds, 10 fall in high priority (SWS-1, 4, 5, 6, 8, 10, 15, 20, 21, 23), 10 fall in medium priority (SWS- 2, 3, 12, 13, 17, 22, 25, 27, 29, 31), and the remaining 11 fall in low priority (SWS- 7, 9, 11, 14, 16, 18, 19, 24, 26, 28, 30). Hence, the high priority subwatersheds are possible for soil erosion. In connection to that, we suggest that the high-priority subwatersheds should take precaution methods to avoid the future hazard.

Keywords Amaravathi · Aster DEM · Subwatersheds

S. Arivazhagan (✉) · A. Karthi · M. Lavanya
Centre for Applied Geology, The Gandhigram Rural Institute – Deemed to be University, Gandhigram, Dindigul, Tamilnadu, India

M. Kirubakaran
Department of Civil Engineering, University VOC College of Engineering, Thoothukudi, Tamil Nadu, India

B. Panneerselvam et al. (eds.), *Climate Change Impact on Groundwater Resources*, https://doi.org/10.1007/978-3-031-04707-7_24

24.1 Introduction

Watershed (WS) is an area of landform, and water-limited by a drainage divide that collects in the surface runoff and exits the WS through the single outlet into a larger river (TNAU). WS is considered to be the most efficient and appropriate unit for resource assessment and subsequent planning for sustainable development of natural water resources because landform and water resources have the best interface and combination effect (PATHSALA). Therefore, the WS approach in India is increasingly used in numerous development projects such as soil management, water resource conservation, erosion control, forest management, desert development program, river valley projects, flood-prone area management, and reclamation of wastelands (TNAU Agritech Portal). WS is a convenient and clearly defined terrain unit that can take the basic erosional terrain element for effective management of the natural land and water resources. Hydrological concern the shape of the WS is most vital because it controls the runoff flow time to concentrate at the outlet (Biswas 2016; Rout 2017; Patel et al. 2019).

Characteristics of WS in terms of climate, drainage, geomorphology, geology, soil, etc., play an important role in natural resource management. As WS is considered a natural system, it has variations in terms of drainage, soil type, etc. To demarcate the most vulnerable subwatersheds (SWSs), they should be arranged based on their surface drainage morphometric characteristics (Patel et al. 2019). Watershed management (WSM) is used to describe the process of implementing landuse/landcover practices and water management practices, and the need to make sustainable use of available land and water resources to protect and improve the quality of water and other natural resources of WS (Rout 2017).

Strahler (1957) provided the quantitative geomorphic methods for measuring the size and shape properties of drainage basins developed over the past few years. Descriptive numbers are present in two general classes such as (i) linear scale measurements and (ii) angles or ratios of length measures (dimensionless numbers).

The quantitative description of basin geometry, such as morphometric analysis (MA), is effectively used with the assistance of integrated remote sensing (RS), and Geographic Information System (GIS) techniques in recent times. Biswas et al. (1999) and Das and Mukherjee (2005) analyzed the drainage and morphometric characteristics of WSs in the Raigad district, Maharashtra, India, using effective techniques such as RS and GIS. The results are proposed that 15 WSs have been identified and a detailed MA concluded that eight of the WSs with low infiltration numbers have high infiltration capacity and a thus good choice for artificial recharge. Vittala et al. (2008) studied the North Pennar basin based on MA and drainage characteristics where the SWSs have been grouped into three classes such as high priority, medium priority, and low priority. These SWSs can immediately implement development and control plans in a sustainable manner, which will eventually lead to water and soil management. Mishra and Nagarajan (2010) investigated the WS of the Bhawanipatna area, Kalahandi district, Odisha, India. The WS area was divided into 12 SWSs such as SWS1 to SWS12. The MA and compound parameter (CP) values are evaluated and the SWS with the lowest CP getting the top priority. The SWS 1 has a minimum CP value which is likely to cause extreme soil erosion and natural hazards in this area. Therefore, immediate soil maintenance and preventive measures should be provided. Biswas et al. (1999) have taken Nayagram Block in the Midnapore District, West Bengal, to investigate the morphometric parameters (MOPs) such as Dd, bifurcation ratio, form factor, stream frequency, texture ratio, and stream length. Nine SWSs have been evaluated through the Sediment Yield Index (SYI) method and also prioritized all the SWSs and found that eight SWSs are having the highest priority because of high erosion intensity which is observed through the above-mentioned MOPs.

The Karawan WS in the Dhasan basin and part of the mega Yamuna basin in Sagar district, Madhya Pradesh, have been studied through MOPs analysis by using the Advanced Space borne Thermal Emission and Reflection Radiometer digital elevation model (ASTER DEM) and GIS techniques. The MOPs such as linear, areal, and relief were used to prioritize and map the SWSs (Pareta and Pareta 2011). Somashekar et al. (2011) studied the Hesaraghatta WS by using RS and GIS techniques. The area is grouped into four SWSs based on drainage patterns and terrain features. IRS–1D LISS III was used to generate thematic layers such as drainage, landuse/landcover, soil, and hydrological soil group. Quantitative MA of (a) linear aspects (b) aerial aspects, and (c) relief aspects were carried out for the entire WS. It is fundamentally a low rainfall region that is characterized by an irregular rainfall along with less structural disturbance.

Magesh et al. (2012) evaluated the Chimmini and Mupily WSs in the Thrissur district, Kerala, India. Quantitative geomorphometric analysis was carried out for the WSs independently by estimating their (a) linear aspects (b) aerial aspects (c) relief aspects. The Chimmini WS was categorized as a sixth-order drainage basin, and the Mupily WS was ordered as a fifth-order basin. The physiographic and structural conditions are predominantly controlled by the stream order of the basin. The local geology and erosion cycles are the important factors to control the slope variation. Wilson et al. (2012) studied the SWSs in the Aiyar and Karai-Pottanar basin of Tamil Nadu, India. Shuttle Radar Topographic Mission (SRTM) data was used to prepare the aspect grid, drainage network (DN), and slope maps. The SWSs are studied based on the different MOPs that were used to evaluate the drainage characteristics and divided into four SWSs such as SWS-I-IV. Magesh et al. (2013) investigated the Bharathapuzha river basin based on MA. The extraction of the river basin and their DNs were generated with the help of GIS techniques. According to Strahler's system of classification DN was extracted and categorized and DN shows dendritic to the subdendritic pattern. The slope of the basin ranges from 0 to 70 and the erosion cycles and local geology are the important factors for control the slope variation. It is observed that the basin is very much helpful for hydrological studies and basin management.

Ali and Ali (2014) studied the five micro-WSs of Romushi-Sasar Catchment using RS and GIS systems. Stream network was extracted from Survey of India

(SOI) toposheets and ASTER DEM. The three aspects of MOPs namely, linear, shape, and relief, and their prioritization were thoroughly investigated. In order to the ranking of each parameter has given based on MA, the SWSs have been divided into three classes as high, medium, and low. The lowest CP has a high priority ranking. Therefore, the SWS RSMW4 and RSMW5 getting very high priority ranking which is possible for maximum soil erosion. Immediate soil conservation measures should be provided in this area. Rai et al. (2014) studied the area of Kanhar basin tributaries of Son River, and the stream network extracted through the ASTER DEM. The detailed study of drainage characteristics shows seven SWSs and MOPs are calculated. The area is mainly controlled by lithological and physiographic conditions. Waikar and Nilawar (2014) analyzed the Charthana drainage basin located in the Parbhani district of Maharashtra, India, and stream order ranges from 1 to 4th order. The drainage basin features characteristics have been studied by three aspects of MOPs such as linear, aerial, and relief aspects with the help of GIS and image processing techniques and investigate about detailed information of the basin.

Salvi et al. (2017) studied that the WSM is based on the MOPs. It is very useful to evaluate morphometry characters such as size, shape, earth surface, and groundwater conditions. Morphometric features are based on three aspects, namely, linear aspects, aerial aspect, and slope of the drainage basin. The study of various observations was made based on WS problems within the drainage basin such as soil erosion, WS evaluation, drought area, and change of landuse/landcover. Kamala and Samynathan (2018) described morphometry of Amaravathi River Basin, Tamil Nadu, based on three aspects of MOPs linear, relief, and arial aspects using the SRTM data. The stream order ranges from 1 to 7th order. The basin drainage system reveals the homogeneous character of the area and helps WSM.

Sushanth and Bhardwaj (2019) studied the Morphometric drainage characteristics of the Patiala-Ki-Rao WS in the Shivalik foot-hills, Punjab. The Dd, drainage texture, mean bifurcation ratio, and stream frequency indicated quick removal and low holding of runoff. The area is classified into three zones such as high, moderate, and low runoff potential zones. The moderate to high runoff prospective zones were typically located near the drainage upstream side, whereas low runoff prospective was toward the downstream side. The study is very helpful to develop water management and planning, rainwater harvesting.

RS and GIS in WSM studies have been accepted as an effective method in recent times to collect facts about the water resources more precisely, proficiently, and rapidly than conventional methods. This investigation deals with the "MA of Amaravathi WS with the help of ArcGIS software around Dindigul, Tirupur, Idukki districts of Tamil Nadu and Kerala, South India" to develop an integrated database and numerically analyze the existing groundwater conditions. It is designed to propose the possible options for successful water management and further development of groundwater resources in the Amaravathi WS, Tamil Nadu, India.

24.2 Study Area

The Amaravathi WS is a semiarid region situated in the district of Tamil Nadu and Kerala states, India, which is the main tributary of the river Cauvery. Amaravathi WS is located at longitude 77°3′31.551″ E and 77°29′43.951″ E and latitude 10°13′27.416" N and 10°40’33.257" N. It has a catchment area of 1546 sq.km constituting of three districts, namely, Dindigul, Tirupur, and Idukki in Tamil Nadu and Kerala, respectively. The Amaravathi drainage basin is mostly characterized by a dendritic drainage pattern. The study area is bounded by Pallatottam in the south, Ramalingapuram in the north, Kumaralingam in the east, and Chinnappanpudur in the west including the Amaravathi reserve forest, Pambadam shola reserve forest, Palani hills, Anaimalai hills. The base map of the Amaravathi WS is shown in Fig. 24.1. The study area experiences a tropical climate that is hot and dry for the most part of the year. The period from March to June is generally warm (temperature ranges from 20° to 38 °C). The study area receives southwest and northeast monsoon rains. The southwest monsoon accounts for 40 percent of the rainfall and the northeast monsoon 50–60 percent. The average annual rainfall in the basin is 720 mm. The Geology of the study is varied in different lithological units such as

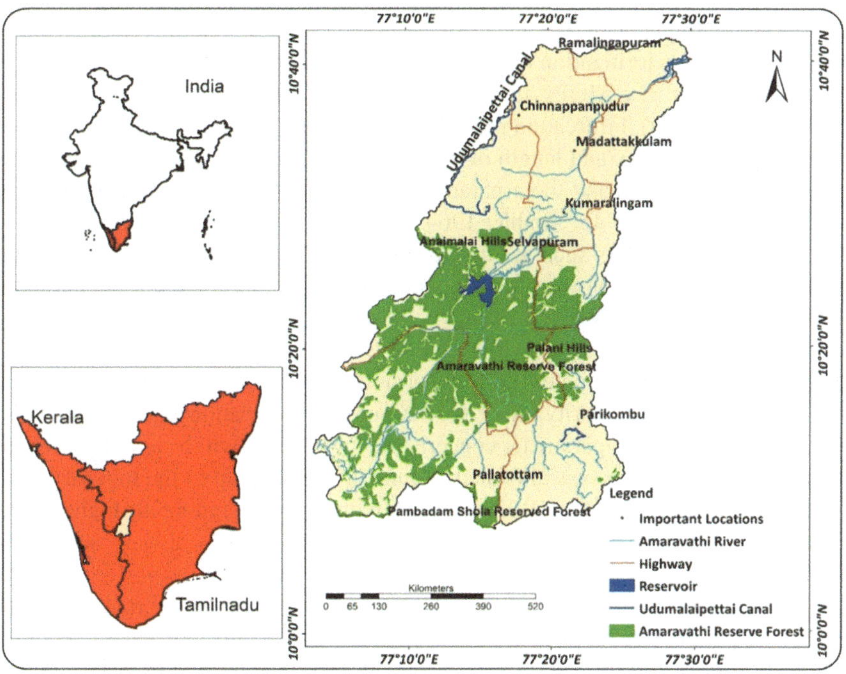

Fig. 24.1 The location base map of the Amaravathi watershed

Calc– granulite, Champion gneiss, Charnockite, Granite, Granitoid gneiss, Migmatite – Gneiss Complex, Peninsular Gneiss, Quartz – Vein, Sand, and Silt, Syenite / Nepheline syenite. The study area is mainly undulated area in nature and the following 17 geomorphological features such as Dome type Denudational Hills, Dome type Residual Hills. Fracture Valley, Hilltop Weathered, Intermontane Valley, Pediment/ Valley Floor, Pediplain Canal Command, Ridge type Structural Hills (Large), Scarp, Shallow Weathered/ Shallow buried pediplain, Stori– Highly dissected Hills and Valleys, Strike Ridge, Valley, Valley Fill/ Filled-in valley, Waterbed Mask, Inselberg are mapped. The soil characteristics of the study area comprise Alfisols, Entisols, Inceptisols, Vertisols, and Forest soil.

24.3 Datasets

In this study, ASTER DEM which is having 30 m spatial resolution in a spectral range of 0.78 to 0.86 mm with 0.6 base-to-height ratio and an intersection angle of 27.7 degrees (Wigglesworth and Alex 2019*)* has been used to generate the thematic maps. It has been downloaded from the USGS Earth Explorer delivery system. *The* Survey of India (SOI) toposheets No. (58 F/2, 58 F/3, 58 F/4, 58 F/6, 58 F/7, 58 F/8) of 1:50,000 scale data were used for digitizing the base and drainage maps.

The MA is implemented by using Horton (1945) and Strahler (1964) classification system. The prioritization of SWSs (PSWS) was approached by the linear and shape parameters. The MOPs, that is, circularity ratio (Rc), elongation ratio (Re), drainage texture (Rt), bifurcation ratio (Rb), Dd, stream frequency (Fs), form factor (Rf), shape factor (Rs), and length of overland flow (Lg) are known as erosion risk assessment parameters and these parameters have been considered for prioritizing SWSs (Horton 1932, 1945; Strahler 1964; Biswas et al. 1999).

The linear parameters such as Rb, Dd, Fs, Rt, and Lg have a direct relationship with erodibility factor where a higher value implies more erosion (Nookaratnam et al. 2005). During PSWSs, the rank 1 was assigned to the highest value linear parameters and the next value will be based on the least rank. Patel et al., (2012) stated that the shape parameters such as form factor (Ff), shape factor (Fs), elongation ratio (Re), and circularity ratio (Rc) have an inverse relationship with erodability criteria. In the shape parameter concern, the lower value will be assigned as rank 1; the next higher value was allocated as rank 2, and thus up to the highest value. Therefore, the ranking of the SWS has been attempted by assigning the highest rank in terms of the highest value in the case of linear parameters, as opposed to the shape parameters.

The compound factor (Fc) will be calculated by summing all the ranks of linear and shape parameters and dividing them by the total number of parameters. On observation-based, SWSs with low Fc value are given the highest priority ranking. The other basic parameters will also be calculated such as area, perimeter, and basin length. The drainage map of the study area is shown in Fig. 24.2.

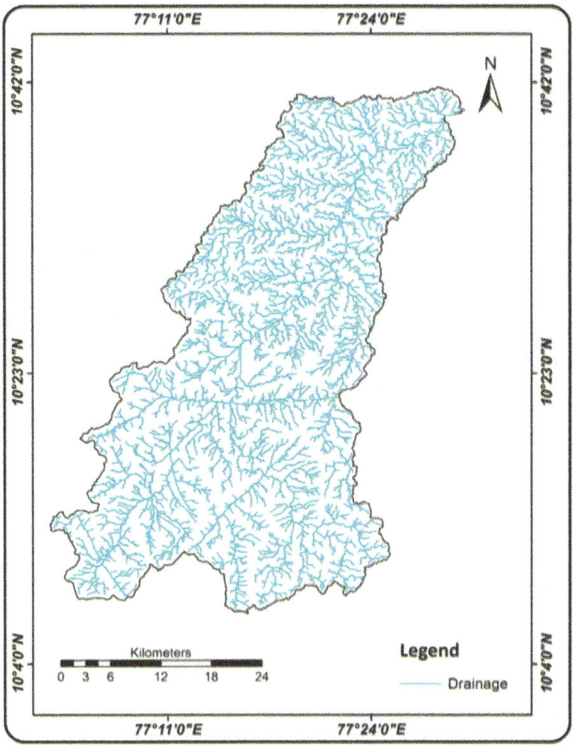

Fig. 24.2 The drainage map of the study area

24.4 Morphometric Analysis (MA)

The MA is done effectively over the measurement of basins linear, aerial, relief aspects. MA of drainage basins is important for suitable planning of WSM and it offers information about the WS characteristics such as slope, topography, soil condition and runoff behavior, etc. Numerical valuation of these parameters has been carried out by using standard mathematical formulae. Area, perimeter, and basin length are the basic parameters used to calculate Ff, Fs, Rc, Re, and Rr (Table 24.1). The Amaravathi SWSs of the study area are shown in Fig. 24.3. The parameters morphometry study are given in Table 24.2.

24.5 Results and Discussion

24.5.1 Linear Aspects

Linear parameters such as U (stream order), Lu (stream length), Lsm (mean stream length), Rl (stream length ratio), Rb (bifurcation ratio), and Rbm (mean bifurcation ratio) were calculated.

Table 24.1 Basic parameters of the study area

S.No.	Subwatershed name	Basic parameters		
		Area of SWS	Perimeter	Basin length
1	SWS-1	11.58	16.85	3.73
2	SWS-2	18.60	24.20	8.62
3	SWS-3	7.65	17.27	5.50
4	SWS-4	26.06	29.44	8.83
5	SWS-5	38.90	43.78	13.07
6	SWS-6	53.73	41.45	8.77
7	SWS-7	6.64	16.74	5.10
8	SWS-8	47.95	42.70	12.30
9	SWS-9	13.22	23.40	4.42
10	SWS-10	120.26	75.97	23.47
11	SWS-11	9.37	16.44	4.72
12	SWS-12	13.99	24.54	7.40
13	SWS-13	126.53	88.69	29.81
14	SWS-14	18.99	23.81	7.65
15	SWS-15	35.13	32.08	9.02
16	SWS-16	7.88	14.18	4.38
17	SWS-17	8.17	22.56	7.15
18	SWS-18	5.47	12.52	4.32
19	SWS-19	17.16	22.04	7.22
20	SWS-20	5.95	12.01	3.73
21	SWS-21	6.38	12.56	3.96
22	SWS-22	30.41	35.11	11.69
23	SWS-23	45.62	31.02	7.55
24	SWS-24	15.30	21.37	7.93
25	SWS-25	40.63	31.41	8.79
26	SWS-26	78.80	55.27	16.26
27	SWS-27	285.94	92.16	28.03
28	SWS-28	19.33	19.82	6.74
29	SWS-29	88.99	49.61	18.53
30	SWS-30	10.66	14.30	4.84
31	SWS-31	133.34	59.42	11.37

24.5.1.1 Stream Order (U)

Stream order (U) is the primary process in drainage MA of the basin. Strahler's (1964) ordering scheme is followed in this study. The sequence of given streams represents the total number of all streams that represent an individual stream segment of the WS (Balasubramanian et al. 2017). The stream order of the study area is shown in Fig. 24.4.

The higher-order stream of the study area is sixth order. Out of 31 SWSs, one SWS represents sixth-order stream (SWS-31). 4 SWS as fifth-order streams

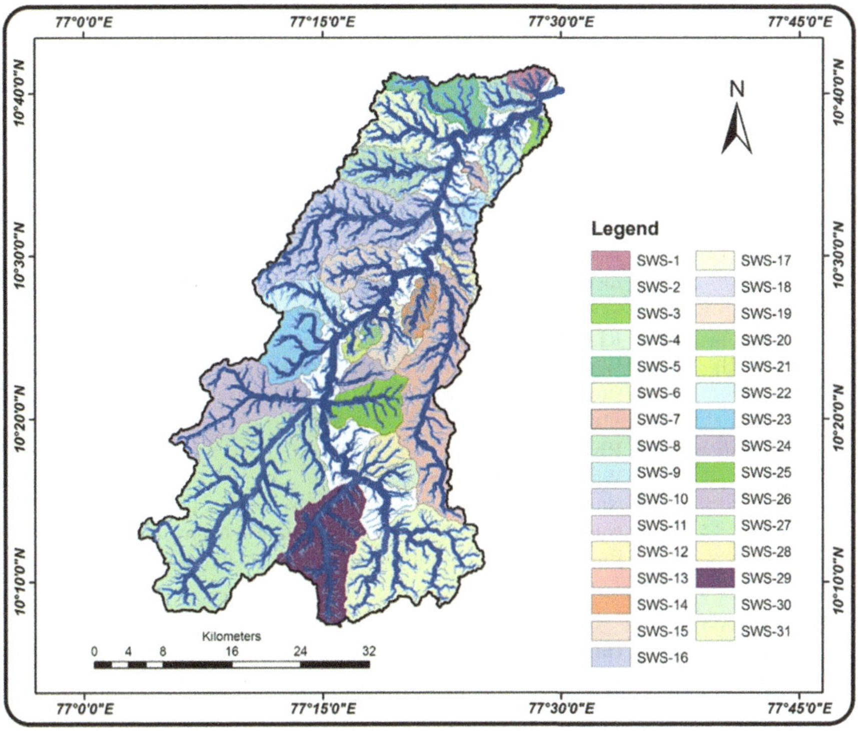

Fig. 24.3 The subwatersheds of the study area

(SWSs -10, 27, 29, 31). 16 SWS as fourth-order streams (SWSs - 1,2,4-6,8,10,13, 15,22,23,25-27,29,31). A total of 3962 streams have been identified in the Amaravathi basin. It represents 2012 as 1st order, 986 as 2nd order, 567 as 3rd order, 264 as of 4th order, 132 as of 5th order, and 1 as 6th order. It may be seen that the highest stream segments are found in SWS-27 (834 km) and the lowest stream segments in SWSs-7 and 18 (15 km). The variation in the stream orders is mainly due to the regions structural conditions. These stream orders were further used for SWS characterization. The stream order values are shown in Table 24.3.

24.5.1.2 Stream Length (Lu)

The Lu was generated based on the law offered by Horton (1945). Lu is measured from the mouth of a river to the drainage divide with the help of Arc GIS software (Horton 1945).

The total stream lengths of various orders have been measured with the help of calculate geometry tool of ArcGIS software. The hilly area is predominant with smaller-length streams (1st and 2nd order) which represent the larger slope area and longer stream lengths (3rd, 4th, 5th, and 6th order) are of a lesser slope with gentle

Table 24.2 Morphometric parameters of Amaravathi watershed with formula

S.No	Parameters	Formula	Reference
1. Linear aspects			
1	Stream order (U)	Hierarchical order	Strahler (1964)
2	Stream length (Lu)	Length of stream	Horton (1945)
3	Mean stream length (Lsm)	**Lsm = Lu/Nu** Where, Lu is the total stream length of order "u" and Nu is the total no. of stream segments of order u	Horton (1945)
4	Stream length ratio (Rl)	**RL = Lu / Lu – 1** Where, RL is the stream length ratio, Lu is the total stream length of the order "u", and Lu – 1 is the total stream length of its next lower order	Horton (1945)
5	Bifurcation Ratio (Rb)	**Rb = Nu/Nu + 1** Where,Rb is the bifurcation ratio, Nu is the total no. of stream segments of order "u" and Nu+1 is the number of the segments of next higher order	Schumn (1956)
6	Mean bifurcation ratio (Rbm)	Rbm is the average of bifurcation ratios of all orders	Strahler (1964)
2. Aerial aspects			
7	Drainage density (Dd)	**Dd = Lu/A** Where, D is the drainage density, Lu is the total stream length of all orders, A is the area of the basin km^2	Horton (1945)
8	Stream frequency (Sf)	Sf = **Nu/A** Where, Sf is the stream frequency, Nu is the total no. of streams of all orders, A is the area of the basin (km^2)	Horton (1932)
9	Drainage texture (Rt)	**Rt = Nu/ P** Where, Rt is the drainage texture, Nu is the total no. of streams of all orders, P is the perimeter km	Horton (1945)
10	Form factor (Ff)	**Ff = A/Lb²** Where, Ff is the form factor, A is the area of the basin km^2 Lb^2 is the square of basin length	Horton (1945)
11	Shape factor (Fs)	**Fs = Lb²/A** Where, Fs is the shape factor, Lb^2 is the square of basin length, A is the area of the basin (km^2)	Horton (1945)

(continued)

Table 24.2 (continued)

S.No	Parameters	Formula	Reference
12	Circularity ratio (Rc)	**Rc = 4Πa/ (P)²** Where, A = Area of WS, Π = 3.14, P = Perimeter of WS	Miller (1953)
13	Elongation ratio (Re)	**Re = √A/π / Lb** Where, A = Area of the basin (km²) Π = 3.14 Lb = Length of the basin	Schuma (1956)
14	Length of overland flow (Lg)	**Lg = 1/D*2** D = Drainage density	Horton (1945)
3. Relief Aspects			
15	Basin relief (R)	**R = H-h** R = Basin relief H = Maximum elevation H = minimum elevation	Schuma (1956)
16	Relief ratio (Rr)	**Rr = H/Lb** Where, "H" is the total relief (relative relief) of basin in km and Lb is the basin length.	Schuma (1956)
17	Ruggedness number (Rn)	**Rn = R*Dd** Where, "R" is the basin relief in meter and Dd is the drainage density	Schuma (1956), Strahler (1957), Melton (1958)

and flat terrain. In the present study, the order-wise stream length for 31 SWSs was measured using GIS techniques. The total length of first- and second-order streams contributes 77.31% of total stream length and the remaining orders comprise 22.64%, which are obviously indicating that the basin Lu is higher in the first-order streams and decreases where the stream order increases. The Lu values are shown in Table 24.4.

24.5.1.3 Mean Stream Length (Lsm)

Lsm divulges the DN components characteristic size and its backing surfaces (Strahler 1964). Lsm has been figured by dividing the total Lu value of order "u" by the number of stream segments (Magesh et al. 2013). The drainage basin represents that the Lsm of a given order is greater than the lower order and lesser than the higher-order.

The value of Lsm varies from 0.30 (SWS-1) to 1.32 (SWS-17). Usually, such a substantial variation of SWSs indicates a change in terrain slope and elevation of the basin. The Lsm values are shown in Table 24.5.

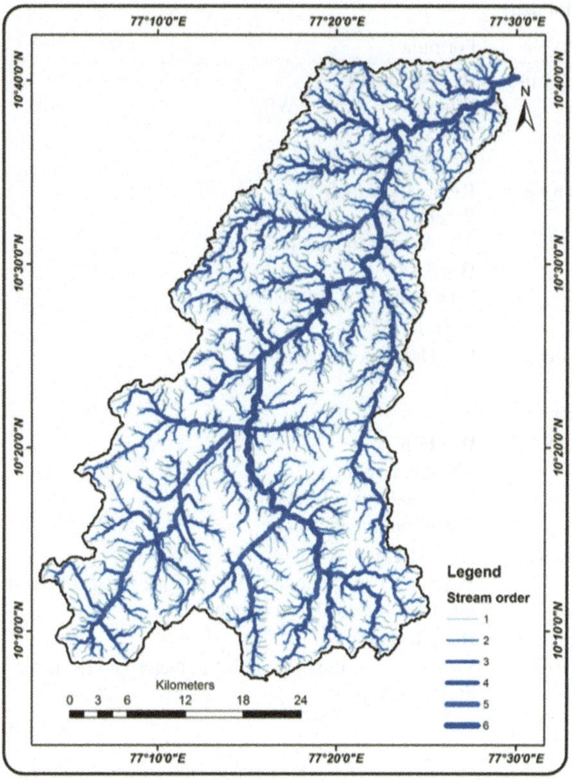

Fig. 24.4 The stream order map of the study

24.5.1.4 Stream Length Ratio (Rl)

Rl value is calculated by dividing the higher-order Lu by the next lower-order stream segments (Horton 1945). Change in Rl between different SWSs showed a tendency to increasing and decreasing in length ratio from the lower order to the higher order. In all SWSs, there was a variation from one order to another order demonstrating the late youth stage of geomorphic development (LYSGD) of streams in the basin area.

The Rl between the streams of different orders of the Amaravathi SWS shows a variation in each SWS. This variation might be attributed to variation in topographic slope and terrain morphology, representing the LYSGD in the streams of the SWS (Singh and Singh 1997). The Rl values are shown in Table 24.6.

24.5.1.5 Bifurcation Ratio (Rb)

Rb is defined as the ratio of the number of stream segments of a given order to the number of segments of the next higher-order (Schumn 1956). The values of Rb are not unique from one order to the next.

Table 24.3 Stream order of the study area

S.No.	Subwatershed name	Area of SWS	Stream order (U) (in km) I	II	III	IV	V	VI	Total (in km)
1	SWS-1	11.58	20	11	6	1	–	–	38
2	SWS-2	18.60	29	13	6	9	–	–	57
3	SWS-3	7.65	13	7	5	–	–	–	25
4	SWS-4	26.06	34	25	6	1	–	–	66
5	SWS-5	38.90	54	31	15	7	–	–	107
6	SWS-6	53.73	82	41	20	19	–	–	162
7	SWS-7	6.64	8	7	–	–	–	–	15
8	SWS-8	47.95	67	34	11	20	–	–	132
9	SWS-9	13.22	19	11	6	–	–	–	36
10	SWS-10	120.26	176	87	52	11	23	–	349
11	SWS-11	9.37	18	7	10	–	–	–	35
12	SWS-12	13.99	24	9	14	–	–	–	47
13	SWS-13	126.53	178	86	33	55	–	–	352
14	SWS-14	18.99	27	12	14	–	–	–	53
15	SWS-15	35.13	52	33	6	10	–	–	101
16	SWS-16	7.88	11	6	4	–	–	–	21
17	SWS-17	8.17	12	4	7	–	–	–	23
18	SWS-18	5.47	8	5	2	–	–	–	15
19	SWS-19	17.16	21	10	9	–	–	–	40
20	SWS-20	5.95	10	6	3	–	–	–	19
21	SWS-21	6.38	14	5	8	–	–	–	27
22	SWS-22	30.41	48	20	10	17	–	–	95
23	SWS-23	45.62	79	36	31	10	–	–	156
24	SWS-24	15.30	26	13	12	–	–	–	51
25	SWS-25	40.63	60	35	11	13	–	–	119
26	SWS-26	78.80	109	46	44	17	–	–	216
27	SWS-27	285.94	421	195	126	25	67	–	834
28	SWS-28	19.33	24	17	5	–	–	–	46
29	SWS-29	88.99	149	62	43	8	32	–	294
30	SWS-30	10.66	17	9	7	–	–	–	33
31	SWS-31	133.34	202	103	41	41	10	1	398
Total		1348.66	2012	986	567	264	132	1	3962
Percentage			50.78	24.88	14.31	6.66	3.33	0.02	99.98

The difference of Rb value might be attributed to the lithological and geological changes of the basin, and further, the higher Rb value (Rb = 10) is noted in SWS 31, representing a robust structural control over the SWS, and a lower Rb value (Rb = 0.25) is noted in SWS 29, representing that the WS is a minor structural disturbance (Table 24.7).

Table 24.4 Stream length of the study area

S.No.	Subwatershed name	Area of SWS	Stream length (Lu) (in km)						Total (in km)
			I	II	III	IV	V	VI	
1	SWS-1	11.58	9.58	10.64	1.84	0.57	–	–	22.63
2	SWS-2	18.60	15.76	9.24	3.86	5.22	–	–	34.08
3	SWS-3	7.65	8.33	3.28	2.37	–	–	–	13.98
4	SWS-4	26.06	22.37	23.10	6.55	0.7	–	–	52.39
5	SWS-5	38.90	42.34	20.96	10.58	3.23	–	–	77.11
6	SWS-6	53.76	42.81	34.15	13.75	9.72	–	–	100.43
7	SWS-7	6.64	7	5	–	–	–	–	12
8	SWS-8	47.95	45.54	29.94	6.69	11.04	–	–	93.21
9	SWS-9	13.22	12.10	9.66	2.97	–	–	–	24.73
10	SWS-10	120.26	111.89	65.39	31.60	6.95	12.73	–	228.56
11	SWS-11	9.37	7.93	4.22	3.50	–	–	–	15.65
12	SWS-12	13.99	12.96	5.11	6.69	–	–	–	24.76
13	SWS-13	126.53	108.57	50.18	17.41	30.85	–	–	207.01
14	SWS-14	18.99	15.31	9.82	8.98	–	–	–	34.11
15	SWS-15	35.13	34.33	21.93	2.98	7.76	–	–	67
16	SWS-16	7.88	8.02	3.51	2.84	–	–	–	14.37
17	SWS-17	8.17	7.33	5.30	3.62	–	–	–	16.25
18	SWS-18	5.47	4.04	4.02	1.39	–	–	–	9.45
19	SWS-19	17.16	17.36	7.37	5.36	–	–	–	30.09
20	SWS-20	5.95	6.05	3.74	2.83	–	–	–	12.62
21	SWS-21	6.38	7.04	2.24	3.18	–	–	–	12.46
22	SWS-22	30.41	29.02	9.24	6.42	8.91	–	–	53.59
23	SWS-23	45.62	38.74	16.31	16.26	5.65	–	–	76.96
24	SWS-24	15.30	10.81	7.50	5.09	–	–	–	23.4
25	SWS-25	40.63	32.76	21.35	7.80	5.66	–	–	67.57
26	SWS-26	78.80	53.63	31.46	19.24	8.06	–	–	12.39
27	SWS-27	285.94	225.94	111.17	68.37	16.03	27.81	–	449.32
28	SWS-28	19.33	14.33	12.74	3.08	–	–	–	30.15
29	SWS-29	88.99	74.0.56	25.20	19.52	3.46	13.07	–	135.81
30	SWS-30	10.66	8.59	5.31	3.84	–	–	–	17.74
31	SWS-31	133.34	103.31	55.98	21.27	23.29	6.62	0.34	210.81
Total		1348.66	1138.3	625.06	309.8	146.7	60.23	0.34	2280.63
Percentage			49.91	27.40	13.58	6.43	2.64	0.01	99.97

24.5.1.6 Mean Bifurcation Ratio (Rbm)

Rbm is defined as the average Rb of all stream orders. The Rbm characteristically ranges from 3 to 5 for the basin when the influence of geological structures on the DN is negligible (Strahler 1964).

The range of values varies from 3.91 to 1.14 for SWSs. SWS-31 (Rbm = 3.91) is indicating that the geological structures on the DN are negligible. SWS-7

Table 24.5 Mean stream length of the study area

S.No.	Subwatershed name	Area of SWS	Mean stream length (Lsm) (in km)					
			I	II	III	IV	V	VI
1	SWS-1	11.58	0.47	0.96	0.30	0.57	–	–
2	SWS-2	18.60	0.54	0.71	0.64	0.58	–	–
3	SWS-3	7.65	0.64	0.46	0.47	–	–	–
4	SWS-4	26.06	0.65	0.92	1.09	0.37	–	–
5	SWS-5	38.90	0.78	0.67	0.70	0.46	–	–
6	SWS-6	53.76	0.52	0.83	0.68	0.51	–	–
7	SWS-7	6.64	0.8	0.71	–	–	–	–
8	SWS-8	47.95	0.67	0.88	0.60	0.55	–	–
9	SWS-9	13.22	0.63	0.87	0.49	–	–	–
10	SWS-10	120.26	0.63	0.75	0.60	0.63	0.55	–
11	SWS-11	9.37	0.44	0.60	0.35	–	–	–
12	SWS-12	13.99	0.54	0.56	0.47	–	–	–
13	SWS-13	126.53	0.60	0.58	0.52	0.56	–	–
14	SWS-14	18.99	0.56	0.81	0.64	–	–	–
15	SWS-15	35.13	0.66	0.66	0.49	0.77	–	–
16	SWS-16	7.88	0.72	0.58	0.71	–	–	–
17	SWS-17	8.17	0.61	1.32	0.51	–	–	–
18	SWS-18	5.47	0.50	0.80	0.69	–	–	–
19	SWS-19	17.16	0.82	0.73	0.59	–	–	–
20	SWS-20	5.95	0.60	0.62	0.94	–	–	–
21	SWS-21	6.38	0.50	0.44	0.39	–	–	–
22	SWS-22	30.41	0.60	0.46	0.64	0.52	0.52	–
23	SWS-23	45.62	0.49	0.45	0.52	0.565	0.565	–
24	SWS-24	15.30	0.41	0.57	0.42	–	–	–
25	SWS-25	40.63	0.54	0.61	0.70	0.43	0.43	–
26	SWS-26	78.80	0.49	0.68	0.43	0.47	0.47	–
27	SWS-27	285.94	0.53	0.57	0.54	0.64	0.64	–
28	SWS-28	19.33	0.59	0.74	0.61	–	–	–
29	SWS-29	88.99	0.50	0.40	0.45	0.43	0.43	–
30	SWS-30	10.66	0.50	0.59	0.54	–	–	–
31	SWS-31	133.34	0.51	0.54	0.51	0.56	0.56	0.34
Average			0.58	0.67	0.55	0.27	0.06	0.01

(Rbm = 1.14) indicating that the geological structures on the DN should be considered.

Table 24.6 Stream length ratio of the study area

S.No.	Subwatershed name	Area of SWS	Stream length ratio (Rl)					
			II/I	III/II	IV/III	V/IV	VI/V	VII/VI
1	SWS-1	11.58	1.11	0.17	0.30	–	–	–
2	SWS-2	18.60	0.58	0.41	1.35	–	–	–
3	SWS-3	7.65	0.39	0.72	–	–	–	–
4	SWS-4	26.06	1.03	0.28	0.05	–	–	–
5	SWS-5	38.90	0.49	0.50	0.30	–	–	–
6	SWS-6	53.76	0.79	0`40	0.70	–	–	–
7	SWS-7	6.64	0.71	–	–	–	–	–
8	SWS-8	47.95	0.65	0.22	1.65	–	–	–
9	SWS-9	13.22	0.79	0.30	–	–	–	–
10	SWS-10	120.26	0.58	0.48	0.21	1.83	–	–
11	SWS-11	9.37	0.53	0.82	–	–	–	–
12	SWS-12	13.99	0.39	1.30	–	–	–	–
13	SWS-13	126.53	0.46	0.34	1.77	–	–	–
14	SWS-14	18.99	0.64	0.91	–	–	–	–
15	SWS-15	35.13	0.63	0.13	2.60	–	–	–
16	SWS-16	7.88	0.43	0.80	–	–	–	–
17	SWS-17	8.17	0.72	0.68	–	–	–	–
18	SWS-18	5.47	0.99	0.34	–	–	–	–
19	SWS-19	17.16	0.42	0.72	–	–	–	–
20	SWS-20	5.95	0.61	0.75	–	–	–	–
21	SWS-21	6.38	0.31	1.41	–	–	–	–
22	SWS-22	30.41	0.31	0.69	1.38	–	–	–
23	SWS-23	45.62	0.42	0.99	0.34	–	–	–
24	SWS-24	15.30	0.69	0.67	–	–	–	–
25	SWS-25	40.63	0.65	0.36	0.72	–	–	–
26	SWS-26	78.80	0.58	0.61	0.41	–	–	–
27	SWS-27	285.94	0.49	0.61	0.23	1.73	–	–
28	SWS-28	19.33	0.88	0.24	–	–	–	–
29	SWS-29	88.99	0.33	0.77	0.17	3.77	–	–
30	SWS-30	10.66	0.61	0.72	–	–	–	–
31	SWS-31	133.34	0.54	0.37	1.09	0.28	0.05	–

24.5.2 Aerial Aspects

Aerial aspects such as Dd (drainage density), Sf (stream frequency), Rt (drainage texture), Rf (form factor), Fs (shape factor), Rc (circulatory ratio), Re (elongation ratio), and Lg (length of overland flow) were calculated using the formulas.

Table 24.7 Bifurcation ratio and mean bifurcation ratio of the study area

S.No.	Subwatershed name	Bifurcation ratio (Rb)					Mean bifurcation ratio (Rbm)
		I/II	II/III	III/IV	IV/V	V/ VI	
1	SWS-1	1.81	1.83	6	–	–	3.21
2	SWS-2	2.23	2.16	0.66	–	–	1.68
3	SWS-3	1.85	1.4	–	–	–	1.62
4	SWS-4	1.36	4.16	6	–	–	3.84
5	SWS-5	1.74	2.06	2.14	–	–	1.98
6	SWS-6	2	2.05	1.05	–	–	1.7
7	SWS-7	1.14	–	–	–	–	1.14
8	SWS-8	1.97	3.09	0.55	–	–	1.87
9	SWS-9	1.72	1.83	–	–	–	1.77
10	SWS-10	2.02	1.67	4.72	0.47	–	2.22
11	SWS-11	2.57	0.7	–	–	–	1.63
12	SWS-12	2.66	0.64	–	–	–	1.65
13	SWS-13	2.06	2.60	0.6	–	–	1.75
14	SWS-14	2.25	0.85	–	–	–	1.55
15	SWS-15	1.57	5.5	0.6	–	–	2.55
16	SWS-16	1.83	1.5	–	–	–	1.66
17	SWS-17	3	0.57	–	–	–	1.78
18	SWS-18	1.6	2.5	–	–	–	2.05
19	SWS-19	2.1	1.11	–	–	–	1.60
20	SWS-20	1.66	2	–	–	–	1.83
21	SWS-21	2.8	0.62	–	–	–	1.71
22	SWS-22	2.4	2	0.58	–	–	1.66
23	SWS-23	2.19	1.16	3.1	–	–	2.15
24	SWS-24	2	1.08	–	–	–	1.54
25	SWS-25	1.71	3.18	0.84	–	–	1.91
26	SWS-26	2.36	1.04	2.58	–	–	1.99
27	SWS-27	2.15	1.54	5.04	0.37	–	2.27
28	SWS-28	1.41	3.4	–	–	–	2.40
29	SWS-29	2.40	1.44	5.37	0.25	–	2.36
30	SWS-30	1.88	1.28	–	–	–	1.58
31	SWS-31	1.96	2.51	1	4.1	10	3.91

24.5.2.1 Drainage Density (Dd)

Dd is the ratio of the total length of all stream orders to the total area of the drainage network of the basin (Horton 1945). The Dd indicates the proximity of the channels. It measures a WS about how well or poorly drained stream channels. Higher Dd is generally associated with the basin characteristics namely weak and impermeable subsurface material, high relief, and rare vegetation cover. Low Dd indicates highly permeable terrain material and low relief with dense vegetation. Low Dd leads to coarse drainage texture while high Dd leads to better high runoff, fine drainage texture, and erosion potential of the basin area (Strahler 1964). The Dd and their textures are shown in Table 24.8.

The Dd in the SWSs of the study area varies from 1.42 (SWS-26) to 2.12 (SWS-20). 31 SWSs indicate less Dd leads to highly permeable terrain material, coarse drainage texture, and low relief with dense vegetation (Table 24.10). The Dd map of the study area is shown in Fig. 24.5.

24.5.2.2 Stream Frequency (Fs)

The drainage frequency or Fs of a drainage basin depends on the Amaravathi basin lithology and is a reflectance to the texture of the drainage. According to Horton, (1932) Fs or Channel Frequency is the total number of streams for all stream orders (U) per unit area.

In this study, the value Fs varies from 2.25 (SWS-7) to 4.23 km/km^2 (SWS-21). A low value of Fs (0–2 km/km^2) indicates low runoff and the value (2–5 km/km2) indicates more runoff in the region. All 31 SWSs indicate a high Fs value, more runoff in the region (Table 24.10).

24.5.2.3 Drainage Texture (Rt)

The texture ratio is the equal spacing of basins drainage lines of all orders of the stream segment. Rt is measured by the number of stream segments of all orders existing as the perimeter of the region (Balasubramanian et al. 2017). Climatic Changes, precipitation, vegetation cover, topsoil types, permeability, rock types, relief, and stage of development are all-natural factors for Rt (Smith 1950). Following Smith (1950), categorized their Dd into five categories: very coarse (< 2), coarse (2–4), moderate (4–6), fine (6–8), and very fine (> 8).

In this study, the Rt value varies from 0.89 (SWS-7) to 9.04 (SWS-27) which indicates that SWSs are very coarse to fine Rt (Table 24.10). Therefore, Rt of SWSs 3, 7, 9, 12, 16, 17, 18, 19, and 20 are very coarse to coarse Rt with less relief, softer and more permeable rock in nature, whereas, SWSs 27 and 31 are fine DN with high relief, hard and impermeable rock in nature. Remaining SWSs 1,2,4,5,6,8,10, 11,1 3,14,15,21,22,23,24,25,26,28,29, and 30 are moderate Rt (Table 24.10).

Table 24.8 Drainage density and textures (Somashekar et al. 2011)

Drainage density (Sq.km)	Textures
<1.24	Very coarse
1.24–2.49	Coarse
2.49–3.73	Moderate
3.73–4.97	Fine
>4.97	Very fine

Table 24.10 Aerial aspects of morphometric parameters

S.No.	Subwatershed name	Drainage density (Dd)	Stream frequency (Fs)	Drainage texture (Rt)	Form factor (Ff)	Shape factor (Fs)	Circularity ratio (Rc)	Elongation ratio (Re)	Length of overland flow (Lg)
1	SWS-1	1.95	3.28	2.25	0.83	1.20	0.51	0.29	0.97
2	SWS-2	1.83	3.06	2.35	0.25	3.99	0.39	0.15	0.91
3	SWS-3	1.82	3.26	1.44	0.25	3.95	0.32	0.16	0.91
4	SWS-4	2.01	2.53	2.24	0.33	2.99	0.37	0.18	1.00
5	SWS-5	1.98	2.75	2.44	0.22	4.39	0.25	0.15	0.99
6	SWS-6	1.86	3.01	3.90	0.69	1.43	0.39	0.26	0.93
7	SWS-7	1.80	2.25	0.89	0.25	3.91	0.29	0.16	0.9
8	SWS-8	1.94	2.75	3.09	0.31	3.15	0.33	0.17	0.97
9	SWS-9	1.87	2.72	1.5	0.67	1.47	0.30	0.26	0.93
10	SWS-10	1.90	2.90	4.59	0.21	4.58	0.26	0.14	0.95
11	SWS-11	1.67	3.73	2.12	0.42	2.37	0.43	0.20	0.83
12	SWS-12	1.76	3.35	1.91	0.25	3.91	0.29	0.16	0.88
13	SWS-13	1.63	2.78	3.96	0.14	7.02	0.20	0.12	0.81
14	SWS-14	1.79	2.79	2.22	0.32	3.08	0.42	0.18	0.89
15	SWS-15	1.90	2.87	3.14	0.43	2.31	0.42	0.20	0.95
16	SWS-16	1.82	2.66	1.48	0.41	2.43	0.49	0.20	0.91
17	SWS-17	1.98	2.81	1.01	0.15	6.25	0.20	0.12	0.99
18	SWS-18	1.72	2.74	1.19	0.29	3.41	0.43	0.17	0.86
19	SWS-19	1.75	2.33	1.81	0.32	3.03	0.44	0.18	0.87
20	SWS-20	2.12	3.19	1.58	0.42	2.33	0.51	0.20	1.06
21	SWS-21	1.95	4.23	2.14	0.40	2.45	0.50	0.20	0.97

Table 24.10 (continued)

S.No.	Subwatershed name	Drainage density (Dd)	Stream frequency (Fs)	Drainage texture (Rt)	Form factor (Ff)	Shape factor (Fs)	Circularity ratio (Rc)	Elongation ratio (Re)	Length of overland flow (Lg)
22	SWS-22	1.76	3.12	2.70	0.22	4.49	0.30	0.15	0.88
23	SWS-23	1.68	3.41	5.02	0.80	1.24	0.59	0.28	0.84
24	SWS-24	1.52	3.33	2.38	0.24	4.11	0.42	0.15	0.76
25	SWS-25	1.66	2.92	3.78	0.52	1.90	0.51	0.23	0.83
26	SWS-26	1.42	2.74	3.90	0.29	3.35	0.32	0.17	0.71
27	SWS-27	1.57	2.91	9.04	0.36	2.74	0.42	0.19	0.78
28	SWS-28	1.55	2.37	2.32	0.42	2.35	0.61	0.20	0.77
29	SWS-29	1.52	3.30	5.92	0.25	3.85	0.45	0.16	0.76
30	SWS-30	1.66	3.09	2.30	0.45	2.19	0.65	0.21	0.83
31	SWS-31	1.58	2.98	6.69	1.03	0.96	0.47	0.32	0.79

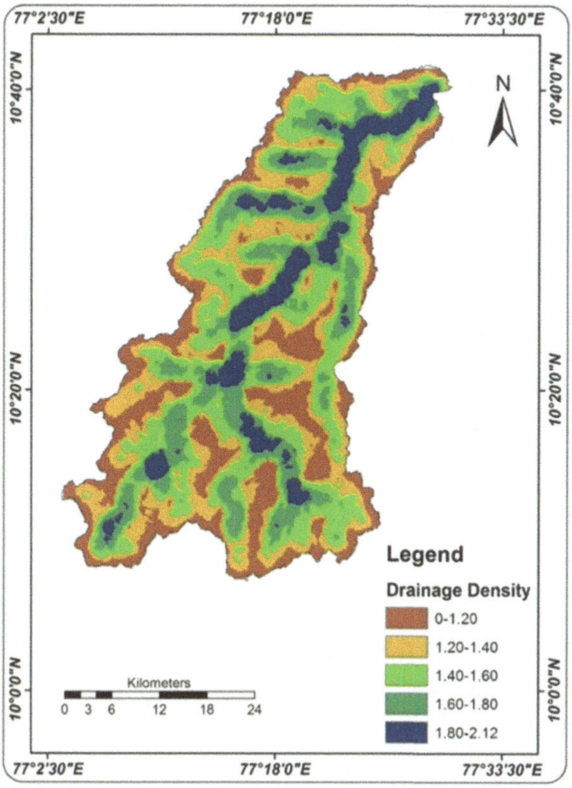

Fig. 24.5 Drainage density map

Table 24.9 Elongation ratio and their shape of the basin (Wilson et al. 2012)

Elongation ratio	Shape of basin
<0.7	Elongated
0.8–0.7	Less elongated
0.9–0.8	Oval
>0.9	Circular

24.5.2.4 Form Factor (Ff)

Schumn (1956) calculated the Ff according to the ratio of the area of the basin (A) to the square of the basin length and also called the shape of the basin. Ff is generally used to refer to various basin shapes. The Ff value is superior than 0.78 for a perfectly circular basin. Smaller Ff value is for the more elongated basin.

The Ff values vary from 0.14(SWS-13) to 0.83(SWS-1) (Table 24.10). The 29 SWSs represent an elongated shape with low peak flows for short periods of time.

Whereas, 0.83 (SWS-1) and 0.80 (SWS-23) are of higher Ff values which indicate a nearly circular WS with a greater peak flow of long periods of time.

24.5.2.5 Shape Factor (Fs)

Fs is the ratio of the square of the basin length (Lb) to A (Horton 1945). The Fs are related to parameters such as basin shape, Fs, Rc, Re, and compactness coefficient.

The study areas Fs value ranges between 0.96 (SWS 31) and 7.02 SWS 13 for the basin (Table 24.10). This study demonstrates that the shape of SWSs varies from circular to elongate form.

24.5.2.6 Circularity Ratio (Rc)

The Rc also called hydraulic radius which is the ratio between the basin area (A) to the area of the circle having the same circumference as the perimeter of the basin (P) (Miller 1953).

In this study, the values of Rc ranges from 0.20 to 0.65 which indicates that SWSs 2, 3, 4, 5, 6, 7, 8, 9, 10, 11, 12, 13, 14, 15, 16, 17, 18, 19, 22, 24, 26, 27, 29, and 31 are elongated in shape; the remaining SWSs (1, 20, 21, 23, 25, 28, 30, and) are more or less circular in shape (Table 24.10). The circular SWS indicate high infiltration over the SWS area, where it takes a longer time to reach the basin outlet.

24.5.2.7 Elongation Ratio (Re)

Re is the ratio between the diameter of the circle of the same area as the drainage basin and the maximum length of the basin (Lb) (Schumm 1956). Therefore, Re is dimensionless which is generally classified into two classes, that is, the low value which has high relief and steep slope where it is most prone to erosion, and high value (circular WS) which has flat land with low relief and low slope. The four-level classifications of Re are Elongated (<0.7), less elongated (0.7–0.8), oval (0.8–0.9), and circular (>0.9). The Re and shape of the study are shown in Table 24.9.

Re for the study area varies from 0.12 (SWS-13, 17) to 0.32 (SWS-14). The SWSs 31 indicates elongated in shape (Table 24.10).

24.5.2.8 Length of Overland Flow (Lg)

The Lg almost equals half the of reciprocal of Dd (Horton 1945). The Lg values of the study area show vary from 0.71 to 1.06. Low Lg values of WSs indicate overall high relief in the study area (Table 24.10).

Table 24.11 Relief aspects of morphometric parameters

S.No.	Subwatershed name	Basin relief (R)	Relief ratio (Rr)	Ruggedness number (Rn)
1	SWS-1	80	21.44	156
2	SWS-2	80	9.28	146.4
3	SWS-3	40	7.27	72.8
4	SWS-4	60	6.79	120.6
5	SWS-5	100	7.65	198
6	SWS-6	100	11.40	186
7	SWS-7	40	7.84	72
8	SWS-8	100	8.13	194
9	SWS-9	60	13.57	112.2
10	SWS-10	560	23.86	1064
11	SWS-11	40	8.47	66.8
12	SWS-12	220	29.72	387.2
13	SWS-13	2080	69.77	3390.4
14	SWS-14	360	47.05	644.4
15	SWS-15	560	62.08	1064
16	SWS-16	540	123.28	982.8
17	SWS-17	800	111.88	1584
18	SWS-18	560	129.62	963.2
19	SWS-19	1080	149.58	1890
20	SWS-20	280	75.06	593.6
21	SWS-21	360	90.90	702
22	SWS-22	980	83.83	1724.8
23	SWS-23	1600	211.92	2688
24	SWS-24	1840	232.03	2796.8
25	SWS-25	1420	161.54	2357.2
26	SWS-26	1960	120.54	2783.2
27	SWS-27	2080	74.20	3265.6
28	SWS-28	1440	213.64	2232
29	SWS-29	1880	101.45	2857.6
30	SWS-30	1360	280.99	2257.6
31	SWS-31	980	86.19	1548.4

24.5.3 Relief Aspects

Relief aspect parameters such as basin relief (R), relief ratio (Rr), and ruggedness number (Rn) were evaluated using the formulas represented in Table 24.11.

24.5.3.1 Basin Relief (R)

R is the elevation difference of the highest and lowest point of the valley floor (Rudraiah et al. 2008). The R is determined as 2080 (SWS-27), indicating excess runoff and sediment transmission. The R value is used to know the geo-morphological process and terrain characteristics of the river basin. The basin relief values are shown in Table 24.11.

24.5.3.2 Relief Ratio (Rr)

The max relief to horizontal distance along the longest dimension of the basin parallel to the principal drainage line is termed as Rr (Schumm 1956). Rr has a direct relationship between the relief and the channel gradient. It measures the overall steepness of the WS which is an indicator of the erosion intensity.

The SWS 30 is having a relatively high Rf (280.99) compared to the other SWSs (Table 24.11). The higher value of Rf is an indicator of a higher degree of slope, high runoff, high soil erosion.

24.5.3.3 Ruggedness Number (Rn)

The Rn is expressed as the product of the R and Dd (Strahler 1957). The higher value of the ruggedness number is 3390.4 (SWS 13) compared to the other SWSs. An extremely high value of ruggedness number occurs the slope is generally steep. The ruggedness number values are shown in Table 24.11.

24.5.4 Prioritization of Subwatersheds

The Fc is an important prime factor in the prioritization of SWS. The prioritized SWSs are identified with the help of the Fc. Estimation of the Fc with priority ranking is shown in Table 24.12.

24.5.4.1 Compound Factor (Fc)

The Fc can be calculated by summing the ranks of linear and shape parameters and then divided by the total number of parameters are called the Fc. The linear parameters such as Rb, Dd, Fs, Rt, and Lg. The shape parameters such as Ff, Fs, Re, and Rc. Rank 1 was assigned the for lowest Fc value and rank 2 was assigned for the highest Fc value as, and so on until the highest value.

The prioritization of 31 SWSs of Amaravathi WS on the basis of MOPs, for each SWSs and the compound value (Cp) based on rank. During SWS prioritization, higher ranks were assigned to SWS with the lowest Cp. In the current study, SWS 10 gets prioritized with rank 1 whose Cp is 8.55 and the SWS with the highest Cp value at 16.22 (SWS 19) was assigned with the lowest rank. The lower value of the Cp was probably possible for soil erosion. The SWSs was categorized as high (8.55–10.88), medium (10.88–13), and low (13–16.22) priority on the final Cp.

Table 24.12 Estimation of compound factor with priority ranking

S.No.	Subwatershed name	Rbm	Dd	Fs	Rt	Ff	Fs	Rc	Re	Lg	Compound factor	Prioritized rank
1	SWS-1	3	4	7	17	19	2	17	13	4	9.55	3
2	SWS-2	21	8	12	14	6	24	9	3	7	11.55	15
3	SWS-3	26	9	8	27	6	23	6	4	7	12.88	19
4	SWS-4	2	2	26	18	10	15	8	6	2	9.88	5
5	SWS-5	12	3	22	12	4	26	2	3	3	9.66	4
6	SWS-6	20	8	13	6	19	4	9	11	6	10.66	7
7	SWS-7	31	10	29	30	6	22	4	4	8	16	30
8	SWS-8	14	5	22	10	8	18	7	5	4	10.33	6
9	SWS-9	17	7	24	25	18	5	5	11	6	13.11	21
10	SWS-10	8	6	17	5	3	28	3	2	5	8.55	1
11	SWS-11	25	17	2	21	14	11	11	8	15	13.77	23
12	SWS-12	24	12	4	22	6	22	4	4	10	12	17
13	SWS-13	18	18	21	7	1	30	1	1	16	12.55	18
14	SWS-14	29	11	20	19	9	17	10	6	9	14.44	24
15	SWS-15	4	6	18	9	15	8	10	8	5	9.22	2
16	SWS-16	23	9	25	26	13	12	15	8	7	15.33	28
17	SWS-17	16	3	19	29	2	29	1	1	3	11.44	13
18	SWS-18	10	14	23	28	7	20	11	5	13	14.55	25
19	SWS-19	27	13	28	23	9	16	12	6	12	16.22	31
20	SWS-20	15	1	9	24	14	9	17	8	1	10.88	9
21	SWS-21	19	4	1	20	12	13	16	8	4	10.77	8
22	SWS-22	22	12	10	11	4	27	5	3	11	11.66	16
23	SWS-23	9	15	3	4	20	3	18	12	14	10.88	10
24	SWS-24	30	22	5	13	5	25	10	3	19	14.66	26
25	SWS-25	13	16	15	8	17	6	17	10	15	13	20
26	SWS-26	11	23	23	6	7	19	6	5	20	13.33	22
27	SWS-27	7	20	16	1	11	14	10	7	17	11.44	14
28	SWS-28	5	21	27	15	14	10	19	8	18	15.22	27
29	SWS-29	6	22	6	3	6	21	13	4	19	11.11	11
30	SWS-30	28	16	11	16	16	7	20	9	15	15.33	29
31	SWS-31	1	19	14	2	21	1	14	14	16	11.33	12

Based on the observation, the prioritization of ten SWSs with a higher degree of slope was severely prone to soil erosion (SWS-10, SWS-15, SWS-1, SWS-5, SWS-4, SWS-8, SWS-6, SWS-21, SWS-20, SWS-23).

24.6 Conclusion

In this study, the prioritization of SWSs of the Amaravthi river WS has been investigated based on MOPs.

- The RS and GIS practices have been used for the identification of morphometric characteristics and the properties of the Amaravathi WS using ASTER DEM with the help of ArcGIS software (10.3).
- The totally 3962 stream segments are identified and they are classified as 1st order, 2nd order, and so on down to the 6th order. The stream segments that fell in different orders are grouped as follows: 1st order (50.78%), 2nd order (24.88%), 3rd order (14.31%), 4th order (6.66%), 5th order (3.33%), and 6th order (0.02%). The first order recorded 49.91% of the stream length whereas the sixth order had only 0.01% of the stream segments.
- The study area is classified into 31 sub-WSs which have been considered for the analysis as well as most of the WSs are elongated in shape.
- The linear, aerial, and relief aspects based on 17 MOPs have been analyzed. The bifurcation results showed that most of the SWSs are good structural control. The Dd varies from 1.42 to 2.12 km^2. This suggests that the 31 SWSs indicate high Dd leads to fine drainage texture, highly impermeable subsurface material, and land covered with sparse vegetation and high relief. The Td is very coarse to coarse in 9 SWSs, moderate in 20 SWSs, and only 2 SWSs show fine Td.
- PSWS on the basis of MA shows that out of 31 SWSs, 10 fall in high priority (SWS-10, SWS-15, SWS-1, SWS-5, SWS-4, SWS-8, SWS-6, SWS-21, SWS-20, SWS-23), 10 fall in medium priority (SWS-29, SWS-31, SWS-17, SWS-27, SWS-2, SWS-22, SWS-12, SWS-13, SWS-3, SWS-25), and 11 fall in low priority (SWS-9, SWS-26, SWS-11, SWS-14, SWS-18, SWS-24, SWS-28, SWS-16, SWS-30, SWS-7, SWS-19). High priority SWSs are possible for soil erosion. These high priority SWSs have the highest Rb, high slope, and deep valley terrain which indicate robust structural control on the drainage and consequently more prone to natural calamities because the existence of structures in hard rock acting as weak planes. Hence, it is possible for extreme soil erosion and natural threats. Therefore, it should be provided with instant soil maintenance measures.
- The quantitative approach was done for the PSWSs most possible for soil erosion conservative measures. These identified SWSs need to have ensued immediate remediation for soil and water conservation for the purpose of efficient WSM planning.
- This SWS prioritization study was quite useful as a resource planner, decision-maker, or any government, private agencies, who effort to take up soil safeguarding measures in the Amaravathi WS of Dindigul and Tirupur districts in Tamil Nadu, and Idukki district in Kerala.

References

U. Ali, S.A. Ali, Analysis of drainage morphometry and watershed prioritization of Romushi - Sasar catchment, kashmir valley, India using remote sensing and GIS technology. Int. J. Adv. Res. **2**(12), 5–23 (2014)

A. Balasubramanian, K. Duraisamy, S. Thirumalaisamy, S. Krishnaraj, R.K. Yatheendradasan, Prioritization of subwatersheds based on quantitative morphometric analysis in lower Bhavani basin, Tamil Nadu, India using DEM and GIS techniques. Arab. J. Geosci. **10**, 552 (2017). https://doi.org/10.1007/s12517-017-3312-6

S.S. Biswas, Analysis of GIS based morphometric parameters and hydrological changes in Parbati river basin, Himachal Pradesh, India. Biswas. J. Geogr. Nat. Disast. **6**, 2 (2016). https://doi.org/10.4172/2167-0587.1000175

S.S. Biswas, S. Sudhakar, V.R. Desai, Prioritisation of subwatersheds based on morphometric analysis of drainage basin: A remote sensing and GIS approach. J. Indian Soc. Remote Sens. **27**, 155–166 (1999)

A.K. Das, S. Mukherjee, Drainage morphometry using satellite data and GIS in Raigad district, Maharashtra. J. Geol. Soc. India **65**, 577–586 (2005)

R.E. Horton, Drainage basin characteristics. Trans. Am. Geophys. Union **13**, 350–361 (1932)

R.E. Horton., (1945). Erosional development of streams and their drainage basins: hydrophysical approach to quantitative morphology. Bull. Geol. Soc. Am. 56; 275–370. https://doi.org/10.1130/0016-760656[275:EDOSAT]2.0.CO;2

M. Kamala, M. Samynathan (2018). Morphometric analysis of drainage basin using GIS techniques. A case study of Amaravathi river basin, Tamilnadu. Vol. 9, Issue, 7(F), pp. 28142–28147

N.S. Magesh, K.V. Jitheshlal, N. Chandrasekar, K.V. Jini, GIS based morphometric evaluation of Chimmini and Mupily watersheds parts of Western Ghats Thrissur District Kerala India. Earth Sci. Inform. **5**(2), 111–121 (2012). https://doi.org/10.1007/s12145-012-0101-3

N.S. Magesh, K.V. Jitheshlal, N. Chandrasekar, K.V. Jini, Geographical information system-based morphometric analysis of Bharathapuzha river basin Kerala India. Appl. Water Sci. **3**(2) 467–477 (2013). https://doi.org/10.1007/s13201-013-0095-0

M.A. Melton, Correlation structure of morphometric properties of drainage systems and their controlling agents. J. Geol. **66**(4), 442–460 (1958). https://doi.org/10.1086/626527

V.C. Miller. A quantitative geomorphic study of drainage basin characteristics in the clinch mountain area, virginia and tennessee. Project NR 389042, Tech. Rept. 3. Columbia University, Department of Geology, ONR, Geography Branch, N Y (1953)

S.S. Mishra, R. Nagarajan, Morphometric analysis and prioritization of subwatersheds using GIS and remote sensing techniques: A case study of Odisha, India. Int. J. Geomatics Geosci. **1**(3), 2010 (2010). ISSN 0976–4380

K. Nookaratnam, Y.K. Srivastava, V. Venkateswarao, E. Amminedu, K.S.R. Murthy, Check dam positioning by prioritization of micro-watersheds using SYI model and morphometric analysis - remote sensing and GIS perspective. J. Indian Soc. Remote Sens **33**(1), 25–38 (2005)

K. Pareta, U. Pareta, Quantitative morphometric analysis of a watershed of Yamuna basin, India using ASTER (DEM) data and GIS. Int. J. Geomatics Geosci. **2**(1), 248–269 (2011)

D.P. Patel, M.B. Dholakia, N. Naresh, P.K. Srivastava, Waterharvesting structure positioning by using geo-visualization concept and prioritization of mini-watersheds through morphometric analysis in the lower Tapi basin. J. Indian Soc. Remote Sens **40**(2), 299–312 (2012)

K.C. Patel, N.K. Gontia, K.M. Gojiya, Morphological parameters and their implications in forest watershed. Int. J. Curr. Microbiol. App. Sci. **8**(04), 2818–2826 (2019). ISSN: 2319-7706

P.K. Rai, K. Mohan, S. Mishra, A. Ahmad, V.N. Mishra, A GIS-based approach in drainage morphometric analysis of Kanhar River Basin, India. Appl. Water Sci. **7**(1), 217–232 (2014)

M. Rout, Water management, Madras student at Banaras Hindu University (2017)

M. Rudraiah, S. Govindaiah, S.S. Vittala, Morphometry using remote sensing and GIS techniques in the sub-basins of Kagna River basin, Gulburga District, Karnataka. J. Indian Soc Remote Sens. **36**, 351–360 (2008)

S.S. Salvi, S. Mukhopadhyay, S.D. Ranade, A. Rajagopalan, (2017). Morphometric analysis of river drainage basin/watershed using GIS and RS: A review. **5**

S.A. Schumn, Evolution of drainage systems and slopes in badlands at Perth Amboy, New Jersey. Geol. Soc. Am. Bull. **67**(5), 597–646 (1956). https://doi.org/10.1130/00167606(1956)67[597:EODSAS]2.0.CO;2

S. Singh, M.C. Singh, Morphometric analysis of Kanhar river basin. Nat. Geo. J. India **43**(1), 31–43 (1997)

K.G. Smith, Standards for grading textures of erosional topography. Am. J. Sci. **248**(9), 655–668 (1950). https://doi.org/10.2475/ajs.248.9.655

R.K. Somashekar, P. Ravikumar, S.V. Sowmya, M.A. Dar, A.S. Ravikumar, Runoff estimation and morphometric analysis for Hesaraghatta watershed using IRS–1D LISS III FCC satellite data. J. Indian Soc. Remote Sens. **39**(1), 95–106 (2011). https://doi.org/10.1007/s12524-011-0074-6

A.N. Strahler, Quantitative analysis of watershed geomorphology. Trans. Am. Geophys. Union **38**(6), 913–920 (1957). https://doi.org/10.1029/TR038i006p00913

A.N. Strahler, Quantitative geomorphology of drainage basin and channel networks, in *Handbook of Applied Hydrology*, ed. by V. T. Chow, (McGraw Hill Book Company, New York, 1964)., section 4

K. Sushanth, A. Bhardwaj. Assessing geomorphologic characteristics and demarcating run-off potential zones using RS and GIS in Patiala-Ki-Rao watershed of Shivalik Foot-Hills, Punjab (2019)

S.S. Vittala, S. Govindaiah, H.H. Gowda, Prioritization of sub-watersheds for sustainable development and management of natural resources: An integrated approach using remote sensing, GIS and socio-economic data. Curr. Sci. **95**(3), 345–354 (2008)

M.L. Waikar, A.P. Nilawar, Morphometric analysis of a drainage basin using Geographical Information System: A case study. Int. J. Multidisciplinary Curr. Res. **2**, 179–184 (2014). ISSN: 2321-3124

Wigglesworth, Alex. Satellite image shows Kincade fire burn scar", Advanced Spaceborne Thermal Emission and Reflection Radiometer (2019)

J.S.J. Wilson, N. Chandrasekar, N.S. Magesh. Morphometric analysis of major sub-watersheds in Aiyar & Karai Pottanar basin, Central Tamil Nadu, India using Remote Sensing & GIS techniques. **2**. ISSN 2277 – 5056 (2012)

Chapter 25
Conclusion

Kirubakaran Muniraj, Balamurugan Panneerselvam, Nagavinothini Ravichandran, and Chaitanya Baliram Pande

25.1 General

The main focus of this chapter is to discuss and summarize the major findings and conclusions of chapters of this book titled Climate Change Impact on Groundwater Resources – Human Health Risk Assessment in Arid and Semi-Arid Regions. Groundwater is a significant freshwater resource and provides essential freshwater supply, chiefly in dry regions where surface water availability is limited. However, groundwater problems are one of the most serious and complex technical problems in today's world. The main purpose of this book is to disseminate advanced technical knowledge to the technology community, and this book seeks to fill the research gap on groundwater resource sustainable management activities.

K. Muniraj (✉)
Department of Civil Engineering, University VOC college of Engineering, Thoothukudi, Tamil Nadu, India

B. Panneerselvam
Department of Civil, Architectural and Environmental Engineering, University of Naples Federico II, Naples, Italy

N. Ravichandran
Department of Structures for Engineering and Architecture, University of Naples Federico II, Naples, Italy

C. B. Pande
GIS Consultant, Primus Partners Private Limited, Delhi, India

B. Panneerselvam et al. (eds.), *Climate Change Impact on Groundwater Resources*, https://doi.org/10.1007/978-3-031-04707-7_25

25.2 Climatic Change Impact on Groundwater

Chapter 1 discusses the importance, availability, and present scenario of groundwater from a local and global perspective. Chapter 2 mainly focuses on the climate change impact on groundwater resources specifically in semi-arid regions. Chapter 2 also examines the general effects of climate change and discusses, in particular, the impact of climate change on the factors of the hydrological cycle. It concludes that climatic change is caused by deforestation, pollution, etc., which affect the groundwater resources in the following forms: increase in atmospheric water vapour content, changes in rainfall patterns, evapotranspiration, soil moisture, runoff, change in groundwater recharge, ice melting change, snow cover reduction, snowfall, and sea-level increase. Cchapter 3 reviews the important factors that influence the hydrological cycle, such as climate change, landuse change, urbanization, land degradation, and disasters. This chapter uses the synoptic view of the satellite data and attempted to exercise certain indices to identify certain classes and classification algorithms to classify landuse classes.

The purpose of Chapter 4 is to examine the spatial variation in long-term rainfall in Maharashtra. The results of the annual and seasonal rainfall analysis using Mann–Kendall test at district and sub-district level for 36 years from 1982 to 2018 indicate the fact that 76% of the districts (26 districts) in Maharashtra consist of 130 subdistricts. There is a significant increase in rainfall over 48.1% of the state.

25.3 Groundwater Hydrochemistry

Chapters 5, 6, 7, 8, 9, 10, and 11 discuss groundwater hydrochemistry in order to evaluate the groundwater quality and groundwater suitability for various purposes and also try to explore the credibility of various suitability indexes and modern assessment techniques. Chapter 5 reviews the impact of climate change on the semi-arid regions' groundwater quality and explains that the major cities in Madhya Pradesh, India, have been affected by groundwater quality issues.

Chapter 6 attempts to assess the groundwater quality of the Kattar Micro-Watershed (~ 245 km^2) located in the Tiruchirappalli and Pudukottai districts of Tamil Nadu by examining 38 samples collected in and around the study area in January 2019. This study is very important in identifying low quality areas of groundwater that require immediate attention to proper water management strategies. Chapter 7 aims to combine hydrochemical and statistical methods to understand saltwater intrusion along the Cuddalore coastal aquifers, south India. The coastal aquifers are affected by different sources from the coast and land, which alter the quality of groundwater in nature.

Chapters 8 and 9 integrate WQI, EWQI, and irrigation indices with GIS using a comprehensive investigation to assess groundwater quality for drinking water and agricultural purposes. These chapters identify the primary sources of pollution such as rock water contact, the nature of the aquifer, and other anthropological functions.

Chapter 10 mainly aims to assess the quality of groundwater and its suitability for drinking and irrigation along the delta areas of the Cauvery River in south India. This chapter suggests that existing norms and regulations be strictly enforced in order to prevent illegal discharge from various industries along the catchment areas. GIS-integrated MCDA has been used for analyzing the groundwater quality of Pattukottai taluk, where major agricultural activities are taking place in Tamil Nadu, India, which are discussed in Chapter 11.

25.4 Heavy Metal Contamination

The presence of heavy metals in the environment is considered to have a negative impact because of their toxic properties, causing various health problems. Chapter 12 is a study to assess the energy and bioavailability of heavy metals at the site of the Durgapur Dam on the Damodar River. This chapter clearly shows that heavy metals can contaminate river water and have a dangerous effect on biota due to their high metal bioavailability.

Chapter 13 describes the study conducted at the landfill in Pallavaram Municipality and various aspects of deep water quality. Significant impact was observed on surface water samples adjacent to perimeter leachate samples, and the same variation was observed in the respective heavy metals and subsequent monsoon seasons. Therefore, the heavy metals in the leak samples play an important role in increasing the concentration of groundwater.

25.5 Human Health Risk Assessment

Global climate change, which threatens human health by altering hydrobiological systems and the quality of their water resources by increasing nitrate concentrations, which are considered to be exemplary pollutants in groundwater, has become a major concern in recent years. Chapter 14 documents the health effects of nitrate exposure on groundwater quality and an integrated management perspective to avoid unplanned actions by private enterprises.

Climate change is the foremost challenge faced by humans in this century. Global climate change and hydrological extremes with high intensity and frequency result in an upsurge of health risks. To develop a systematic adaptation strategy, it is necessary to integrate climate change, flood risk, and health risk. Chapter 15 seeks to identify the severity and health risk of floods caused by climate change. Health implications for different ages are analyzed in this chapter.

Chapter 16 examines the important climate change factors that add to the toxic levels, conventional therapies, ideas, and possibilities of current measures to mitigate these pollutants. This chapter focuses on understanding the groundwater vulnerability to both fluoride and nitrate by which studying the various forms in which these contaminations occur is important to mitigate these hazards impacts on human health. The findings and recommended measures are expected to benefit the state and local government in regulating systems that ensure the health and safety of residents.

25.6 Hydraulic Modeling

Chapter 17 gives a detailed study of the development of hydraulic models over the decades and the reliability of different models.

25.7 Groundwater Prospective Zone

Chapters 18, 19, 20, and 21 of this book discuss the groundwater prospect zone delineation methods using integrated remote sensing, GIS, and various Multi-criteria decision making (MCDA) techniques. Assessment of ground water potential zone and suggested the soil and water conservation structure is a very challenging task especially in arid region. In this century, remote sensing is a very useful technique to provide recent data on surface and subsurface with less time and cost.

In Chap. 18, geological methods are used to identify and map future regions to estimate groundwater availability in the Liddar sub-basin. In contrast, low potential zones are found in the upstream area of the bed, due to steeper slopes and lower penetration rates due to massive hard rocks. Chapter 19 deals with the identification and delineation of groundwater potential zones in the region through the combined use of remote sensing, Geographic Information System and Analysis Hierarchy Process (AHP).

Chapter 20 discusses the identification of groundwater potential zones in PTK4 watersheds in Akola district, Maharashtra. Many criteria are approached based on remote sensing and GIS using six parameters, namely, geography, LULC, drainage density, line density, slope, and soil. Chapter 21 discusses the fact that these factors have been conducted in Adigrat Town and its surrounding area, Ethiopia, to identify suitable groundwater potential zones and methods of proper groundwater management and find suitable groundwater potential zones using recent scientific approaches to remote sensing and GIS techniques.

25.8 Other Important Topics

Soil infiltration acts a main role in the freshwater intrusion as a downward direction of the ground. Thus, the basic source of water resources is the rainfall that occurs in a tropical climate. Chapter 22 focuses on the study of the spatial and temporal variation of soil infiltration off the coast of Andhra Pradesh. Finally, coastal vulnerability zones have been predicted in order to maintain coastal sustainability. Chapter 23 evaluates the feasibility of a low-cost filter for use in roof-top rainwater harvesting. Prioritization of subwatersheds in a river basin using morphometric parameters is one of the elegant approaches for the development of watersheds and its management. The concept of prioritization plays a key role in identifying areas which requisite immediate intervention for natural resource conservation plans at the sub-watershed level. In Chap. 23, an attempt is made to prioritize sub-waters based on morphometric parameters using remote sensing and GIS techniques to understand the role of the channel network in the geo-hydrological behavior of the Amravati watershed.

Index

B. Panneerselvam et al. (eds.), *Climate Change Impact on Groundwater Resources*, https://doi.org/10.1007/978-3-031-04707-7

The manufacturer's authorised representative in the EU is Springer Nature Customer Service Centre GmbH, Europaplatz 3, 69115 Heidelberg, Germany. If you have any concerns regarding our products, please contact ProductSafety@springernature.com

Printed and bound by CPI Group (UK) Ltd, Croydon, CR0 4YY
30/09/2026
02225250-0017